Frida Android SO
逆向深入实践

陈佳林 著

清华大学出版社
北京

内 容 简 介

本书主要介绍使用Frida辅助分析SO进行逆向工程项目开发。首先从基础开始介绍NDK编译套件，开发编译包含SO的应用软件并进行动静态分析调试；接着介绍如何将算法移植到SO中保护App，并使用模拟执行框架加载SO运行算法，使用GDB、HyperPwn、Objection、Frida联合调试SO中的算法；此外，还将介绍ARM/ELF的文件格式和反编译工具IDA，Frida/Jnitrace Hook/Invoke JNI，JNI与反射及简单风控案例设计分析，onCreate的Native化，分析Android源码追踪JNI的绑定流程。最后介绍哈希和对称加解密算法的核心原理与实现细节，以及使用Frida辅助逆向分析的工作流程。

本书按照由易到难、由浅入深的方式进行讲解，适合Native层的初、中级读者阅读。

本书封面贴有清华大学出版社防伪标签，无标签者不得销售。
版权所有，侵权必究。举报：010-62782989，beiqinquan@tup.tsinghua.edu.cn。

图书在版编目（CIP）数据

Frida Android SO 逆向深入实践 / 陈佳林著. —北京：清华大学出版社，2023.9（2024.7重印）
ISBN 978-7-302-64559-7

I. ①F… II. ①陈… III. ①移动终端－应用程序－程序设计 IV. ①TN929.53

中国国家版本馆 CIP 数据核字（2023）第 178229 号

责任编辑：王金柱
封面设计：王　翔
责任校对：闫秀华
责任印制：杨　艳

出版发行：清华大学出版社
　　　网　　址：https://www.tup.com.cn，https://www.wqxuetang.com
　　　地　　址：北京清华大学学研大厦A座　　邮　编：100084
　　　社 总 机：010-83470000　　　　　　　　邮　购：010-62786544
　　　投稿与读者服务：010-62776969，c-service@tup.tsinghua.edu.cn
　　　质 量 反 馈：010-62772015，zhiliang@tup.tsinghua.edu.cn
印 装 者：小森印刷霸州有限公司
经　　销：全国新华书店
开　　本：185mm×235mm　　　印　张：32.5　　　字　数：780千字
版　　次：2023年10月第1版　　　　　　　　印　次：2024年7月第2次印刷
定　　价：149.00元

产品编号：097962-01

前　　言

熟悉笔者的读者都知道，2021年出版的《安卓Frida逆向与抓包实战》和2022年出版的《安卓Frida逆向与协议分析》主要介绍了Frida框架在Android平台Java层的工作和作用，包括但不限于Hook、抓包、算法还原、协议分析等领域。Frida框架独特的自动化和规模化逆向功能大大提高了安全从业者的工作效率。它可以使用Python、JavaScript等语言搭建综合性应用安全审计和隐私监控框架。据笔者所知，老牌的安全工具MobSF和Inspackage都将其主要核心动态分析监控代码由Xposed转移到了Frida，足以说明Frida已经成为事实上Java层分析的第一工具。

与此同时，安卓移动端还有很多应用使用C/C++语言进行开发，将逻辑和算法打包进应用包中。这样做带来的好处是：

（1）运行效率和速度的显著提升。相比Java语言的虚拟机解释执行，C/C++语言编译生成的SO文件是直接运行在CPU上的，减少了解释执行的效率损耗，从而使运行速度得到质的提升。例如，我们常用的微信语音和视频功能都是由C/C++代码开发的，这样在较旧的手机上也能有不错的性能表现。

（2）更加安全。Java语言生成的Dex几乎没有安全性可言，一旦被反编译器反编译，其源码几乎等同于完全暴露。相比之下，C/C++语言反编译起来难度较大，即使生成的伪代码也晦涩难懂。此外，经过多年的发展，C/C++代码保护工具和技术更为成熟，例如花指令、强大的混淆、虚拟机等，将算法与逻辑实现放到C/C++代码中可以劝退大多数新手。

因此，安全工具和安全产品几乎全都是由C/C++代码开发的，先将代码编译成SO文件，再在应用中加载这个SO文件，以提供相应的安全功能。作为安全或逆向工程师，学习以C/C++为代表的本地Native开发和JNI（Java与C之间的转换层）开发，就成为必须掌握的技能栈之一。

由此衍生的反编译、动静态调试分析、算法开发、模拟执行、JNI开发、Frida Hook C/C++或JNI、汇编与ELF文件格式、Android源码中的JNI绑定、Native层算法开发及结合Frida

的逆向分析技巧等技术，均在本书中有详细的介绍、原理阐述和案例的综合实战。

虽然对于大多数新手来说，SO 中的 ARM、JNI、算法等已经足够复杂，但这些实际上都还没有涉及更加深入的算法保护机制。在加入 OLLVM、VMProtect 等机制后的算法分析逆向还原才是真正的噩梦。此外，本书还介绍了忽略算法实现细节而直接调用算法获取结果的 Frida RPC 机制，帮助只需要结果而不需要细节的读者来翻越强混淆的大山。

Frida 不仅在 Java 平台如鱼得水，而且在 Native 原生开发的 ARM 架构上也表现出色。实际上，Frida 已经渗透到 SO 生命中的各个方面，包括进程、线程、指令集、链接器、数据结构、参数传递、内存布局、读写执行、块和执行流追踪等。Frida 提供了独立而完善的 API，学好 Frida 对掌握 SO 的逆向工程也会起到事半功倍的效果。

本书内容介绍

本书共 26 章，主要内容如下：

第 1 章介绍基本开发环境配置。

第 2 章介绍安卓 SO 动态调试入门。

第 3 章介绍静态分析工具安装和基本使用。

第 4 章介绍 C 算法开发及模拟执行。

第 5 章介绍动态调试：GDB 动态调试、HyperPwn/（内存）断点/栈帧。

第 6 章介绍汇编开发：ARM 汇编原理/流程/调用约定/动态调试。

第 7 章介绍逆向分析：ELF 文件结构、节/区/表/段/符号/链接器。

第 8 章介绍反编译工具 IDA。

第 9 章介绍 JNI 接口初识。

第 10 章介绍 JNI 特性：Java/Native 互相调用、反射/全局/局部引用。

第 11 章介绍 onCreate 进行 Native 化和引用。

第 12 章介绍 JNI 动静态绑定和追踪。

第 13 章介绍 MD5 算法分析和魔改。

第 14 章介绍对称加密算法逆向分析。

第 15 章介绍读懂 Dex 并了解 DexDump 解析过程。

第 16 章介绍 ELF 文件格式解读及其生成过程。

第 17 章介绍高版本 Android 函数地址索引彻底解决方案。

第 18 章介绍从 findExportByName 源码分析到 anti-frida 的新思路。

第 19 章介绍 PLT 表和 GOT 表 Hook。

第 20 章介绍番外篇——另类方法寻找首地址。

第 21 章介绍 Java Hook 原理。

第 22 章介绍 inline Hook 中用到的汇编指令。

第 24 章介绍基于 capstone 处理特殊指令。

第 25 章介绍通杀的检测型框架 r0Invoke。

第 26 章介绍 SO 加载流程分析与注入实战。

本书按照由易到难、由浅入深的方式编写，定位于 Native 层的初、中级内容。希望能够抛砖引玉，对读者学习 SO 逆向工程有所帮助。

配书资源

为方便读者使用本书，本书还提供了代码源文件，读者需要用微信扫描下面的二维码获取。如果阅读中发现问题或疑问，请联系 booksaga@126.com，邮件主题写"Frida Android SO 逆向深入实践"。

读者对象

- Android 应用安全工程师
- Android 逆向分析工程师
- 爬虫工程师
- 大数据收集和分析工程师

书本是静止的，知识是流动的。在本书写作、编撰、排版的过程中，Frida 已经升级到了 15/16 版本，Android 也升级到了 13 版本。但本书中的代码都是可以在特定版本的 Frida 和 Android 中成功运行的。同时，Android 逆向实践性极强，读者在动手实践的过程中难免会产生各式各样的疑问。为此，笔者特地准备了 GitHub 仓库更新和勘误，读者有疑问可以在仓库的 issue 页面提出，笔者会尽力解答和修正。

希望可以与读者一起学习和进步。

编　者

2023 年 8 月

目 录

第 1 章 基本开发环境配置 1
- 1.1 虚拟机环境搭建 1
- 1.2 逆向环境搭建 3
  - 1.2.1 Android Studio 安装 NDK 编译套件 3
  - 1.2.2 ADB 的配置和使用 5
  - 1.2.3 Python 版本管理 6
  - 1.2.4 移动设备环境准备 7
  - 1.2.5 Frida 版本管理 7
  - 1.2.6 Objection 的安装和使用 8
- 1.3 Frida 基本源码开发环境搭建 10
- 1.4 初识 NDK 12
- 1.5 其他工具 16
- 1.6 本章小结 18

第 2 章 Android SO 动态调试入门 19
- 2.1 Android SO 基本动态分析调试 19
  - 2.1.1 第一个 NDK 程序 19
  - 2.1.2 动态调试 NDK 程序 20
  - 2.1.3 交叉编译 24
- 2.2 LLDB 动态调试（三方）Android SO 27
- 2.3 Capstone/Keystone/Unicorn（反）汇编器 30
- 2.4 Frida 动态调试 Android Native 部分 32
- 2.5 Frida Instruction 模块动态反汇编 32
- 2.6 本章小结 34

第 3 章 静态分析工具的安装和基本使用 35
- 3.1 使用 objdump 反汇编目标文件命令显示二进制文件信息 35
- 3.2 使用 010 Editor 解析 SO 文件显示二进制基本信息 38

3.3 Ghidra/JEB/IDA 高级反汇编器 ··· 40

3.4 Binary Ninja 新晋反汇编器 ··· 47

3.5 本章小结 ··· 51

第 4 章 C 算法开发及模拟执行 ··· 52

4.1 Native 层密码学套件移植开发 ··· 52

4.2 Frida Hook/主动调用执行算法 ··· 62

4.3 使用 AndroidNativeEmu 模拟执行算法 ··· 63

4.4 本章小结 ··· 64

第 5 章 动态调试：GDB 动态调试、Hyperpwn/（内存）断点/栈帧 ···················· 65

5.1 GDB C/S 的调试架构 ··· 65

5.2 将 App 编译成带调试符号的 SO 文件 ··· 67

5.3 使用 Android 调试模式来启动 App ··· 69

5.4 Hyperpwn 调试入门 ··· 73

5.5 Objection+Frida+Hyperpwn 联合调试 ·· 79

5.6 本章小结 ··· 82

第 6 章 汇编开发：ARM 汇编原理/流程/调用约定/动态调试 ······························· 83

6.1 Android 和 ARM 处理器 ·· 83

6.2 ARM 原生程序的生成过程 ··· 84

6.3 汇编语言简介 ··· 87

6.3.1 汇编程序组成 ·· 87

6.3.2 ARM 处理器的工作模式与寻址方式 ·· 91

6.4 ARM 汇编指令及动态调试分析 ··· 94

6.5 多功能 CPU 模拟器：Unicorn ·· 109

6.6 本章小结 ··· 110

第 7 章 逆向分析：ELF 文件结构、节/区/表/段/符号/链接器 ····························· 111

7.1 操作系统 ELF 文件动态加载的基础知识 ··· 111

7.1.1 从几个问题入手 ·· 111

7.1.2 操作系统的核心概念 ·· 112

7.2 可执行文件的加载过程 ··· 114

7.3 使用 Unidbg 模拟执行 SO 文件中的函数 ································· 116
    7.3.1 Unidbg 框架的基本运作原理 ··································· 117
    7.3.2 Unidbg 各组件的基本功能 ······································ 119
    7.3.3 追踪 SO 文件的加载与解析流程 ································ 123
7.4 本章小结 ··· 126

第 8 章 反编译工具 IDA ··· 127

8.1 IDA 入门 ··· 127
    8.1.1 IDA 的安装与使用 ··· 127
    8.1.2 IDA 插件的使用 ··· 129
    8.1.3 IDA 反汇编介绍 ··· 130
    8.1.4 IDA 分析与 Frida Hook 结合 ·································· 136
8.2 动静态 SO 算法还原与脱机 ······································· 137
    8.2.1 IDA 动态调试 SO 算法 ·· 137
    8.2.2 Keypatch 原理/实战硬改算法逻辑 ······························ 149
8.3 本章小结 ··· 150

第 9 章 JNI 接口初识 ··· 151

9.1 JNI 及其工作原理 ··· 151
    9.1.1 NDK 简介 ·· 151
    9.1.2 JNI——NDK 具体的实现接口 ·································· 153
9.2 Frida 手动追踪 JNI 接口 ··· 163
9.3 jnitrace 自动追踪 JNI ·· 167
9.4 JNI 接口大横评 ··· 168
    9.4.1 Frida Hook 并主动调用 ·· 169
    9.4.2 jnitrace ··· 179
    9.4.3 ExAndroidNativeEmu ·· 179
9.5 本章小结 ··· 183

第 10 章 JNI 的特性：Java/Native 互相调用、反射/全局/局部引用 ········· 184

10.1 反射"滥用"类和对象的基本属性 ································· 184
    10.1.1 反射的概念与相关的 Java 类 ·································· 184
    10.1.2 实例：Xposed 刷机和编译使用的插件 ·························· 186

	10.1.3	反射设置/调用类和对象的域和方法 ································ 190
	10.1.4	来自 Native 层的反射调用追踪 ······································ 192
10.2	设计简单风控 SDK 并主动调用观察效果 ·· 194
	10.2.1	收集设备关键信息的常见方向和思路 ·································· 194
	10.2.2	Native 层使用反射调用 Java 层 API 获取设备信息 ··················· 195
10.3	本章小结 ··· 208

第 11 章 onCreate 进行 Native 化和引用 ·························· 209

11.1	将 onCreate 函数 Native 化 ·· 209
11.2	Java 内存管理 ·· 220
	11.2.1 C 和 Java 内存管理的差异 ·· 220
	11.2.2 JNI 的三种引用 ·· 220
11.3	本章小结 ·· 221

第 12 章 JNI 动静态绑定和追踪 ·································· 222

12.1	Dalvik 下动静态注册流程追踪 ·· 222
12.2	ART 下动静态注册流程追踪 ·· 228
12.3	本章小结 ·· 237

第 13 章 MD5 算法分析和魔改 ·································· 238

13.1	MD5 算法的描述 ·· 240
13.2	MD5 工程实现 ·· 246
13.3	哈希算法逆向分析 ·· 248
	13.3.1 Findcrypt/Signsrch 源码剖析 ·· 248
	13.3.2 算法识别插件的核心原理与改进方向 ································ 254
	13.3.3 使用 findhash 插件检测哈希算法 ······································ 255
	13.3.4 SHA1 算法逆向分析实战 ·· 258
13.4	哈希算法的扩展延伸 ·· 264
	13.4.1 哈希算法的特征 ·· 264
	13.4.2 大厂最爱：HMAC-MD5/SHA1 详解 ································ 266
13.5	Frida MemoryAccessMonitor 的使用场景 ·· 270
13.6	本章小结 ·· 277

第 14 章 对称加密算法逆向分析 278

14.1 DES 详解 278
14.1.1 分组密码的填充与工作模式 286
14.1.2 三重 DES 287

14.2 AES 289
14.2.1 AES 初识 289
14.2.2 深入了解 AES 296
14.2.3 Unicorn 辅助分析 302
14.2.4 AES 的工作模式 305

14.3 本章小结 312

第 15 章 读懂 DEX 并了解 DexDump 解析过程 313

15.1 环境及开发工具 313

15.2 认识 DEX 文件结构 314
15.2.1 DEX 文件格式概貌 314
15.2.2 DEX 文件格式项目搭建 315
15.2.3 DEX 文件详细分析 316

15.3 DexDump 解析 330
15.3.1 ULEB128 格式讲解 330
15.3.2 DexDump 解析过程 333

15.4 本章小结 338

第 16 章 ELF 文件格式解读及其生成过程 339

16.1 ELF 文件头 339
16.1.1 分析环境搭建 340
16.1.2 elf_header 341
16.1.3 program_header_table 344
16.1.4 section_header_table 344

16.2 ELF 可执行文件的生成过程与执行视图 352
16.2.1 ARM 可执行文件的生成过程 352
16.2.2 执行视图 357
16.2.3 GOT 和 PLT 359

16.3 本章小结 365

第 17 章　高版本 Android 函数地址索引彻底解决方案 ... 366

17.1　不同版本对于动态链接库的调用对比 ... 366
17.2　高版本加载 SO 文件 ... 367
17.2.1　自定义库查看库函数的偏移 ... 367
17.2.2　自定义库实现的背景 ... 368
17.2.3　自定义库 findsym 的实现 ... 370
17.3　SO 符号地址寻找 ... 372
17.3.1　通过节头获取符号地址 ... 372
17.3.2　模仿 Android 通过哈希寻找符号 ... 374
17.4　本章小结 ... 377

第 18 章　从 findExportByName 源码分析到 anti-frida 新思路 ... 378

18.1　两种模式下 anti-frida 的演示 ... 378
18.1.1　Frida attach 模式下的 anti-frida ... 378
18.1.2　Frida spawn 模式下的 anti-frida ... 380
18.2　源码分析 ... 383
18.2.1　Frida 编译 ... 384
18.2.2　源码追踪分析 ... 385
18.3　本章小结 ... 388

第 19 章　PLT 和 GOT 的 Hook ... 389

19.1　GOT 的 Hook ... 389
19.1.1　根据节头实现 Hook ... 392
19.1.2　根据程序头来实现 Hook ... 395
19.2　PLT 的 Hook ... 397
19.2.1　根据节头来实现 Hook ... 397
19.2.2　根据程序头来实现 Hook ... 403
19.3　从 GOT 和 PLT 的 Hook 到 xHook 原理剖析 ... 406
19.3.1　xHook 的优点 ... 406
19.3.2　源码赏析 ... 406
19.4　本章小结 ... 407

第 20 章 番外篇——另类方法寻找 SO 文件首地址 408

20.1 项目搭建 408
20.2 封装成库 413
20.3 通过 soinfo 的映射表遍历符号 415
20.4 dlopen 和 dlsym 获取符号地址 417
20.5 本章小结 418

第 21 章 Java Hook 的原理 419

21.1 Java 函数源码追踪 419
21.1.1 什么是 Java Hook 419
21.1.2 源码追踪 420
21.2 Java Hook 实践 424
21.3 Frida 中 Java Hook 的实现 429
21.3.1 Frida perform 源码追踪 429
21.3.2 Frida implementation 源码追踪 435
21.4 本章小结 437

第 22 章 inline Hook 中用到的汇编指令 438

22.1 两种 Hook 方式的介绍 438
22.2 定向跳转 441
22.3 寄存器保存 445
22.3.1 寄存器选择 445
22.3.2 3 种寄存器赋值的方案 446
22.4 本章小结 449

第 23 章 基于 Capstone 处理特殊指令 450

23.1 编译 Capstone 并配置测试环境 450
23.2 Capstone 官方测试案例演示 453
23.3 自定义汇编翻译函数 456
23.4 基于 Capstone 修正指令 459
23.4.1 指令修复的目的 459
23.4.2 修复指令的原理 460
23.4.3 指令修复的种类 460

23.5 本章小结 ... 461

第 24 章 inline Hook 框架集成 ... 462

24.1 inline Hook 框架测试 ... 462
24.2 结合 Capstone 框架 ... 467
24.3 本章小结 ... 469

第 25 章 通杀的检测型框架 r0Invoke ... 470

25.1 r0Invoke 牛刀小试脱壳 fulao2 ... 471
    25.1.1 APK 静态分析 ... 471
    25.1.2 使用 r0Invoke 脱壳 ... 471
    25.1.3 脱壳操作 ... 472
25.2 r0Invoke 进阶：跟踪所有运行在 ART 下的 Java 函数 ... 477
25.3 r0Invoke 主动调用 Native 函数并且修改参数 ... 479
25.4 r0Invoke Trace 高度混淆 OLLVM ... 482
25.5 本章小结 ... 484

第 26 章 SO 文件加载流程分析与注入实战 ... 485

26.1 SO 文件的加载方式 ... 485
26.2 SO 文件加载流程 ... 487
26.3 Frida Hook dlopen 和 android_dlopen_ext ... 491
26.4 编译 AOSP 注入 SO ... 496
    26.4.1 直接加载 sdcard 中的 SO 文件 ... 497
    26.4.2 加载私有目录的 SO 文件 ... 498
    26.4.3 编译 AOSP 注入 SO 文件 ... 500
26.5 注入优化 ... 502
26.6 本章小结 ... 504

第 1 章

基本开发环境配置

工欲善其事，必先利其器。本章将会介绍 Android 应用安全技术所需的环境配置，包括主机和测试机的基础环境。

一个良好的系统能给工作人员带来很多便利，不必因为环境问题焦头烂额。建议 Android 应用安全学习的新手，准备好与本书相同的环境，这样更有利于之后的学习，以免因环境配置问题而导致实验无法复现的可能性。

1.1 虚拟机环境搭建

推荐使用虚拟机而不是真机。

首先，虚拟机自带"系统时光机"功能——"快照"，虚拟机的这个特性让用户能够随时得到一个全新的真机，避免由于一个配置失误导致系统崩溃，最终只能选择重装系统而懊恼。图 1-1 所示为笔者在日常工作过程中开发 FART 脱壳机时创建的诸多虚拟机快照。

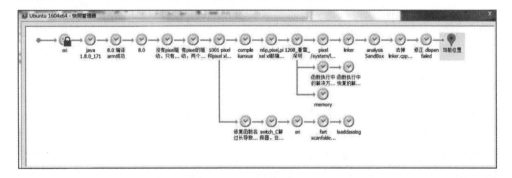

图 1-1 带快照功能的虚拟机

其次，虚拟机在工作环境中具有良好的隔离特性，可以确保在实验过程中不会"污染"真机，是测试全新功能的天然"沙盘"环境。推荐使用 VMware 公司出品的虚拟机软件系列。VMware 具有良好的跨平台特性，不同系统的虚拟机文件都能做到复制即用。

对于虚拟机环境的选择，笔者推荐使用 Ubuntu 系列的 Linux 操作系统。经过笔者测试，不论是编译 Android 源码，还是使用 Frida、GDB、Ollvm 等重要工具，这个操作系统病毒总是表现出更少受系统环境影响的特性。

在笔者的工作中，主要使用的是 Kali Linux 操作系统，Kali Linux 是基于 Debian 的 Linux 发行版，与 Ubuntu 师出同门，专门设计用于数字取证。Kali Linux 预装了许多渗透测试软件，包括 Metasploit、Burp Suite、SQLmap、Nmap 等，提供了一整套开箱即用的专业渗透测试工具。

Kali Linux 自带 VMware 镜像版本，下载并解压该镜像文件，然后双击打开.vmx 文件即可启动虚拟机。

图 1-2 为 Kali Linux 的界面。

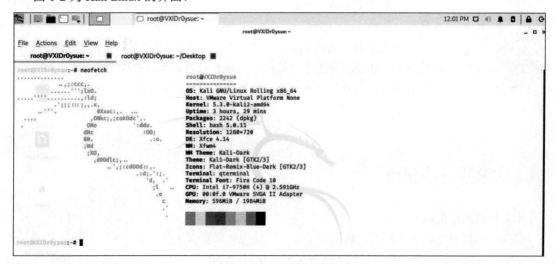

图 1-2 Kali Linux 的界面

因为虚拟机本身的时间可能与东八区时间不一致，所以需要进行时区设置。在启动虚拟机后，首先打开终端 Terminal，然后使用如下命令设置时区：

```
root@VXIDr0ysue:~# dpkg-reconfigure tzdata

Current default time zone: 'Asia/Shanghai'
Local time is now:      Sat Nov 14 12:11:05 CST 2020.
Universal Time is now:  Sat Nov 14 04:11:05 UTC 2020.
```

在弹出的窗口中选择 Asia→Shanghai 即可，读者可以根据自己所在的国家或地区进行选择。

另外，Kali Linux 默认不带中文支持，如果需要在中文网页上浏览或者进行抓包时解析包含中文的数据包，则还需要对 Kali 系统进行中文支持的配置，具体可执行如下命令：

```
root@VXIDr0ysue:~# apt update
```

```
root@VXIDr0ysue:~# apt install xfonts-intl-chinese

root@VXIDr0ysue:~# apt install ttf-wqy-microhei
```

> **注意** 一定不要把系统切换为纯中文环境，切换为纯中文环境会出现意想不到的问题。

1.2 逆向环境搭建

在准备好虚拟机环境之后，为了进行后面的逆向开发工作，还需要安装一些基础的开发工具。本节将介绍 Android Studio、ADB 工具、Python/Frida 环境以及 Objection 的安装配置和使用方法。

1.2.1 Android Studio 安装 NDK 编译套件

作为 Android 逆向开发人员，Android Studio 是一款必不可少的开发工具。在 Eclipse 退出 Android 开发历史舞台后，作为 Google 官方的 Android 应用开发 IDE，笔者首先推荐的就是这款软件。在虚拟机中，从官网下载和解压 Android Studio 之后，切换到 android-studio/bin 目录下，运行当前目录下的 studio.sh 即可启动 Android Studio。

首次启动 Android Studio 时，该软件会自动下载一些插件，比如 Android SDK 等工具，这些配置是后续开发所必需的，因此保持默认设置一直单击 Next 按钮即可。在插件下载完毕后，Android Studio 的界面将如图 1-3 所示。

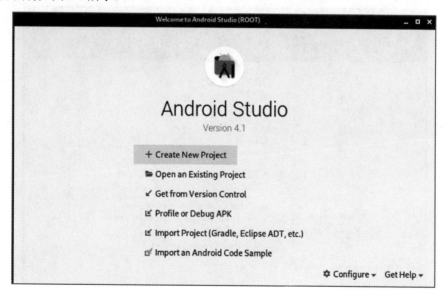

图 1-3 Android Studio 主界面

单击 Create New Project 选项，创建新工程并选择模板 Native C++，如图 1-4 所示。

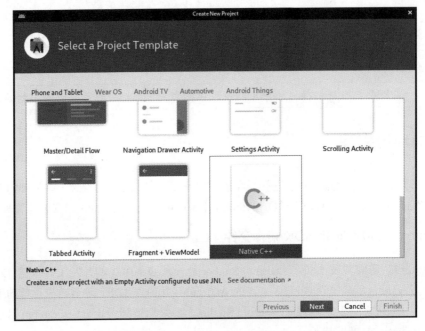

图 1-4 Native C++模板

下一步配置项目。项目的名称、包名以及保存位置可根据个人喜好自定义，语言选择 Java，最小 SDK 版本保持默认即可。单击 Next 按钮进入下一步，如图 1-5 所示。

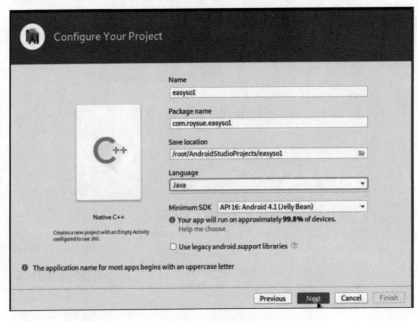

图 1-5 项目配置

在配置好这些选项后，单击 Finish 按钮，等待 Android Studio 完成配置和依赖的同步。在第一次创建 Project 时，Android Studio 会下载编译系统，下载完成之后进行编译时会发现没有 NDK（Native Development Kit），如图 1-6 所示。单击 Install latest NDK and sync project 安装最新版 NDK 即可。同步环节通常耗时比较久，这个时候只需要去喝杯茶静静等待即可。

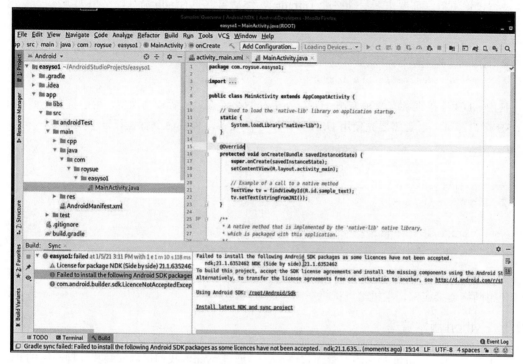

图 1-6　单击 Install latest NDK and sync project 以安装最新版 NDK

1.2.2　ADB 的配置和使用

ADB（Android Debug Bridge，Android 调试桥）是一种功能多样的命令行工具，用于与移动设备进行通信。由于 Android Studio 预先安装了 ADB，因此我们可以通过以下命令进入 ADB 的安装目录并查看其路径：

```
root@VXIDr0ysue:~# cd Android/Sdk/platform-tools
```

```
root@VXIDr0ysue:~/Android/Sdk/platform-tools# pwd
/root/Android/Sdk/platform-tools
```

这里使用 nano 命令将 ADB 工具所在目录加入环境变量，以便在任意目录下都能执行 adb 命令。

```
root@VXIDr0ysue:~/Android/Sdk/platform-tools# nano ~/.zshrc
```

文件打开后，将光标移动到文件末尾，添加一行代码：

```
export PATH="/root/Android/Sdk/platform-tools:$PATH"
```

读者可参考终端窗口最下方的提示，保存修改并退出文件。在将 adb 命令加入环境变量后，为了使得设置生效，需要重新启动终端 Terminal，然后再次执行 adb 命令，结果如下：

```
root@VXIDr0ysue:~# adb shell
* daemon not running; starting now at tcp:5037
* daemon started successfully
adb: no devices/emulators found
```

使用 ADB 连接设备有两种方法：USB 连接和网络连接。前者较为简单，用手机数据线接上主机 USB 接口即可。后者需要先使用 USB 数据线连接设备，再在命令行执行以下操作：

```
root@VXIDr0ysue:~# adb tcpip 5555

root@VXIDr0ysue:~# adb connect 10.203.171.25:5555
failed to authenticate to 10.203.171.25:5555

root@VXIDr0ysue:~# adb connect 10.203.171.25:5555
already connected to 10.203.171.25:5555
```

第一次尝试连接的时候，在手机上单击"允许调试"，再次尝试即可连接成功。需要注意的是，使用网络连接应保证主机与测试机在同一局域网。

1.2.3 Python 版本管理

笔者推荐使用 Python 版本的包管理软件 pyenv。通过 pyenv，我们可以安装和管理不同的 Python 版本，每个由 pyenv 包管理软件安装的 Python 版本都是相互隔离的。换句话说，无论在这个 Python 中安装了多少依赖包，对于另一个 Python 版本都是不可见的。如果你觉得自己的 Python 环境不纯或者需要一个新的环境，随时可以安装一个新的纯净的 Python，这也可以看作是一种特殊的虚拟机环境。

需要注意的是，在安装 pyenv 之前，建议读者先对虚拟机进行一次快照。这样做是为了防止在安装 pyenv 的最后一步依赖时，可能导致整个系统无法进入桌面环境。下面是安装 pyenv 的具体过程：

```
root@VXIDr0ysue:~# git clone https://github.com/pyenv/pyenv.git ~/.pyenv
Cloning into '/root/.pyenv'

done.
Resolving deltas: 100% (12507/12507), done.

root@VXIDr0ysue:~# echo 'export PYENV_ROOT="$HOME/.pyenv"' >> ~/.zshrc
```

```
root@VXIDr0ysue:~# echo 'export PATH="$PYENV_ROOT/bin:$PATH"' >> ~/.zshrc

root@VXIDr0ysue:~# echo -e 'if command -v pyenv 1>/dev/null 2>&1; then\n  eval "$(pyenv init -)"\nfi' >> ~/.zshrc

root@VXIDr0ysue:~# exec "$SHELL"

root@VXIDr0ysue:~# sudo apt-get update; sudo apt-get install --no-install-recommends make build-essential libssl-dev zlib1g-dev libbz2-dev libreadline-dev libsqlite3-dev wget curl llvm libncurses5-dev xz-utils tk-dev libxml2-dev libxmlsec1-dev libffi-dev liblzma-dev
```

如果安装后重启能够正常进入桌面环境，接下来就可以方便地使用 pyenv install 命令安装不同版本的 Python 了，在安装完毕后还需要运行 pyenv local 命令切换到对应的版本。例如安装 Python 3.8.5 的过程如下：

```
root@VXIDr0ysue:~# pyenv install 3.8.5

root@VXIDr0ysue:~# pyenv local 3.8.5

root@VXIDr0ysue:~# python -V
Python 3.8.5
```

使用以上方法，可以再安装一个 3.8.0 版本的 Python，在之后的实验过程中，我们将会用到这两个版本的 Python 环境。

1.2.4　移动设备环境准备

在讲解后续虚拟机上的逆向环境配置之前，我们先准备移动设备。对于手机环境的搭建，建议参考以下文章，文章内容从硬件准备到刷机流程都讲解得非常详细，本书的实验环境也是按照下述文章进行配置的，参考网址 1（参见配书资源文件）。

另外，笔者推荐安装投屏软件 QtScrcpy，参考网址 2（参见配书资源文件）安装使用。

1.2.5　Frida 版本管理

移动设备环境准备好以后，就可以安装我们的主角——重要的逆向工具 Frida。在本书的实验中，Frida 安装了两个版本：一个是最新版，另一个是 12.8.0 版，后者运行一些相对较旧的代码比较管用。官网安装 Frida 的命令如下：

```
root@VXIDr0ysue:~# pip install frida-tools
```

安装成功后，可以在终端 Terminal 看到 Frida 安装的版本号（笔者在进行实验时，Frida 最新版本为 14.2.13）。在网址 3（参见配书资源文件）的 Frida 网站找到与之对应的版本号的 frida-server 进行下载。例如，安装版本号为 14.2.13 的 Frida，其对应的 frida-server 是 frida-server-14.2.13-android-arm64.xz。

下载 frida-server 后，使用 7z 解压并通过 adb push 命令将解压后的 frida-server 文件上传到 /data/local/tmp 目录下。接着，进入手机 Shell 并切换到 root 用户下的 /data/local/tmp 目录，也就是 frida-server 文件所在的位置，在该目录下，为 frida-server 文件赋予执行权限并运行它。

注意，在运行 14.2.13 版本的 frida-server 之前，应保证此时虚拟机中的 Python 环境为 3.8.5。

具体过程如下：

```
root@VXIDr0ysue:~/Desktop # 7z x frida-server-14.2.13-android-arm64.xz

root@VXIDr0ysue:~/Desktop # adb push frida-server-14.2.13-android-arm64 /data/local/tmp

root@VXIDr0ysue:~/Desktop # adb shell
bullhead:/ $ su
bullhead:/ # cd /data/local/tmp/
bullhead:/ # chmod 777 frida-server-14.2.13-android-arm64
bullhead:/ # ./frida-server-14.2.13-android-arm64
```

另外，可以使用以下命令查看手机的进程：

```
root@VXIDr0ysue:~/Desktop # frida-ps -U
```

1.2.6 Objection 的安装和使用

Objection 是基于 Frida 的命令行 Hook 工具，可以不写代码，简单地使用命令就可以对 Java 函数进行 Hook。安装 Objection 的命令如下：

```
root@VXIDr0ysue:~# pip install objection
```

如果使用基于特定 Frida 版本的 Objection，则需要先安装特定版本的 Frida 和 frida-tools。然后，在 Objection 的 Releases 页面中找到在该 Frida 版本之后发布的 Objection 版本。在本文的实验中，使用的是 12.8.0 版本的 Frida。首先切换到 Python 环境 3.8.0，再执行如下指令：

```
root@VXIDr0ysue:~/Desktop # pyenv local 3.8.0

root@VXIDr0ysue:~/Desktop # pip install frida==12.8.0

root@VXIDr0ysue:~/Desktop # pip install frida-tools==5.3.0

root@VXIDr0ysue:~/Desktop # pip install objection==1.8.4
```

按照这个顺序安装 Objection 时，系统会直接显示 Requirement already satisfied（需求已满足），不会再下载新的 Frida 来安装了。完成后，下载 12.8.0 版本的 frida-server，具体过程可参照 1.2.5 节的步骤。接下来，在手机 Shell 中运行 12.8.0 版本的 frida-server，并使用 Objection 打开 Settings，如图 1-7 所示。

第 1 章　基本开发环境配置　　9

图 1-7　使用 Objection 打开 Settings

```
root@VXIDr0ysue:~/Desktop # pyenv local 3.8.0

root@VXIDr0ysue:~/Desktop # objection -g com.android.settings explore
Using USB device `LGE Nexus 5X`
Agent injected and responds ok!

     _         _
 ___ | |_  _ ___ ___ | |_  _ ___ ___
| . || . || | | -_|  _||  _|| . || . |
|___||___||___|___||___||___||___||_|
      |___|(object)inject(ion) v1.8.4

     Runtime Mobile Exploration
        by: @leonjza from @sensepost

[tab] for command suggestions
com.android.settings on (google: 8.1.0) [usb] #
```

可以看到手机的设置打开了。

查看所有的类：

```
om.android.settings on (google: 8.1.0) [usb] # android hooking list classes
```

同理，如果使用 14.2.13 版本的 Frida，则在手机上运行 14.2.13 版本的 frida-server，还要将虚拟机上的 Python 环境切换至 3.8.5 版本，否则会报错。之后，再次使用 Objection 打开 Settings，我们将看到不同的 Objection 版本号。

1.3　Frida 基本源码开发环境搭建

本节继续讲解 Frida 基本源码开发环境的搭建。在这个过程中，我们需要用到带有包管理器的 Node.js。

打开网址 4（参见配书资源文件），选择 Installation instructions，然后执行以下命令：

```
root@VXIDr0ysue:~/Desktop# curl -fsSL https://deb.nodesource.com/setup_15.x | bash -

root@VXIDr0ysue:~/Desktop# apt-get install -y nodejs

root@VXIDr0ysue:~/Desktop# node -V
v15.12.0
```

安装好 Node 后，就可以开发项目了：

```
root@VXIDr0ysue:~# cd Desktop

root@VXIDr0ysue:~/Desktop# mkdir 20210105
```

启动 VS Code→20210105 文件夹，创建文件 demo.js。此时，还没有代码提示功能，我们需要安装 Gum，以便在任意工程中获得 Frida 代码提示、补全和 API 查看功能。

```
root@VXIDr0ysue:~/Desktop# npm install --save @types/frida-gum
```

重启 VS Code，即可获得代码提示。VS Code 是指 Visual Studio Code，是微软公司向开发者们提供的一款跨平台源代码编辑器。

接下来，在 demo.js 中编写代码实现基本的 Hook 项目 easyso1 中的 onCreate 函数，内容见代码清单 1-1。

代码清单 1-1　demo.js

```
setImmediate(function(){
    Java.perform(function(){
        Java.use("com.r0ysue.easyso1.MainActivity").onCreate.implementation = function(x){
            console.log("Entering onCreate!");
            return this.onCreate(x);
        }
    })
})
```

打开终端 Terminal，使用以下指令运行脚本：

```
root@VXIDr0ysue:~/Desktop/20210105 # frida -U -f com.roysue.easyso1 -l demo.js
--no-pause
```

结果如图 1-8 所示，说明我们 Hook 上了。

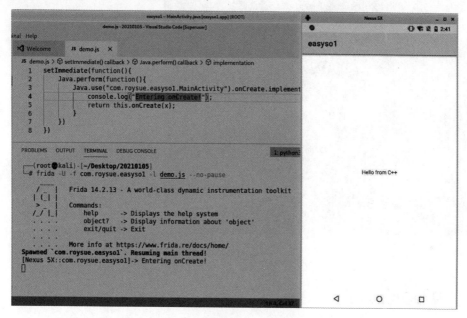

图 1-8　Hook onCreate 函数

继续尝试 Hook 项目 easyso1 中的 stringFromJNI 方法，内容见代码清单 1-2。

代码清单 1-2　demo.js

```
setImmediate(function(){
    Java.perform(function(){
        Java.use("com.r0ysue.easyso1.MainActivity").onCreate.implementation = function(x){
            console.log("Entering onCreate!");
            return this.onCreate(x);
        }
        Java.use("com.r0ysue.easyso1.MainActivity").stringFromJNI.implementation = function(){
            var result = this.stringFromJNI();
            console.log("Return value of stringFromJNI is => ", result);
            return result;
        }
    })
})
```

结果如图 1-9 所示,说明我们 Hook 成功了。

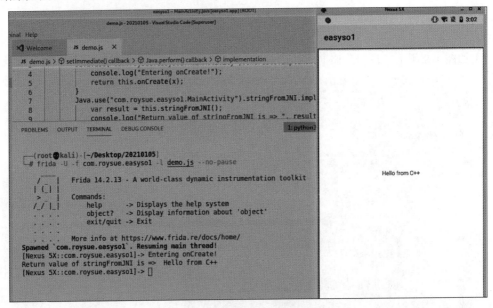

图 1-9 Hook stringFromJNI 方法

1.4 初识 NDK

笔者在这里推荐网址 5(参见配书资源文件)的 NDK 入门指南:读者可以参阅此文档来了解 NDK 开发的相关知识。

NDK 开发使用编译型语言 C++,相较于脚本型语言 Java,并不适合初学者。C/C++代码被编译成 CPU 代码,可以直接在 CPU 上运行,而 Java 代码会被编译成中间语言,在解释器中进行解释和执行。很明显,编译型语言执行起来效率更高,而脚本型语言的优势在于代码只需要编译一次,无须修改便可在任意平台上运行。因此,为了提高性能、保护程序的逻辑和代码,可以使用 C++来开发,C++的语言特性更加难以被反编译和分析。

在 NDK 入门指南文档中,比较关键的内容是"创建 CMake 构建脚本"。对于 CMake 如何对原生源文件进行编译,由 CMake 的构建脚本中的一些重要选项来指定。如果原生库已有 CMakeLists.txt 构建脚本,则可直接使用;否则需要自己创建一个编译脚本。CMakeLists.txt 脚本文件在项目的 app/src/main/cpp 路径下,如图 1-10 所示。

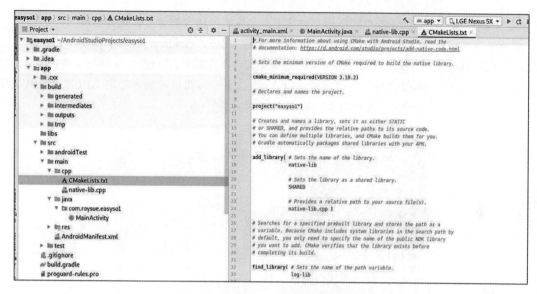

图 1-10　CMakeLists.txt 所在的路径

由于不同设备使用不同的 CPU，不同的 CPU 支持的指令集也不同，因此需要提供对应的二进制接口交互规则才能进行交互。Android ABI（Application Binary Interface，应用程序二进制接口）相当于 CPU 指令集的一种模式和调用规范。读者可参阅文档了解 Android ABI，在后续的章节中，我们还会进一步学习 Android ABI 的知识。了解 CPU 和架构对于开发非常重要。因为代码是直接在 CPU 上运行的，如果不了解 CPU 和架构，就无法理解不同的寄存器和指令代表的含义。不同的 CPU 架构所支持的 ABI 如图 1-11 所示。

支持的 ABI

表 1. ABI 和支持的指令集。

ABI	支持的指令集	备注
armeabi-v7a	• armeabi • Thumb-2 • VFPv3-D16	与 ARMv5/v6 设备不兼容。
arm64-v8a	• AArch64	
x86	• x86 (IA-32) • MMX • SSE/2/3 • SSSE3	不支持 MOVBE 或 SSE4。
x86_64	• x86-64 • MMX • SSE/2/3 • SSSE3 • SSE4.1、4.2 • POPCNT	

★ 注意：NDK 以前支持 ARMv5 (armeabi) 以及 32 位和 64 位 MIPS，但 NDK r17 已不再支持。

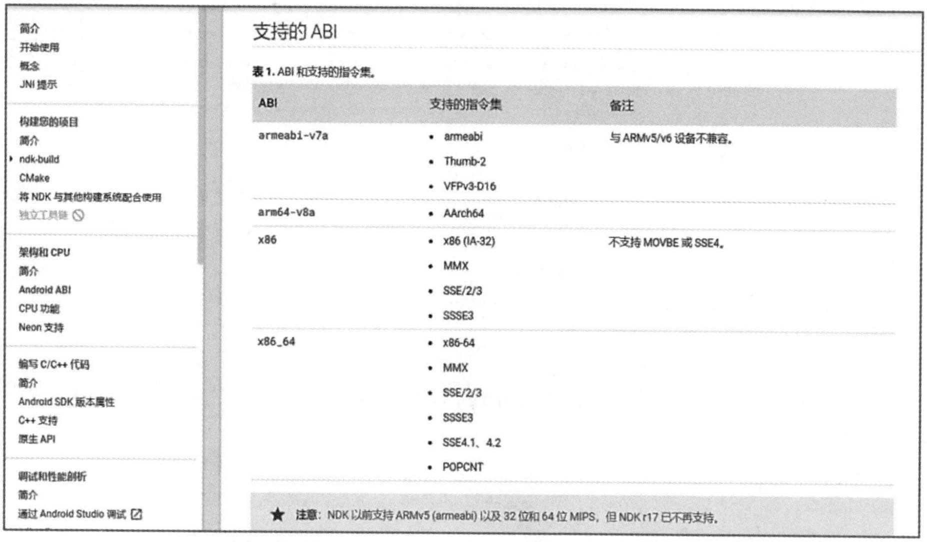

图 1-11　不同的 CPU 架构支持的 ABI

通常情况下，我们从 APK 中解压出文件夹时，文件夹中包含的子文件夹会出现如图 1-11 所示的文件名，文件夹的命名表示对应的 ABI 架构。默认情况下，Gradle 会针对手机支持的 ABI 进行构建。我们可以从终端 Terminal 进入 app-debug.apk 文件所在的位置，使用 7z 解压文件，再从文件管理器打开 app-debug.apk 文件所在的目录，进入 lib 文件夹，可以看到为特定 ABI 生成的代码文件夹，其命名即为手机支持的 ABI 版本，如图 1-12 所示。

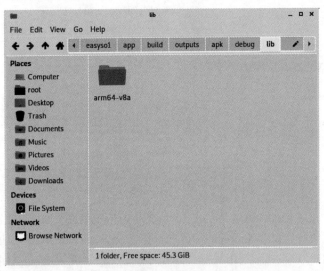

图 1-12　手机支持的 ABI 版本

如果需要为特定 ABI 生成代码，可通过在项目的 Gradle 中更改配置来针对不同的 ABI 进行构建，如图 1-13 所示。

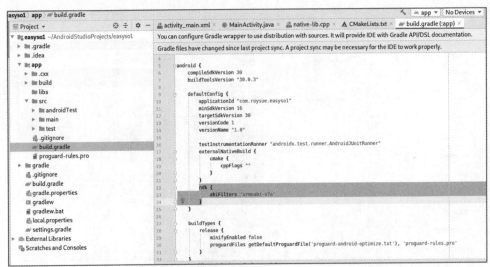

图 1-13　修改 Gradle 配置

在修改完成后，保存文件并重新运行 App，解压 app-debug.apk 文件后，打开 lib 文件夹，我们可以看到为特定 ABI 生成的代码文件夹的命名发生了变化，如图 1-14 所示。

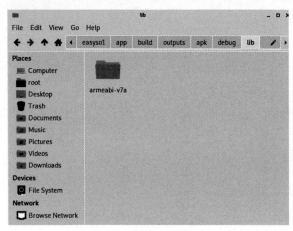

图 1-14　armeabi-v7a

查看系统主要的 ABI 接口需要用到 Objection。首先，安装 Wallbreaker 插件，启动 Objection：

```
root@VXIDr0ysue:~/Desktop # git clone git clone https://github.com/hluwa/
Wallbreaker.git

root@VXIDr0ysue:~/Desktop# objection -g com.android.settings explore

com.android.settings on (google: 8.1.0) [usb] # plugin load /root/Desktop/Wallbreaker
Loaded plugin: wallbreaker
com.android.settings on (google: 8.1.0) [usb] # plugin wallbreaker classdump
android.os.Build
```

执行上述命令后，终端显示出系统主要的 ABI 接口，如图 1-15 所示。

```
static String[] SUPPORTED_32_BIT_ABIS; => armeabi-v7a,armeabi
static String[] SUPPORTED_64_BIT_ABIS; => arm64-v8a
static String[] SUPPORTED_ABIS; => arm64-v8a,armeabi-v7a,armeabi
```

图 1-15　系统主要的 ABI 接口

接下来，将 arm64-v8a 添加到 Gradle 配置文件的 NDK 中，运行 App，这时优先以 64 位运行（因为当前系统本来就是 64 位的）。

再次启动 Objection：

```
root@VXIDr0ysue:~/Desktop # objection -g com.roysue.easyso1 explore

com.roysue.easyso1 on (google: 8.1.0) [usb] # frida
-------------------  -----------
Frida Version        14.2.13
```

```
Process Architecture    arm64
Process Platform        linux
Debugger Attached       False
Script Runtime          QJS
Script Filename         /script1.js
Frida Heap Size         7.5 MiB
-------------------     -----------
```

从 Process Architecture 表项可以看到进程架构是 64 位。此外，我们也可以把 App 强制安装成 32 位。卸载 App，并重新解压 app-debug.apk 文件，可以发现 lib 中有两个架构，如图 1-16 所示。

使用 ADB 强制安装指定架构：

```
root@VXIDr0ysue:~/…/build/outputs/apk/debug # adb install -r -t --abi armeabi-v7a app-debug.apk
Performing Streamed Install Success
```

然后用 Objection 查看。如果 Objection 报错，则可切换 Python 和 Frida 版本再次尝试。如图 1-17 所示，我们可以看到 Process Architecture 变成了 32 位的。

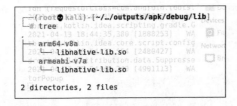
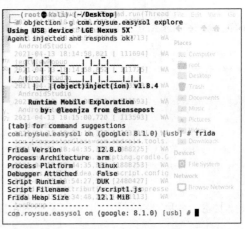

图 1-16　tree 查看两个架构　　　　图 1-17　强制安装成 32 位架构

1.5　其他工具

至此，我们基本完成了环境准备工作。另外，笔者还要推荐一些在日常工作中使用的小工具。尽管这些工具可能不会直接对工作产生影响，但一旦拥有了这些工具，用户在日常工作中可以更准确地掌握自己的工具环境。

首先，推荐使用 htop 这款加强版 top 工具。htop 可以动态地查看当前活跃的、占用资源较多的进程，如图 1-18 所示。这一点在编译 Android 源码时非常好用，当我们执行 make 命令之后，可以肉眼可见的速度看到内存 Mem 跑到底之后，开始侵占 Swp 的进度条。Uptime 表示系统的运行时间，

Load average 表示平均负载，比如我们的系统有一个四核CPU，在平均负载跑到4时说明系统满载了。有关 htop 的其他操作教程，读者可以在网上搜索。

另外，还要推荐一款实时查看系统网络负载的工具 jnettop。在安装和使用软件（比如 Frida）的过程中，jnettop 可以实时查看其下载和安装进度。甚至在 AOSP 编译过程中，仍然可以观察到它连接到国外服务器进行依赖包下载等操作。除此之外，在抓包时打开这个工具往往会有奇效，比如实时查看对方的 IP 等信息。jnettop 界面如图 1-19 所示，我们可以清楚地看到主机连接的远程 IP、端口、速率以及协议等信息。

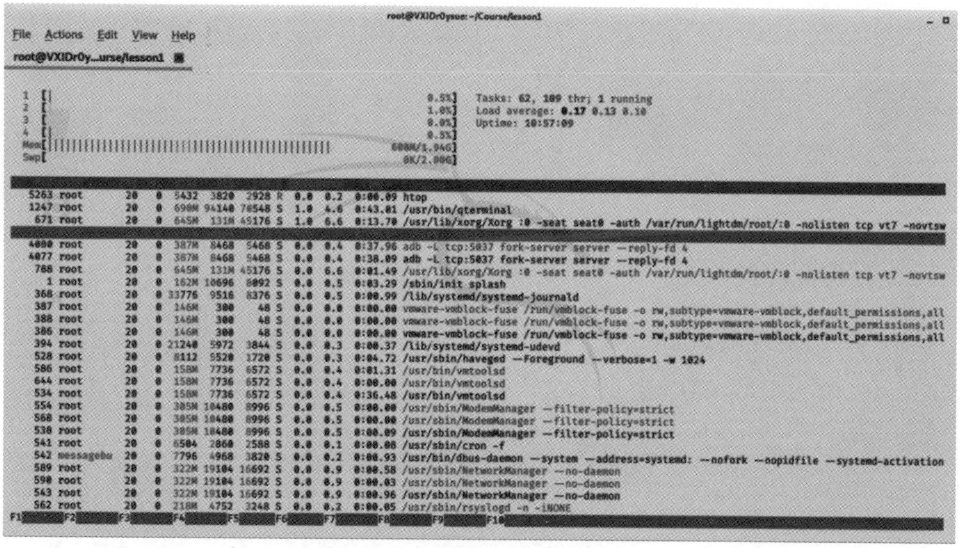

图 1-18　htop 界面

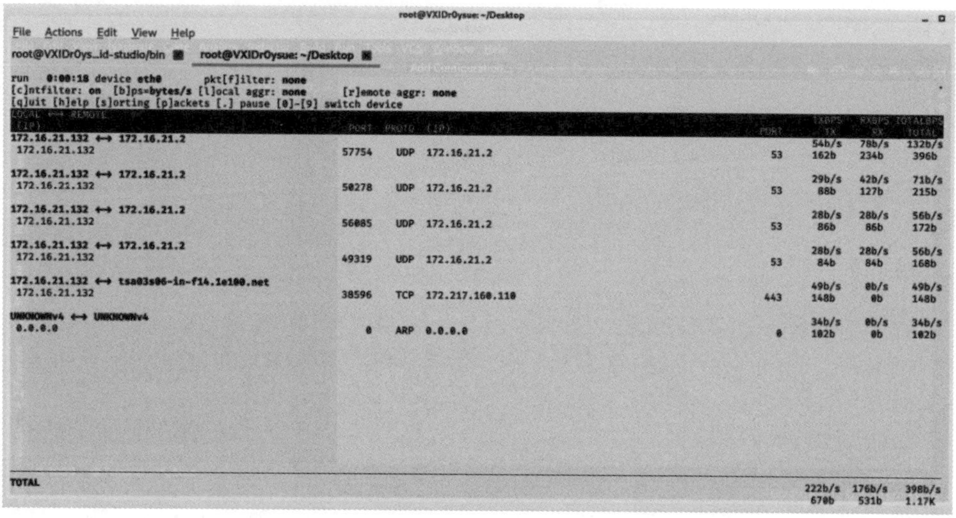

图 1-19　jnettop 界面

1.6 本章小结

本章主要介绍了 Android 应用安全技术中常用的一些基础环境配置，虽然这一章没有过多的技术内容，但它却是整本书的基石。良好的环境配置能够节省学习过程中的时间，从而大大提高学习效率。

第 2 章

Android SO 动态调试入门

通过第 1 章的学习,我们完成了基本的环境配置。在本章中,我们将介绍 NDK 程序的动态调试、交叉编译以及汇编/反汇编工具的简单应用。

2.1 Android SO 基本动态分析调试

SO(Shared Object,共享库)是机器可以直接运行的二进制代码,它是 Android 上的动态链接库,类似于 Windows 上的 DLL。Android SDK 编译出来的 SO 文件有好几种,以适配不同的 CPU 架构。一般人并不需要了解 SO 文件,但如果要做逆向分析工作,就必须了解 SO 文件,并且具备对 SO 文件进行分析和调试的能力。

2.1.1 第一个 NDK 程序

ndk-samples 中提供了非常多的案例,我们选择其中最简单的 hello-jni,打开它来安装一下,结果如图 2-1 所示。

在 hello-jni.c 文件中,以下代码决定了显示内容:

```
return (*env)->NewStringUTF(env, "Hello from JNI !
Compiled with ABI " ABI ".");
```

其中,ABI 是预定义的。如果想要更改显示的内容,则需要在项目的 build.gradle 文件中添加如图 2-2 所示的代码。

图 2-1 hello.jni 运行结果

提示　什么是 ABI？

每一种 CPU 架构都定义了一种 ABI（Application Binary Interface，应用程序二进制接口），ABI 定义了对应的 CPU 架构能够执行的二进制文件（如 SO 文件）的格式规范，决定了二进制文件如何与系统进行交互。

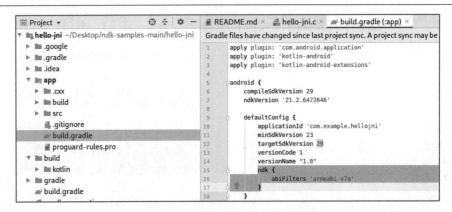

图 2-2　改变 hello-jni 显示的内容

删除 productFlavors 中的 arm64-v8a，保留 armeabi-v7a，以免配置被覆盖，如图 2-3 所示。

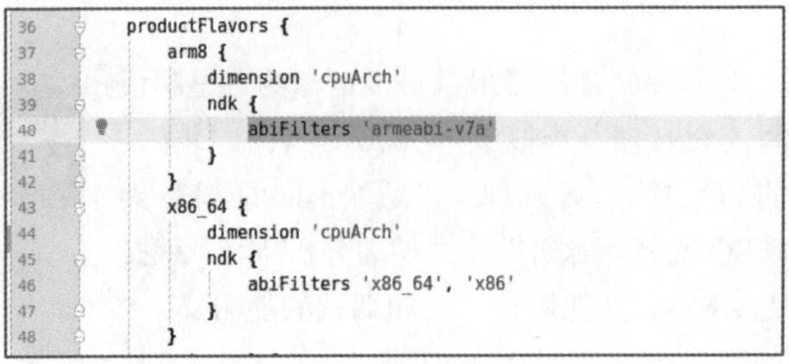

图 2-3　保留 armeabi-v7a

在弹出的提示中单击 Sync Now，再次进行编译并运行，可以看到应用界面显示出 armeabi-v7a。

2.1.2　动态调试 NDK 程序

通过运行 hello-jni，我们对 NDK 开发有了一个初步的印象。接下来，将通过编写一个简单的 C 语言程序来学习 SO 的基本动态分析调试方法。

首先，在 app/src/main/cpp 路径下新建一个文件 roysue.c，右击 cpp 文件夹选择 New→C/C++ Source File。我们可以复制 ndk-samples 中的程序 hello-jni，并在此基础之上进行改动，修改后的代码如下：

```
#include <string.h>
#include <jni.h>
```

```c
/* This is a trivial JNI example where we use a native method
 * to return a new VM String. See the corresponding Java source
 * file located at:
 *
 *   hello-jni/app/src/main/java/com/example/hellojni/HelloJni.java
 */
JNIEXPORT jstring JNICALL
Java_com_roysue_easyso1_MainActivity_stringFromJNI( JNIEnv* env, jobject thiz )
{
#if defined(__arm__)
   #if defined(__ARM_ARCH_7A__)
     #if defined(__ARM_NEON__)
       #if defined(__ARM_PCS_VFP)
         #define ABI "armeabi-v7a/NEON (hard-float)"
       #else
         #define ABI "armeabi-v7a/NEON"
       #endif
     #else
       #if defined(__ARM_PCS_VFP)
         #define ABI "armeabi-v7a (hard-float)"
       #else
         #define ABI "armeabi-v7a"
       #endif
     #endif
   #else
    #define ABI "armeabi"
   #endif
#elif defined(__i386__)
#define ABI "x86"
#elif defined(__x86_64__)
#define ABI "x86_64"
#elif defined(__mips64)   /* mips64el-* toolchain defines __mips__ too */
#define ABI "mips64"
#elif defined(__mips__)
#define ABI "mips"
#elif defined(__AArch64__)
#define ABI "arm64-v8a"
#else
#define ABI "unknown"
#endif

    return (*env)->NewStringUTF(env, "Hello from JNI !  Compiled with ABI " ABI ".");
}
```

此时，我们会在代码区域上方看到一行提示，如图2-4所示。

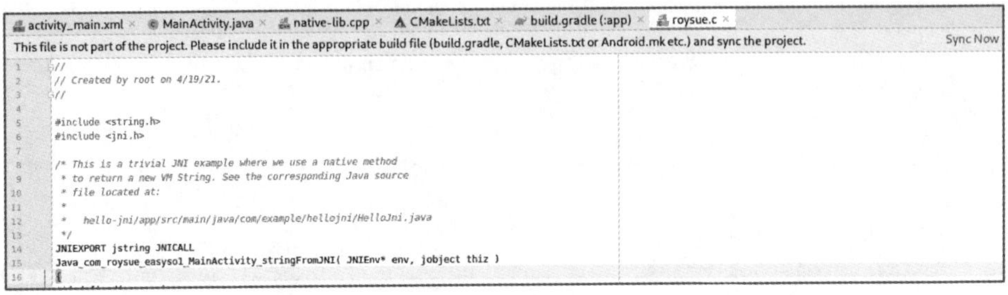

图 2-4　Sync 提示

根据提示，打开 build.gradle，在 externalNativeBuild 中可以看到项目使用 CMake 的编译系统生成 CMakeLists.txt。因此，当我们新建文件之后，需要把它加入 CMakeLists.txt 中，否则编译系统无法识别新建的文件。打开 CMakeLists.txt，将 native-lib 全部替换为 roysue，再将 add_library 中的 roysue.cpp 改为 roysue.c，然后把不需要的 native-lib.cpp 文件删掉。

回到 roysue.c，在提示中单击 Sync Now 进行同步。除此之外，还需要在 MainActivity 中修改加载模块为 roysue，如图 2-5 所示。

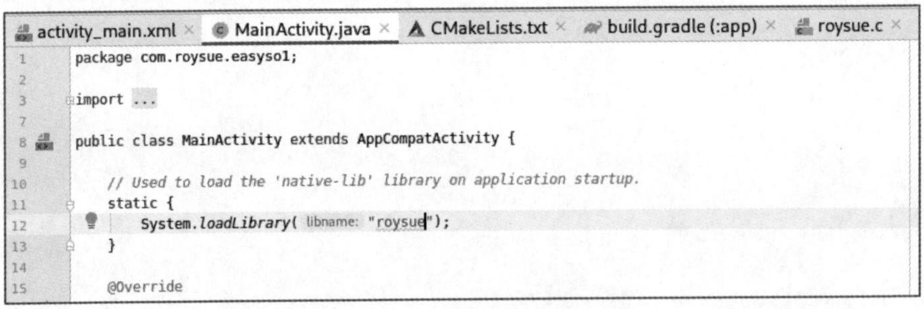

图 2-5　修改加载模块

这时就可以运行项目了。

接下来，我们调试一个简单的 C 语言程序来实现数字求和，并以 info 的日志权限打印出结果。首先，在 roysue.c 中包含头文件<android/log.h>，在函数体中添加一段 C 循环代码如下：

```
int i=1, sum=0;
    while(i<=10){
        sum+=i;
        i++;
        __android_log_print(ANDROID_LOG_INFO, "r0ysue", "now sum is %d", sum);
    }
```

运行程序，可以在日志输出中看到迭代求和结果，如图 2-6 所示。

第 2 章　Android SO 动态调试入门

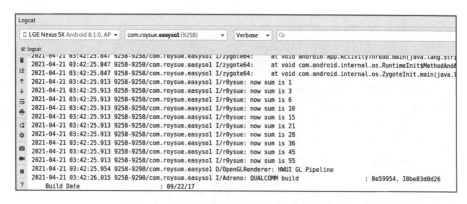

图 2-6　以 info 的日志权限打印出求和结果

对于使用 Android Studio 调试应用的具体流程，网址 6（参见配书资源文件）的 Android 官方文档提供了详细说明。

这里，我们以 roysue.c 为例进行演示。

首先，在"sum+=1;"一行处加上断点，单击调试键开始调试，程序遇到断点后，进入 Debugger，就可以开始调试（Debug）了。调试主要使用图 2-7 中框出的前 4 个按钮，从左至右依次是：Step Over、Step Into、Force Step Into、Step Out，如图 2-7 所示。

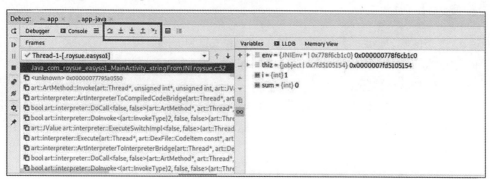

图 2-7　Debug 的 5 个按钮（常用前 4 个）

- Step Over：前进到代码中的下一行，如果下一行是方法调用，则不进入方法。
- Step Into：前进到代码中的下一行，如果下一行是方法调用，则进入方法（仅限于自定义的方法）。
- Force Step Into：前进到代码中的下一行，如果下一行是方法调用，则进入方法（强制进入，包括系统方法）。
- Step Out：前进到当前方法之外的下一行（跳出方法）。

主要就是通过操作上述 4 个按钮来进行调试，观察图 2-7 右边 Variables 区域所显示的变量信息的变化，以便对程序进行修改。

从图 2-7 中可以看到，除了 Variables 标签页以外，还有一个标签页 LLDB，进入后可以看到 LLDB 的命令行。在命令行中，输入 LLDB 命令可以实现很多强大的功能，比如打印变量、寻址、调用堆栈等，通过这些命令可以有效地帮助我们调试 NDK 程序。

例如，在 LLDB 的命令行输入 dis（dissemble）执行反汇编操作，如图 2-8 所示。

```
(lldb) script exec("if libstdcxx_printers_available: import printers")
(lldb) script exec("if libstdcxx_printers_available: printers.register_libstdcxx_printers(None)")
(lldb) script exec("if lldb.debugger.GetCategory('libstdc++-v6').IsValid(): lldb.debugger.GetCategory('gnu-libstdc++')
(lldb) dis
libroysue.so`Java_com_roysue_easyso1_MainActivity_stringFromJNI:
    0x777935f63c <+0>:   sub    sp, sp, #0x30             ; =0x30
    0x777935f640 <+4>:   stp    x29, x30, [sp, #0x20]
    0x777935f644 <+8>:   add    x29, sp, #0x20            ; =0x20
    0x777935f648 <+12>:  mov    w8, #0x1
    0x777935f64c <+16>:  stur   x0, [x29, #-0x8]
    0x777935f650 <+20>:  str    x1, [sp, #0x10]
    0x777935f654 <+24>:  str    w8, [sp, #0xc]
    0x777935f658 <+28>:  str    wzr, [sp, #0x8]
    0x777935f65c <+32>:  ldr    w8, [sp, #0xc]
    0x777935f660 <+36>:  cmp    w8, #0xa                  ; =0xa
    0x777935f664 <+40>:  cset   w8, gt
    0x777935f668 <+44>:  tbnz   w8, #0x0, 0x777935f6a8    ; <+108> at roysue.c:57
->  0x777935f66c <+48>:  ldr    w8, [sp, #0xc]
    0x777935f670 <+52>:  ldr    w9, [sp, #0x8]
```

图 2-8　LLDB 反汇编

2.1.3　交叉编译

通过 2.1.2 节的学习，我们基本了解了如何在 Android Studio 上调试一个 NDK 程序。然而在实际应用中，如何将一个 C 语言程序移植到已有的 NDK 工程中呢？接下来，我们将通过演示一个稍微复杂一些的案例来介绍交叉编译。

以 MD5 算法（消息-摘要算法）为例，MD5 是计算机安全领域广泛使用的一种散列函数，用以提供消息的完整性保护，可作为数字签名入门算法。首先，在 GitHub 上找一个 C 语言实现的 MD5 算法，复制到本地：

```
root@VXIDr0ysue:~/Desktop # mkdir 20210110

root@VXIDr0ysue:~/Desktop # cd 20210110

root@VXIDr0ysue:~/Desktop/20210110 # git clone https://github.com/pod32g/MD5.git
```

用 VS Code 打开 md5.c，可以在"扩展"搜索 C/C++ 插件并进行安装，以便查看函数实现。安装好插件后，按住 Ctrl 键，单击函数即可查看函数实现。编译并运行 md5.c：

```
root@VXIDr0ysue:~/Desktop/20210110/md5 # gcc -o md5 md5.c

root@VXIDr0ysue:~/Desktop/20210110/md5 # ls
md5  md5.c  README.md
```

```
root@VXIDr0ysue:~/Desktop/20210110/md5 # ./md5 r0ysue
8e2b0b38f557578f578ef21b33e4df0d
```

通过访问网址 7（参见配书资源文件），将计算结果在 CyberChef 网站上与标准结果进行比对，看是否相同，若相同，则说明这个程序是一个标准的 MD5 算法的实现。

查看编译后的 MD5 可执行文件格式：

```
root@VXIDr0ysue:~/Desktop/20210110/md5 # file md5
md5: ELF 64-bit LSB pie executable, x86-64, version 1 (SYSV), dynamically linked, interpreter /lib64/ld-linux-x86-64.so.2, BuildID[sha1]=14a627a2ddebaa8f4b86b02bb9b259922ed34255, for GNU/Linux 3.2.0, not stripped
```

如果我们要把它编译到手机上，该如何操作呢？由于手机与虚拟机指令级架构不同，因此需要借助 NDK 提供的工具链对现有的 C 语言代码进行构建，即进行交叉编译。关于交叉编译的具体用法，可参考网址 8（参见配书资源文件）文档。

下面，我们以 md5.c 为例进行交叉编译的演示。

首先，进入 ndk 目录，选择最新版本：

```
root@VXIDr0ysue:~/Desktop/20210110/md5 # cd ~/Android/Sdk/ndk

root@VXIDr0ysue:~/Android/Sdk/ndk # ls
21.1.6352462  21.2.6472646  23.0.7123448

root@VXIDr0ysue:~/Android/Sdk/ndk # cd 23.0.7123448

root@VXIDr0ysue:~/Android/Sdk/ndk/23.0.7123448 # cd toolchains

root@VXIDr0ysue:~/Android/Sdk/ndk/23.0.7123448/toolchains # cd llvm

root@VXIDr0ysue:~/Android/Sdk/ndk/23.0.7123448/toolchains/llvm # cd prebuilt

root@VXIDr0ysue:~/Android/Sdk/ndk/23.0.7123448/toolchains/llvm/prebuilt # cd linux-x86_64

root@VXIDr0ysue:~/Android/Sdk/ndk/23.0.7123448/toolchains/llvm/prebuilt/linux-x86_64 # cd bin

root@VXIDr0ysue:~/Android/Sdk/ndk/23.0.7123448/toolchains/llvm/prebuilt/linux-x86_64/bin # pwd
/root/Android/Sdk/ndk/23.0.7123448/toolchains/llvm/prebuilt/linux-x86_64/bin
```

复制 pwd 返回的路径，回到 VS Code，打开终端 Terminal，执行以下命令：

```
root@VXIDr0ysue:~/Desktop/20210110/md5 # /root/Android/Sdk/ndk/23.0.7123448/toolchains/llvm/prebuilt/linux-x86_64/bin/clang -target AArch64-linux-android21 md5.c
```

```
root@VXIDr0ysue:~/Desktop/20210110/md5 # ls
a.out  md5  md5.c  README.md

root@VXIDr0ysue:~/Desktop/20210110/md5 # file a.out
a.out: ELF 64-bit LSB pie executable, ARM AArch64, version 1 (SYSV), dynamically linked,
interpreter /system/bin/linker64, not stripped
```

可以看到，生成的 a.out 文件是为在 ARM 架构的 CPU 上运行而生成的。接下来，将 a.out 文件推送到手机上：

```
root@VXIDr0ysue:~/Desktop/20210110/md5 # adb push a.out /data/local/tmp
a.out: 1 file pushed, 0 skipped. 0.3 MB/s (8304 bytes in 0.029s)
```

在手机中查看 a.out 文件，并加上权限：

```
root@VXIDr0ysue:~# adb shell
bullhead:/ $ su
bullhead:/ # cd /data/local/tmp
bullhead:/data/local/tmp # chmod 777 a.out
bullhead:/data/local/tmp # ./a.out r0ysue
8e2b0b38f557578f578ef21b33e4df0d
```

以上就是交叉编译到手机上直接运行的流程。可以将之前复制的 clang 路径加入 zsh 的路径，这样使用的时候更方便一些：

```
root@VXIDr0ysue:~ # nano ~/.zshrc
```

在打开的文件末尾添加：

```
export PATH="/root/Android/Sdk/ndk/23.0.7123448/toolchains/llvm/prebuilt/linux-x86_64/bin:$PATH"
```

验证：

```
root@VXIDr0ysue:~ # clang --version
Debian clang version 11.0.1-2
Target: x86_64-pc-linux-gnu
Thread model: posix
InstalledDir: /usr/bin
```

再把它移植到我们开发的工程中。

首先在 roysue.c 中添加 md5.c 的头文件，再将除 main 函数之外的其他部分代码复制到 roysue.c 中。

按照原 md5.c 的算法逻辑，在 Java_com_roysue_easyso1_MainActivity_stringFromJNI 中添加以下代码实现 MD5 算法并在日志（log）中打印出结果。

```
char* msg = "r0ysue";
size_t len = strlen(msg);
uint8_t result[16];
```

```
char* r = (char*)malloc(16);
memset(r, 0x00, sizeof(char) *16);

char* final = (char*)malloc(16);
memset(final, 0x00, sizeof(char) *16);

md5((uint8_t*)msg, len, result);

for (int i = 0; i < 16; i++) {
    sprintf(r, "%2.2x", result[i]);
    __android_log_print(ANDROID_LOG_INFO, "r0ysue", "Hi,now i is %s", r);
    sprintf(final, "%s%s", final, r);
}

jstring jresult = (*env)->NewStringUTF(env, final);
return jresult;
```

编写 MD5 算法的代码时,需要注意将输出结果作为一个完整的字符串返回给前端显示,而不能按照原程序逐个输出。此外,还要遵循 C 语言的编写规范。例如,调用 memset() 方法将存放字符串的数组清空后再赋值。

修改配置文件 build.gradle,在 NDK 中只保留 32 位的 ABI,即 armeabi-v7a,然后运行即可。

2.2 LLDB 动态调试(三方)Android SO

LLDB 调试第三方 App 可以使用以下两个方法之一:

- App 处于调试(Debug)模式、APK 包中的 debuggable==true(需要重新打包,或者 Xposed/Frida 去 Hook)。
- 手机是 AOSP 系统,编译成 userdebug 模式(N5X、Sailfish)。

第一种方法容易失败,笔者在这里推荐第二种方法,使用第二种方法可以调试手机中所有的 App。

对于 LLDB 调试工具的使用可参考网址 9 和网址 10 中的文章(参见配书资源文件)。

在 2.1 节中,我们使用了 Android Studio 中自带的 LLDB 调试。因为是首次调试,lldb-server 会被自动推送(push)到手机/data/local/tmp 目录下,Android Studio 调试控制台记录如图 2-9 所示。

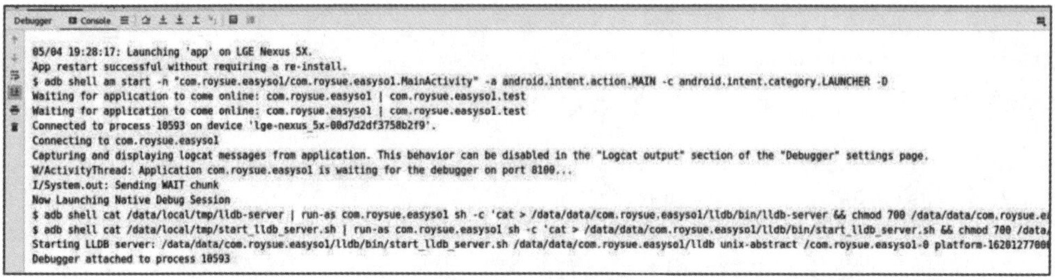

图 2-9　ADB push lldb-server

如果没有调试过，则可通过以下途径查看 lldb-server 的位置：

```
root@VXIDr0ysue:~ # cd ~/Android/Sdk/ndk

root@VXIDr0ysue:~/Android/Sdk/ndk # cd 23.0.7123448

root@VXIDr0ysue:~/Android/Sdk/ndk/23.0.7123448 # tree -NCfhl | grep -i lldb-server
│   │        │       │       │       └──
[ 28M]  ./toolchains/llvm/prebuilt/linux-x86_64/lib64/clang/11.0.5/lib/linux/AArch64/lldb-server
│   │        │       │       │       └──
[ 25M]  ./toolchains/llvm/prebuilt/linux-x86_64/lib64/clang/11.0.5/lib/linux/arm/lldb-server
│   │        │       │       │       └──
[ 26M]  ./toolchains/llvm/prebuilt/linux-x86_64/lib64/clang/11.0.5/lib/linux/i386/lldb-server
│   │        │       │       │       └──
[ 29M]  ./toolchains/llvm/prebuilt/linux-x86_64/lib64/clang/11.0.5/lib/linux/x86_64/lldb-server
```

我们把 64 位的和 32 位的 lldb-server 复制出来：

```
root@VXIDr0ysue:~/Android/Sdk/ndk/23.0.7123448 # cp toolchains/llvm/prebuilt/linux-x86_64/lib64/clang/11.0.5/lib/linux/AArch64/lldb-server ~/Desktop/20210110/ls64

root@VXIDr0ysue:~/Android/Sdk/ndk/23.0.7123448 # cp toolchains/llvm/prebuilt/linux-x86_64/lib64/clang/11.0.5/lib/linux/arm/lldb-server ~/Desktop/20210110/ls
```

64 位和 32 位的 lldb-server 并不是混用的，至于如何选择，需要与进程相匹配。将 ls 与 ls64 用 ADB 推送（push）到/data/local/tmp 目录下：

```
root@VXIDr0ysue:~ # cd Desktop/20210110

root@VXIDr0ysue:~ # ls
ls  ls64  MD5
```

```
root@VXIDr0ysue:~ # adb push ls* /data/local/tmp
ls: 1 file pushed, 0 skipped. 12.5 MB/s (26343620 bytes in 2.018s)
ls64: 1 file pushed, 0 skipped. 14.4 MB/s (28965712 bytes in 1.924s)
2 files pushed, 0 skipped. 13.0 MB/s (55309332 bytes in 4.062s)
```

进入手机目录给 ls 和 ls64 添加权限，然后启动 ls64：

```
bullhead:/data/local/tmp # ls
bullhead:/data/local/tmp # chmod 777 * ls*
bullhead:/data/local/tmp # ./ls64 platform --listen "0.0.0.0:10086" --server
```

查看 lldb 命令在哪里：

```
root@VXIDr0ysue:~/Android/Sdk/ndk/23.0.7123448 # tree -NCfhl | grep -i lldb
─ [  91]  ./ndk-lldb
│   │            │   ├─ [229K]  ./toolchains/llvm/prebuilt/linux-x86_64/bin/lldb
│   │            │   ├─
[126K]  ./toolchains/llvm/prebuilt/linux-x86_64/bin/lldb-argdumper
│   │            │   ├─ [  98]  ./toolchains/llvm/prebuilt/linux-x86_64/bin/lldb.sh

root@VXIDr0ysue:~/Android/Sdk/ndk/23.0.7123448 # cd

root@VXIDr0ysue:~ # lldb
(lldb) platform select remote-android
  Platform: remote-android
 Connected: no
(lldb) platform connect connect://00d7d2df3758b2f9:10086
  Platform: remote-android
    Triple: AArch64-unknown-linux-android
OS Version: 27 (3.10.73-g89fd15db99aa)
  Hostname: localhost
 Connected: yes
WorkingDir: /data/local/tmp
    Kernel: #1 SMP PREEMPT Thu Oct 11 19:31:31 UTC 2018
(lldb)
```

进入 platform 使用 connect 连接时，如果手机是使用 USB 连接的，则要在 connect:// 后输入 adb devices 显示的设备名称；如果是使用网络连接的，则需要输入手机的 IP 地址。

然后，根据需要执行以下操作。

查看 easyso 进程号：

```
bullhead:/ # ps -e |grep easy
u0_a102      16289   569 4310276  63632 SyS_epoll_wait 78120c83f8 S com.roysue.easyso1
```

回到 LLDB，attach easyso 进程：

```
(lldb) attach -p 16289
```

```
Process 16289 stopped
```

查看当前线程:

```
(lldb) thread list
```

查看当前执行的函数:

```
(lldb) dis
libc.so`__epoll_pwait:
```

关于调试更具体的内容，我们会在后续章节展开讲解。

2.3　Capstone/Keystone/Unicorn（反）汇编器

在之前的内容中，我们了解到使用 LLVM 可以进行一定的汇编与反汇编。本节介绍 Capstone/Keystone（反）汇编器。

首先，我们简单回顾一些基本概念。

- 汇编（编译）：指将源码转换为机器码的过程，编译的过程用的是 LLVM。
- 反汇编（反编译）：指将机器码转换为汇编语言代码的操作，用的是 LLVM 的 LLDB 调试器的 Disassembler 功能，将当前的机器码反汇编成更容易理解的指令。

汇编与反汇编是互逆的过程。

我们可以用 Capstone、Keystone 和 Unicorn 实现同样的效果。为什么使用这些工具呢？因为 LLVM 工具链使用起来并不是很方便，尤其是当我们对 SO 文件进行自定义修改的时候。

接下来，我们将对 Capstone、Keystone 以及 Unicorn 进行简单的介绍。这 3 款工具是由同一个团队开发的，其中，最先发布的是 Capstone，而 Unicorn 将是我们课程的重点。模拟执行是指对于一个 SO 文件，我们可以在不用知道其实现细节的情况下直接运行它并获取结果。

Capstone 和 Keystone 的实现基于 LLVM。LLVM 项目是一个模块化、可重用的编译器和工具链技术集合。例如，LLVM 将我们编写的 C 语言代码编译成中间状态代码 LLVM IR，然后通过后端将中间代码编译成各个平台的机器码，从而可以在不同的目标机器上运行。LLVM 的架构如图 2-10 所示。

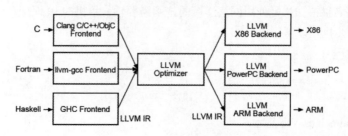

图 2-10　LLVM 的架构

反编译器 Capstone 是一个精准裁剪的 LLVM 的子模块，可以将指定的 SO 文件中的二进制数据转换成指令并返回结果。相对应的是汇编器 Keystone，可以将一个或一组指令转换成二进制数据并返回结果。

Keystone 相比于 LLVM，是非常轻量化的中间件。一个基于 Keystone 的工具 Keypatch，其操作界面如图 2-11 所示，可以通过输入指令得到相应的机器码。

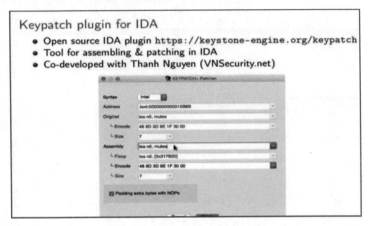

图 2-11　Keypatch 工具的操作界面

我们可以用 Keypatch 来进行硬编码，即打开 SO 文件并直接修改其中的二进制数据。与 Capstone 和 Keystone 不同，Unicorn 是经过精准裁剪的 QEMU。QEMU 本身是一个完整的模拟器。Android Studio 自带的模拟器也是基于 QEMU 实现的。Unicorn 是一个基于 QEMU 的 CPU 模拟器，它可以直接运行机器码并忽略设备之间的差异。在某些场景下，我们可能只需要模拟代码的执行，而不需要一个真实的 CPU 来执行这些操作。例如，我们当前使用的这个虚拟机是 x86 架构的，如图 2-12 所示，我们可以在此虚拟机上使用 Unicorn 运行 ARM 架构的二进制文件。

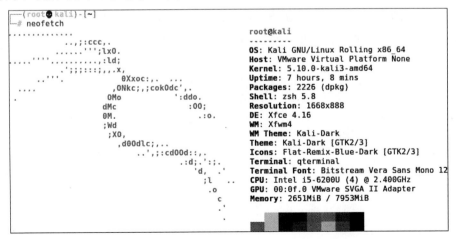

图 2-12　使用 neofetch 查看系统信息

2.4　Frida 动态调试 Android Native 部分

本节将介绍一款调试工具 Dwarf，读者简要了解即可。

Dwarf 是一个基于 Frida 开发的调试器，它将 Capstone、Keystone 和 Unicorn 集成到自己的框架中。在使用 Dwarf 调试器之前，需要先启动 frida-server：

```
bullhead:/data/local/tmp # ./frida-server-12.8.0-android-arm64
```

安装 Dwarf：

```
root@VXIDr0ysue:~ # cd Desktop/20210110

root@VXIDr0ysue:~/Desktop/20210110 # git clone https://github.com/iGio90/Dwarf

root@VXIDr0ysue:~/Desktop/20210110 # cd Dwarf

root@VXIDr0ysue:~/Desktop/20210110/Dwarf # pip3 install -r requirements.txt

root@VXIDr0ysue:~/Desktop/20210110/Dwarf # python3 dwarf.py
```

本质上，Dwarf 使用了 Frida 断点功能，然后使用 Capstone 将当前的指令进行反编译，再用 Keystone 将用户修改的内容写入内存区域。

我们也可以直接使用 Frida 的 Capstone 引擎，因为该引擎已经内置在 Frida 中。

2.5　Frida Instruction 模块动态反汇编

Frida 的 Instruction 模块也是通过集成 Capstone 实现对目标内存地址中指令的解析。本节将演示 Frida Instruction 模块的用法。

首先，在手机上运行 12.8.0 版本的 frida-server，并切换到相对应的 Python 环境 3.8.0。然后，在~/Desktop/20210110 目录下新建文件 instruction.js。使用 Instruction 模块时，可以通过 Objection 找到想要查看的地址：

```
root@VXIDr0ysue:~ # objection -g com.roysue.easyso1 explore

com.roysue.easyso1 on (google: 8.1.0) [usb] # memory list modules

com.roysue.easyso1 on (google: 8.1.0) [usb] # memory list exports libroysue.so
```

libroysue.so 导出的信息如图 2-13 所示。

```
com.roysue.easysol on (google: 8.1.0) [usb] # memory list exports libroysue.so
Save the output by adding `--json exports.json` to this command
Type       Name                                                    Address
--------   --------------------------------------------------      ------------
variable   r                                                       0x7776beb580
function   to_bytes                                                0x7776beab0c
variable   k                                                       0x7776beb480
function   to_int32                                                0x7776beab5c
function   md5                                                     0x7776beaba4
function   Java_com_roysue_easysol_MainActivity_stringFromJNI      0x7776beb064
com.roysue.easysol on (google: 8.1.0) [usb] #
```

图 2-13　libroysue.so 导出的信息

以 stringFromJNI 函数为例编写如下代码：

```
setImmediate(function(){
    var stringFromJniaddr = Module.findExportByName("libroysue.so","Java_com_roysue_easysol_MainActivity_stringFromJNI")
    var ins = Instruction.parse(stringFromJniaddr);
    console.log("ins =>", ins.toString())
})
```

保存并运行，可以看到打印出的指令：

```
root@VXIDr0ysue:~/Desktop/20210110 # frida -UF -l instruction.js

ins => sub sp, sp, #0x130
[LGE Nexus 5X::easysol]->
```

修改程序，打印 10 条指令：

```
function dis(address, number) {
    for(var i = 0; i < number; i++) {
        var ins = Instruction.parse(address);
        console.log("address:" + address + "--dis:" + ins.toString());
        address = ins.next;
    }
}

setImmediate(function(){
    var stringFromJniaddr = Module.findExportByName("libroysue.so","Java_com_roysue_easysol_MainActivity_stringFromJNI")
    dis(stringFromJniaddr, 10);
})
```

打印结果如图 2-14 所示。

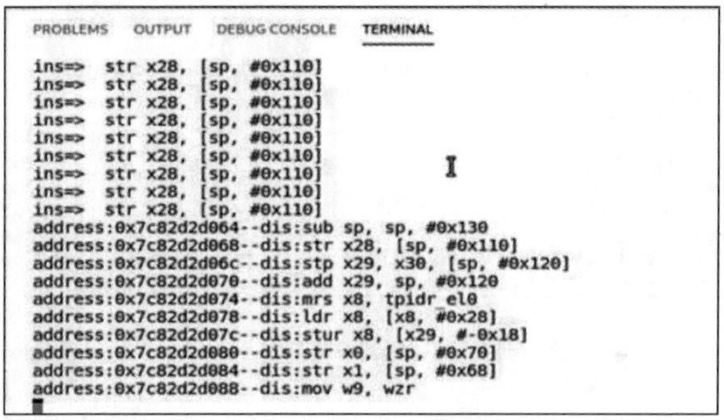

图 2-14 打印结果

结果与 LLDB 的调试结果相同，如图 2-15 所示。

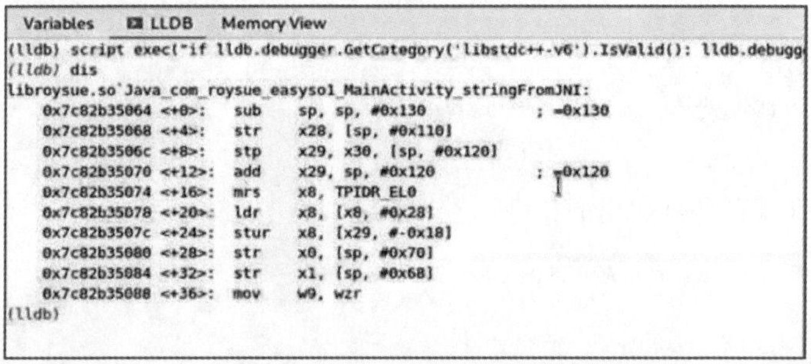

图 2-15 LLDB 的调试结果

2.6 本章小结

本章主要介绍了在 Android Studio 中使用 Debugger 进行调试的方法，以及 LLDB 动态调试 NDK 程序的过程。通过一个简单的 MD5 案例，我们学习了交叉编译的过程，即如何将一个 C 语言程序移植到现有的 NDK 项目中。之后介绍了反汇编器 Capstone 和汇编器 Keystone，并演示了集成 Capstone 的 Frida Instruction 模块的基本用法。在后续的章节中，我们将更加深入地学习应用调试的技术并介绍 Unicorn CPU 模拟器的使用。

第 3 章

静态分析工具的安装和基本使用

前两章我们学习了 NDK 开发的基本环境配置以及一些动态调试的内容。在动态调试中，主要使用了 LLDB 和 GDB 等主流的系统级调试器。相比之下，基于 Frida 的动态调试实际使用较少，因此只需简单了解即可。本章将介绍一些静态分析工具的安装和基本使用方法。

3.1 使用 objdump 反汇编目标文件命令显示二进制文件信息

本节将带领读者了解 objdump 命令的使用方法。

objdump 命令是 Linux 下的反汇编目标文件或者可执行文件的命令，它以一种可阅读的格式帮助读者了解二进制文件可能带有的附加信息。

我们先来看最终编译成的 SO 文件的位置。在 Android Studio 中打开项目 easyso1，右击 build 目录，单击 Show in Visual Studio Code，可以在文件资源管理器中显示 build 文件夹（此处 Android Studio 显示可能会有问题）。在路径 build/outputs/apk/debug 下找到 app-debug.apk，并将其复制到 Desktop/20210112/app 目录下以备后续使用。

执行以下命令解压 app-debug.apk 文件：

```
root@VXIDr0ysue:~/Desktop/20210112/app # 7z x app-debug.apk
```

打开 lib/armeabi-v7a，右击 Open Terminal Here。由于 objdump 在 /root/Android/Sdk/ndk/21.1.6352462/toolchains/llvm/prebuilt/linux-x86_64/bin 路径下，因此现在可以直接使用。

我们可以检查配置文件：

```
root@VXIDr0ysue:~/Desktop/20210112/app # cat ~/.zshrc

export PATH="/root/Android/Sdk/platform-tools:$PATH"
```

```
    export
PATH="/root/Android/Sdk/ndk/21.1.6352462/toolchains/llvm/prebuilt/linux-x86_64/bin:$PAT
H"
```

通过 which 命令查看 objdump 的位置，可以看到系统中自带了 objdump。此外，在上述路径中也包括 64 位、32 位、LLVM 等不同版本的 objdump：

```
root@VXIDr0ysue:~/Desktop/20210112/app # which objdump
/usr/bin/objdump

root@VXIDr0ysue:~/Desktop/20210112/app # cd /root/Android/Sdk/ndk/21.1.6352462/toolchains/llvm/prebuilt/linux-x86_64/bin

root@VXIDr0ysue:~/…/llvm/prebuilt/linux-x86_64/bin # ls |grep objdump
AArch64-linux-android-objdump
arm-linux-androideabi-objdump
i686-linux-android-objdump
llvm-objdump
x86_64-linux-android-objdump
```

至此，我们可以在任意目录下使用 objdump 了。

也有带 GUI（图形界面）的 objdump，例如 ObjGui，它实际上属于 GNU 的工具集 binutils。GNU 是指一整套 Linux 的二进制工具的集合。

接下来演示 objdump 命令及其部分参数的使用方法。

首先，使用 objdump 命令查看 libroysue.so 的导出表：

```
root@VXIDr0ysue:~/Desktop/20210112 # cd app/lib/armeabi-v7a

root@VXIDr0ysue:~/…/20210112/app/lib/armeabi-v7a # arm-linux-androideabi-objdump -tT libroysue.so

libroysue.so:     file format elf32-littlearm

SYMBOL TABLE:
no symbols

DYNAMIC SYMBOL TABLE:
00000000      DF *UND*  00000000 LIBC        __cxa_atexit

00000c44 g    DF .text  0000013c Base        Java_com_roysue_easyso1_MainActivity_stringFromJNI
0000250c g    DO .rodata        00000100 Base        k
000009cc g    DF .text  00000278 Base        md5
000009c0 g    DF .text  0000000c Base        to_int32
```

还可以用 objdump 命令进行反编译，可以使用参数-D(--disassemble-all)表示反汇编所有 section，同时在 Java_com_roysue_easyso1_MainActivity_stringFromJNI 的偏移 0c44 处指定操作行数，带-D 参数表示返回结果包含操作码、指令等信息：

```
root@VXIDr0ysue:~/…/20210112/app/lib/armeabi-v7a # arm-linux-androideabi-objdump -D libroysue.so|grep 0c44
    00000c44 <Java_com_roysue_easyso1_MainActivity_stringFromJNI@@Base>:

root@VXIDr0ysue:~/…/20210112/app/lib/armeabi-v7a # arm-linux-androideabi-objdump -D libroysue.so|grep -100 0c44

    00000c44 <Java_com_roysue_easyso1_MainActivity_stringFromJNI@@Base>:
        c44:     b580              push    {r7, lr}
        c46:     466f              mov     r7, sp
```

还可以使用-g 参数来显示调试信息。

注意调试信息一般在源码被编译成 debug 时才会有，而在 release 库中则会被去掉。例如，用 LLDB 可以专门生成一个在调试模式下的 SO 文件，然后就会有调试信息。有了调试信息，我们在调试时就可以用 Android Studio 将调试的信息打印出来。

使用 GNU 的工具集 binutils 提供的一些工具也可以查看二进制信息。例如，通过 nm 命令获取导出表：

```
root@VXIDr0ysue:~/…/20210112/app/lib/armeabi-v7a # nm -D libroysue.so
         U abort@LIBC
         U __aeabi_memclr
         U __aeabi_memcpy

00000c45 T Java_com_roysue_easyso1_MainActivity_stringFromJNI
0000250c R k
         U malloc@LIBC
000009cd T md5
```

使用 strings 命令打印字符串：

```
root@VXIDr0ysue:~/…/20210112/app/lib/armeabi-v7a # strings libroysue.so
td$.
Android
r21b
6352462
```

使用 readelf 命令显示基本的 symbols：

```
root@VXIDr0ysue:~/…/20210112/app/lib/armeabi-v7a # readelf -s libroysue.so

Symbol table '.dynsym' contains 29 entries:
   Num:    Value  Size Type    Bind   Vis      Ndx Name
```

```
0: 00000000     0 NOTYPE  LOCAL  DEFAULT  UND
1: 00000000     0 FUNC    GLOBAL DEFAULT  UND __cxa_atexit@LIBC (2)
2: 00000000     0 FUNC    GLOBAL DEFAULT  UND __cxa_f[...]@LIBC (2)
```

可见，我们并不需要使用 IDA，通过 binutils 默认自带的工具就可以看到二进制文件中大量的信息。

3.2 使用 010 Editor 解析 SO 文件显示二进制基本信息

本节将介绍一款功能强大的代码编辑器——010 editor。首先，在官网下载 010 Editor，解压到同级目录下，进入文件夹执行以下命令进行安装：

```
root@VXIDr0ysue:~/Downloads # ./010EditorLinux64Installer
```

安装过程中保持默认设置，单击 Next 按钮即可。010 Editor 启动后，将 libroysue.so 文件拖曳进去以打开该文件。如果不安装任何插件，则只能看到原始的十六进制数据。此外，在 010 Editor 中可以显示的字符串同样可以通过上文提到的 strings 命令提取，如图 3-1 所示。

图 3-1　010 Editor

接下来，将演示在 010 Editor 中直接对二进制数据进行修改（patch）。

什么是 patch？

许多时候，我们不能获得程序源码，只能直接对二进制文件进行修改，这就是所谓的 patch，用户可以使用十六进制编辑器直接修改文件的字节，也可以利用一些半自动化的工具。

patch 有很多种形式：

● patch 二进制文件（程序或库）。

- 在内存中 patch（利用调试器）。
- 预加载库替换源库文件中的函数。
- triggers（Hook，然后在运行时 patch）。

例如，将字符串 now 修改为 ooo，如图 3-2 所示。

```
26C0h: 0B 00 00 00 10 00 00 00 17 00 00 00 06 00 00 00   ................
26D0h: 0A 00 00 00 0F 00 00 00 15 00 00 00 06 00 00 00   ................
26E0h: 0A 00 00 00 0F 00 00 00 15 00 00 00 06 00 00 00   ................
26F0h: 0A 00 00 00 0F 00 00 00 15 00 00 00 06 00 00 00   ................
2700h: 0A 00 00 00 0F 00 00 00 15 00 00 00 72 30 79 73   ............r0ys
2710h: 75 65 00 25 32 2E 32 78 00 48 69 2C 6F 6F 6F 20   ue.%2.2x.Hi,ooo
2720h: 69 20 69 73 20 25 73 00 25 73 25 73 2E 6C 69 62   i is %s.%s%s.lib
2730h: 75 6E 77 69 6E 64 3A 20 25 73 20 25 73 3A 25 64   unwind: %s %s:%d
2740h: 20 2D 20 25 73 0A 00 5F 55 6E 77 69 6E 64 5F 52    - %s.._Unwind_R
2750h: 65 73 75 6D 65 00 2F 62 75 69 6C 64 62 6F 74 2F   esume./buildbot/
2760h: 73 72 63 2F 61 6E 64 72 6F 69 64 2F 6E 64 6B 2D   src/android/ndk-
```

图 3-2 将字符串 now 修改为 ooo

我们的目标是使用修改后的 libroysue.so 文件替换掉手机中原有的 libroysue.so 文件，然后观察运行结果是否发生了改变。

首先，将修改后的 libroysue.so 文件推送（push）到手机的/sdcard/Download 目录下：

```
root@VXIDr0ysue:~/…/20210112/app/lib/armeabi-v7a # adb push libroysue.so /sdcard/Download
    libroysue.so: 1 file pushed, 0 skipped. 32.2 MB/s (13904 bytes in 0.000s)
```

查看 App 原有的 libroysue.so 在哪个目录下。可以先启动 easyso1，显示进程信息，并根据进程号找到 libroysue.so 文件所在的目录，进入：

```
bullhead:/ # ps -e | grep -i easyso
u0_a100       6181   566 1653804  50776 SyS_epoll_wait e798ca74 S com.roysue.easyso1

bullhead:/ # cat /proc/6181/maps |grep -i libroysue
cdda6000-cdda9000 r-xp 00000000 fd:00 376200  /data/app/com.roysue.easyso1-xkpi8IZIwqCHRQ9Oh_L7og==/lib/arm/libroysue.so
cdda9000-cddaa000 r--p 00002000 fd:00 376200  /data/app/com.roysue.easyso1-xkpi8IZIwqCHRQ9Oh_L7og==/lib/arm/libroysue.so
cddaa000-cddab000 rw-p 00003000 fd:00 376200  /data/app/com.roysue.easyso1-xkpi8IZIwqCHRQ9Oh_L7og==/lib/arm/libroysue.so

bullhead:/ # cd /data/app/com.roysue.easyso1-xkpi8IZIwqCHRQ9Oh_L7og==/lib/arm/
bullhead:/data/app/com.roysue.easyso1-xkpi8IZIwqCHRQ9Oh_L7og==/lib/arm # ls
libroysue.so
```

在检查权限和用户后，我们将/sdcard/Download 目录下的 libroysue.so 复制到当前目录，完成 SO 文件的替换。在复制之前，我们可以先检查 libroysue.so 的 MD5 值：

```
bullhead:/data/app/com.roysue.easyso1-xkpi8IZIwqCHRQ9Oh_L7og==/lib/arm # ls -alit
total 24
376199 drwxr-xr-x 2 system system  4096 2021-05-15 20:06 .
376198 drwxr-xr-x 3 system system  4096 2021-05-15 20:06 ..
376200 -rwxr-xr-x 1 system system 13904 1981-01-01 01:01 libroysue.so
bullhead:/data/app/com.roysue.easyso1-xkpi8IZIwqCHRQ9Oh_L7og==/lib/arm # cp
/sdcard/Download/libroysue.so ./
```

当我们再次运行 easyso1 的时候，输出的日志发生了变化，如图 3-3 所示。

```
bullhead:/data/app/com.roysue.easyso1-xkpi8IZIwqCHRQ9Oh_L7og==/lib/arm # logcat
|grep -i r0ysue
```

```
05-19 09:17:04.247  6181  6181 I r0ysue  : Hi,now i is f2
05-19 09:17:04.247  6181  6181 I r0ysue  : Hi,now i is 1b
05-19 09:17:04.247  6181  6181 I r0ysue  : Hi,now i is 33
05-19 09:17:04.247  6181  6181 I r0ysue  : Hi,now i is e4
05-19 09:17:04.248  6181  6181 I r0ysue  : Hi,now i is df
05-19 09:17:04.248  6181  6181 I r0ysue  : Hi,now i is 0d
05-19 09:38:20.779  6737  6737 I r0ysue  : Hi,ooo i is 8e
05-19 09:38:20.779  6737  6737 I r0ysue  : Hi,ooo i is 2b
05-19 09:38:20.779  6737  6737 I r0ysue  : Hi,ooo i is 0b
05-19 09:38:20.779  6737  6737 I r0ysue  : Hi,ooo i is 38
05-19 09:38:20.779  6737  6737 I r0ysue  : Hi,ooo i is f5
05-19 09:38:20.779  6737  6737 I r0ysue  : Hi,ooo i is 57
05-19 09:38:20.779  6737  6737 I r0ysue  : Hi,ooo i is 57
05-19 09:38:20.779  6737  6737 I r0ysue  : Hi,ooo i is 8f
05-19 09:38:20.779  6737  6737 I r0ysue  : Hi,ooo i is 57
05-19 09:38:20.779  6737  6737 I r0ysue  : Hi,ooo i is 8e
05-19 09:38:20.779  6737  6737 I r0ysue  : Hi,ooo i is f2
05-19 09:38:20.779  6737  6737 I r0ysue  : Hi,ooo i is 1b
05-19 09:38:20.779  6737  6737 I r0ysue  : Hi,ooo i is 33
05-19 09:38:20.779  6737  6737 I r0ysue  : Hi,ooo i is e4
05-19 09:38:20.779  6737  6737 I r0ysue  : Hi,ooo i is df
05-19 09:38:20.779  6737  6737 I r0ysue  : Hi,ooo i is 0d
```

图 3-3 now 变为 ooo

以上过程是最简单的 patch。我们可以对二进制数据进行修改，修改完之后就可以把结果体现到运行过程中。

3.3 Ghidra/JEB/IDA 高级反汇编器

首先，下载并安装 JEB。如果 JEB 是从物理机复制到虚拟机中的，则需要清除缓存：

```
root@VXIDr0ysue:~ # cd ~/.cache/vmware

root@VXIDr0ysue:~/.cache/vmware # rm -rf *
zsh: sure you want to delete the only file in /root/.cache/vmware [yn]? y
```

进入 jeb 目录，运行 jeb_linux.sh：

```
root@VXIDr0ysue:~/Desktop/20210112/jeb # ls
bin         doc           jeb_linux.sh  jeb_wincon.bat  scripts   typelibs
coreplugins jebKeygen.py  jeb_macos.sh  jvmopt.txt                siglibs
```

```
root@VXIDr0ysue:~/Desktop/20210112/jeb # ./jeb_linux.sh
```

需要输入 JEB 解密密码。成功解密后，弹出的窗口如图 3-4 所示。

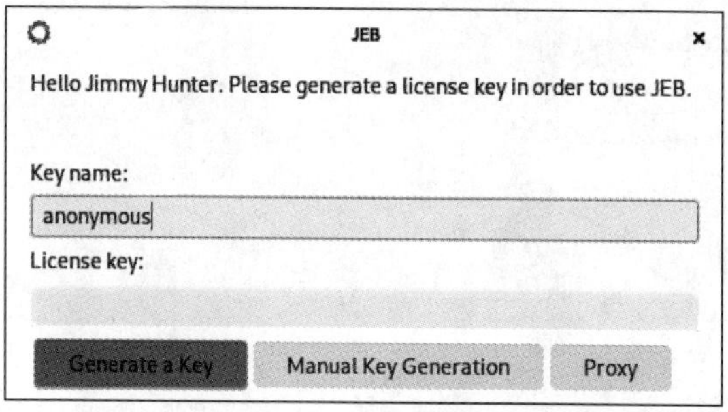

图 3-4　Key name

> **注意**　不要使用自己的 ID。

单击 Manual Key Generation 按钮并复制 License data。我们要利用 jebKeygen.py 计算 License key，由于 jebKeygen.py 需要用 Python 2 来运行，因此切换成系统自带的 Python 2：

```
root@VXIDr0ysue:~/Desktop/20210112/jeb # pyenv versions
  system
* 3.8.0 (set by /root/Desktop/.python-version)
  3.8.5

root@VXIDr0ysue:~/Desktop/20210112/jeb # pyenv local system

root@VXIDr0ysue:~/Desktop/20210112/jeb # python -V
Python 2.7.18

root@VXIDr0ysue:~/Desktop/20210112/jeb # python jebKeygen.py
Input License Data:
```

在 Input License Data 后粘贴 License data，按回车键即可得到 License key。将 License key 填入表单，如图 3-5 所示。

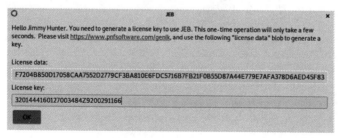

图 3-5　License key

使用 JEB 反编译 libroysue.so 文件，将 libroysue.so 拖曳进 JEB。高级反编译器的优势在于能够完整地显示反编译信息，如图 3-6 所示。

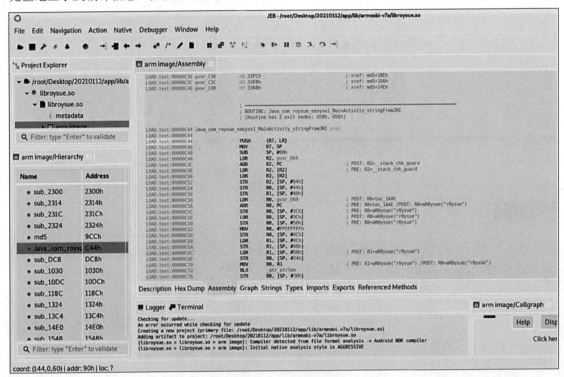

图 3-6　JEB 界面

左侧是符号表，右侧是反汇编的结果以及日志等。对照 objdump，机器码的反汇编结果应该是一模一样的，如图 3-7 所示。

```
0000ec44 <Java_com_roysue_easysol_MainActivity_stringFromJNI@@Base>:
    c44:   b580        push    {r7, lr}
    c46:   466f        mov     r7, sp
    c48:   b0a2        sub     sp, #136        ; 0x88
    c4a:   4a46        ldr     r2, [pc, #280]  ; (d64 <Java_com_roysue_easysol_MainActivity_stringFromJNI@@Bas
e+0x120>)
    c4c:   447a        add     r2, pc
    c4e:   6812        ldr     r2, [r2, #0]
    c50:   6812        ldr     r2, [r2, #0]
    c52:   9221        str     r2, [sp, #132]  ; 0x84
    c54:   9011        str     r0, [sp, #68]   ; 0x44
    c56:   9110        str     r1, [sp, #64]   ; 0x40
    c58:   4843        ldr     r0, [pc, #268]  ; (d68 <Java_com_roysue_easysol_MainActivity_stringFromJNI@@Bas
e+0x124>)
```

图 3-7　objdump 反汇编结果

从 metadata 可以看到 ELF 原信息，如图 3-8 所示。

第 3 章 静态分析工具的安装和基本使用

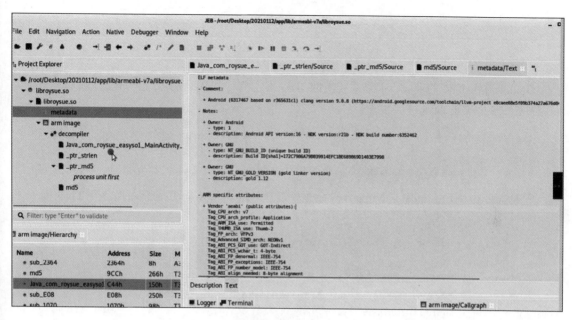

图 3-8　metadata

按 Tab 键可以反编译到 Java 伪代码，如图 3-9 所示。

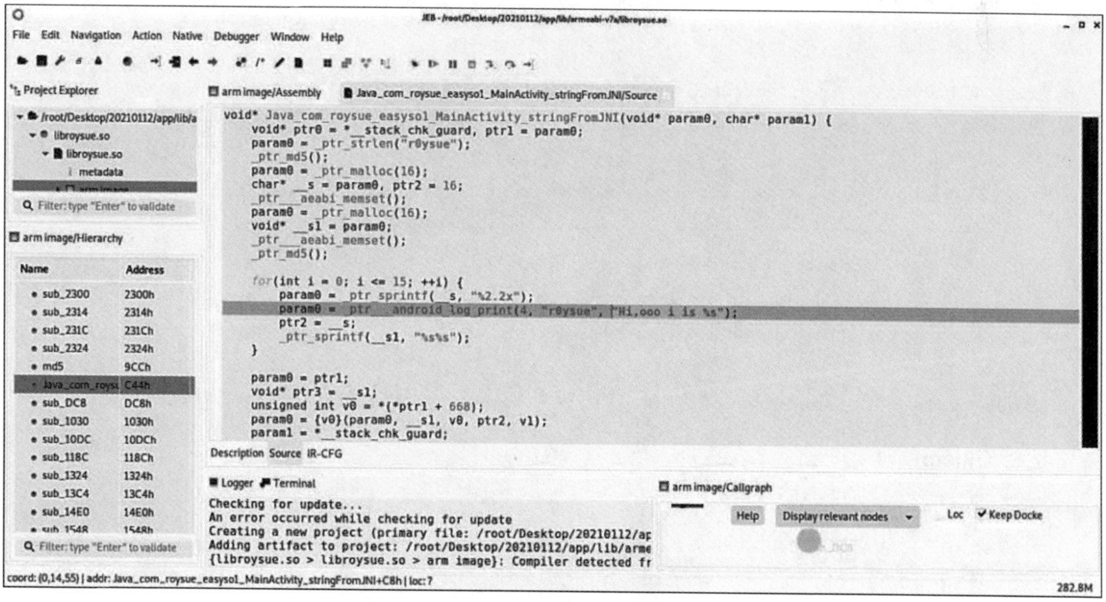

图 3-9　反编译到 Java 伪代码

除了 JEB 外，Ghidra 也是时下流行的反汇编器。从 Ghidra 官网把 Ghidra 下载到本地：

```
root@VXIDr0ysue:~/Downloads # 7z x ghidra_9.2.2_PUBLIC_20201229.zip

root@VXIDr0ysue:~/Downloads # cd ghidra_9.2.2_PUBLIC
```

Ghidra 需要 JDK 的环境，但是系统上的 JDK 不完整。由于系统中缺少 Java 汇编器，因此我们需要先安装完整的 Java。

```
root@VXIDr0ysue:~/Downloads/ghidra_9.2.2_PUBLIC # apt search openjdk|grep openjdk

root@VXIDr0ysue:~/Downloads/ghidra_9.2.2_PUBLIC # apt install openjdk-11-jdk
```

一般情况下，使用 update-alternatives 命令来选择 JDK 的版本。更换 Java 版本至 11：

```
root@VXIDr0ysue:~/Downloads/ghidra_9.2.2_PUBLIC # update-alternatives --config java
There are 2 choices for the alternative java (providing /usr/bin/java).

  Selection    Path                                            Priority   Status
------------------------------------------------------------
  0            /usr/lib/jvm/java-17-openjdk-amd64/bin/java      1711      auto mode
* 1            /usr/lib/jvm/java-11-openjdk-amd64/bin/java      1111      manual mode
  2            /usr/lib/jvm/java-17-openjdk-amd64/bin/java      1711      manual mode

Press <enter> to keep the current choice[*], or type selection number: 1

root@VXIDr0ysue:~/Downloads/ghidra_9.2.2_PUBLIC # update-alternatives --config javac
There are 2 choices for the alternative javac (providing /usr/bin/javac).

  Selection    Path                                             Priority   Status
------------------------------------------------------------
  0            /usr/lib/jvm/java-17-openjdk-amd64/bin/javac      1711      auto mode
* 1            /usr/lib/jvm/java-11-openjdk-amd64/bin/javac      1111      manual mode
  2            /usr/lib/jvm/java-17-openjdk-amd64/bin/javac      1711      manual mode

Press <enter> to keep the current choice[*], or type selection number: 1
```

运行 Ghidra：

```
root@VXIDr0ysue:~/Downloads/ghidra_9.2.2_PUBLIC # ./ghidraRun
Picked up _JAVA_OPTIONS: -Dawt.useSystemAAFontSettings=on -Dswing.aatext=true
Picked up _JAVA_OPTIONS: -Dawt.useSystemAAFontSettings=on -Dswing.aatext=true
```

Ghidra 运行起来后，新建项目，依次单击 File→New Project→Non-shared Project，如图 3-10 所示。

第 3 章 静态分析工具的安装和基本使用　　45

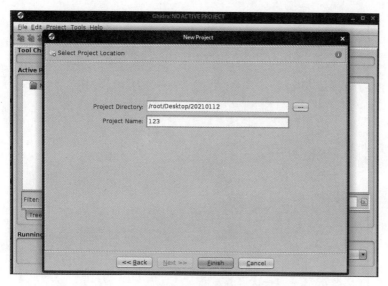

图 3-10　新建项目

完成后，单击 Tool Chest 中的绿色按键，如图 3-11 所示。

为了防止两个 SO 文件重叠，在 libroysue.so 同级目录下将其复制一份。Ghidra 启动后，我们把 libroysue(copy1).so 拖进去以打开该文件。Ghidra 分析 SO 文件的结果中包含完整的基本信息，如图 3-12 所示，因此，我们就不需要再用 Binutils 提供的工具了。

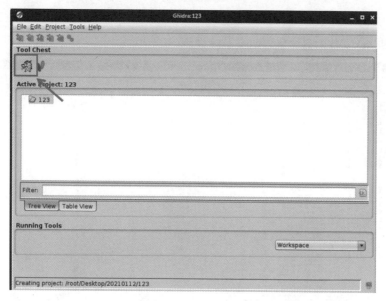

图 3-11　启动 Ghidra

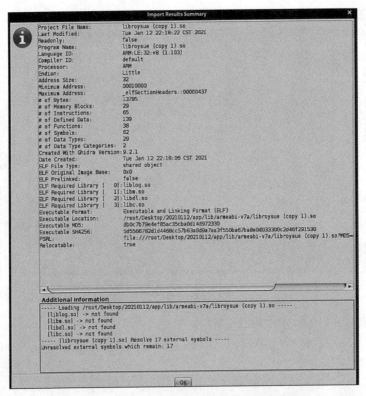

图 3-12 Ghidra 分析 SO 文件的基本信息

Ghidra 分析 SO 文件的具体信息如图 3-13 所示。

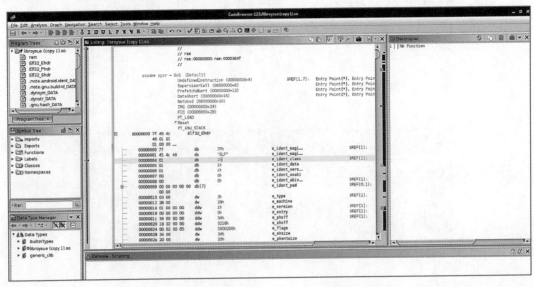

图 3-13 Ghidra 分析 SO 文件的具体信息

中间窗口起始为程序入口点，从 magic 开始（7f 45 4c），显示的是最原始的信息。左侧 Program Trees 窗口列举出了程序的各个段，后续章节我们会学习各个段的基本信息。使用 Symbol Tree 窗口可以查看程序中的所有符号，例如导出符号、导入符号、类、函数、标签等。

我们可以选择导出符号中的 Java_com_roysue_easyso1_MainActivity_stringFromJNI，可以看到右侧反编译视图与中间反汇编列表同步。因此，在反编译器视图中浏览时，我们可以在列表窗口中看到相应的反汇编行，如图 3-14 所示。

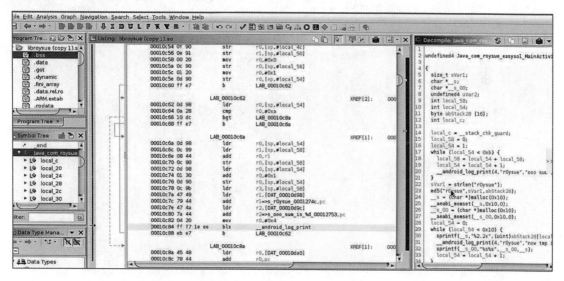

图 3-14　相应的反汇编行

IDA 是 Windows 下的软件，在 Kali 上运行是为了演示方便。首先，我们需要在 Kali 上安装 Wine 运行 Windows 应用。

```
root@VXIDr0ysue:~/Downloads/ghidra_9.2.2_PUBLIC # dpkg --add-architecture i386
root@VXIDr0ysue:~/Downloads/ghidra_9.2.2_PUBLIC # apt update
root@VXIDr0ysue:~/Downloads/ghidra_9.2.2_PUBLIC # apt-get install wine32
```

反汇编器主要的功能就是以上这些，之后可以基于代码对二进制进行一些修改和处理。无论是 Ghidra 还是 IDA，它们都支持自己的中间码，在中间码的基础之上可以用脚本语言对二进制进行更加深入的分析。尽管 JEB 不是主流的二进制反编译工具，但作为一份交叉对照的参考也是不错的选择。另外，环境配置问题往往是最难解决的问题之一，有的时候环境配置到位了，很多问题就会迎刃而解，而当环境配置不到位时，就会经常浪费我们的时间和精力。

还有很多其他的反汇编工具可供选择，读者可以自己编写源码比较各个工具反编译出来的效果。

3.4　Binary Ninja 新晋反汇编器

除了 3.3 节讲述的高级反汇编器外，还有一些新晋反汇编器，如 Binary Ninja 等。

Binary Ninja 的界面如图 3-15 所示，包含符号（Symbols）、交叉引用（Cross References）、CFG、热力图等。

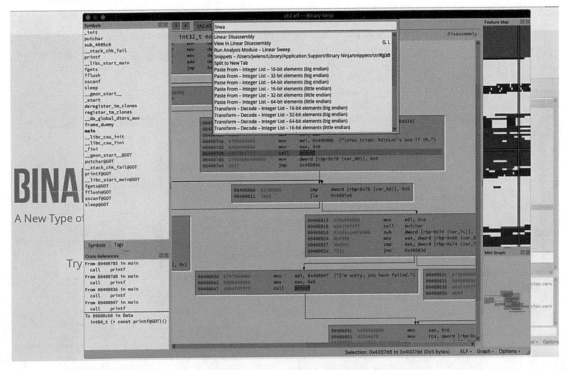

图 3-15　Binary Ninja 的界面

单击首页的 Cloud 可以在线试用，在线试用限制文件大小为 100MB。注册并登录后，上传 libroysue.so，分析完成后在左侧选择函数可以看到具体的函数信息，如图 3-16 所示。

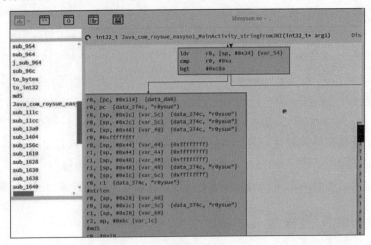

图 3-16　Binary Ninja 分析 SO 文件

字符串表以及引用分析如图 3-17 所示。

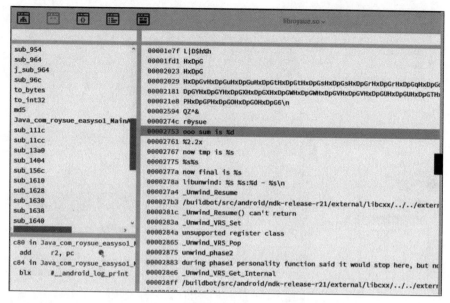

图 3-17　Binary Ninja 字符串表

反汇编显示的指令如图 3-18 所示。

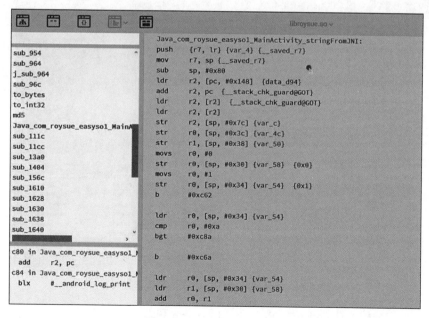

图 3-18　反汇编显示的指令

选择 High Level IL 可以显示反编译优化后的伪代码，如图 3-19 所示，双击 md5 可以进入并查看其使用的一系列函数。

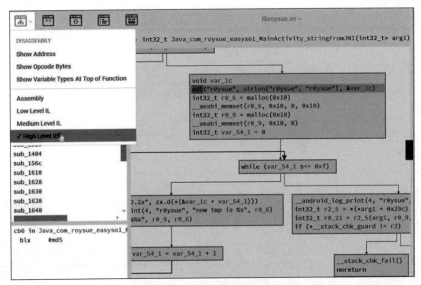

图 3-19　反汇编显示的指令

查看导入函数、导出函数、节区表等，如图 3-20 所示。

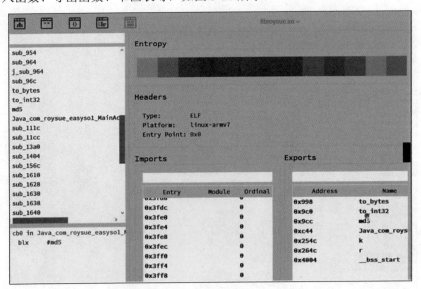

图 3-20　导入函数、导出函数、节区表等信息

还有一款逆向分析工具 Radare2，该工具与 Keystone 和 Capstone 比较接近。Radare2 提供了一组库和工具，用于处理二进制文件，它的功能包括分析二进制文件、调试程序、修补二进制文件、修改代码或数据等。

另外，Radare2 可以和 Frida 结合使用，网址 11（参见配书资源文件）中有一篇详细介绍 r2frida（Radare2 与 Frida 结合起来的一个工具）的文章。

其中许多功能 Frida 也提供了，不过 r2frida 的命令行交互模式能够更加清晰地显示结果。例如，我们可以用 r2frida 在内存中进行实时的 patch，具体内容将在后续章节中介绍。

3.5 本章小结

本章主要介绍了一些静态分析工具的安装和基本使用，包括使用 objdump 反汇编目标文件来显示基本信息，常用的 Binutils 二进制工具包命令如 nm、strings 和 readelf，使用 010 Editor 解析 SO 文件以显示二进制基本信息，常见的高级反汇编器 Ghidra/JEB/IDA，以及新的反汇编器 Binary Ninja 等。这些工具的最终结果大同小异，读者可以自行编写源码来比较各个工具反编译出来的效果，并根据个人喜好选用不同的反编译工具。

第 4 章

C 算法开发及模拟执行

第 3 章学习了静态分析工具的安装和基本使用，包括一些基础的反编译工具、高级反汇编器以及新的反汇编器等。其中，高级反汇编器具有更为完整的功能，可以完全代替 objdump 工具来查询信息。然而，静态分析始终是靠不住的，在实际工作中，往往需要使用多种不同的工具进行交叉确认。本章将重点介绍 C 算法开发及模拟执行，特别是在 Native 层进行密码学套件的移植开发，以及对 AES 算法进行静态分析。

4.1　Native 层密码学套件移植开发

常见的密码学方法中，用得比较多的是编码、哈希、对称加解密等。具体算法的实现可以参考网址 12 或网址 13（参见配书资源文件）。

在实际应用中，我们通常不需要自己去实现算法，可以直接在以上网站找到相应的代码并使用即可。下面分别执行以下命令：

```
root@VXIDr0ysue:~/Desktop/20210117 # git clone https://github.com/WaterJuice/WjCryptLib.git

root@VXIDr0ysue:~/Desktop/20210117 # cd WjCryptLib

root@VXIDr0ysue:~/Desktop/20210117/WjCryptLib # apt install cmake

root@VXIDr0ysue:~/Desktop/20210117/WjCryptLib # cmake ./

root@VXIDr0ysue:~/Desktop/20210117/WjCryptLib # make
```

编译成功后，我们可以查看源码的位置。启动 VS Code 打开 Desktop/20210117/WjCrytLib 文件夹，可以看到源码位于 lib 文件夹下。

读者可以参考第 2 章的内容，将 C 语言的加密算法实现移植到 Android 系统中。

下面演示移植 AES 算法，在此推荐网址 14（（参见配书资源文件）的一篇 AES 文章。

该文章主要介绍了在 Native 代码中使用 AES CBC PKCS5Padding 进行加密，并对密钥 Key 和偏移量 IV 进行保护的方法。

对于花指令，可以自行编写，越复杂越好，目的是尽可能地让反编译器无法将其解析为可读性较强的代码。

参考上述文章，我们进行下面的实践。执行以下命令：

```
root@VXIDr0ysue:~/Desktop/20210117 # git clone https://github.com/luck-apple/aesTool
```

从 VS Code 打开 AESTOOL，看一下代码的基本逻辑。源代码文件位于路径 AESTOOL/app/jniLibrary/src/main/cpp 下。将 java/FooTools.java 中的 native 方法调用复制到 easyso1 项目的 MainActivity.java 文件中，如图 4-1 所示。

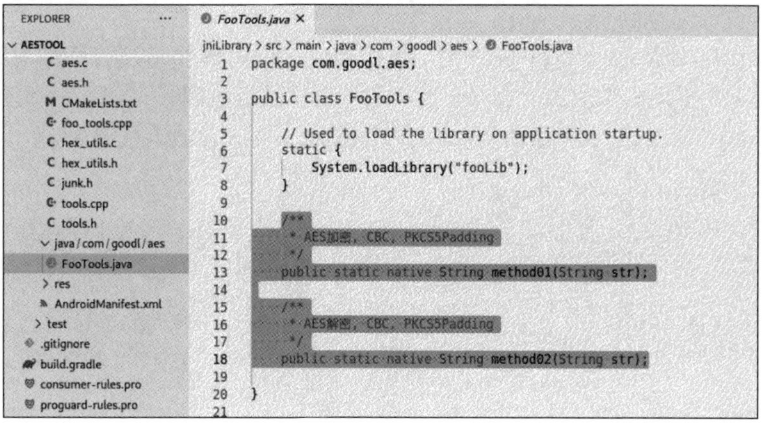

图 4-1　native 方法调用

此时会报错，因为现在还没有 native 层的实现，如图 4-2 所示。

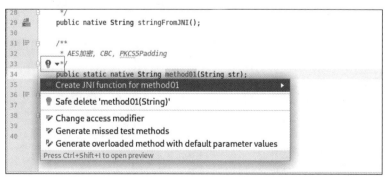

图 4-2　为 method01 创造 JNI 方法

在提示符的下拉选项中选择 Create JNI function for method01 后，界面会自动跳转到方法的实现代码。method02 同理，结果如图 4-3 所示。

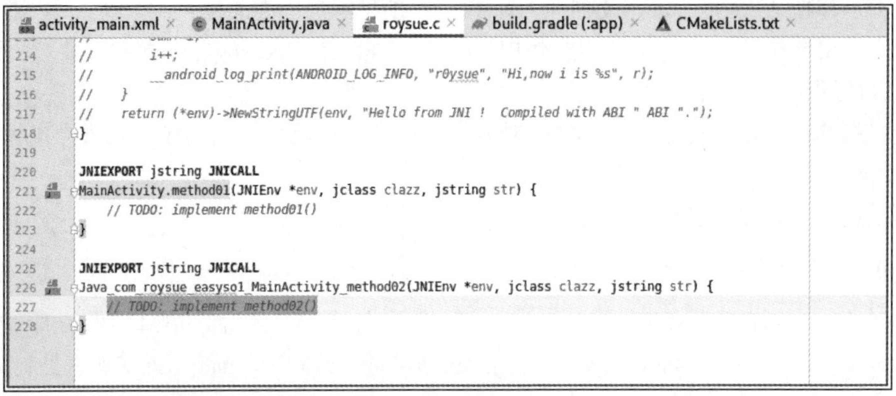

图 4-3　native 方法的实现（空）

源程序的实现位于 foo_tools.cpp 文件中，除了 method01、method02 方法的实现之外，还包括动态注册的实现，关于动态注册的内容将在后续的章节中进行讲解。

先把 method01、method02 方法的实现复制到 roysue.c 中，修改后代码如下：

```c
#include <string.h>
#include <jni.h>
#include <android/log.h>
#include <stdio.h>
#include <stdlib.h>
#include <stdint.h>

#include "aes_utils.h"
#include "tools.h"
#include "junk.h"

#define NELEM(x) ((int) (sizeof(x) / sizeof((x)[0])))

const uint32_t k[64] = {
        0xd76aa478, 0xe8c7b756, 0x242070db, 0xc1bdceee,
        0xf57c0faf, 0x4787c62a, 0xa8304613, 0xfd469501,
        0x698098d8, 0x8b44f7af, 0xffff5bb1, 0x895cd7be,
        0x6b901122, 0xfd987193, 0xa679438e, 0x49b40821,
        0xf61e2562, 0xc040b340, 0x265e5a51, 0xe9b6c7aa,
        0xd62f105d, 0x02441453, 0xd8a1e681, 0xe7d3fbc8,
        0x21e1cde6, 0xc33707d6, 0xf4d50d87, 0x455a14ed,
        0xa9e3e905, 0xfcefa3f8, 0x676f02d9, 0x8d2a4c8a,
        0xfffa3942, 0x8771f681, 0x6d9d6122, 0xfde5380c,
        0xa4beea44, 0x4bdecfa9, 0xf6bb4b60, 0xbebfbc70,
```

```c
            0x289b7ec6, 0xeaa127fa, 0xd4ef3085, 0x04881d05 ,
            0xd9d4d039, 0xe6db99e5, 0x1fa27cf8, 0xc4ac5665 ,
            0xf4292244, 0x432aff97, 0xab9423a7, 0xfc93a039 ,
            0x655b59c3, 0x8f0ccc92, 0xffeff47d, 0x85845dd1 ,
            0x6fa87e4f, 0xfe2ce6e0, 0xa3014314, 0x4e0811a1 ,
            0xf7537e82, 0xbd3af235, 0x2ad7d2bb, 0xeb86d391 };

    // r specifies the per-round shift amounts
    const uint32_t r[] = {7, 12, 17, 22, 7, 12, 17, 22, 7, 12, 17, 22, 7, 12, 17, 22, 5,
9, 14, 20, 5,  9, 14, 20, 5,  9, 14, 20, 5,  9, 14, 20, 4, 11, 16, 23, 4, 11, 16, 23, 4, 11,
16, 23, 4, 11, 16, 23, 6, 10, 15, 21, 6, 10, 15, 21, 6, 10, 15, 21, 6, 10, 15, 21};

    // leftrotate function definition
    #define LEFTROTATE(x, c) (((x) << (c)) | ((x) >> (32 - (c))))

    void to_bytes(uint32_t val, uint8_t *bytes)
    {
        bytes[0] = (uint8_t) val;
        bytes[1] = (uint8_t) (val >> 8);
        bytes[2] = (uint8_t) (val >> 16);
        bytes[3] = (uint8_t) (val >> 24);
    }

    uint32_t to_int32(const uint8_t *bytes)
    {
        return (uint32_t) bytes[0]
            | ((uint32_t) bytes[1] << 8)
            | ((uint32_t) bytes[2] << 16)
            | ((uint32_t) bytes[3] << 24);
    }

    void md5(const uint8_t *initial_msg, size_t initial_len, uint8_t *digest) {

        // These vars will contain the hash
        uint32_t h0, h1, h2, h3;

        // Message (to prepare)
        uint8_t *msg = NULL;

        size_t new_len, offset;
        uint32_t w[16];
        uint32_t a, b, c, d, i, f, g, temp;

        // Initialize variables - simple count in nibbles
        h0 = 0x67452301;
        h1 = 0xefcdab89;
```

```
    h2 = 0x98badcfe;
    h3 = 0x10325476;

    //Pre-processing:
    //append "1" bit to message
    //append "0" bits until message length in bits ≡ 448 (mod 512)
    //append length mod (2^64) to message

    for (new_len = initial_len + 1; new_len % (512/8) != 448/8; new_len++)
        ;

    msg = (uint8_t*)malloc(new_len + 8);
    memcpy(msg, initial_msg, initial_len);
    msg[initial_len] = 0x80; // append the "1" bit; most significant bit is "first"
    for (offset = initial_len + 1; offset < new_len; offset++)
        msg[offset] = 0; // append "0" bits

    // append the len in bits at the end of the buffer
    to_bytes(initial_len*8, msg + new_len);
    // initial_len>>29 == initial_len*8>>32, but avoids overflow
    to_bytes(initial_len>>29, msg + new_len + 4);

    // Process the message in successive 512-bit chunks
    //for each 512-bit chunk of message:
    for(offset=0; offset<new_len; offset += (512/8)) {

        // break chunk into sixteen 32-bit words w[j], 0 ≤ j ≤ 15
        for (i = 0; i < 16; i++)
            w[i] = to_int32(msg + offset + i*4);

        // Initialize hash value for this chunk
        a = h0;
        b = h1;
        c = h2;
        d = h3;

        // Main loop
        for(i = 0; i<64; i++) {

            if (i < 16) {
                f = (b & c) | ((~b) & d);
                g = i;
            } else if (i < 32) {
                f = (d & b) | ((~d) & c);
                g = (5*i + 1) % 16;
            } else if (i < 48) {
```

```
                f = b ^ c ^ d;
                g = (3*i + 5) % 16;
            } else {
                f = c ^ (b | (~d));
                g = (7*i) % 16;
            }

            temp = d;
            d = c;
            c = b;
            b = b + LEFTROTATE((a + f + k[i] + w[g]), r[i]);
            a = temp;

        }

        // Add this chunk's hash to result so far:
        h0 += a;
        h1 += b;
        h2 += c;
        h3 += d;

    }

    // cleanup
    free(msg);

    //var char digest[16] := h0 append h1 append h2 append h3 //(Output is in little-endian)
    to_bytes(h0, digest);
    to_bytes(h1, digest + 4);
    to_bytes(h2, digest + 8);
    to_bytes(h3, digest + 12);
}


/* This is a trivial JNI example where we use a native method
 * to return a new VM String. See the corresponding Java source
 * file located at:
 *
 *   hello-jni/app/src/main/java/com/example/hellojni/HelloJni.java
 */
JNIEXPORT jstring JNICALL
Java_com_roysue_easyso1_MainActivity_stringFromJNI( JNIEnv* env, jobject thiz )
{
#if defined(__arm__)
    #if defined(__ARM_ARCH_7A__)
```

```
      #if defined(__ARM_NEON__)
        #if defined(__ARM_PCS_VFP)
          #define ABI "armeabi-v7a/NEON (hard-float)"
        #else
          #define ABI "armeabi-v7a/NEON"
        #endif
      #else
        #if defined(__ARM_PCS_VFP)
          #define ABI "armeabi-v7a (hard-float)"
        #else
          #define ABI "armeabi-v7a"
        #endif
      #endif
    #else
    #define ABI "armeabi"
    #endif
#elif defined(__i386__)
#define ABI "x86"
#elif defined(__x86_64__)
#define ABI "x86_64"
#elif defined(__mips64)  /* mips64el-* toolchain defines __mips__ too */
#define ABI "mips64"
#elif defined(__mips__)
#define ABI "mips"
#elif defined(__AArch64__)
#define ABI "arm64-v8a"
#else
#define ABI "unknown"
#endif

    char* msg = "r0ysue";
    size_t len = strlen(msg);
    uint8_t result[16];
    md5((uint8_t*)msg, len, result);

    char* r = (char*)malloc(16);
    memset(r, 0x00, sizeof(char) *16);

    char* final = (char*)malloc(16);
    memset(final, 0x00, sizeof(char) *16);

    md5((uint8_t*)msg, len, result);

// display result
    for (int i = 0; i < 16; i++) {
        sprintf(r, "%2.2x", result[i]);
```

```c
        __android_log_print(ANDROID_LOG_INFO, "r0ysue", "Hi,now i is %s", r);
        sprintf(final, "%s%s", final, r);
    }

    jstring jresult = (*env)->NewStringUTF(env, final);
    return jresult;
}

JNIEXPORT jstring JNICALL
Java_com_roysue_easyso1_MainActivity_method01(JNIEnv *env, jclass clazz, jstring str_)
{
    // TODO: implement method01()
    //if (str_ == nullptr) return nullptr;

    const char *str = (*env)->GetStringUTFChars(env, str_, JNI_FALSE);
    char *result = AES_128_CBC_PKCS5_Encrypt(str);

    (*env)->ReleaseStringUTFChars(env, str_, str);

    jstring jResult = getJString(env, result);
    free(result);

    return jResult;
}

JNIEXPORT jstring JNICALL
Java_com_roysue_easyso1_MainActivity_method02(JNIEnv *env, jclass clazz, jstring str_)
{
    // TODO: implement method02()
    //if (str_ == nullptr) return nullptr;

    const char *str = (*env)->GetStringUTFChars(env, str_, JNI_FALSE);
    char *result = AES_128_CBC_PKCS5_Decrypt(str);

    (*env)->ReleaseStringUTFChars(env, str_, str);

    jstring jResult = getJString(env, result);
    free(result);

    return jResult;
}
```

将源程序中如图4-4所示的文件复制到我们项目的/root/AndroidStudioProjects/easyso1/app/src/main/cpp/目录下。

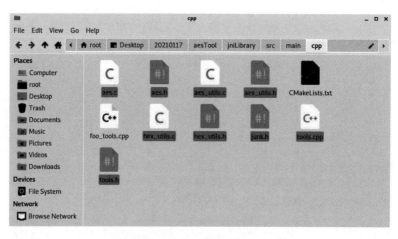

图 4-4 复制文件

之后，需要包含头文件，并把复制的文件添加到 CMakeLists 中，可以参考源程序中的 CMakeLists。修改 onCreate 中的对应语句为：

```
tv.setText(method01("r0ysue"));
```

修改完成后运行即可，结果如图 4-5 所示。

再次修改代码，对加密结果进行解密，得到明文，结果如图 4-6 所示。

```
tv.setText(method02(method01("r0ysue")));
```

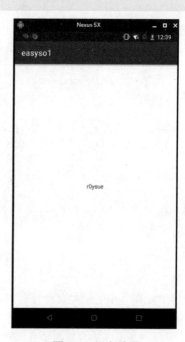

图 4-5 method01 的运行结果　　　　图 4-6 解密结果

接下来，拦截 native 层的 method01 和 method02 两个函数，java 层函数可以通过 objection 进行拦截。因为是 C 语言，所以我们使用 Ghidra 来看一下它的导出函数。

将/root/AndroidStudioProjects/easyso1/app/build/outputs/apk/debug/目录下的 app-debug.apk 复制到/root/Desktop/20210117/apk/目录下，以便后续使用。

解压 app-debug.apk，运行 Ghidra：

```
root@VXIDr0ysue:~/Desktop/20210117/apk # 7z x app-debug.apk

root@VXIDr0ysue:~/Downloads # cd ghidra_9.2.2_PUBLIC

root@VXIDr0ysue:~/Downloads/ghidra_9.2.2_PUBLIC # ./ghidraRun
Picked up _JAVA_OPTIONS: -Dawt.useSystemAAFontSettings=on -Dswing.aatext=true
Picked up _JAVA_OPTIONS: -Dawt.useSystemAAFontSettings=on -Dswing.aatext=true
```

在/root/Desktop/20210117 目录下创建 project123，启动并打开/root/Desktop/20210117/apk/lib/armeabi-v7a/目录下的 SO 文件，反编译完成后，可以从导出表看到，除了 method01、method02 和 stringFromJNI 外，其他函数都变成了我们不认识的字符串，如图 4-7 所示。

我们可以从反汇编后的伪代码中看到一些逻辑，但可能看不懂其中的内容，如图 4-8 所示。

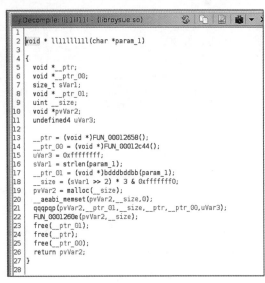

图 4-7　导出表　　　　　　　　　图 4-8　导出表

但是如果 MD5 函数没有进行代码混淆，我们仍然可以看懂其反汇编后的伪代码。在源码中，最核心的函数是 AES_128_CBC_PKCS5_Encrypt，然而这个函数在反汇编后的伪代码中很难看出它的具体逻辑。

4.2 Frida Hook/主动调用执行算法

前面介绍的是基本的静态分析。针对模拟执行，我们不需要分析它所有的流程，只需要用 Frida 主动 Hook 即可，或者使用 AndroidNativeEmu 模拟执行算法。

下面我们来演示一下 Frida 如何实现模拟执行。

启动 VS Code，在/root/Desktop/20210117 目录下创建 FridaInvokeNative.js 文件。

首先通过导出函数找到 method01 的地址，通过附加调用 Java.perform(fn)，将当前线程附加到 Java VM 并且调用 fn 方法。然后定义变量 method01 为 NativeFunction，需要传入的参数如下：

- address：method01 的地址。
- returnType：返回值 jstring（在 C 语言中是指针）。
- argTypes[]：传入的 3 个参数类型也是指针。

调用 method01 方法，由于返回值是 jstring，不能直接打印，因此需要调用 getStringUtfChars 函数。编写代码如下：

```
setImmediate(function(){
    console.log("invoking...")
    var method01_addr = Module.findExportByName("libroysue.so", "Java_com_roysue_easyso1_MainActivity_method01")
    console.log("method01 address is => ", method01_addr)

    Java.perform(function(){
        var jstring = Java.vm.getEnv().newStringUtf("roysue")
        var method01 = new NativeFunction(method01_addr, "pointer", ["pointer", "pointer", "pointer"])
        var result = method01(Java.vm.getEnv(), jstring, jstring)
        console.log("Final result is => ", Java.vm.getEnv().getStringUtfChars(result, null).readCString());
    })
})
```

将虚拟机上的 Python 环境切换至 3.8.0 版本，在手机上启动 frida-server 12.8.0，在 Frida 上运行 FridaInvokeNative.js：

```
root@VXIDr0ysue:~/Desktop/20210117 # frida -UF -l FridaInvokeNative.js

invoking...
method01 address is => 0xd30073c9
Final result is => 47fcda3822cd10a8e2f667fa49da783f
```

我们可以对经过简单混淆的 SO 文件进行静态分析，但如果 SO 文件经过混淆膨胀后体积过大，则无法进行静态分析，这时可以通过 Frida 主动调用导出函数得到结果。

类似地，可以调用 method02 进行解密，读者可以自行尝试。

通过主动调用来执行算法可以解决这一类问题，而不需要分析出每一个细节。

4.3 使用 AndroidNativeEmu 模拟执行算法

AndroidNativeEmu 是一个基于 Unicorn 的完整的模拟器。我们先将它复制到本地，执行以下命令：

```
root@VXIDr0ysue:~/Desktop/20210117 # git clone https://github.com/maiyao1988/
ExAndroidNativeEmu.git
```

另外，AndroidNativeEmu 需要 Python 3.7 的环境，执行以下命令安装 Python 3.7：

```
root@VXIDr0ysue:~ # pyenv install 3.7.0
```

使用 VS Code 打开 AndroidNativeEmu，安装 Python 扩展后，切换到 Python 3.7 环境，再安装依赖包：

```
root@VXIDr0ysue:~/Desktop/20210117/ExAndroidNativeEmu # pyenv local 3.7.0
root@VXIDr0ysue:~/Desktop/20210117/ExAndroidNativeEmu # pip install -r
requirements.txt
```

安装成功后，需要对 example_jni.py 做一些修改。把 libroysue.so 复制到 /root/Desktop/20210117/ExAndroidNativeEmu/tests/bin/ 目录下，编辑 load_library，如图 4-9 所示。

```
81    # Load all libraries.
82    lib_module = emulator.load_library("tests/bin/libroysue.so")
83
```

图 4-9　load_library

直接调用算法：

```
    try:
        # Run JNI_OnLoad
        # JNI_OnLoad will call 'RegisterNatives'
        # emulator.call_symbol(lib_module, 'JNI_OnLoad', emulator.java_vm.address_ptr, 0x00)

        # Do native stuff
        # main_activity = MainActivity()
        # logger.info("Response from JNI call: %s" % main_activity.string_from_jni(emulator))
        logger.info("Response from JNI call: %s" % emulator.call_symbol(lib_module,
"Java_com_roysue_easyso1_MainActivity_method01", emulator.java_vm.jni_env.address_ptr,
0x00, String('roysue')))
```

由于 String 是模拟的，因此导入 String：

```
from androidemu.java.classes.string import String
```

运行结果如图 4-10 所示。

```
(              libc.so[0xCBBD8000])[FB FF FF EA   ]0x0000E844:   B      #0XCBBE6838        ;(PC=0xCBBE6844 )
(              libc.so[0xCBBD8000])[08 D0 8D E2   ]0x0000E838:   ADD    SP, SP, #8         ;(SP=0x100FFFB8 )
(              libc.so[0xCBBD8000])[70 80 BD E8   ]0x0000E83C:   POP    {R4, R5, R6, PC}   ;(SP=0x100FFFC0 )
(              libc.so[0xCBBD8000])[08 BD         ]0x0000DC0C:   POP    {R3, PC}           ;(SP=0x100FFFD0 )
(          libroysue.so[0xCBBCB000])[02 98         ]0x0000341A:   LDR    R0, [SP, #8]       ;(SP=0x100FFFD8 )
(          libroysue.so[0xCBBCB000])[68 B0         ]0x0000341C:   ADD    SP, #0X20          ;(SP=0x100FFFD8 )
(          libroysue.so[0xCBBCB000])[80 BD         ]0x0000341E:   POP    {R7, PC}           ;(SP=0x100FFFF8 )
22313 - 2021-05-29 20:44:12,626 - INFO - Response from JNI call: JavaString(47fcda3822cd10a8e2f667fa49da783f)
22313 - 2021-05-29 20:44:12,627 - INFO - Exited EMU
22313 - 2021-05-29 20:44:12,627 - INFO - Native methods registered to MainActivity:
```

图 4-10 直接调用 method01 后的运行结果

将结果 47fcda3822cd10a8e2f667fa49da783f 回填，调用 method02，添加如下代码：

```
logger.info("Response from JNI call: %s" % emulator.call_symbol(lib_module,
"Java_com_roysue_easyso1_MainActivity_method02", emulator.java_vm.jni_env.address_ptr,
0x00, String('47fcda3822cd10a8e2f667fa49da783f')))
```

打印的内容包含模拟器执行的所有（汇编）指令。我们强调的是定位最细粒度的算法逻辑，最关心的是第三方的调用，然后我们再模拟这个系统。

注释掉部分代码，查看对系统调用的指令有哪些。

```
# Register Java class.
emulator.java_classloader.add_class(MainActivity)
# emulator.mu.hook_add(UC_HOOK_CODE, hook_code, emulator)

# emulator.mu.hook_add(UC_HOOK_MEM_WRITE, hook_mem_write)
# emulator.mu.hook_add(UC_HOOK_MEM_READ, hook_mem_read)
```

在打印输出的结果中，最后两个 Response from JNI call 之间都是调用的系统的 API。这些 API 是 AndroidNativeEmu 已经实现的。我们可以在以下代码段加断点来进行调试：

```
logger.info("Response from JNI call: %s" % emulator.call_symbol(lib_module,
"Java_com_roysue_easyso1_MainActivity_method01", emulator.java_vm.jni_env.address_ptr,
0x00, String('roysue')))
```

方法执行的具体过程这里就不演示了。

4.4 本章小结

本章介绍并演示了在 native 层进行密码学套件的移植开发。通过使用 Ghidra 对 AES 算法进行静态分析，了解其特征及简单的混淆技术，并通过添加花指令增加破解难度。

最后，本章还延伸了两个部分，即 Frida Hook 和主动调用执行算法，以及使用 AndroidNativeEmu 模拟执行算法，不过这两部分不是本章的重点，并且具有一定的难度。然而，模拟执行是一项很重要的技术，需要读者在后续的学习中逐步掌握。

第 5 章

动态调试：GDB 动态调试、Hyperpwn/（内存）断点/栈帧

在之前的章节中，我们学习了基本的静态分析以及模拟执行的思想。接下来，将带领读者学习动态调试。动态调试可以分为源码级调试和汇编级调试。对于源码级调试，我们之前介绍了通过断点调试项目 easyso1 以及 Android Studio 内置的 LLDB 实现方式，本章将进一步介绍以 GDB 为主的其他动态调试方法。

5.1 GDB C/S 的调试架构

我们之前介绍了通过断点调试项目 easyso1 实现调试，但用这种方法调试第三方 APK 存在问题。这是因为 SO 库并不是在 App 启动时加载的，而是在调用 loadLibrary 时加载的，如图 5-1 中所示选中的部分。

```
1   package com.roysue.easyso1;
2
3   import ...
7
8   public class MainActivity extends AppCompatActivity {
9
10      // Used to load the 'native-lib' library on application startup.
11      static {
12          System.loadLibrary( libname: "roysue");
13      }
14
15      @Override
16      protected void onCreate(Bundle savedInstanceState) {
17          super.onCreate(savedInstanceState);
18          setContentView(R.layout.activity_main);
19
```

图 5-1　loadLibrary 加载 SO 库

这意味着我们必须先 Hook 加载器。在本项目中，App 一经启动，onCreate 函数立即执行。当执行完 onCreate 函数的最后一行语句后，App 就运行结束了。因此，在 roysue.c 中设置断点将不再生效。也就是说，当我们调试第三方 App 时，需要 Hook SO 的解析器，也就是以下语句：

```
System.loadLibrary("roysue");
```

在 SO 库加载完成而函数还未开始执行的一个时机点，我们需要将它暂停下来，这对于程序员的技术水平要求较高。

然而，以上方法并不是本书的重点。我们需要将项目改造一下，让它多次调用 method01 和 method02 函数。这样做的好处是，当 SO 库加载完成后，我们依旧可以在 method01 和 method02 函数中设置断点，并观察它前后的行为。

我们可以使用 Runnable 接口创建一个线程来循环执行 method01 函数。在 onCreate 函数中添加以下代码：

```
class MyThread extends Thread {
    @Override
    public void run() {
        while (true) {
            try {
                Thread.sleep(2000);
            } catch (InterruptedException e) {
                e.printStackTrace();
            }
            Log.i("roysue easyso1", method01("r0ysue"));
        }
    }
}

new MyThread().start();
```

运行程序，在 Logcat 中可以看到打印的日志，如图 5-2 所示。

```
22:15:46.566 7513-7513/com.roysue.easyso1 I/roysue easyso1: 4e8de2f3c674d8157b4862e50954d81c
22:15:48.569 7513-7513/com.roysue.easyso1 I/roysue easyso1: 4e8de2f3c674d8157b4862e50954d81c
22:15:50.571 7513-7513/com.roysue.easyso1 I/roysue easyso1: 4e8de2f3c674d8157b4862e50954d81c
22:15:58.579 7513-7513/com.roysue.easyso1 I/chatty: uid=10100(com.roysue.easyso1) identical 4 lines
22:16:00.582 7513-7513/com.roysue.easyso1 I/roysue easyso1: 4e8de2f3c674d8157b4862e50954d81c
22:16:02.583 7513-7513/com.roysue.easyso1 I/roysue easyso1: 4e8de2f3c674d8157b4862e50954d81c
22:16:04.585 7513-7513/com.roysue.easyso1 I/roysue easyso1: 4e8de2f3c674d8157b4862e50954d81c
22:16:06.587 7513-7513/com.roysue.easyso1 I/roysue easyso1: 4e8de2f3c674d8157b4862e50954d81c
22:16:08.589 7513-7513/com.roysue.easyso1 I/roysue easyso1: 4e8de2f3c674d8157b4862e50954d81c
22:16:24.609 7513-7513/com.roysue.easyso1 I/chatty: uid=10100(com.roysue.easyso1) identical 8 lines
22:16:26.611 7513-7513/com.roysue.easyso1 I/roysue easyso1: 4e8de2f3c674d8157b4862e50954d81c
```

图 5-2　打印的日志：method01("r0ysue")

完成以上准备工作之后，接下来正式开始学习 GDB C/S 的调试架构。首先，我们要明确 GDB 的客户端（Client）和服务端（Server）。GDB 的客户端跟 Android Studio 的角色比较接近，它运行在宿主机上；GDB 的服务端则运行在目标机上，此处是移动设备端。因此，我们需要将 gdbserver 推送（push）到手机上。在之前的章节中提到过，使用 Android Studio 的调试功能时，debug 将 lldb-server 推送到用户的目录下。GDB 也是一样的过程，具体介绍可以参考网址 15（参见配书资源文件）中的文章。

建议调试用的手机系统是 Android 8.1.0。

不同目标系统的 gdbserver 位于/root/Android/Sdk/ndk/21.1.6352462/prebuilt 目录下；GDB 位于/root/Android/Sdk/ndk/21.1.6352462/prebuilt/linux-x86_64/bin/目录下。可以将包含 gdbserver 的两个文件夹 android-arm 和 android-arm64 复制到/root/Desktop/20210120/目录下以备后续使用。

```
root@VXIDr0ysue:~/Desktop/20210120 # ls
android-arm  android-arm64

root@VXIDr0ysue:~/Desktop/20210120 # adb push android-* /data/local/tmp
android-arm/: 1 file pushed, 0 skipped. 4.2 MB/s (1006532 bytes in 0.229s)
android-arm64/: 1 file pushed, 0 skipped. 5.6 MB/s (1343712 bytes in 0.230s)
2 files pushed, 0 skipped. 4.6 MB/s (2350244 bytes in 0.491s)
```

选择 32 位的 arm 还是 64 位的 arm64 取决于 App 的进程，我们将在本章后续的内容中讲解如何根据 App 的进程来选择 arm 或 arm64。

5.2 将 App 编译成带调试符号的 SO 文件

Android 编译生成的 SO 文件默认会去除（strip）所有符号，这是我们不愿意看到的。本节将学习如何将 app 编译成带调试符号的 SO 文件。

首先，Android Studio 使用的是 CMake 编译工具。

CMake 主要编写 CMakeLists.txt 文件，然后通过 cmake 命令将 CMakeLists.txt 文件转化为 make 所需要的 Makefile 文件，最后用 make 命令编译源码以生成可执行程序或者库文件。

读者可以参阅网址 16（参见配书资源文件）的文章进一步了解 CMake。

我们先来看一下项目 easyso1 的 SO 文件大小：

```
root@VXIDr0ysue:~/AndroidStudioProjects/easyso1/app/build/outputs/apk/debug # 7z x app-debug.apk

root@VXIDr0ysue:~/AndroidStudioProjects/easyso1/app/build/outputs/apk/debug/lib/armeabi-v7a # du -h -d 1 *
28K     libroysue.so
```

结果显示，libroysue.so 只有 28KB。为了添加符号和调试信息，需要在 Android Studio 项目配置文件 build.gradle 中添加以下代码，添加位置如图 5-3 所示。

```
packagingOptions {
    doNotStrip "*/armeabi/*.so"
```

```
        doNotStrip "*/armeabi-v7a/*.so"
        doNotStrip "*/x86/*.so"
}
```

有关添加调试信息的内容，可参考网址 17（参见配书资源文件）的文章。

```
31          }
32       }
33       externalNativeBuild {
34          cmake {
35              path "src/main/cpp/CMakeLists.txt"
36              version "3.10.2"
37          }
38       }
39       compileOptions {
40          sourceCompatibility JavaVersion.VERSION_1_8
41          targetCompatibility JavaVersion.VERSION_1_8
42       }
43       packagingOptions {
44          doNotStrip "*/armeabi/*.so"
45          doNotStrip "*/armeabi-v7a/*.so"
46          doNotStrip "*/x86/*.so"
47       }
48  }
49
50  dependencies {
51
```

图 5-3　packagingOptions 添加位置

Sync Now 同步后重新运行，在菜单栏中按以下顺序选择：Build→Clean Project，Build→Refresh Linked C++ Projects，Build→Rebuild Project，运行项目。

从以上过程可以看到，我们在运行第三方 App 的时候是无法通过修改项目配置文件 build.gradle 来获取调试信息的。检查一下调试信息是否成功添加了：在/root/AndroidStudioProjects/easyso1/app/build/outputs/apk/debug/路径下打开终端，解压此目录下的 app-debug.apk：

```
root@VXIDr0ysue:~/…/build/outputs/apk/debug # 7z x app-debug.apk

root@VXIDr0ysue:~/…/build/outputs/apk/debug # cd lib/armeabi-v7a

root@VXIDr0ysue:~/…/build/outputs/apk/debug/lib/armeabi-v7a # du -h -d 1 *
212K    libroysue.so
```

libroysue.so 文件大小由 28KB 变为 212KB，包含非常多的调试信息。使用 objdump 命令查看符号表，使用选项-t 查看所有符号：

```
root@VXIDr0ysue:~/…/build/outputs/apk/debug/lib/armeabi-v7a # objdump -t libroysue.so

libroysue.so:     file format elf32-little

DYNAMIC SYMBOL TABLE:
00000000      DF *UND*  00000000  LIBC        __cxa_atexit
```

```
00000000        DF *UND*    00000000    LIBC       __cxa_finalize
00000000        DF *UND*    00000000               __aeabi_memcpy
```

使用 nm 命令同样能看到符号的偏移，结果如图 5-4 所示。

```
root@VXIDr0ysue:~/…/build/outputs/apk/debug/lib/armeabi-v7a # nm -s libroysue.so
```

```
File  Actions  Edit  View  Help
00003fb5 t  _ZN9libunwind12UnwindCursorINS_17LocalAddressSpaceENS_13Registers_armEE24setInfoBasedOnIPRegisterEb
00003f13 t  _ZN9libunwind12UnwindCursorINS_17LocalAddressSpaceENS_13Registers_armEE4stepEv
00003eeb t  _ZN9libunwind12UnwindCursorINS_17LocalAddressSpaceENS_13Registers_armEE6getRegEi
00003f75 t  _ZN9libunwind12UnwindCursorINS_17LocalAddressSpaceENS_13Registers_armEE6jumptoEv
00003ef1 t  _ZN9libunwind12UnwindCursorINS_17LocalAddressSpaceENS_13Registers_armEE6setRegEij
00003f67 t  _ZN9libunwind12UnwindCursorINS_17LocalAddressSpaceENS_13Registers_armEE7getInfoEP15unw_proc_info_t
00003edd t  _ZN9libunwind12UnwindCursorINS_17LocalAddressSpaceENS_13Registers_armEE8validRegEi
00003edb t  _ZN9libunwind12UnwindCursorINS_17LocalAddressSpaceENS_13Registers_armEED0Ev
00004029 t  _ZN9libunwind13Registers_arm11getRegisterEi
000040a1 t  _ZN9libunwind13Registers_arm11setRegisterEij
00004a48 t  _ZN9libunwind13Registers_arm12restoreVFPv3EPy
000045e1 t  _ZN9libunwind13Registers_arm15getRegisterNameEi
00004115 t  _ZN9libunwind13Registers_arm16getFloatRegisterEi
00003d0c t  _ZN9libunwind13Registers_arm16saveVFPWithFSTMDEPy
00003d14 t  _ZN9libunwind13Registers_arm16saveVFPWithFSTMXEPy
000041c1 t  _ZN9libunwind13Registers_arm16setFloatRegisterEiy
00004a38 t  _ZN9libunwind13Registers_arm19restoreVFPWithFLDMDEPy
00004a40 t  _ZN9libunwind13Registers_arm19restoreVFPWithFLDMXEPy
00004a24 t  _ZN9libunwind13Registers_arm20restoreCoreAndJumpToEv
000042dd t  _ZN9libunwind13Registers_arm26restoreSavedFloatRegistersEv
000042c9 t  _ZN9libunwind13Registers_arm6jumptoEv
00003d1c t  _ZN9libunwind13Registers_arm9saveVFPv3EPy
00004311 t  _ZN9libunwind17LocalAddressSpace16findFunctionNameEjPcjPj
000070c0 b  _ZN9libunwind17LocalAddressSpace17sThisAddressSpaceE
00004371 t  _ZN9libunwind17LocalAddressSpace18findUnwindSectionsEjRNS_18UnwindInfoSectionsE
00003ed9 t  _ZN9libunwind20AbstractUnwindCursorD2Ev
0000455d t  _ZNSt6__ndk11upper_boundIN9libunwind20EHABISectionIteratorINS1_17LocalAddressSpaceEEEjNS_6__lessIjjEEEET_S7
7_RKT0_T1
00004599 t  _ZNSt6__ndk113__upper_boundIRNS_6__lessIjjEEN9libunwind20EHABISectionIteratorINS4_17LocalAddressSpaceEEEjEET
S8_S8_RKT1_T
00006de8 d  _ZTVN9libunwind12UnwindCursorINS_17LocalAddressSpaceENS_13Registers_armEEE
```

图 5-4   使用 nm 命令查看符号的偏移

5.3 使用 Android 调试模式来启动 App

Native 的调试参考一些比较旧的方法会更加精准有效。让我们一起来学习如何使用 Android 调试模式来启动 App。首先，arm/arm64 取决于 App 的进程。如果 App 正在运行，则可使用 objection 命令来查看：

```
root@VXIDr0ysue:~ # pyenv local 3.8.0

bullhead:/data/local/tmp # ./frida-server-12.8.0-android-arm64

root@VXIDr0ysue:~ # objection -g com.roysue.easyso1 explore
com.roysue.easyso1 on (google: 8.1.0) [usb] # frida
--------------------  -----------
Frida Version         12.8.0
Process Architecture  arm
Process Platform      linux
Debugger Attached     False
Script Runtime        DUK
```

```
Script Filename       /script1.js
Frida Heap Size       12.1 MiB
-------------------   -----------
```

从 Process Architecture 表项为 arm 可以看出进程是 32 位的。另一种方法更直观,先找到自己的进程:

```
bullhead:/data/local/tmp # ps -e |grep roysue
u0_a100      2001   567 1631732  44900 futex_wait_queue_me ef199ccc S
com.roysue.easyso1
```

进程号在第二列是 2001,第三列是父进程号 567,即 Zygote 进程。

在 Android 中,负责孵化新进程的进程叫作 Zygote,Android 上其他的应用进程都是由它孵化的。众所周知,Android 是 Linux 内核,Android 系统上运行的一切程序都是放在 Dalvik 虚拟机上的,Zygote 也不例外,事实上,它是 Android 运行的第一个 Dalvik 虚拟机进程。既然 Zygote 负责孵化其他的 Android 进程,那么它自己是由谁孵化的呢?既然 Android 是基于 Linux 内核的,Zygote 当然就是 Linux 内核启动的用户级进程 Init 创建的了。

进程 1 是 Init 进程:

```
bullhead:/data/local/tmp # ps -e |grep 567
u0_a100      2001   567 1631636  41572 futex_wait_queue_me ef199ccc S
com.roysue.easyso1
bullhead:/data/local/tmp # ps -e |grep 567
root          567     1 1552136  40988 poll_schedule_timeout ef1c9c4c S zygote
u0_a100      2001   567 1632148  41708 futex_wait_queue_me ef199ccc S
com.roysue.easyso1
```

孵化器进程 Zygote 不止一个:

```
bullhead:/data/local/tmp # ps -e |grep -i zygote
root          566     1 4213108  61252 poll_schedule_timeout 7b9ed79518 S zygote64
root          567     1 1552136  40920 poll_schedule_timeout ef1c9c4c S zygote
webview_zygote 1078    1 1333580  21448 poll_schedule_timeout f6b13c4c S
webview_zygote32
```

我们也可以通过父进程是 64 位的 Zygote64 还是 32 位的 Zygote 来区分。当然,这种方法只能通过动态判断。除此之外,我们还需要获取包名和 activity 名字。可以用 APK 反汇编,打开 App 后获得顶层 activity:

```
root@VXIDr0ysue:~ # adb shell dumpsys activity top

ACTIVITY com.roysue.easyso1/.MainActivity e5ea38 pid=3799
root@VXIDr0ysue:~ # adb shell am start -D -n com.roysue.easyso1/.MainActivity
Starting: Intent { cmp=com.roysue.easyso1/.MainActivity }
```

可以看到，显示与 Android Studio 中的调试效果是一样的，如图 5-5 所示。

重新获取进程 PID，并 attach 进程，默认端口是 23946，并在此端口上进行监听：

```
bullhead:/data/local/tmp/android-arm # ps -e |grep -i roysue
u0_a100    7735   567 1625028  31584 futex_wait_queue_me
ef199ccc S com.roysue.easyso1

bullhead:/data/local/tmp/android-arm # cd gdbserver

bullhead:/data/local/tmp/android-arm/gdbserver # ./gdbserver
0.0.0.0:23946 --attach 7735
Attached; pid = 7735
Listening on port 23946
```

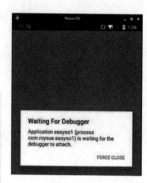

图 5-5 调试

然后，启动 GDB：

```
root@VXIDr0ysue:~ # cd /root/Android/Sdk/ndk/21.1.6352462/prebuilt/linux-x86_64/bin/

root@VXIDr0ysue:~/…/21.1.6352462/prebuilt/linux-x86_64/bin # ./gdb
GNU gdb (GDB) 8.3

(gdb) target remote 192.168.0.183:
23946
```

此时就连接上了虚拟机。我们把远程的 SO 文件拿过来，看它有没有符号。可以用 GDB 的命令查看已经加载了哪些 Shared Library 共享库：

```
(gdb) info shared
```

结果(*)表示没有调试信息。按理来说，我们的 SO 文件应该是有调试信息的，但是结果中为什么没有我们的 SO 文件呢？刚才也看到了，这个时候断点在整个 MainActivity 程序执行之前，即进程刚开始的时候。所以说，这个时候连 MainActivity 都还没有开始执行，更不用谈加载一个静态库了。因此，我们需要让程序恢复执行。

此外，因为我们没有用 GDB 设置断点，程序只要恢复执行，就会直接运行到结束。当然，现在可以设置断点，但是这个时候没有调试信息，只能加在地址上。然后，我们让它执行起来。命令如下：

```
root@VXIDr0ysue:~ # cd ~/Desktop/android-studio/jre/bin
root@VXIDr0ysue:~/Desktop/android-studio/jre/bin # jdb -connect
com.sun.jdi.SocketAttach:hostname=127.0.0.1,port=8700
```

如果出现 Connection refuse，可能是因为 Android Studio 把 8700 端口占用了，可以把 Android Studio 关闭。我们也可以用 DDMS 检查 8700 端口是否打开。如果 Kali Linux 中 as4 的 DDMS 启动

失败，则要用 Android Studio 自带的 JRE 来启动。直接运行 ./monitor 使用的是 Kali 系统的 JDK，发现版本太高了，需要先设置一个软链接：

```
root@VXIDr0ysue:~/Desktop/android-studio/jre/bin # ln -s /root/Desktop/android-studio/jre/ /root/Android/Sdk/tools/lib/monitor-x86_64/
```

注意目录一致。然后可以运行 ./monitor 查看 8700 端口是否开启：

```
root@VXIDr0ysue:~/Desktop/android-studio/jre/bin # cd ~/Android/Sdk/tools

root@VXIDr0ysue:~/Android/Sdk/tools # ./monitor
```

结果显示 8700 端口已开启，如图 5-6 所示。

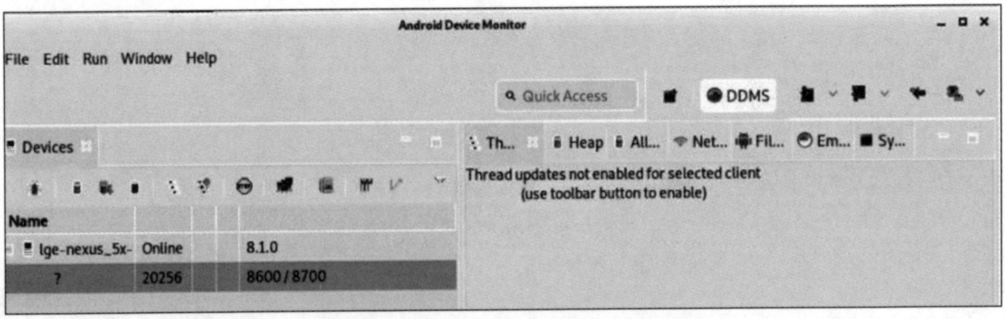

图 5-6　运行 monitor 检查 8700 端口

如果没有以上信息，那么用 JDB 是恢复不了的。先执行 GDB 的 c 命令让程序运行起来，再执行 jdb 命令，手机就会进入被调试状态。

```
(gdb) c
Continuing.
root@VXIDr0ysue:~/Android/Sdk/tools # cd ~/Desktop/android-studio/jre/bin
root@VXIDr0ysue:~/Desktop/android-studio/jre/bin # jdb -connect com.sun.jdi.SocketAttach:hostname=127.0.0.1,port=8700

Initializing jdb
>
```

从图 5-7 可以看到，GDB 已经开始读取 ODEX 文件和 libroysue.so。这个时候 App 开始执行了，从 Monitor 可以看到它打印的信息。

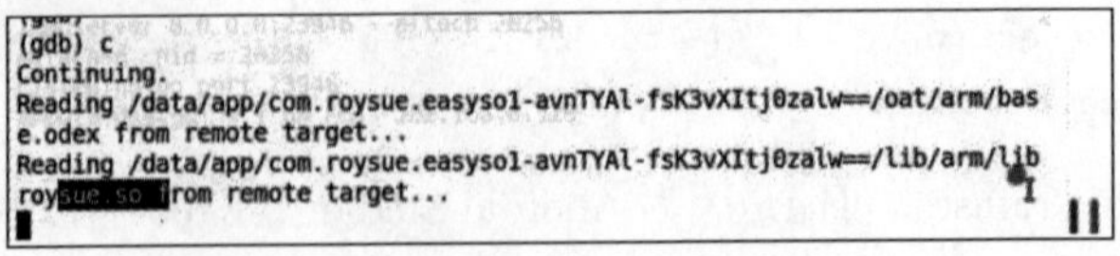

图 5-7　开始读取 ODEX 文件和 libroysue.so 文件

5.4 Hyperpwn 调试入门

无论是学习 GDB 还是 LLDB，重要的是掌握原理，即 SO 文件从加载、寻址、执行到解析的一整套流程，这才是我们研究的关键。

为什么选择 GDB？原因如下：

（1）可以调内核（更传统 legend）。
（2）支持全平台（binary）。
（3）QEMU 内置 GDB。
（4）内核硬件断点，开启内存断点（IDA 没有）。
（5）历史悠久。

我们在本章的前几节已经搭建好了 GDB 调试环境。关于 Hyperpwn，可以参考网址 18（参见配书资源文件）的文章。

Hyperpwn 是一个基于 Hyper 实现的 GDB 调试插件，用于改善调试过程的结果显示。Hyper 本质上是一个终端（Terminal），但可以用前端语言描述它的行为。Hyperpwn 在前辈们的基础上能够进一步将调试过程中的 gef、pwndbg、peda 等调试插件显示的结果进行窗口自动化布局，从而让整个调试过程更人性化。

Hyperpwn 引入了一个重要的功能：记录调试状态，即对动态调试过程中的每一个状态进行保存，从而避免逆向调试人员手动记录每个调试状态的寄存器信息、内存信息等烦琐过程。在调试过程中，只需要使用 Ctrl+Shift+Page Up 快捷键即可查看上一个调试状态的信息，使用 Ctrl+Shift+Page Down 即可查看下一个调试状态的信息，非常方便，从而真正做到解放双手，提高效率。

GDB 更多地用于漏洞调试。漏洞调试与系统息息相关。一般情况下，系统漏洞分析人员对工具的要求是比较高的。虽然分析 Android App 对工具的要求没有分析系统漏洞那么高，但我们仍然可以借鉴优秀的工具，以提升调试工作的效率。

参考网址 19（参见配书资源文件）的内容。

按照提示下载 Hyper-3.1.0-canary.4：

```
root@VXIDr0ysue:~/Desktop/20210124 # wget
https://github.com/vercel/Hyper/releases/download/v3.1.0-canary.4/Hyper_3.1.0-canary.4_amd64.deb
```

在等待下载完成的过程中，我们可以浏览一下官网。在官网推荐的 3 个增强插件 gef、pwndbg、peda 中选择一个来使用即可。笔者使用的是 pwndbg。在 pwndbg 官网中，官方推荐安装在 Ubuntu 16.04 中，Ubuntu 16.04 是一个长期支持（Long Time Support，LTS）的版本。因此，在 Ubuntu 16.04 中安装使用 pwndbg 应该可以获得最佳的效果。现阶段，在 Kali 中使用 pwndbg 也是没有问题的。如果读者在使用过程中发现问题或想深入学习 pwndbg，再切换回 Ubuntu 16.04 系统也不迟。

下载完成后，解压：

```
root@VXIDr0ysue:~/Desktop/20210124 # dpkg -i Hyper_3.1.0-canary.4_amd64.deb
```

如果报错，可以参考网址 20（参见配书资源文件）中的解答。
也可以尝试运行以下命令：

```
root@VXIDr0ysue:~/Desktop/20210124 # apt --fix-broken install
```

注意 运行 Hyper 时需要以普通用户权限来运行：

```
root@VXIDr0ysue:~/Desktop/20210124 # su kali
kali@kali:/root/Desktop/20210124 $ cd
kali@kali:~ $ Hyper -v
No protocol specified

(Hyper:3918): Gtk-WARNING **: 18:35:29.523: cannot open display: :0.0
```

出现以上问题是因为 Kali 用户没有启动。单击屏幕右上角的图标，选择 Switch User，进入 Kali，如图 5-8 所示。

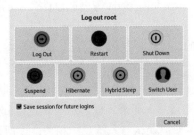

图 5-8　Switch User

在终端（Terminal）运行 Hyper，运行成功后如图 5-9 所示。

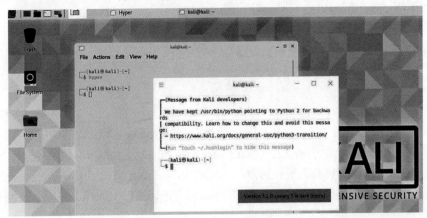

图 5-9　Hyper 运行成功

Hyper 安装好了之后，我们再切换用户并安装 Hyper 的插件：

```
kali@kali:/root/Desktop/20210124 $ cd /opt/Hyper
kali@kali:/opt/Hyper $ Hyper i Hyperinator
Hyperinator installed successfully!
kali@kali:/opt/Hyper $ Hyper i Hyperpwn
Hyperpwn installed successfully!
```

切换回 Kali，重新启动 Hyper：

```
kali@kali:~ $ Hyper
```

依次单击 Plugins→Update，显示更新成功后，再依次单击 Plugins→Reload。
之前我们使用的是源码中的 GDB。其实，也可以使用 Android Studio 中的 GDB：

```
root@VXIDr0ysue:~/Android/Sdk/ndk/23.0.7123448 # adb push prebuilt/ /data/local/tmp
root@VXIDr0ysue:~/Android/Sdk/ndk/23.0.7123448 # apt search gdb-multiarch
Sorting... Done
Full Text Search... Done
gdb-multiarch/unstable 10.1-2 amd64
  GNU Debugger (with support for multiple architectures)
root@VXIDr0ysue:~/Android/Sdk/ndk/23.0.7123448 # apt install gdb-multiarch
```

NDK 中的 GDB 只支持 ARM 架构，而 gdb-multiarch 可以支持所有架构。这样在任意目录下就可以运行 gdb：

```
root@VXIDr0ysue:~ # gdb
GNU gdb (Debian 10.1-2) 10.1.90.20210103-git
(gdb)
```

输入 q 退出。然后安装一个 pwndbg。pwndbg 是 GDB 的一个插件，主要是为了让 GDB 更易于使用。安装 pwndbg 前需要先切换到 Kali，因为 Hyper 在 Kali 中才能够启动：

```
kali@VXIDr0ysue:~ $ git clone https://github.com/pwndbg/pwndbg
kali@VXIDr0ysue:~ $ cd pwndbg
kali@VXIDr0ysue:~ $ ./setup.sh
```

我们已经安装好了 gdb-multiarch 和 pwndbg，运行 gdb-multiarch：

```
kali@VXIDr0ysue:/root $ gdb-multiarch
GNU gdb (Debian 10.1-2) 10.1.90.20210103-git
pwndbg>
```

输入 q 退出 pwndbg。这个时候切换到 Kali 用户运行 Hyper，可以看到如图 5-10 所示的界面。为了节省时间，我们可以使用 tmux 来分窗。查看进程并在手机中启动 gdbserver：

```
bullhead:/data/local/tmp/android-arm/gdbserver # ps -e | grep roysue
bullhead:/data/local/tmp/android-arm/gdbserver # ./gdbserver 0.0.0.0:23946 --attach 8687
```

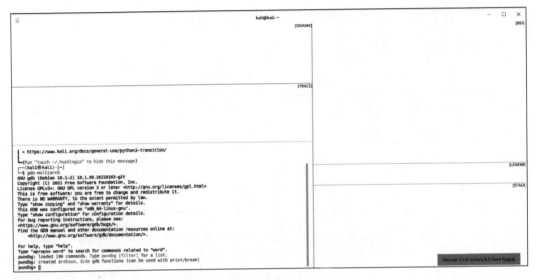

图 5-10　pwndbg

8687 为正在运行的项目 easyso1 的进程号。进入 Hyper 中：

```
pwndbg> target remote 10.203.190.13:23946
```

开始从手机上抽取 SO 文件到本地。抽取完成后，我们可以看到 Logcat 日志输出被暂停了。这是因为 GDB 正在附加（attach）到进程上。一旦附加完成，这个信息就会显示出来，如图 5-11 所示。

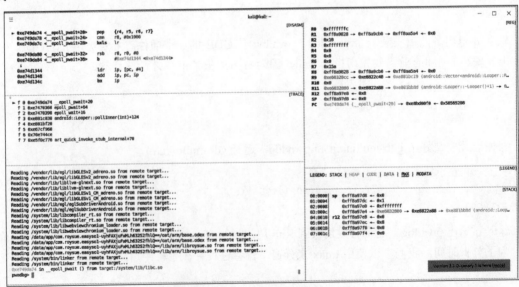

图 5-11　SO 文件抽取结束，显示信息

第 5 章　动态调试：GDB 动态调试、Hyperpwn/（内存）断点/栈帧

继续运行：

```
pwndbg> set arch arm
pwndbg> set arm fallback-mode thumb
pwndbg> c
Continuing...
```

输入 c 是运行到下一个断点的意思。Logcat 继续输出。查看加载的 SO 文件：

```
pwndbg> info share
```

结果如图 5-12 所示，左边的地址不是基址。

```
0xde5880e0  0xde627ca4  Yes (*)  target:/system/lib/libvixl-arm.so
0xde699460  0xde71b540  Yes (*)  target:/system/lib/libvixl-arm64.so
0xdbc131a4  0xdbc179a9  Yes (*)  target:/system/lib/libsoundpool.so
0xdbbcdca8  0xdbbbdf228 Yes (*)  target:/system/lib/libjavacrypto.so
0xdbb1b550  0xdbb1ea00  Yes (*)  target:/vendor/lib/hw/android.hardware.graphics.mapper@2.0-impl.so
0xdba90580  0xdba98724  Yes (*)  target:/vendor/lib/egl/libEGL_adreno.so
0xdbae12a8  0xdbae574c  Yes (*)  target:/vendor/lib/libadreno_utils.so
0xdb95b720  0xdb9d85c8  Yes (*)  target:/vendor/lib/libgsl.so
0xd82d63a0  0xd842c3c4  Yes (*)  target:/vendor/lib/egl/libGLESv2_adreno.so
0xd84cd140  0xd8cc05bc  Yes (*)  target:/vendor/lib/libllvm-glnext.so
0xdb9189c0  0xdb92e1bc  Yes (*)  target:/vendor/lib/egl/libGLESv1_CM_adreno.so
0xdb8ea380  0xdb8efc3c  Yes (*)  target:/vendor/lib/egl/eglSubDriverAndroid.so
0xdb89f848  0xdb8a357c  Yes (*)  target:/system/lib/libcompiler_rt.so
0xdb826a6c  0xdb828a18  Yes (*)  target:/system/lib/libwebviewchromium_loader.so
0xccaf1000  0xccb2f6dc  Yes (*)  target:/data/app/com.roysue.easyso1-uyhFAXjuFxMLh632S2fhlQ==/oat/arm/base.odex
0xcc9b3f40  0xcc9b7b40  Yes      target:/data/app/com.roysue.easyso1-uyhFAXjuFxMLh632S2fhlQ==/lib/arm/libroysue.so
(*): Shared library is missing debugging information.
pwndbg>
```

图 5-12　info share 查看加载的 SO 文件

可以通过以下方式查看基址：

```
bullhead:/ # cat /proc/8687/maps | grep libroysue
```

使用 nm 查看符号偏移，结果如图 5-13 所示。

```
┌──(root㉿kali)-[/home/.../debug/tmp/lib/armeabi-v7a]
└─# nm -s libroysue.so | grep method
000033c9 T Java_com_roysue_easyso1_MainActivity_method01
00003421 T Java_com_roysue_easyso1_MainActivity_method02
00002399 W _ZN7_JNIEnv9NewObjectEP7_jclassP10_jmethodIDz
```

图 5-13　查看偏移

root@VXIDr0ysue:~/home/.../debug/tmp/lib/armeabi-v7a # nm -s libroysue.so | grep method

通过基址与偏移相加得到 method01 的内容，如图 5-14 所示。

```
pwndbg> x/20i 0xcc990000+0x33c9
```

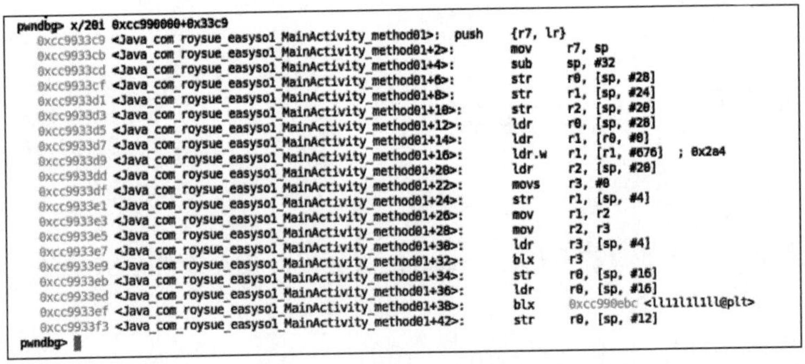

图 5-14　method01 的内容

然后我们给它设置一个断点，结果如图 5-15 所示。

```
pwndbg> b *(0xcc990000+0x33c9)
pwndbg> info b
```

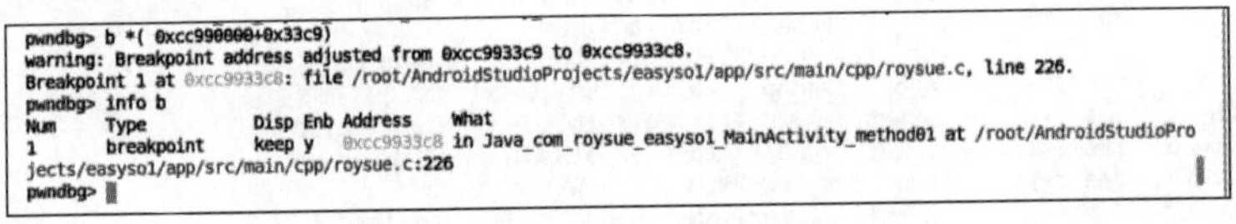

图 5-15　method01 的断点信息

这样断点就设置好了。从结果中可以看到源码，原因是编译的时候符号没有抽掉（其实是我们主动加上的）。因为线程在不断地调用 method01，输入 c 就会运行到断点的位置。disassemble、trace、寄存器、内存指向的内容都已经更新，如图 5-16 所示。

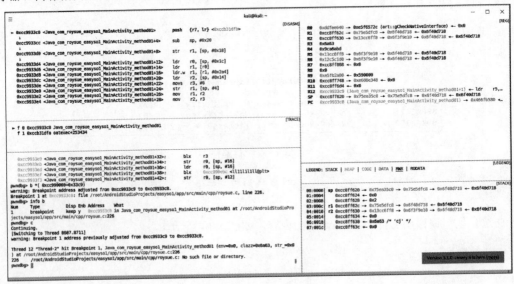

图 5-16　程序运行到断点

此时，我们可以使用断点调试的两个快捷键：

- 步入：F7 键。
- 步进/步过（遇到下一条指令不会进去）：F8 键。

5.5 Objection+Frida+Hyperpwn 联合调试

本节将尝试使用 Objection+Frida+Hyperpwn 联合调试。首先，启动手机上的 frida-server 12.8.0。运行 objection 命令列举所有的模块，如图 5-17 所示。

```
┌──(root㉿kali)-[~/Desktop/20210124]
└─# objection -g com.roysue.easyso1 explore
Using USB device `LGE Nexus 5X`
Agent injected and responds ok!

     |___|(object)inject(ion) v1.8.4

     Runtime Mobile Exploration
        by: @leonjza from @sensepost

[tab] for command suggestions
com.roysue.easyso1 on (google: 8.1.0) [usb] #

A newer version of objection is available!
You have v1.8.4 and v1.11.0 is ready for download.

Upgrade with: pip3 install objection --upgrade
For more information, please see: https://github.com/sensepost/objection/wiki/Updating

com.roysue.easyso1 on (google: 8.1.0) [usb] #
com.roysue.easyso1 on (google: 8.1.0) [usb] # memory list modules
Save the output by adding `--json modules.json` to this command
Name                        Base        Size                 Path
--------------------------  ----------  -------------------  ------------------------------
app_process32               0xab161000  32768 (32.0 KiB)     /system/bin/app_process32
libandroid_runtime.so       0xe672e000  1359872 (1.3 MiB)    /system/lib/libandroid_runtime.so
libbinder.so                0xe7011000  434176 (424.0 KiB)   /system/lib/libbinder.so
libcutils.so                0xe9123000  61440 (60.0 KiB)     /system/lib/libcutils.so
libhwbinder.so              0xe8714000  126976 (124.0 KiB)   /system/lib/libhwbinder.so
liblog.so                   0xe6f22000  81920 (80.0 KiB)     /system/lib/liblog.so
libnativeloader.so          0xe8d31000  32768 (32.0 KiB)     /system/lib/libnativeloader.so
libutils.so                 0xe880c000  98304 (96.0 KiB)     /system/lib/libutils.so
libwilhelm.so               0xe6e4b000  192512 (188.0 KiB)   /system/lib/libwilhelm.so
libc++.so                   0xe6ad1000  598016 (584.0 KiB)   /system/lib/libc++.so
libc.so                     0xe7455000  634880 (620.0 KiB)   /system/lib/libc.so
libm.so                     0xe7258000  139264 (136.0 KiB)   /system/lib/libm.so
libdl.so                    0xe615a000  24576 (24.0 KiB)     /system/lib/libdl.so
libmemtrack.so              0xe6bf1000  24576 (24.0 KiB)     /system/lib/libmemtrack.so
```

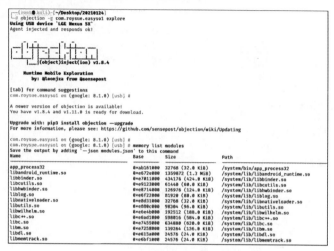

图 5-17　运行 Objection 命令列举所有的模块

可以看到 libroysue.so 的基址，也可以直接查看 libroysue.so 所有符号的地址，如图 5-18 所示。

```
com.roysue.easyso1 on (google: 8.1.0) [usb] # memory list exports libroysue.so
Save the output by adding `--json exports.json` to this command
Type      Name                                                    Address
--------  ------------------------------------------------------  ----------
function  md5                                                     0xcc9b9015
function  to_bytes                                                0xcc9b8fe1
function  qpppqp                                                  0xcc9b7261
function  qqqpqp                                                  0xcc9b73a1
variable  r                                                       0xcc9bb764
function  qppqqp                                                  0xcc9b7471
function  qqppqp                                                  0xcc9b6f85
function  _ZN7_JNIEnv9NewObjectEP7_jclassP10_jmethodIDz           0xcc9b8399
function  qqppqq                                                  0xcc9b71c9
function  _ZN7_JNIEnv12NewStringUTFEPKc                           0xcc9b8377
function  _ZN7_JNIEnv9FindClassEPKc                               0xcc9b8329
function  to_int32                                                0xcc9b9009
function  bdddbddbb                                               0xcc9b80c1
function  _ZN7_JNIEnv18SetByteArrayRegionEP11_jbyteArrayiiPKa     0xcc9b8239
function  _ZN7_JNIEnv11GetMethodIDEP7_jclassPKcS3_                0xcc9b8347
function  lllllll1l                                               0xcc9b8571
function  Java_com_roysue_easyso1_MainActivity_method01           0xcc9b93c9
function  lll1l1l1ll                                              0xcc9b842d
function  Java_com_roysue_easyso1_MainActivity_method02           0xcc9b9421
variable  k                                                       0xcc9bb664
function  _ZN7_JNIEnv14DeleteLocalRefEP8_jobject                  0xcc9b840d
function  bbbddbdbb                                               0xcc9b8031
function  bbddbbdbb                                               0xcc9b7fa9
function  _ZN7_JNIEnv12NewByteArrayEi                             0xcc9b8219
function  o0000O0O                                                0xcc9b81d1
function  o0OoO000                                                0xcc9b8279
function  Java_com_roysue_easyso1_MainActivity_stringFromJNI      0xcc9b928d
com.roysue.easyso1 on (google: 8.1.0) [usb] #
```

图 5-18　查看 libroysue.so 所有符号的地址

我们使用之前的脚本 FridaInvokeNative.js，Frida 主动调用 method02，如图 5-19 所示。

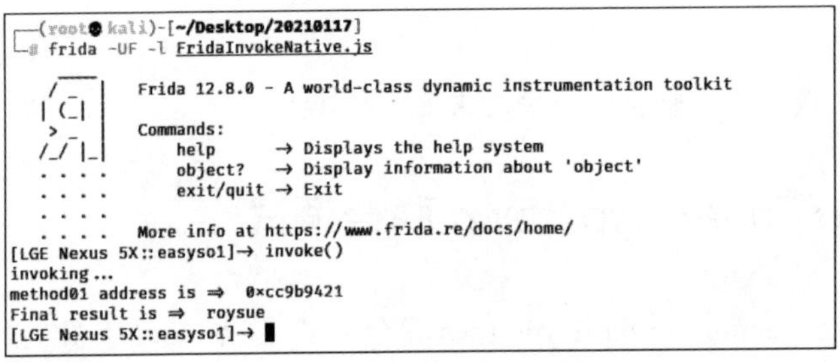

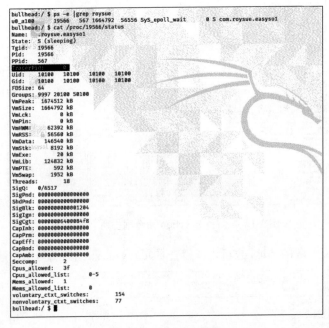

图 5-19　Frida 主动调用 method02

用 Frida 进行 attach 并主动调用之后，TracerPid 是 0，如图 5-20 所示。

图 5-20　TracerPid 是 0

因为注入器只有在注入 SO 文件的阶段需要 ptrace，注入器注入完 SO 文件之后，就会 detach，接下来就不需要 ptrace 了。所以它在 detach 之后的 TracerPid 就是 0。这个时候可以使用 gdb attach 进程，而 Frida 不能再调用 invoke()。gdb attach 结束后，执行 GDB 的 c 命令继续运行程序，invoke() 才能显示结果。

接下来，按 Ctrl+C 快捷键停止并设置断点，我们要主动调用 method02，也是通过基址加上偏移实现的。首先查看基址：

```
bullhead:/ # cat /proc/20840/maps | grep libroysue
```

查看偏移：

```
root@VXIDr0ysue:~/home/.../debug/tmp/lib/armeabi-v7a # nm -s libroysue.so | grep
method
```

查看并确认 method02 的内容，结果如图 5-21 所示。

```
pwndbg> x/20i 0xd0ea7000+0x3421
```

```
pwndbg> x/20i 0xd0ea7000+0x3421
   0xd0eaa421 <Java_com_roysue_easysol_MainActivity_method02>:  push   {r7, lr}
   0xd0eaa423 <Java_com_roysue_easysol_MainActivity_method02+2>:    mov    r7, sp
   0xd0eaa425 <Java_com_roysue_easysol_MainActivity_method02+4>:    sub    sp, #32
   0xd0eaa427 <Java_com_roysue_easysol_MainActivity_method02+6>:    str    r0, [sp, #28]
   0xd0eaa429 <Java_com_roysue_easysol_MainActivity_method02+8>:    str    r1, [sp, #24]
   0xd0eaa42b <Java_com_roysue_easysol_MainActivity_method02+10>:   str    r2, [sp, #20]
   0xd0eaa42d <Java_com_roysue_easysol_MainActivity_method02+12>:   ldr    r0, [sp, #28]
   0xd0eaa42f <Java_com_roysue_easysol_MainActivity_method02+14>:   ldr    r1, [r0, #0]
   0xd0eaa431 <Java_com_roysue_easysol_MainActivity_method02+16>:   ldr.w  r1, [r1, #676]  ; 0x2a4
   0xd0eaa435 <Java_com_roysue_easysol_MainActivity_method02+20>:   ldr    r2, [sp, #20]
   0xd0eaa437 <Java_com_roysue_easysol_MainActivity_method02+22>:   movs   r3, #0
   0xd0eaa439 <Java_com_roysue_easysol_MainActivity_method02+24>:   str    r1, [sp, #4]
   0xd0eaa43b <Java_com_roysue_easysol_MainActivity_method02+26>:   mov    r1, r2
   0xd0eaa43d <Java_com_roysue_easysol_MainActivity_method02+28>:   mov    r2, r3
   0xd0eaa43f <Java_com_roysue_easysol_MainActivity_method02+30>:   ldr    r3, [sp, #4]
   0xd0eaa441 <Java_com_roysue_easysol_MainActivity_method02+32>:   blx    r3
   0xd0eaa443 <Java_com_roysue_easysol_MainActivity_method02+34>:   str    r0, [sp, #16]
   0xd0eaa445 <Java_com_roysue_easysol_MainActivity_method02+36>:   ldr    r0, [sp, #16]
   0xd0eaa447 <Java_com_roysue_easysol_MainActivity_method02+38>:   blx    0xd0ea7ed4 <ll11lll11l@plt>
   0xd0eaa44b <Java_com_roysue_easysol_MainActivity_method02+42>:   str    r0, [sp, #12]
pwndbg>
```

图 5-21　查看并确认 method02 的内容

然后设置断点：

```
pwndbg> b *(0xd0ea7000+0x3421)
```

输入 c，程序继续运行，但此时不会命中断点，因为还没有调用 method02。在源码中，不断调用的是 method01，而没有调用 method02。method02 只在 App 启动的时候调用过一次。但是，可以通过 Frida 将它运行起来。Frida 主动调用 invoke()后，命中断点。再按 C 键就返回了。

在混淆的方法 ll11ll111 处再设置一个断点。Frida 主动调用 invoke()进入第一个断点，输入 c 进入第二个断点，如图 5-22 所示。

```
pwndbg> c
Continuing.
warning: Breakpoint 1 address previously adjusted from 0xcc987421 to 0xcc987420.

Thread 18 "Thread-8" hit Breakpoint 1, Java_com_roysue_easysol_MainActivity_method02 (env=0x6764, clazz=0x1, str_=0xc
b2dbee0) at /root/AndroidStudioProjects/easysol/app/src/main/cpp/roysue.c:242
242     /root/AndroidStudioProjects/easysol/app/src/main/cpp/roysue.c: No such file or directory.
pwndbg> c
Continuing.
warning: Breakpoint 2 address previously adjusted from 0xcc986571 to 0xcc986570.

Thread 18 "Thread-8" hit Breakpoint 2, ll11lll11l (input=0xcb2dbda8 "\270\275-\313Gu", <incomplete sequence \313>) at
 /root/AndroidStudioProjects/easysol/app/src/main/cpp/aes_utils.c:156
156     /root/AndroidStudioProjects/easysol/app/src/main/cpp/aes_utils.c: No such file or directory.
```

图 5-22　命中两个断点

Hyperpwn 最大的创新在于保存状态。在两个断点之间切换状态的快捷键是 Ctrl+Shift+Page Up，可以切回上一个断点的状态。

在源代码中，我们已经将 Key 和 IV 混淆了，且在静态分析时只能看到混淆后的内容。但动态调试时，可以直接得到 Key 和 IV。动态调试所有寄存器的结果都是真实可见的。

05

GEF 也是 GDB 的一个增强插件，功能与 pwndbg 类似。安装 GEF 的命令如下：

```
kali@VXIDr0ysue:/root $ wget -O ~/.gdbinit-gef.py -q http://gef.blah.cat/py
```

5.6 本章小结

本章主要介绍了动态调试技术，包括 GDB 的 C/S 架构与 GDB 动态调试、Hyperpwn 的使用。之后完整演示了 Objection+Frida+Hyperpwn 联合调试。动态调试非常重要，相比于静态调试，动态调试所有寄存器的结果都是真实可见的。

第 6 章

汇编开发：ARM 汇编原理/流程/调用约定/动态调试

本章将继续介绍 SO 文件的相关内容，进一步学习 SO 文件的结构等。提到 SO 文件的内容就离不开 ARM 汇编。因为 SO 文件是一个可以在 CPU 上直接执行的二进制文件，我们需要了解它在 CPU 上执行的过程及其所有的细节。因此，理解 ARM 基本的汇编原理、开发和调试的基础知识是非常重要的。

6.1 Android 和 ARM 处理器

对于手机来讲，处理器应该称作 SoC（System-on-Chip，片上系统），因为它的整个系统都集成在一块芯片之上，因此不像计算机，因为计算机的主板、CPU、内存等都可以定制，而 SoC 是无法定制的。

我们更多关注的是与 Android 系统相关的 ARM 处理器架构。ARM 产品可以按照架构来区分，也可以按照内核来区分。编写本书时，最后一个版本的内核是 ARM11，后来以 Cortex 对应的序列号重新命名。目前，移动智能终端常见的为 ARM11 和 Cortex 内核。Android 系统一般都使用 Cortex 内核。Cortex 系列属于 ARMv7 架构，Android 系统以基于 v7A 的 Cortex-A 系列为主。现在用得最多的就是 ARMv7，而之前的 MIPS 架构的处理器基本上见不到了。每一个处理器都关联着一个相应的 ABI（Application Binary Interface，应用程序二进制接口），现在的 ABI 以 armeabi-v7a 和 armeabi-v8a 为主，其他的非常少见。

我们之前用 Clang 演示过不同编译工具的一些使用方法。Clang 和 GCC 不一样，Clang 属于 LLVM 套件家族的一个前端工具，它在汇编过程中的转化比 GCC 多得多。因为它有自己的中间语言，会将代码转换成 LLVM 所特有的中间代码（ByteCode 字节码），然后将 ByteCode 字节码转换并生成与机器相关的汇编指令。

需要注意，v7a 不支持 ARMv5、ARMv6。因此，在使用 Android NDK 编译原生代码时，设置

ABI 支持列表不兼容 ARMv5、ARMv6，必须单独指定一个 armeabi。ARMv5、ARMv6 基本上只会在 Android 4.0 以下版本的手机中使用。ARM64 的 v8a 从 Android 5 以后开始使用，目前已经成为主流。ARM64 的 v8a 支持最新的 AArch64 架构，同时也兼容 AArch32 架构。

ARM64 的 v8a 处理器支持 32 位和 64 位两种运行模式。在 64 位模式下，它所有的代码寻址采用 64 位的地址空间，寄存器也采用 64 位；在 32 位模式下，它可以与之前的 v7a 一样，代码寻址、寄存器都采用 32 位。因此，可以说 v8a 对 v7a 是完全兼容的。在进行 APK 文件体积优化时，可以考虑牺牲 64 位代码的运行效率而只编译一个 v7a 的 SO 文件。这样，我们编译出来的 APK 可以在 v7a 和 v8a 处理器上同时运行。

6.2 ARM 原生程序的生成过程

使用 VS Code 打开之前编译过的 MD5 项目。回顾编译过程，我们用 Clang 编译时是一步到位的，即直接编译成可执行文件，如图 6-1 所示。

```
┌──(root㉿kali)-[~/Desktop/20210202/MD5]
└─# file *
a.out:      ELF 64-bit LSB pie executable, ARM aarch64, version 1 (SYSV), dynamically linked, interpreter /system/bin/linker64, not
 stripped
md5:        ELF 64-bit LSB pie executable, x86-64, version 1 (SYSV), dynamically linked, interpreter /lib64/ld-linux-x86-64.so.2, B
uildID[sha1]=14a627a2ddebaa8f4b86b02bb9b259922ed34255, for GNU/Linux 3.2.0, not stripped
md5.c:      C source, UTF-8 Unicode text
README.md:  ASCII text
```

图 6-1 MD5 直接编译成可执行文件

下面编译一个在 Android 手机上原生执行的可执行文件：

```
root@VXIDr0ysue:~/Desktop/20210202 # /root/Android/Sdk/ndk/23.0.7123448/toolchains/llvm/prebuilt/linux-x86_64/bin/clang -target AArch64-linux-android21 md5.c -o md5AArch64

root@VXIDr0ysue:~/Desktop/20210202 # adb push md5AArch64 /data/local/tmp
md5AArch64: 1 file pushed, 0 skipped. 18.9 MB/s (8304 bytes in 0.000s)
```

在手机上运行 md5aarch：

```
bullhead:/data/local/tmp # ./md5AArch64 r0ysue
8e2b0b38f557578f578ef21b33e4df0d
```

这个过程是可以拆分的，比如拆分成预处理、编译、汇编、链接 4 个过程。我们以简单的 hello.c 为例，该文件的代码如下：

```
#include <stdio.h>

int main()
{
    /* 我的第一个 C 程序 */
    printf("Hello, World! \n");
    return 0;
```

第 6 章　汇编开发：ARM 汇编原理/流程/调用约定/动态调试

```
}
```

我们把它编译成 32 位的可执行文件：

```
root@VXIDr0ysue:~/Desktop/20210202/hello # /root/Android/Sdk/ndk/23.0.7123448/
toolchains/llvm/prebuilt/linux-x86_64/bin/clang -target arm-linux-android21 hello.c -o
helloARM

root@VXIDr0ysue:~/Desktop/20210202/hello # adb push helloARM /data/local/tmp
helloARM: 1 file pushed, 0 skipped. 2.9 MB/s (3992 bytes in 0.001s)
```

在手机上运行 helloARM：

```
bullhead:/data/local/tmp # ./helloARM
Hello, World!
```

接下来，进行分步处理。可以通过-E 选项查看 Clang 在预处理这步做了什么。执行以下命令进行预处理：

```
root@VXIDr0ysue:~/Desktop/20210202 # /root/Android/Sdk/ndk/23.0.7123448/toolchains/
llvm/prebuilt/linux-x86_64/bin/clang -target arm-linux-android21 -E hello.c -o hello.i
```

生成的 hello.i 文件是可以查看内容的。预处理会把一些依赖展开，包括宏的替换、头文件的导入以及类似#if 的处理都会在这个阶段进行。预处理的下一个阶段是编译。

我们使用参数-S 进行编译：

```
root@VXIDr0ysue:~/Desktop/20210202 # /root/Android/Sdk/ndk/23.0.7123448/toolchains/
llvm/prebuilt/linux-x86_64/bin/clang -target arm-linux-android21 -S hello.i
```

打开生成的 hello.s 文件可以看到如下的 ARM 汇编代码：

```
        .text
        .syntax unified
        .eabi_attribute 67, "2.09"  @ Tag_conformance
        .eabi_attribute 6, 10       @ Tag_CPU_arch
        .eabi_attribute 7, 65       @ Tag_CPU_arch_profile
        .eabi_attribute 8, 1 @ Tag_ARM_ISA_use
        .eabi_attribute 9, 2 @ Tag_THUMB_ISA_use
        .fpu neon
        .eabi_attribute 34, 1       @ Tag_CPU_unaligned_access
        .eabi_attribute 15, 1       @ Tag_ABI_PCS_RW_data
        .eabi_attribute 16, 1       @ Tag_ABI_PCS_RO_data
        .eabi_attribute 17, 2       @ Tag_ABI_PCS_GOT_use
        .eabi_attribute 20, 1       @ Tag_ABI_FP_denormal
        .eabi_attribute 21, 0       @ Tag_ABI_FP_exceptions
        .eabi_attribute 23, 3       @ Tag_ABI_FP_number_model
        .eabi_attribute 24, 1       @ Tag_ABI_align_needed
        .eabi_attribute 25, 1       @ Tag_ABI_align_preserved
        .eabi_attribute 38, 1       @ Tag_ABI_FP_16bit_format
```

```asm
        .eabi_attribute 18, 4       @ Tag_ABI_PCS_wchar_t
        .eabi_attribute 26, 2       @ Tag_ABI_enum_size
        .eabi_attribute 14, 0       @ Tag_ABI_PCS_R9_use
        .file   "hello.c"
        .globl  main                        @ -- Begin function main
        .p2align 2
        .type   main,%function
        .code   32                          @ @main
main:
        .fnstart
@ %bb.0:
        .save   {r11, lr}
        push {r11, lr}
        .setfp  r11, sp
        mov r11, sp
        .pad #8
        sub sp, sp, #8
        ldr r0, .LCPI0_0
.LPC0_0:
        add r0, pc, r0
        movw r1, #0
        str r1, [sp, #4]
        bl  printf
        movw r1, #0
        str r0, [sp]                        @ 4-byte Spill
        mov r0, r1
        mov sp, r11
        pop {r11, pc}
        .p2align 2
@ %bb.1:
.LCPI0_0:
        .long   .L.str-(.LPC0_0+8)
.Lfunc_end0:
        .size   main, .Lfunc_end0-main
        .cantunwind
        .fnend
                                    @ -- End function
        .type   .L.str,%object              @ @.str
        .section .rodata.str1.1,"aMS",%progbits,1
.L.str:
        .asciz  "Hello, World! \n"
        .size   .L.str, 16


        .ident  "Android (6875598, based on r399163b) clang version 11.0.5 (https://android.googlesource.com/toolchain/llvm-project 87f1315dfbea7c137aa2e6d362dbb457e388158d)"
```

```
.section ".note.GNU-stack","",%progbits
```

这时编译已经完成了，可以看到代码节、数据节等内容，但还没有转换成二进制形式。在这些内容中，以"."开头的汇编指令是汇编器提供的伪指令，不同的伪指令用于描述汇编程序的不同部分。

接下来，将汇编输出的文件 hello.s 编译成 hello.o 二进制文件：

```
root@VXIDr0ysue:~/Desktop/20210202 # /root/Android/Sdk/ndk/23.0.7123448/toolchains/
llvm/prebuilt/linux-x86_64/bin/clang -target arm-linux-android21 -c hello.s -o hello.o
```

最后链接，生成 helloARM2 可执行文件，并推送（push）到手机上运行：

```
root@VXIDr0ysue:~/Desktop/20210202 # /root/Android/Sdk/ndk/23.0.7123448/
toolchains/llvm/prebuilt/linux-x86_64/bin/clang -target arm-linux-android21 hello.o -o
helloARM2

root@VXIDr0ysue:~/Desktop/20210202/hello # adb push helloARM2 /data/local/tmp
helloARM: 1 file pushed, 0 skipped. 2.9 MB/s (3992 bytes in 0.001s)

bullhead:/data/local/tmp # ./helloARM2
Hello, World!
```

经过以上步骤，生成的文件如图 6-2 所示。

图 6-2　预处理、编译、汇编、链接生成的文件

可以从 hello.s 中得到汇编代码，对 hello.s 进行修改势必会影响 hello.o，进而影响最终运行的结果。一门程序语言一般都有自己的关键字、代码规范、子程序调用、注释等内容，汇编语言也不例外。

6.3　汇编语言简介

6.3.1　汇编程序组成

汇编语言其实就是将一系列与处理器相关的汇编指令、语法和结构组织在一起的程序语言形式。手写汇编语言的意义不大。一般情况下，只有在效率要求非常高的地方，比如做 shellcode（通常理解为一段用于利用软件漏洞而执行的代码）、逆向相关的时候才会用到手写汇编语言。其实，

C 语言的效率已经非常高了。但是我们在做逆向分析的时候，对汇编语言的了解也是需要非常深入的。

使用汇编语言规范编写的汇编代码可以被完整地编译并嵌入一般高级语言中。汇编程序由以下几个部分组成：

1. 指定 CPU 架构

指定 CPU 架构如图 6-3 所示。

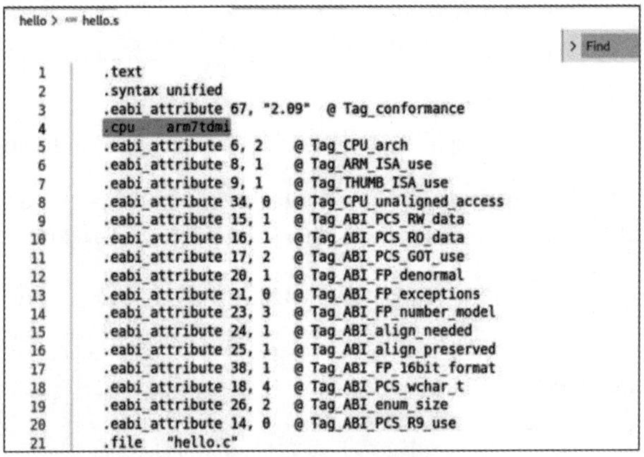

如果使用了浮点指令，还会使用 .fpu 指定协处理器的类型。

2. 代码段与数据段声明

一个完整的汇编程序在编译后都有用于存放代码的地方，称为代码段，它的段名为 .text，以及用于存放数据的数据段，段名为 .rodata，如图 6-4 所示。

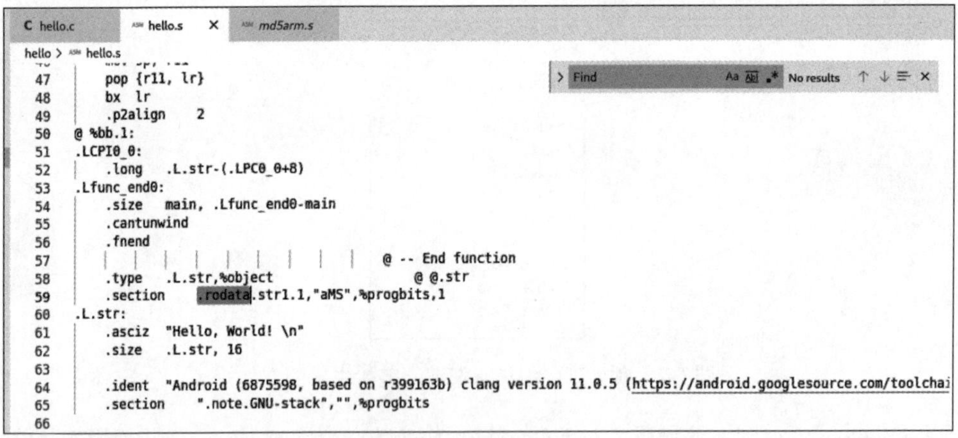

图 6-3 指定 CPU 架构

图 6-4 数据段的声明

在一些特别的程序中可能没有数据段，但是都会有代码段。在汇编代码中声明段的时候一般要用到 .section。

那么，如何查看这些数据段和代码段的内容呢？之前讲过，在编译完程序之后，可以用 readelf 命令查看这些段的内容，比如，这里通过 llvm-readelf 命令查看可执行文件 helloARM 的段内容：

```
root@VXIDr0ysue:~/Desktop/20210202/hello # /root/Android/Sdk/ndk/23.0.7123448/
toolchains/llvm/prebuilt/linux-x86_64/bin/llvm-readelf --sections helloARM
```

结果如图 6-5 所示。

```
┌──(root㊣kali)-[~/Desktop/20210202/hello]
└─# /root/Android/Sdk/ndk/23.0.7123448/toolchains/llvm/prebuilt/linux-x86_64/bin/llvm-readelf --sections helloARM
There are 27 section headers, starting at offset 0xb60:

Section Headers:
  [Nr] Name              Type            Address  Off    Size   ES Flg Lk Inf Al
  [ 0]                   NULL            00000000 000000 000000 00      0   0  0
  [ 1] .interp           PROGBITS        00000194 000194 000013 00   A  0   0  1
  [ 2] .note.android.ident NOTE          000001a8 0001a8 000098 00   A  0   0  4
  [ 3] .dynsym           DYNSYM          00000240 000240 000040 10   A  8   1  4
  [ 4] .gnu.version      VERSYM          00000280 000280 000008 02   A  3   0  2
  [ 5] .gnu.version_r    VERNEED         00000288 000288 000020 00   A  8   1  4
  [ 6] .gnu.hash         GNU_HASH        000002a8 0002a8 000018 00   A  3   0  4
  [ 7] .hash             HASH            000002c0 0002c0 000028 04   A  3   0  4
  [ 8] .dynstr           STRTAB          000002e8 0002e8 000037 00   A  0   0  1
  [ 9] .rel.dyn          REL             00000320 000320 000008 08   A  3   0  4
  [10] .ARM.exidx        ARM_EXIDX       00000340 000340 000020 00  AL 13   0  4
  [11] .rel.plt          REL             00000360 000360 000018 08   A  3  20  4
  [12] .rodata           PROGBITS        00000378 000378 000010 01 AMS  0   0  1
  [13] .text             PROGBITS        00001388 000388 0000cc 00  AX  0   0  4
  [14] .plt              PROGBITS        00001460 000460 000050 00  AX  0   0 16
  [15] .preinit_array    PREINIT_ARRAY   000024b0 0004b0 000008 00  WA  0   0  4
  [16] .init_array       INIT_ARRAY      000024b8 0004b8 000008 00  WA  0   0  4
  [17] .fini_array       FINI_ARRAY      000024c0 0004c0 000008 00  WA  0   0  4
  [18] .dynamic          DYNAMIC         000024c8 0004c8 0000e8 08  WA  8   0  4
```

图 6-5　通过 llvm-readelf 命令查看可执行文件 helloARM 的段内容

一个 .s 汇编文件会包含各种符号，这些符号是汇编文件可以检索和引用的目标。它可以是一个变量、一个常量或者一个函数。在汇编代码中，可以直接引用外部的符号，比如 printf 函数，这个函数并不是我们自己实现的，它是 libc 中的函数。之前也提到过，可以用 llvm-nm 命令查看所有符号的信息：

```
root@VXIDr0ysue:~/Desktop/20210202/MD5 # /root/Android/Sdk/ndk/23.0.7123448/
toolchains/llvm/prebuilt/linux-x86_64/bin/llvm-nm md5v7a
```

结果如图 6-6 所示。

```
┌──(root㊣kali)-[~/Desktop/20210202/MD5]
└─# /root/Android/Sdk/ndk/23.0.7123448/toolchains/llvm/prebuilt/linux-x86_64/bin/llvm-nm md5v7a
00002d58 d _DYNAMIC
00002d50 D __FINI_ARRAY__
00002d48 D __INIT_ARRAY__
00002d40 D __PREINIT_ARRAY__
000016bc t __atexit_handler_wrapper
         U __cxa_atexit
00003e7c b __dso_handle
         U __libc_init
00001660 T _start
00001668 t _start_main
000016c8 t atexit
         U free
00000444 R k
00001b8c T main
         U malloc
00001780 T md5
         U memcpy
00000200 r ndk_build_number
000001c0 r ndk_version
000001a8 r note_android_ident
000001bc r note_data
00000240 r note_end
000001b4 r note_name
```

图 6-6　通过 llvm-nm 命令查看可执行文件 md5v7a 的符号信息

其中，全局符号使用大写 T（Text）来表示，意为在代码段中定义的符号；外部符号使用大写

U（Undefined）表示，在程序运行的时候进行动态链接；大写字母 D（Data）的含义为数据段定义的符号；大写字母 A（Absolute）表示绝对符号；小写字母 r（read only）是只读数据段中的符号。如果想了解更多的符号，可以用以下命令查看：

```
root@VXIDr0ysue:~/Desktop/20210202 # man nm
```

一个 .s 文件中还包含子程序。高级语言中的函数在汇编语言中叫作子程序。汇编语言中的子程序可由 type 伪指令声明为 function 符号。子程序由 .fnstart 伪指令开始，.fnend 结束。比如子程序 to_bytes 由 .type 声明，如图 6-7 所示。

```
24          .type    to_bytes,%function
25          .code    32                       @ @to_bytes
26      to_bytes:
27          .fnstart
28      @ %bb.0:
```

图 6-7　子程序 to_bytes

其中，类似于 push {r11, lr} 的汇编指令是最小的执行单元，它代表单条 CPU 执行的指令。每条汇编指令完成一项特定的工作，将多条汇编指令组织在一起就形成了各种各样的应用程序。

在汇编代码中调用程序没有像高级语言中的语法糖（语言中的一个构件，去掉该构件后并不影响语言的功能和表达能力），很多数据与代码的引用甚至跳转执行都依赖标号来完成。在声明标号的时候，只需要在声明为标号的内容后加上冒号即可。为了便于阅读，声明的标号通常从首行开始，中间的内容用 Tab 键进行缩进，如图 6-8 所示。

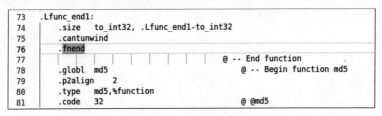

图 6-8　声明标号

汇编代码也支持注释，注释用 @ 标识，@ 后的内容将被解释成注释。注意，编译器不支持多行注释。

汇编代码能实现什么功能完全由它的子程序的功能决定，而子程序中全部都是处理器的命令。所以说，在一个程序中，除了里面包含的重要数据之外，最重要的就是一条条汇编指令。ARM 处理器使用的是 RISC（Reduced Instruction Set Computer，精简指令集计算机），所有指令的长度都是一样的。比如说 ARM 指令采用 32 位指令，Thumb 统一采用 16 位指令，这与 x86 的变长指令集不同，x86 用的是复杂指令集，指令的长度是不一样的。ARM 指令集采用统一长度的指令带来的好处是程序执行时处理器的效率更快，执行速度更快。ARM 中定义的每条汇编指令都有特定的含义，比如 add r0 r0 #1 表示把 r0 寄存器中的值和立即数 1 相加，结果存到 r0 寄存器中；再比如 str r1, [r2, #4] 表示将寄存器 r1 中的值存到 r2+4 指定的地址中；ldr 是数据读取指令，操作与 str 相反。很多汇编指令需要指定源操作对象与目的操作对象。操作的对象很多时候就是寄存器。寄存器是处理器特有的高速存储部件,可用于暂存指令、数据和地址。在高级语言中使用的变量、常量、结构体、类的数据到了 ARM 汇编语言中就是使用寄存器保存的值或者内存地址,也就是指针。寄存器的数量是有限的,32 位的 ARM 微处理器共有 37 个 32 位的寄存器,其中 31 个为通用寄存器,6 个为状态寄存器。

6.3.2 ARM 处理器的工作模式与寻址方式

1. ARM 处理器的工作模式与指令

1)ARM 处理器的工作模式

ARM 处理器支持的运行模式:用户模式(正常的程序执行状态)、快速中断模式、中断模式、管理模式(操作系统使用的模式)、数据访问终止模式、未定义指令终止模式。

不同工作模式可以通过软件进行切换,也可以通过中断进入相应的模式。除用户模式之外,其他 5 种工作模式都属于特权模式。

特权模式中除了系统模式以外的其他 5 种模式又称为异常模式。大多数程序运行于用户模式,通过 svc(管理模式)可以进入特权模式。

2)ARM 处理器的工作状态

我们再来看 ARM 处理器的工作状态。

ARM 体系的 CPU 有两种工作状态,不同状态下的寄存器的工作模式也不同。

(1)ARM

ARM 有 31 个通用的 32 位寄存器,6 个程序状态寄存器,共分为 7 组,有些寄存器是所有工作模式共用的,还有一些寄存器专属于每一种工作模式,例如:

- R13——栈指针寄存器,用于保存堆栈指针。
- R14——程序连接寄存器,当执行 BL 子程序调用指令时,R14 中得到 R15 的备份,而当发生中断或异常时,R14 保存 R15 的返回值。
- R15——程序计数器。

(2)Thumb

Thumb 状态下的寄存器集是 ARM 状态下的寄存器集的一个子集,程序可以直接访问 8 个通用寄存器(R7~R0)、程序计数器(PC)、堆栈指针(SP)、链接寄存器(LR)和当前程序状态寄存器(CPSR)。同时,在每一种特权模式下都有一组 SP、LR 和 SPSR。

在 Thumb 状态下,高位寄存器 R8~R15 并不是标准寄存器集的一部分,而是跟 ARM 指令集下的一些寄存器存在对应关系,如图 6-9 所示。

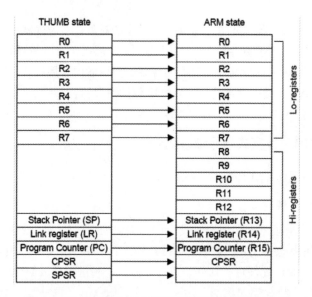

图 6-9　Thumb 与 ARM 状态下的寄存器的对应关系

其中，R8~R12 是 ARM 特有的，R13 是栈指针，R14 是 LR，R15 是 PC。

3）Thumb 指令和 ARM 指令的跳转模式

由于 Thumb 指令集不包括 MSR（对状态寄存器 CPSR 和 SPSR 进行读操作）和 MRS（对状态寄存器 CPSR 和 SPSR 进行写操作）指令，如果用户需要修改 CPSR 的任何标志位，则必须回到 ARM 模式。通过 BX 和 BLX 指令来改变指令集模式，而且当完成复位（Reset）或者进入异常模式时，将会被自动切换到 ARM 模式。这就是所谓的 Thumb 指令集会加 1 这样的情况的由来。

CPSR 中有一个标志位 T：

- T 位为 1——CPU 处于 Thumb 状态。
- T 位为 0——CPU 处于 ARM 状态。

> **注意** 所有的异常处理都是在 ARM 状态下处理的。在 Thumb 状态下，如果有任何中断或者异常标志出现，则处理器会自动切换到 ARM 状态下进行异常处理，完成后又自动切换到 Thumb 状态。

至于 arm64-v8a 则又有一些新的特点。而 AArch64 使用 32 位固定长度的指令集，其有如下特性：

- 引入异常等级的概念，有 EL0 ~ EL3 四种异常等级。
- 提供基于 5 位寄存器说明符的简洁解码表。
- 指令语义与 AArch32 中大致相同。
- 提供 31 个可随时访问的通用 64 位寄存器 x0 ~ x30。

- 提供无模式 GP 寄存器组。
- 提供程序计数器（PC）和堆栈指针（SP）非通用寄存器。
- 提供可用于大多数指令的专用零寄存器 XZR/WZR。

AArch64 和 AArch32 的主要差异在于：

- AArch64 支持 64 位操作数的指令，大多数指令可具有 32 位或 64 位的参数。
- AArch64 地址假定为 64 位，主要的目标数据模型是 P64（指 Pointer 是 64-bit 的）和 LLP64（指 Long Long 和 Pointer 是 64-bit 的）。
- AArch64 的条件指令远少于 AArch32 架构。
- AArch64 没有任意长度的加载和存储多重指令，增加了用于处理寄存器对的 LD/ST 指令。

2. 处理器的寻址方式

寻址是指通过指令中给出的地址码字段寻找真实操作数地址的方式。虽然 ARM 采用精简指令集，但指令间组合的灵活度却比 x86 高。x86 支持 7 种寻址方式，ARM 支持 9 种。下面简要介绍几种常见的寻址方式。

1）立即寻址

立即寻址是最简单的寻址方式，多用于给寄存器赋初值，并且大多数处理器都支持这种方式。在立即寻址指令中，后面的地址码部分为立即数（常量或常数，只能用于源操作数字段）。例如 MOV R0, #123 指令执行后，R0=123。立即数以 "#" 为前缀，表示十六进制数值时以 "0x" 开头，如 "#0x10"。

2）寄存器寻址

寄存器寻址指令操作数的值在寄存器中，指令执行时直接从寄存器中取值进行操作。例如 MOV R0, R1 指令执行后，R0=R1。

3）寄存器移位寻址

寄存器移位寻址是 ARM 指令集特有的寻址方式。与寄存器寻址类似，寄存器移位寻址只是在操作前要对源寄存器操作数进行移位操作，包括逻辑左移（LSL）、逻辑右移（LSR）、算数右移（ASR）、循环右移（ROR）和带扩展的循环右移（RRX）。例如 MOV R0, R1, LSL #2 指令的功能是将 R1 寄存器中的值左移 2 位，后赋值给 R0 寄存器，等价于 R0=R1*4。

4）寄存器间接寻址

在寄存器间接寻址的指令中，由地址码给出的寄存器是操作数的地址指针，所需的操作数保存在由寄存器指定的地址的存储单元中。例如 MOV R0, [R1] 指令的功能是将 R1 寄存器的值作为地址，取出此地址中的值赋给 R0 寄存器。

5）基址寻址

基址寻址是指将地址码给出的基址寄存器与偏移量相加，形成操作数的有效地址，并将所需的操作数保存在有效地址指向的存储单元中。例如 LDR R0, [R1, #-4]指令的功能是将 R1 寄存器的值减 4 作为地址,取出此地址中的值赋给 R0 寄存器。

还有多寄存器寻址、堆栈寻址、块拷贝寻址、相对寻址等寻址方式,由于本书侧重介绍安全应用与逆向实战,这里就不一一讲解了,读者如果感兴趣,可以通过阅读相关书籍来了解更详细的内容。

6.4 ARM 汇编指令及动态调试分析

我们已经学习了基本的 GDB 调试并搭建好了环境。本节将介绍程序的动态生成过程和通过 gdbserver+Hyperpwn 动态调试分析 ARM 汇编指令,这是本章的重点内容。

如第 5 章所述,使用传统方法安装使用 Hyperpwn,需要切换到普通用户权限。这里讲解 Hyperpwn 不切换用户的使用方法,即直接在 root 用户下使用。

在网址 21(参见配书资源文件)按照提示下载 Hyper-3.1.0-canary.4.AppImage。

执行以下命令赋予权限:

```
root@VXIDr0ysue:~/Desktop/20210207 # wget https://github.com/vercel/Hyper/releases/download/v3.1.0-canary.4/Hyper-3.1.0-canary.4.AppImage
root@VXIDr0ysue:~/Desktop/20210207 # chmod 777 Hyper-3.1.0-canary.4.AppImage
```

在手机上运行 gdbserver:

```
bullhead:/data/local/tmp/android-arm/gdbserver # ./gdbserver 0.0.0.0:23946 ./hello
Process ./hello created; pid = 7089
Listening on port 23946
```

执行 Hyper-3.1.0-canary.4.AppImage:

```
root@VXIDr0ysue:~/Desktop/20210207 # ./Hyper-3.1.0-canary.4.AppImage --no-sandbox
```

为了方便使用,可以在配置文件~/.zshrc 中加上 alias,以便在任意目录下都可以打开 Hyper,如图 6-10 和图 6-11 所示。

```
# enable color support of ls, less and man, and also add handy aliases
if [ -x /usr/bin/dircolors ]; then
    test -r ~/.dircolors && eval "$(dircolors -b ~/.dircolors)" || eval "$(dircolors -b)"
    alias ls='ls --color=auto'
    alias hyper2='/root/Desktop/20210207/Hyper-3.1.0-canary.4.AppImage --no-sandbox'
    #alias dir='dir --color=auto'
    #alias vdir='vdir --color=auto'

    alias grep='grep --color=auto'
    alias fgrep='fgrep --color=auto'
    alias egrep='egrep --color=auto'
    alias diff='diff --color=auto'
    alias ip='ip --color=auto'
```

图 6-10 在配置文件~/.zshrc 中加上 alias

```
root@VXIDr0ysue:~ # source ~/.zshrc
root@VXIDr0ysue:~ # Hyper2
```

图 6-11 启动 Hyper

退出 Hyper，安装插件：

```
root@VXIDr0ysue:~ # npm install Hyperinator
root@VXIDr0ysue:~ # npm install Hyperpwn
```

下载并安装 pwndbg：

```
root@VXIDr0ysue:~ # git clone https://github.com/pwndbg/pwndbg
root@VXIDr0ysue:~ # cd pwndbg
root@VXIDr0ysue:~/pwndbg # ./setup.sh
```

使用这种方法就不用频繁切换 Kali 用户了。以 hello.s 为例，我们复制一个副本，并在其基础上稍作修改，作为手写汇编的样例。删除一些不必要的伪指令后，可以看到：基本的文件名 .file"hello.c"；注释 @ -- Begin function main，表示从这里开始是一个新的函数；.type main,%function，这是一个函数；.code 32，是指 32 位。

真正发挥作用的是从以下代码 push {r11, lr} 开始的部分。

```
main:
    .fnstart
@ %bb.0:
    .save   {r11, lr}
    push{r11, lr}
    .setfp  r11, sp
    mov r11, sp
    .pad #8
    sub sp, sp, #8
    ldr r0, .LCPI0_0
.LPC0_0:
    add r0, pc, r0
```

```
movw r1, #0
str  r1, [sp, #4]
bl   printf
movw r1, #0
str  r0, [sp]                    @ 4-byte Spill
mov  r0, r1
mov  sp, r11
pop  {r11, pc}
.p2align 2
```

各个指令的含义如下。

- push {r11, lr}：将 r11 和 lr 压入栈中。
- .setfpr11, sp：（提示）设置栈底，r11 就是 fp，fp 是栈底的一个指针。
- mov r11, sp：将 sp 存入 r11。
- sub　sp, sp, #8：将 sp 减 8。
- ldr　r0, .LCPI0_0：取指，将标号.LCPI0_0 的地址取出放到 r0。
- add　r0, pc, r0：r0 加上 pc 赋给 r0。
- movw　r1, #0：把立即数 0 赋值给 r1。

对于剩下的汇编指令，可以随着调试过程详细了解。

我们还需要修改配置文件~/.Hyperjs，如图 6-12 所示。

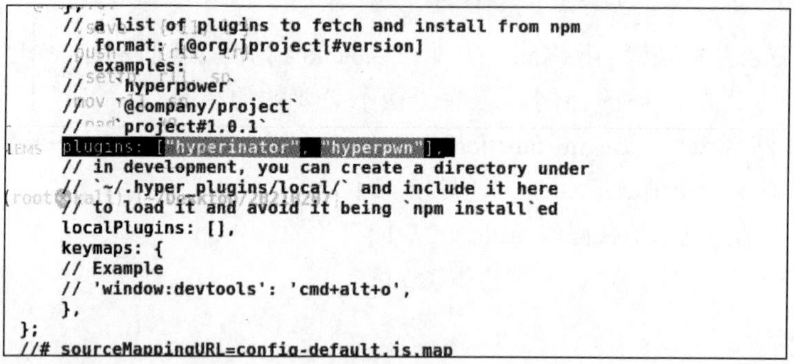

```
    // a list of plugins to fetch and install from npm
    // format: [@org/]project[#version]
    // examples:
    //   `hyperpower`
    //   `@company/project`
    //   `project#1.0.1`
    plugins: ["hyperinator", "hyperpwn"],
    // in development, you can create a directory under
    // `~/.hyper_plugins/local/` and include it here
    // to load it and avoid it being `npm install`ed
    localPlugins: [],
    keymaps: {
    // Example
    // 'window:devtools': 'cmd+alt+o',
    },
};
//# sourceMappingURL=config-default.js.map
```

图 6-12　修改配置文件~/.Hyperjs

启动 Hyper2，更新插件：依次单击 Plugins→Update，然后依次单击 View→Reload 即可。退出并重启 Hyper2，运行 gdb-multiarch：

```
root@VXIDr0ysue:~ # gdb-multiarch
```

又看到了熟悉的 pwndbg 界面，如图 6-13 所示。

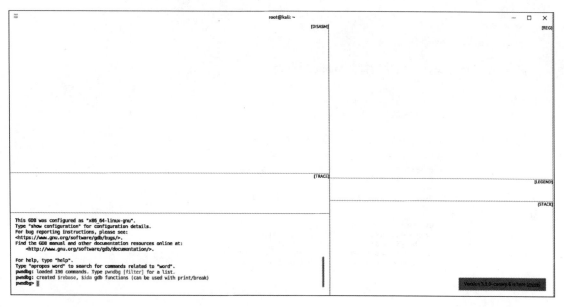

图 6-13 pwndbg 界面

```
pwndbg> target remote 10.203.183.234:23946
```

target remote 完成后的结果如图 6-14 所示。

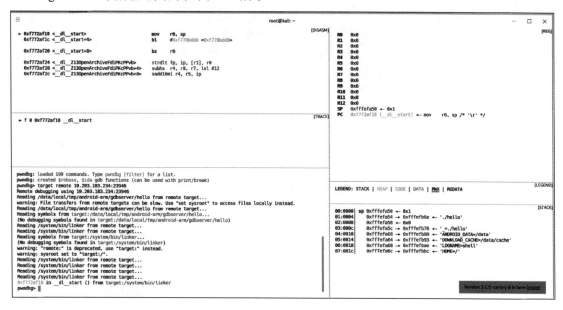

图 6-14 target remote 完成

下一步还是在 main 函数设置断点：

```
pwndbg> b main
Breakpoint 1 at 0xaab4941c
```

输入 c（Continue），继续运行到 main 断点处：

```
pwndbg> c
Breakpoint 1, 0xaab4941c in main ()
```

断点在 0xaab4941c 地址处，对应指令如图 6-15 所示。

```
► 0xaab4941c <main+12>    ldr    r0, [pc, #0x20]
  0xaab49420 <main+16>    add    r0, pc, r0
  0xaab49424 <main+20>    movw   r1, #0
  0xaab49428 <main+24>    str    r1, [sp, #4]
  0xaab4942c <main+28>    bl     #0xaab49490 <0xaab49490>
  0xaab49430 <main+32>    movw   r1, #0
  0xaab49434 <main+36>    str    r0, [sp]
  0xaab49438 <main+40>    mov    r0, r1
  0xaab4943c <main+44>    mov    sp, fp
  0xaab49440 <main+48>    pop    {fp, pc}
```

图 6-15　0xaab4941c 地址处的指令

0xaab4941c 地址处为 ldr 取指令，这明显是不符合常规的。常规的函数从入口到开始执行应该是先保存现场，即保存原来的一些参数，如图 6-16 所示。

```
 9    main:
10        .fnstart
11    @ %bb.0:
12        .save   {r11, lr}
13        push    {r11, lr}
14        .setfp  r11, sp
15        mov r11, sp
16        .pad    #8
17        sub sp, sp, #8
18        ldr r0, .LCPI0_0
```

图 6-16　保存现场

push {r11, lr} 指令中的 lr 寄存器保存返回地址，最后 pop {r11, pc} 指令会将返回地址取出，传给 pc 后就会回到上一层函数（pc 指向下一条指令的地址）。断点地址显示直接到了 main+12 的位置。为什么会发生这样的情况呢？我们先来反编译一下 main 函数：

```
pwndbg> disassemble main
Dump of assembler code for function main:
   0xaab49410 <+0>:    push    {r11, lr}
   0xaab49414 <+4>:    mov     r11, sp
```

直接反编译 main 函数，我们发现，正确的地址为 0xaab49410，通过 objdump 验证：

```
root@VXIDr0ysue:~/Desktop/20210207 # objdump -t hello
...
00001410 g    F .text  00000038              main
```

第 6 章 　汇编开发：ARM 汇编原理/流程/调用约定/动态调试

结果显示，正确的位置在 410，这里却断在了 41c 处，很明显是有问题的。再通过基址加上偏移获取真实地址：

```
bullhead:/ # ps -e |grep hello
root         14468 14465    4216   1236 ptrace_stop  aab4941c t hello
bullhead:/ # cat /proc/14468/maps |grep -i hello
aab48000-aab49000 r--p 00000000 fd:00 539734                             /data/local/tmp/android-arm/gdbserver/hello
```

基址为 0xaab48000，偏移为 0x1410，相加得到的结果为 0xaab49410。所以说 0xaab49410 才是 main 函数的真实地址。但是为什么 pc 指向的是 main+12 的地址呢？这是因为 ARM 三级流水线的设定。这一点很重要，之后我们经常会遇到这类事情。

如果只是一级流水线，CPU 按照取指、译码、执行三个步骤执行每条指令，效率是非常低的。为了提高 CPU 的效率，现采用三级流水线技术。三级流水线的本质是利用指令运行的不同阶段使用的 CPU 硬件互不相同，并发地运行多条指令，从而提高时间效率，如图 6-17 所示。

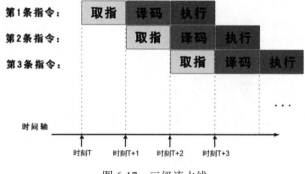

图 6-17 　三级流水线

因此，当执行一条指令的时候，pc 已经指到三条之后，即 pc 的值等于当前正在执行的指令在内存中的地址加 8。这一点非常关键，请务必牢记。这也是为什么我们目前执行的是 410，但 pc 的值已经指向 41c 了。所以断点应该设置在我们计算出的地址处，即 0xaab49410。

此时，输入 c，程序将执行到结束。重新启动 gdbserver 并在 target remote 结束后反编译 main 函数以取得地址 0xaab06410，然后设置断点：

```
pwndbg> b*0xaab06410
Breakpoint 1 at 0xaab06410
pwndbg> c
Continuing.

Breakpoint 1, 0xaab06410 in main ()
```

断点在 410 之后，如图 6-18 所示，第一步是 push，将当前寄存器 fp、lr（fp 和 lr 分别对应 r11 和 r14）的值压入栈。

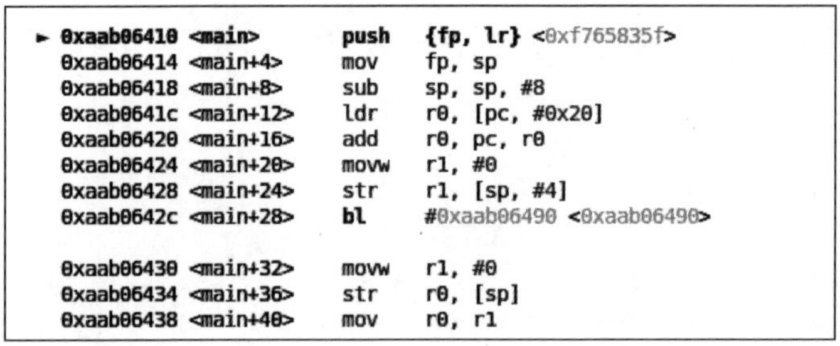

```
▶ 0xaab06410 <main>        push    {fp, lr}    <0xf765835f>
  0xaab06414 <main+4>      mov     fp, sp
  0xaab06418 <main+8>      sub     sp, sp, #8
  0xaab0641c <main+12>     ldr     r0, [pc, #0x20]
  0xaab06420 <main+16>     add     r0, pc, r0
  0xaab06424 <main+20>     movw    r1, #0
  0xaab06428 <main+24>     str     r1, [sp, #4]
  0xaab0642c <main+28>     bl      #0xaab06490 <0xaab06490>

  0xaab06430 <main+32>     movw    r1, #0
  0xaab06434 <main+36>     str     r0, [sp]
  0xaab06438 <main+40>     mov     r0, r1
```

图 6-18 main 函数的断点

我们可以通过以下指令查看寄存器的值：

```
pwndbg> info registers
r0              0x1                 1


r11             0xfffefa48          4294900296
r12             0xf766f630          4150720048
sp              0xfffefa20          0xfffefa20
lr              0xf765835f          -144342177
pc              0xaab06410          <main>
cpsr            0x800d0010          -2146631664
fpscr           0x60000000          1610612736
```

ni 是汇编级别的步过（Step Over），n 是源码级别的步过。执行 ni 后可以看到 fp、lr 的值分别被压入栈中。最后程序执行结束时，这两个值会被弹出（恢复现场）。在右上方的视图[REG]中，*表示寄存器中的值发生了改变，如图 6-19 所示。

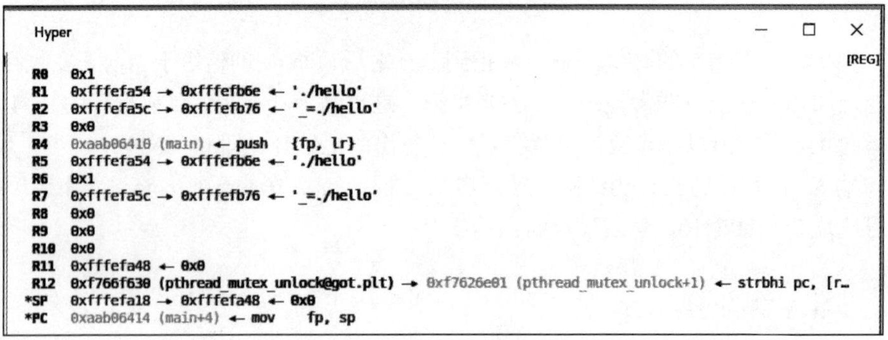

```
Hyper                                                                     —  □  ×
                                                                              [REG]
R0   0x1
R1   0xfffefa54 → 0xfffefb6e ← './hello'
R2   0xfffefa5c → 0xfffefb76 ← '_=./hello'
R3   0x0
R4   0xaab06410 (main) ← push    {fp, lr}
R5   0xfffefa54 → 0xfffefb6e ← './hello'
R6   0x1
R7   0xfffefa5c → 0xfffefb76 ← '_=./hello'
R8   0x0
R9   0x0
R10  0x0
R11  0xfffefa48 ← 0x0
R12  0xf766f630 (pthread_mutex_unlock@got.plt) → 0xf7626e01 (pthread_mutex_unlock+1) ← strbhi pc, [r…
*SP  0xfffefa18 → 0xfffefa48 ← 0x0
*PC  0xaab06414 (main+4) ← mov    fp, sp
```

图 6-19 视图[REG]

mov fp, sp 是把 sp 的值赋给 fp，ni 步过，fp 的值从 0xfffefa48 变为 0xfffefa18，如图 6-20 所示。

第 6 章　汇编开发：ARM 汇编原理/流程/调用约定/动态调试

```
Hyper                                                                    — □ ×
                                                                           [REG]
R0   0x1
R1   0xfffefa54 → 0xfffefb6e ← './hello'
R2   0xfffefa5c → 0xfffefb76 ← '_=./hello'
R3   0x0
R4   0xaab06410 (main) ← push {fp, lr}
R5   0xfffefa54 → 0xfffefb6e ← './hello'
R6   0x1
R7   0xfffefa5c → 0xfffefb76 ← '_=./hello'
R8   0x0
R9   0x0
R10  0x0
*R11 0xfffefa18 → 0xfffefa48 ← 0x0
R12  0xf766f630 (pthread_mutex_unlock@got.plt) → 0xf7626e01 (pthread_mutex_unlock+1) → strbhi pc, [r...
SP   0xfffefa18 → 0xfffefa48 ← 0x0
*PC  0xaab06418 (main+8) ← sub    sp, sp, #8
```

图 6-20　fp 的值从 0xfffefa48 变为 0xfffefa18

下一条指令 sub sp, sp, #8 可直观地理解为 sp=sp-8。结果是 sp 从 0xfffefa18 变为 0xfffefa10，如图 6-21 所示。

```
Hyper                                                                    — □ ×
                                                                           [REG]
R0   0x1
R1   0xfffefa54 → 0xfffefb6e ← './hello'
R2   0xfffefa5c → 0xfffefb76 ← '_=./hello'
R3   0x0
R4   0xaab06410 (main) ← push {fp, lr}
R5   0xfffefa54 → 0xfffefb6e ← './hello'
R6   0x1
R7   0xfffefa5c → 0xfffefb76 ← '_=./hello'
R8   0x0
R9   0x0
R10  0x0
R11  0xfffefa18 → 0xfffefa48 ← 0x0
R12  0xf766f630 (pthread_mutex_unlock@got.plt) → 0xf7626e01 (pthread_mutex_unlock+1) → strbhi pc, [r...
*SP  0xfffefa10 → 0xfffefb76 ← '_=./hello'
*PC  0xaab0641c (main+12) ← ldr    r0, [pc, #0x20] /* ' ' */
```

图 6-21　sp 从 0xfffefa18 变为 0xfffefa10

ldr r0, [pc, #0x20] 表示把当前 pc 指针指向的地址加上立即数 0x20，取此内存地址上的值赋给 r0。此操作又涉及三级流水线。其实只要用到 pc 上的值就无法避免这个问题。我们可以用 Python 计算一下 pc 当前显示的值 0xaab0641c 加上 0x20 的结果：

```
>>> hex(0xaab0641c+0x20)
'0xaab0643c'
```

尝试读取内存地址 0xaab0643c 中的值：

```
pwndbg> x/4xw 0xaab0643c
0xaab0643c <main+44>:    0xe1a0d00b    0xe8bd8800    0xffffef50    0x00000000
```

由于暂时看不出什么内容，因此执行 ni 步过，可以看到 R0 的值变为 0xffffef50。这说明 pc 当前显示的值 0xaab0641c 加上 0x20 的结果是不对的，实际上还需要再加 0x8。验证一下：

```
>>> hex(0xaab0641c+0x20+0x8)
'0xaab06444'
```

```
pwndbg> x/4xw 0xaab06444
0xaab06444 <main+52>:    0xffffef50    0x00000000    0x00000000    0xe52de004
```

继续下一条指令 add r0, pc, r0 <0xaab06420>，计算方法与上述步骤类似：

```
>>> hex(0xffffef50+0xaab06420+0x8)
'0x1aab05378'
```

去掉溢出位，查看内存地址 0xaab05378 中存放的内容，结果如图 6-22 所示。

```
pwndbg> x/40c 0xaab05378
```

```
pwndbg> x/40c 0xaab05378
0xaab05378:    72 'H'  101 'e'  108 'l'  108 'l'  111 'o'  44 ','  32 ' '  87 'W'
0xaab05380:    111 'o' 114 'r'  108 'l'  100 'd'  33 '!'  32 ' '  10 '\n' 0 '\000'
0xaab05388:    13 '\r' 0 '\000'          -96 '\240'       -31 '\341'      -1 '\377'    -1 '\
377'     -1 '\377'          -22 '\352'
0xaab05390:    0 '\000'         72 'H'   45 '-'   -23 '\351'       13 '\r' -80 '\260'      -96 '
\240'    -31 '\341'
0xaab05398:    16 '\020'        -48 '\320'        77 'M'   -30 '\342'      48 '0'  16 '\020'    -
97 '\237'         -27 '\345'
```

图 6-22　x/40c 查看地址 0xaab05378 存放的内容

也可以使用 pwndbg 提供的命令更方便地查看内存地址中的内容，结果如图 6-23 所示。

```
pwndbg> hexdump 0xaab05378
```

```
pwndbg> hexdump 0xaab05378
+0000 0xaab05378  48 65 6c 6c  6f 2c 20 57  6f 72 6c 64  21 20 0a 00  │Hell│o,.W│orld│!...│
+0010 0xaab05388  0d 00 a0 e1  ff ff ff ea  00 48 2d e9  0d b0 a0 e1  │....│....│.H-.│....│
+0020 0xaab05398  10 d0 4d e2  30 10 9f e5  04 30 8d e2  01 10 9f e7  │..M.│0...│.0..│....│
+0030 0xaab053a8  0c 10 8d e5  24 10 9f e5  01 10 9f e7  08 10 8d e5  │....│$...│....│....│
```

图 6-23　使用 hexdump 命令查看地址 0xaab05378 存放的内容

执行 add r0, pc, r0 <0xaab06420>，结果如图 6-24 所示，R0 存放的地址正是我们所计算出的地址，其内容是 Hello, World! \n。

```
 *R0   0xaab05378 ◂— 'Hello, World! \n'
  R1   0xfffefa54 —▸ 0xfffefb6e ◂— './hello'
  R2   0xfffefa5c —▸ 0xfffefb76 ◂— '_=./hello'
  R3   0x0
  R4   0xaab06410 (main) ◂— push    {fp, lr}
  R5   0xfffefa54 —▸ 0xfffefb6e ◂— './hello'
  R6   0x1
  R7   0xfffefa5c —▸ 0xfffefb76 ◂— '_=./hello'
  R8   0x0
  R9   0x0
  R10  0x0
  R11  0xfffefa18 —▸ 0xfffefa48 ◂— 0x0
  R12  0xf766f630 (pthread_mutex_unlock@got.plt) —▸ 0xf7626e01 (pthread_mutex_unlock+1) ◂— strbhi pc, [r-
  SP   0xfffefa10 —▸ 0xfffefa5c —▸ 0xfffefb76 ◂— '_=./hello'
 *PC   0xaab06424 (main+20) ◂— movw    r1, #0
```

图 6-24　R0 存放的地址变化

movw r1, #0 将 0 赋给 r1。str r1, [sp, #4]表示把 r1 的值保存到 sp 加 4 内存地址处的地方。

对照动态执行的指令与我们之前编译出来的汇编代码，可以发现两者是相同的，如图 6-25 所示。

第 6 章　汇编开发：ARM 汇编原理/流程/调用约定/动态调试

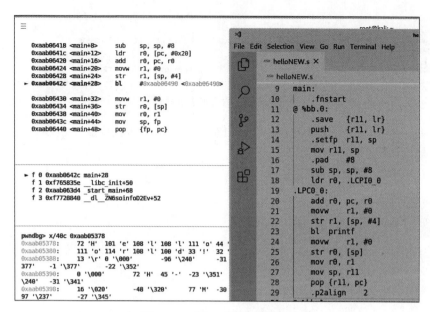

图 6-25　动态与静态汇编对照

注意静态汇编代码中的 ldr r0, .LCPI0_0 被动态翻译成了 ldr r0, [pc, #0x20]，说明从标号 .LPC0_0 开始的指令就是在 pc 下一条指令加 0x20 的地方。所以说动态执行的效果才是最准确的。

bl printf 被直接转化为地址，bl 是带返回值的跳转指令，返回值写在 lr 中。lr 存放的值为 0xf765835f，是上一个函数的地址。此处是 main 函数执行完之后要跳转回去的位置。如果现在再用一个 bl 指令，就会进入另一个子程序。我们可以用 si 进入 bl，此时 lr 中存放的值就会变为 bl 的下一条指令的地址，即 0xaab06430：

```
pwndbg> si
pwndbg> info registers
lr              0xaab06430          -1431280592
```

调用约定：r0~r3 四个寄存器用来传参数。finish 回到上一层的 main 函数。movw r1, #0 再次将 0 赋给 r1，目的是把参数清零。str r0, [sp] 将 r0 的值（0xf）保存到 sp 栈顶的地址上（0xfffefa10），如图 6-26 所示。

```
00:0000 | sp  0xfffefa10 ← 0xf
01:0004 |     0xfffefa14 ← 0x0
02:0008 | r11 0xfffefa18 → 0xfffefa48 ← 0x0
03:000c |     0xfffefa1c 0xf765835f (__libc_init+51) ← ldrshteq r4, [sl], #0x10
04:0010 |     0xfffefa20 ← 0x0
... ↓       3 skipped
```

图 6-26　将 r0 的值（0xf）保存到 sp 栈顶的地址上（0xfffefa10）

mov r0, r1 把 r1 的值赋给 r0，即清空。然后把 fp 的值赋给 sp，即弹栈操作。这样原来保存在栈中的内容就恢复了，如图 6-27 所示。

```
00:0000│ r11 sp  0xfffefa18 → 0xfffefa48 ← 0x0                                                    [STACK]
01:0004│         0xfffefa1c → 0xf765835f (__libc_init+51) ← ldrshteq r4, [sl], #0x10
02:0008│         0xfffefa20 ← 0x0
... ↓            4 skipped
07:001c│         0xfffefa34 → 0xaab063d4 (_start_main+68) ← strdeq r1, r2, [r0], -r4
```

图6-27　从栈中恢复数据

最后一步 pop {fp, pc} <0xaab06440>是将 a18 和 a1c 处的值取出来放到 fp 和 pc 中。这样 pc 就恢复了最开始保存的值。最终回到 libc。

在 hello.c 中添加一行代码：

```
#include <stdio.h>
int main()
{
    /* 我的第一个 C 程序 */
    getchar();
    printf("Hello, World! \n");

    return 0;
}
```

编译并推送（push）到手机上：

```
root@VXIDr0ysue:~/Desktop/20210207 # /root/Android/Sdk/ndk/23.0.7123448/toolchains/llvm/prebuilt/linux-x86_64/bin/clang -target armv7a-linux-android21 -mthumb -S hello.c -o hellogetchar.s

root@VXIDr0ysue:~/Desktop/20210207 # /root/Android/Sdk/ndk/23.0.7123448/toolchains/llvm/prebuilt/linux-x86_64/bin/clang -target armv7a-linux-android21 -mthumb hello.c -o hellogetchar

root@VXIDr0ysue:~/Desktop/20210207 # adb push hellogetchar /data/local/tmp
hellogetchar: 1 file pushed, 0 skipped. 20.0 MB/s (4092 bytes in 0.000s)
```

在手机上运行 hellogetchar，一定要按 Enter 键，才会输出 Hello, World!：

```
bash
bullhead:/data/local/tmp # ./hellogetchar
Hello, World!
```

接下来，使用 Frida 进行联合调试。为什么要在 hellogetchar 中设置中断呢？因为不设置中断，Frida 脚本必须编写 Hook 才能让它暂停，否则 Frida 就直接执行到结束了。当然，可以试一下加 --no-pause 参数。在手机上启动 frida-server-12.8.0-android-arm64：

```
bullhead:/data/local/tmp # ./frida-server-12.8.0-android-arm64
```

第 6 章　汇编开发：ARM 汇编原理/流程/调用约定/动态调试

```
root@VXIDr0ysue:~/Desktop/20210207 # pyenv local 3.8.0

root@VXIDr0ysue:~/Desktop/20210207 # frida -U -f "/data/local/tmp/hellogetchar"
--no-pause
...
[LGE Nexus 5X::hellogetchar]->
```

可以找到进程，如图 6-28 所示。

```
root     27159     2        0     0 worker_th+            0 S [kworker/u12:3]
root     27233 25920   140628 75328 poll_sche+     7e3a9ca518 S frida-server-12.8.0-android-arm64
root     27235 27233     9060  1576 __skb_rec+     70c0bc0f80 S logcat
root     27252 27233    10048  2228 poll_sche+     f0b8bc4c  S frida-helper-32
root     27266 27252    27244  5960 n_tty_read     f2a65a54  S hellogetchar
root     27296     2        0     0 worker_th+            0 S [kworker/u12:1]
root     27302     2        0     0 worker_th+            0 S [kworker/0:1]
root     27306 27036    10488  1932 0              71fbcadf20 R ps
bullhead:/data/local/tmp #
```

图 6-28　进程 hellogetchar 的启动

然后可以通过一些指令查看 Frida 的版本、当前进程等信息，如图 6-29 所示。

```
[LGE Nexus 5X::hellogetchar]-> Frida
{
    "_javaSourceMap": {},
    "_objcSourceMap": {},
    "heapSize": 1553240,
    "sourceMap": {},
    "version": "12.8.0"
}
[LGE Nexus 5X::hellogetchar]-> Process
{
    "arch": "arm",
    "codeSigningPolicy": "optional",
    "id": 27266,
    "pageSize": 4096,
    "platform": "linux",
    "pointerSize": 4
}
```

图 6-29　查看 Frida 的版本、当前进程等信息

还可以找到当前的 hellogetchar 模块：

```
[LGE Nexus 5X::hellogetchar]-> Process.findModuleByName("hellogetchar")
{
    "base": "0xab299000",
    "name": "hellogetchar",
    "path": "/data/local/tmp/hellogetchar",
    "size": 16384
}
```

然后用 GDB attach：

```
bullhead:/data/local/tmp/android-arm/gdbserver # ./gdbserver 0.0.0.0:23946 --attach 27266
Attached; pid = 27266
```

```
Listening on port 23946

pwndbg> target remote 10.203.183.234:23946

0xf2a65a50 in read () from target:/system/lib/libc.so
```

此时停在了 getchar 处,如图 6-30 所示。

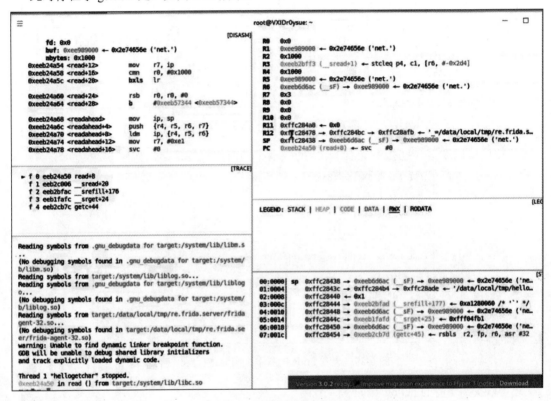

图 6-30 停在了 getchar 处

但实际上,由于没有输入,因此无法运行。而对于另一个程序 hello 就不存在这样的问题。对于 hello 来讲,它的问题是恢复(resume)之后,程序就运行结束了。因此,用 Frida 进行联合调试并没有完美的方案。

接下来,手写一段汇编代码调试一下,输入一个字符串并打印:

```
        .text
        .syntax unified
        .fpu neon
        .file   "hello.c"
        .globl  main                    @ -- Begin function main
        .p2align 2
```

```
        .type   main,%function
        .code   32                          @ @main
main:
        push    {lr}
        ldr     r0, [r1, #8]
        bl      printf
        mov     r0, #0
        pop     {lr}
        bx      lr
```

首先，保存当前环境，将 lr 压入栈。一般来说，我们会将所需要的值保存在 r0、r1、r2 上。将内存地址 r1 加 8 后所在位置的值读取到 r0 中。然后打印寄存器 r0 的内容。将立即数 0 赋值给 r0，最后弹栈并跳转到 lr 中存放的地址所指向的位置处。

将输入的 helloAST 进行编译链接生成可执行文件，并推送（push）到手机上运行：

```
root@VXIDr0ysue:~/Desktop/20210202 #
/root/Android/Sdk/ndk/23.0.7123448/toolchains/llvm/prebuilt/linux-x86_64/bin/clang
-target arm-linux-android21 hello.s -o helloLAST

root@VXIDr0ysue:~/Desktop/20210202/hello # adb push helloLAST /data/local/tmp
helloARM: 1 file pushed, 0 skipped. 2.9 MB/s (3992 bytes in 0.001s)
```

输入两个参数：1、123，打印第二个参数 123：

```
bullhead:/data/local/tmp # ./helloLAST 1 123
123bullhead:/data/local/tmp #
```

启动 gdbserver 调试：

```
bullhead:/data/local/tmp # ./gdbserver 0.0.0.0:23946 ./helloLAST 1 abc
warning: Found...

Listening on port 23946

pwndbg> target remote 192.168.3.13:23946
```

还是在 main 处设置断点：

```
pwndbg> disassemble main
Dump of assembler code for function main:
```

运行结果如图 6-31 所示。

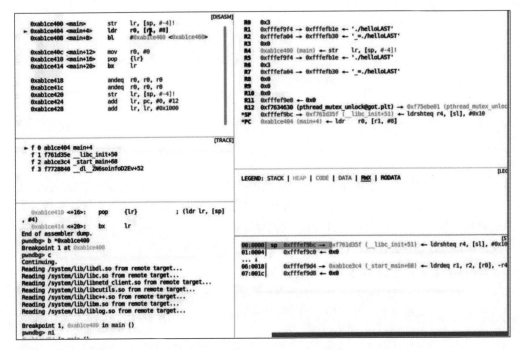

图 6-31 停在了 main 断点的位置

查看寄存器中的值：

```
pwndbg> info registers
r0        0x1        1

lr        0xf763d35f ...
```

首先，把 lr 寄存器中的值保存到栈中，我们看到的 lr 就是上一层的返回值。ni 步过，[r1,#8] 中应该就是我们输入的参数 abc，如图 6-32 所示。

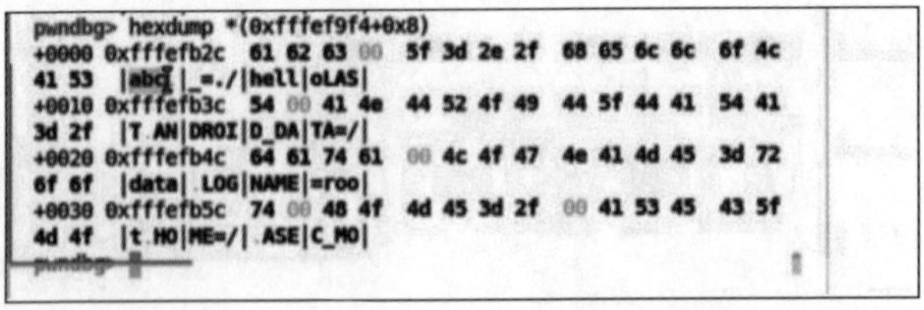

图 6-32 hexdump 查看[r1,#8]处的值

然后，传递给 bl 打印并将 r0 归零，因为 r0 要传递给上一层函数。最后 pop lr, bx 跳转到上一层的指令中。bx 和 bl 不一样，bl 是带返回值的跳转，bx 是状态切换的跳转，从 ARM 切换到 Thumb 或从 Thumb 切换到 ARM。

很多指令这里没有提到，如寻址方式、数据处理指令等，读者可以阅读《Android 软件安全权威指南》一书来学习这些指令。其实，对这些指令有一个基本的了解便可，不必死记硬背。实际上，不需要手写汇编程序代码，只要理解汇编程序代码的大致意思就可以了。

6.5 多功能 CPU 模拟器：Unicorn

笔者之前介绍过的 AndroidNativeEmu 底层用的就是 Unicorn。本节学习用 AndroidNativeEmu 执行单个指令。

```
root@VXIDr0ysue:~/Desktop/20210207 # /root/Android/Sdk/ndk/23.0.7123448/toolchains/llvm/prebuilt/linux-x86_64/bin/clang -target armv7a-linux-android21 -mthumb -S hello.c -o helloTHUMB.s

root@VXIDr0ysue:~/Desktop/20210207 # /root/Android/Sdk/ndk/23.0.7123448/toolchains/llvm/prebuilt/linux-x86_64/bin/clang -target armv7a-linux-android21 -mthumb helloTHUMB.s -o helloTHUMB

root@VXIDr0ysue:~/Desktop/20210207 # adb push helloTHUMB /data/local/tmp
helloTHUMB: 1 file pushed, 0 skipped. 13.4 MB/s (4092 bytes in 0.000s)
```

用 objdump 反编译 helloTHUMB：

```
root@VXIDr0ysue:~/Desktop/20210207 # arm-linux-androideabi-objdump -d helloTHUMB
```

我们以几个简单的指令为例，实现 mov r0, r1 和 add r1, pc 的功能：

```
import unicorn
import capstone
import binascii

# 1452:     4608        mov    r0, r1
# 144e:     4479        add    r1, pc

def testthumb():
    CODE = b'\x08\x46\x79\x44'
    CP = capstone.Cs(capstone.CS_ARCH_ARM, capstone.CS_MODE_THUMB)
    for i in CP.disasm(CODE, 0, len(CODE)):
        print("[addr:%x]:%s %s\n"%(i.address, i.mnemonic, i.op_str))

    mu = unicorn.Uc(unicorn.UC_ARCH_ARM, unicorn.UC_MODE_THUMB)
    ADDRESS = 0x1000
    SIZE = 1024
    mu.mem_map(ADDRESS, SIZE)
    mu.mem_write(ADDRESS, CODE)
    bytes = mu.mem_read(ADDRESS, 10)
    print("ADDRESS:0x%x,contents:%s"%(ADDRESS, binascii.b2a_hex(bytes)))
```

```
        mu.reg_write(unicorn.arm_const.UC_ARM_REG_R0, 0x100)
        mu.reg_write(unicorn.arm_const.UC_ARM_REG_R1, 0x200)
        mu.reg_write(unicorn.arm_const.UC_ARM_REG_R2, 0x300)
        mu.emu_start(ADDRESS+1, ADDRESS+4)
        print("result is 0x%x"%(mu.reg_read(unicorn.arm_const.UC_ARM_REG_R1)))

if __name__ == '__main__':
    print("test")
    testthumb()
```

模拟执行:

```
root@VXIDr0ysue:~/Desktop/20210207 # python emulator.py
test
[addr:0]:mov r0, r1

[addr:2]:add r1, pc

ADDRESS:0x1000,contents:b'0846794400000000000000'
result is 0x1206
```

6.6 本章小结

本章首先介绍了 ARM 处理器的一些基础知识, 并且通过简单的案例学习了 ARM 原生程序的生成过程, 最重要的是, 使用 gdbserver+Hyperpwn 进行了动态调试, 分析了 ARM 汇编指令。最后, 使用多功能 CPU 模拟器 Unicorn 模拟执行了几个简单的指令。

第 7 章 逆向分析：ELF 文件结构、节/区/表/段/符号/链接器

在本章中，将介绍 Linux 系统的内存、进程、ELF 文件结构等相关内容。需要明确的是，使用 Linux 系统进行程序的 Native 层开发的主要思想从 Linux 诞生之初起一直延续至今。因此，Android 作为基于 Linux 的系统，情况也类似。通过学习源码，可以了解 Android 的工作内容和过程。

7.1 操作系统 ELF 文件动态加载的基础知识

本节介绍 ELF 文件动态加载，首先从与 SO 文件有关的 4 个问题入手，然后介绍操作系统的核心概念。

7.1.1 从几个问题入手

本小节从以下 4 个问题着手，介绍 ELF 文件动态加载的相关内容。

（1）为什么要学 ELF Parser（ELF 解析器）？
（2）为什么我们要 dump SO 文件？
（3）为什么 dump 下来的 SO 文件无法直接反汇编？
（4）dump 下来的 SO 文件如何进行修复？

1. 为什么要学 ELF Parser

ELF 解析器是对 ELF 文件结构进行解析的一个软件。文件结构的定义都是为了解析器而生的，解析器就是为了解析文件，把文件按照相应的定义放到对应的对象中，然后就可以使用这个对象。这就好比 JSON 的序列化和反序列化，我们同样可以理解为 ELF 的序列化和反序列化。首先把对象保存到 ELF 文件中，然后从 ELF 文件中把对象取出来。这就是为什么要把 ELF 文件结构和解析

器放到一起学——结合解析器可以更直观地理解文件格式及其对应的含义。

2. 为什么我们要 dump SO 文件

对 SO 文件加密之后，我们从 APK 中拿到的 SO 文件其实是经过加密的。但是，我们在内存中运行的 SO 文件才是程序真正的代码。我们把内存中的 SO 文件 dump 下来就可以得到真正的、解密之后的 SO 文件。这其实跟 DEX 的 dump 非常接近。APK 包直接反汇编得到的是被某个壳加固之后的 DEX，由于进行了加密，因此看不到它里面的内容。但如果从内存中 dump，例如用 Frida 的 DEX dump，就可以得到真正的 DEX。静态文件是可以进行加密的，在加载 ELF 文件的过程中就进行了解密。这个加载的过程可以在代码中进行控制。了解加载过程就相当于学习加固 SO 文件的方法。因为加固 SO 文件可以对节区进行加密，然后在 ELF 被加载的过程中进行解密。解密之后再链接和执行。所以，由为什么要 dump SO 文件的这个问题可以引申出对静态文件到动态内存的映射过程的大致理解。我们需要清楚的是，这个过程中进行了加密文件的解密，之后就可以通过内存 dump 得到明文。

3. 为什么 dump 下来的 SO 文件无法直接反汇编

为什么 dump 下来的 SO 文件无法反汇编？原因是它在加载过程中会丢失信息，这些丢失的信息无法直接从内存中得到，因此无法直接反汇编。如果要使它能够反汇编，就需要修复 SO 文件，把丢失的信息重新建立起来。

4. 如何修复 dump 后的 SO 文件

我们一般会使用 last yang 的脚本进行自动化的 dump 和修复，参考网址 22（参见配书资源文件）。

这样修复后的 SO 文件就恢复了原来的节区表等信息。有了这些信息，自然可以进行反汇编了。

以上介绍了一些 SO 文件的相关知识，建议读者了解这些基础知识，基础知识扎实了，学习本章内容就会轻松很多。笔者在这里推荐《程序员的自我修养》一书。这本书主要介绍系统软件的运行机制及原理，有助于读者对整个计算机是如何运行的有个整体的理解。它涉及 Windows 和 Linux 两个系统平台，因为它们本质上是相似的，实现是相近的。

我们之所以研究操作系统，是为了对抗应用层的。因为从操作系统出发，应用层是无法对抗的。本章的重点是 ELF 文件的链接、加载以及库概念的理解与操作，本质上是文件格式如何定义，定义了之后如何解析，解析之后安排在内存中的哪些区域，然后与一些库进行交互，交互完之后一起运行的整个过程。

7.1.2 操作系统的核心概念

对于现代计算机来讲，有 3 个部件最为重要：CPU、内存和 I/O 控制芯片（硬盘），三者不可或缺。比如，一个 8 核 16 线程的处理，这里就涉及核和线程的概念。8 核很好理解，可以认为是将 8 块 CPU 放到一起。此处的线程与相对进程的线程是完全不一样的。此处的线程是 CPU 内部中的实现，具体的含义不必深究，我们简单地理解为 16 个核即可。

与一般的计算机软件体系结构相同，Android 系统自底向上的分层中也包含硬件、内核、运行库和各种软件。每个相邻层之间都要相互通信，通信需要协议，这个协议一般称为接口。接口的下面是接口的提供者（Server），由接口定义；接口的上面是接口的使用者/消费者（Client），使用接口来实现特定的功能。

分析 Android 应用，最关心的是运行时库（Runtime Library）和操作系统内核。运行时库使用操作系统提供的系统调用接口，系统调用接口在实现中往往采用软件中断，比如 Linux 的中断号 0x80。操作系统内核是硬件接口的使用者——驱动程序如何操作硬件。操作系统为应用软件提供服务的同时还需要管理硬件。

需要理解层次的概念，首先是分层，然后是每一层对上一层提供服务。跨层基本上是不可能的，能跨到最远的层是 SSVC，直接调用内核的系统调用即可。而操作系统主要是为应用软件提供服务，同时管理硬件。

进程的总体目标是希望每个进程从逻辑上来看都可以独占计算机的资源，但是从现实意义上讲是不可能的。由于操作系统的多任务功能，从进程的角度来看，似乎是它独占了 CPU，并且操作系统也实现了硬盘设备的抽象和共享。关键是进程内存和真实内存之间的关系。对操作系统不了解的话，肯定会有这样的疑问：为什么内存只有 1GB，但是进程的空间比 1GB 要大？一开始，程序是运行在物理内存上的，程序在运行时访问的是物理地址。这样一台计算机只运行一个程序是没有问题的。但是为了更有效地利用硬件资源，我们必须同时运行多个程序，如何将计算机上有限的物理内存分配给多个程序就成为一个问题。解决这一问题的方法就是中间层，即使用虚拟地址进行间接访问。每个进程都有自己独立的虚拟地址空间，而且每个进程只能访问自己的地址空间。至于虚拟地址和物理地址之间如何转化，进程与进程之间如何相互隔离，就是操作系统要考虑的事情。

分段的思想非常关键。我们的 ELF 文件就是分段的，内存也是分段的。分段的思想是把一段程序所需的内存空间大小的虚拟空间映射到某个地址空间。分段解决了进程间的地址隔离、连续性等问题。但是分段没有解决内存使用效率的问题。分段以程序为单位对内存区域进行映射，如果内存空间不足，则整个程序都将被换入和换出到磁盘，从而造成大量的磁盘访问操作，严重影响效率。

实际上，根据程序的局部性原理，在一段时间内，一个程序运行时只会频繁地用到一小部分数据，其他部分的数据很有可能不会被用到。因此，就有了更小粒度的内存分割方法——分页。为了提高内存使用率，分页是利用程序的局部性原理，将内存地址空间更细地等分成固定大小的页，页的大小由硬件决定。一般计算机上的操作系统使用的页大小为 4KB。以上都是关于操作系统虚拟内存和物理内存的映射方法，与硬件关系比较密切，还没有涉及 ELF 文件如何加载到页中。

除了进程之外，线程也是比较关键的概念。线程有时被称为轻量级进程，是程序执行流的最小单元。通常，一个进程可以由一个或多个线程组成。在大多数软件应用中，线程的数量都不止一个，并且可以互不干扰地并发执行，共享进程的全局变量和堆的数据。线程也拥有自己的私有存储空间，包括栈、线程局部存储和寄存器。在 Linux 中，所有执行实体都称为任务。不同的任务之间可以选择共享内存空间。因此，共享了同一个内存空间的多个任务构成了一个进程，这些任务也就成了这个进程中的线程。通过系统调用 fork 方法可以复制当前进程。调用 fork 方法之后，新的任务将启动并和本任务一起从 fork 方法返回。举个例子，我们安装一个 APK，很多时候都会进行 dex to oat 的转化以提高运行效率。dex to oat 实际上就是 fork 一个子进程，然后在子进程上进行单独的转化。如何保证线程安全？多个线程同时访问一个变量是非常危险的，必须给这块内存区域加个锁，才能保证它在某一时刻只被一个线程访问。

7.2 可执行文件的加载过程

如何把可执行 ELF（Executable Linkable Format，ELF）文件加载到进程中？所谓加载，其实就是一个解析的过程。解析完之后，在内存中变成一个对象。

可执行文件只有加载到内存中才能被 CPU 执行。而随着多进程、多用户、虚拟存储的操作系统出现以后，可执行文件的加载过程变得非常复杂。

程序和进程有以下区别：

- 程序是一个静态的概念，是一些预先编译好的数据和指令的集合。
- 进程是一个动态的概念，是程序运行时的一个过程。

每个程序运行起来之后，都有自己独立的虚拟地址空间。虚拟空间的大小由 CPU 位数决定。在有虚拟内存的情况下，创建一个进程，加载并执行相应的可执行文件，需要先完成以下操作：

（1）创建一段独立的虚拟地址空间。
（2）读取可执行文件头，并且建立虚拟地址空间与可执行文件的映射关系。
（3）将 CPU 的指令寄存器设置成可执行文件的入口地址，并启动执行。

对于一个正常的进程来说，可执行文件中往往包含多个段——代码段、数据段、BSS 等，都会被映射到进程的虚拟内存空间中。我们以手机上的 wifiadb 进程为例，如图 7-1 所示。

```
bullhead:/ # ps -e |grep wifi
wifi           441     1  17392   1456 binder_thread_read 77a63b24e8 S android.hardware.wifi@1.0-service
u0_a98       20794 29687 4328504 56964 SyS_epoll_wait 7b4032a3f8 S com.ttxapps.wifiadb
wifi         29694     1  15164   1436 SyS_epoll_wait 76827bf3f8 S wificond
wifi         30205     1  17020         poll_schedule_timeout 7848081530 S wpa_supplicant
bullhead:/ # ps -e |grep wifiadb
u0_a98       20794 29687 4328504 56964 SyS_epoll_wait 7b4032a3f8 S com.ttxapps.wifiadb
```

图 7-1 wifiadb 进程

常用的查看进程内存布局的方法如下：

```
bullhead:/ # cat /proc/pid/maps
```

可以看到它里面一些详细的内存布局，如图 7-2 所示。

```
7b440a4000-7b440a7000 r--s 00000000 103:09 1301         /system/framework/arm64/boot-android.hidl.base-V1.0-j
ava.vdex
7b440a7000-7b440c7000 r--s 00000000 00:0c 13818         /dev/__properties__/u:object_r:vold_prop:s0
7b440c7000-7b440e7000 r--s 00000000 00:0c 13847         /dev/__properties__/u:object_r:config_prop:s0
7b440e7000-7b440e8000 rw-p 00000000 00:00 0             [anon:linker_alloc]
7b440e8000-7b440e9000 r--p 00002000 fd:00 539663        /data/dalvik-cache/arm64/system@framework@boot-androi
d.hidl.base-V1.0-java.art
7b440e9000-7b440ea000 r--p 00026000 fd:00 539662        /data/dalvik-cache/arm64/system@framework@boot-org.ap
ache.http.legacy.boot.art
7b440ea000-7b440f3000 r--s 00000000 103:09 1328         /system/framework/arm64/boot-legacy-test.vdex
7b440f3000-7b440f4000 rw-p 00000000 00:00 0             [anon:linker_alloc]
7b440f4000-7b440f5000 r--p 0000a000 fd:00 539661        /data/dalvik-cache/arm64/system@framework@boot-ims-co
mmon.art
7b440f5000-7b440f6000 r--p 0000b000 fd:00 539660        /data/dalvik-cache/arm64/system@framework@boot-voip-c
ommon.art
7b440f6000-7b440f7000 r--p 00000000 00:00 0             [anon:linker_alloc]
7b440f7000-7b44117000 r--s 00000000 00:0c 13846         /dev/__properties__/u:object_r:dalvik_prop:s0
7b44117000-7b44137000 r--s 00000000 00:0c 13835         /dev/__properties__/u:object_r:log_tag_prop:s0
7b44137000-7b44138000 rw-p 00000000 00:00 0             [anon:linker_alloc_small_objects]
7b44138000-7b44139000 rw-p 00000000 00:00 0             [anon:linker_alloc_vector]
7b44139000-7b4413a000 r--p 00042000 fd:00 539659        /data/dalvik-cache/arm64/system@framework@boot-teleph
ony-common.art
```

图 7-2 查看进程的内存布局

图 7-2 第一列显示了虚拟内存区域的地址范围；第二列表示权限（r：可读，w：可写，x：可执行，p：私有，s：共享）；第三列表示 Segment 在源文件中的偏移；第四列表示源文件所在设备的主设备号和次设备号；第五列表示源文件的节点号；最后一列表示源文件所在路径。我们也可以看到线程的栈区、栈区的保护，如图 7-3 所示。

```
7aa8468000-7aa8563000 rw-p 00000000 00:00 0             [stack:20815]
7aa8563000-7aa8564000 ---p 00000000 00:00 0             [anon:thread stack guard page]
7aa8564000-7aa8565000 ---p 00000000 00:00 0
7aa8565000-7aa8660000 rw-p 00000000 00:00 0             [stack:20814]
7aa8660000-7aa875e000 r--p 00000000 00:0c 14008         /dev/binder
7aa8b52000-7aa8b53000 ---p 00000000 00:00 0             [anon:thread stack guard page]
7aa8b53000-7aa8b54000 ---p 00000000 00:00 0
7aa8b54000-7aa8c4f000 rw-p 00000000 00:00 0             [stack:32292]
7aa8c4f000-7aa8c50000 ---p 00000000 00:00 0             [anon:thread stack guard page]
7aa8c50000-7aa8c51000 ---p 00000000 00:00 0
7aa8c51000-7aa8d54000 rw-p 00000000 00:00 0             [stack:20807]
7aa8d54000-7aa8d55000 ---p 00000000 00:00 0             [anon:thread stack guard page]
7aa8d55000-7aa8d56000 ---p 00000000 00:00 0
```

图 7-3 线程的栈区、栈区的保护

操作系统通过给进程空间划分出一个个 VMA（虚拟内存区域）来管理进程的虚拟空间，基本原则是将相同权限属性的、有相同映像的文件映射成一个个 VMA。

一个进程的 VMA 基本上可以分为代码 VMA、数据 VMA、堆 VMA、栈 VMA。栈相当于寄存器的扩展，堆相当于代码和数据的扩展。

由于操作系统在加载可执行文件的时候主要关心段的权限问题（可读、可写、可执行），因此对于权限相同的段，操作系统把它们合并到一起当作一个段进行映射，以此来减少内存浪费的问题。

ELF 文件引入了 Segment 这个概念。一个 Segment 包含一个或多个属性相似的 Section。它从加载的角度重新划分了 ELF 的各个段。例如，将权限相同的.text 段和.init 段合并在一起看作一个 Segment，那么加载时就可以对这个 Segment 进行映射。这种映射方式减少了内存空间的浪费。

进程刚开始启动时，我们要知道一些进程运行的环境，最基本的就是系统环境变量和进程的运行参数。

下面介绍 Linux 系统加载 ELF 文件的过程。

用户在 bash 输入执行 ELF 程序的命令后，bash 进程会调用 fork() 创建一个新的进程，然后新的进程调用 execve() 执行指定的 ELF 文件。

```
int execve(const char *filename, char *const argv[], char *const envp[]);
```

使用 Android 地址空间布局随机化（Address Space Layout Randomization，ASLR）技术后，每次 ELF 文件加载的起始地址都是不一样的，包括内核加载的起始地址也是不一样的。

将 ELF 文件加载到内存中只是第一步，接下来还需要进行链接才能执行。

运行时库就比较好理解了，它也被加载到了虚拟页上。之所以用动态链接取代静态链接，是因为静态链接的方式会浪费较多的内存空间。比如 libc.so 这样的共享库是通用的，使用频率非常高。因此，把它预置在操作系统上，而不需要每个程序都静态链接到自己的进程中。

要解决空间浪费和更新困难这两个问题，最简单的办法就是把程序的模块相互分割开来，形成独立的文件，而不再将它们静态链接在一起。简单来说，就是不对那些组成程序的目标文件进行链接，等到程序要运行时再进行链接。也就是说，把链接这个过程推迟到了运行时再进行，这就是动态链接的基本思想。

共享库 libc.so 在系统空间的一段地址上，每个进程需要用到 libc.so 的时候都不会再次加载。

静态链接只有可执行文件本身需要被映射，而动态链接还需要它所依赖的共享目标文件。在这种情况下，进程的地址空间又是怎样分布的呢？Linux 下的动态链接器与共享对象同样被映射到了进程的地址空间。系统开始运行程序之前，首先会把控制权交给动态链接器，由它完成所有的动态链接工作之后，再把控制权交还给程序，然后开始执行。

偏移在加载时进行重新定位。

为了优化动态链接的性能，ELF 文件采用了一种延迟绑定的方法，其基本思想是当函数第一次被调用时才进行绑定（符号查找、重定位等）。

对于 ELF 文件的加载、链接，我们就学习到这里。至于一些更深层次的、更细致的内容，读者可参阅《程序员的自我修养》一书。

7.3 使用 Unidbg 模拟执行 SO 文件中的函数

Unidbg 是一个基于 Unicorn 的逆向工具，可以使用黑盒调用 Android 和 iOS 中的 SO 文件。它不需要直接运行 App，也无须逆向 SO 文件，而是通过在 App 中找到对应的 JNI 接口，然后用 Unicorn 引擎直接执行这个 SO 文件，所以效率比较高。我们需要使用 IDEA 来运行 Unidbg。

IDEA 由 JetBrains 公司开发，是 Java 语言开发的集成环境，也是目前业界公认的最好的 Java 开发工具之一。Android Studio 就是基于 IDEA 开发的。我们从官网下载 Community 版本的 IDEA 即可。

7.3.1 Unidbg 框架的基本运作原理

Unidbg 的项目地址参考网址 23（参见配书资源文件）。
Unidbg 有以下特点：

- 实现了所有的 JNI 的 API，因此可以调用 JNI_OnLoad。
- 支持 JavaVM、JNIEnv。
- 模拟系统调用指令。
- 支持 ARM32 和 ARM64（Unicorn 自带的功能）。
- inline Hook（对于逆向分析来说，有了 Frida，我们也不需要其他的 inline Hook 框架了）。
- Android import Hook。
- Unicorn 后端支持简单的控制台调试、指令跟踪、内存读写等（Unicorn 自带的功能）。
- 支持 4 种模拟器，包括 Unicorn、Hypervisor、KVM、Dynamic。

Unidbg 框架可以简单理解为：Unicorn 加上各种各样的 Android 特有的系统库的实现。Unicorn 和 Unidbg 的区别如下：

- Unicorn 直接执行 ARM 汇编。
- Unidbg = Unicorn + libc.so + libart.so + liblog.so。

虽然 SO 文件是由各种 ARM 汇编组成的，但是它还需要使用各种各样的 API，包括使用 C 的函数、JNI 函数（例如之前的案例中，读取从 Java 层传过来的字符串）。最起码，想要执行一个 SO 文件，需要在 Unicorn 的基础上加上 libc.so 库和 libart.so 库。所以 Unidbg 的功能可以实现模拟 art.so 中的对象和函数，甚至可以模拟系统调用。在 App 中，可以通过 libc 中的 open 函数来打开一个文件，也可以通过系统调用打开一个文件。所以说，既可以通过 libc 打开文件，又可以直接通过系统调用进入内核打开文件。

下载并打开 IDEA：

```
root@VXIDr0ysue:~/Desktop # wget https://download.jetbrains.com/idea/ideaIC-2021.1.3.tar.gz

root@VXIDr0ysue:~/Desktop # tar zxvf ideaIC-2021.1.3.tar.gz

root@VXIDr0ysue:~/Desktop # cd idea-IC-211.7628.21/bin

root@VXIDr0ysue:~/Desktop/idea-IC-211.7628.21/bin # ./idea.sh
```

初始界面如图 7-4 所示。

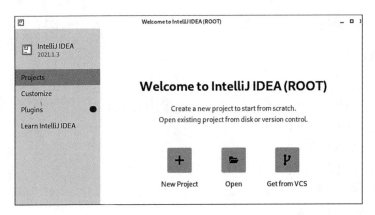

图 7-4　IDEA 初始界面

下载 Unidbg：

```
root@VXIDr0ysue:~/Desktop # git clone https://github.com/zhkl0228/unidbg.git
```

由于一打开 Unidbg 就会从网上下载依赖，因此建议下载完成后对 Unidbg 进行备份。大概的流程要心里有数，可以把 Unicorn 当成一个精简的模拟 Android 系统的实现，毕竟它连 SO 文件都可以运行。Android 系统需要百万、千万级别的代码，但像 Unidbg 这样的框架，上万行代码就可以实现百万、千万级别的代码所实现的功能，这是一件非常了不起的成就。所以从 Unidbg 框架的基本运作原理就可以管中窥豹，Android 系统对于 SO 文件的支持有哪些是必需的、不可取代的组件，比如文件系统，Android 的文件系统更加复杂，但 Unidbg 实现的是最精简的文件系统，可以把 APK 文件或者 SO 文件加载起来。接下来看一下 Unidbg 对于文件系统如何模拟。

从 IDEA 中打开备份的 unidbg2，等待下载完成。我们首先来看 unidbg-android/src/test/java/com/kanxue.test2/MainActivity 下的构造函数 MainActivity：

```java
    private MainActivity() {
        emulator = AndroidEmulatorBuilder
                .for32Bit()
                .addBackendFactory(new DynarmicFactory(true))
                .build();
        Memory = emulator.getMemory();
        LibraryResolver resolver = new AndroidResolver(23);
        memory.setLibraryResolver(resolver);

        vm = emulator.createDalvikVM();
        vm.setVerbose(false);
        DalvikModule dm = vm.loadLibrary(new
File("unidbg-android/src/test/resources/example_binaries/armeabi-v7a/libnative-lib.so"),
false);
        dm.callJNI_OnLoad(emulator);
    }
```

在构造函数中做了什么呢？首先，需要选择架构来构造模拟器 Emulator，此处选择 32 位，之后通过 emulator.getMemory() 获取模拟器的内存，再把 libc 填充到 LibraryResolver 中。这样做是为了将 libc 库加载到模拟器中。如果不填充 libc 库，那么无法执行 SO 文件中的任何 C 函数，因为没有一个 SO 文件可以独立于 C 库而存在，毕竟 SO 库是用 C 语言编写的。

要查看 setLibraryResolver 的实现，右击 Implementation(s)，如图 7-5 所示。

```
private MainActivity() {
    emulator = AndroidEmulatorBuilder
            .for32Bit()
            .addBackendFactory(new DynarmicFactory(fallbackUnicorn: true))
            .build();
    Memory memory = emulator.getMemory();
    LibraryResolver resolver = new AndroidResolver(sdk: 23);
    memory.setLibraryResolver(resolver);

    vm = emulator.createDalvikVM();
    vm.setVerbose(false);
    DalvikModule dm = vm.loadLibrary(new File(pathname: "unidbg-android/src/test/resources/example_binaries/armeab
    dm.callJNI_OnLoad(emulator);
}
```

图 7-5　setLibraryResolver 的实现

包括 AbstractLoader 和两个在不同平台上的实现：Android 平台的 AndroidElfLoader 和 macOS 平台的 MachOLoader。虽然平台不同，但是最终加载还是在 AbstractLoader 中：

```
public void setLibraryResolver(LibraryResolver libraryResolver) {
    this.libraryResolver = libraryResolver;
}
```

SO 文件加载到内存的主要流程就在 AbstractLoader 中。例如 mmap、allocateMapAddress 函数等都是关于开辟内存的最关键的函数。所以说，通过最简单的追踪就可以得到一些重要的信息。

另外，libc.so 等库文件在 unidbg-android/src/main/resources/android/sdk23/lib 的文件夹中，如图 7-6 所示。

其中，liblog.so 是 <android/log.h> 的实现。所以我们调用 log 函数会结果。libssl.so 支持网络功能，因此在进行远程通信使用 SO 文件也没有问题。我们可以通过输出 libssl 相关接口的内容来获取信息，而不需要进行抓包。这样相当于实现了 Native 层的抓包功能。另外，libz.so 是用于解压缩的。

7.3.2　Unidbg 各组件的基本功能

7.3.1 节介绍了一个加载 SDK 组件的基本功能 AndroidResolver。而在 unidbg-android/src/main/resources/android/lib/armeabi-v7a 目录下的都是用来 Hook 的 SO 库，它们都必须注入目标进程中，Hook

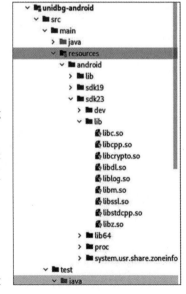

图 7-6　SDK 中引入的 lib 库

才能生效。如果需要用到其他的 SO 库，而 Unidbg 中没有提供，就需要自己把它加进来，放到相应目录下并在 AndroidResolver 中加载。现实世界中的 SO 文件要比 DEMO 复杂得多，SO 文件与 SO 文件之间互相依赖，互相调用对方 SO 文件中的函数，这时需要知道怎么处理。

回到 MainActivity，通过 emulator.createDalvikVM 创建一个 ART 的虚拟机 VM。进入 AndroidEmulator：

```java
public interface AndroidEmulator extends ARMEmulator<AndroidFileIO> {

    VM createDalvikVM();

    /**
     * @param apkFile 可为 null
     */
    VM createDalvikVM(File apkFile);

    /**
     * jar as apk
     */
    VM createDalvikVM(Class<?> callingClass);

    @SuppressWarnings("unused")
    VM getDalvikVM();

}
```

AndroidEmulator 本质上是一个接口。我们需要了解 AndroidEmulator 是如何实现的，才能够在此基础上进行修改。在如图 7-7 所示的界面中右击，选择 Go To Implemetation(s)。

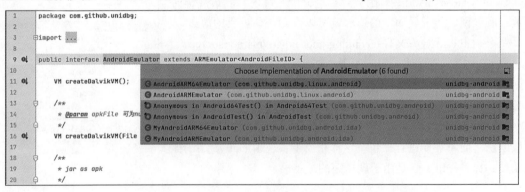

图 7-7　查看 AndroidEmulator 的实现

分别看到 AndroidEmulator32 位的实现 AndroidARMEmulator 和 64 位的实现 AndroidARM64Emulator。这里我们需要自定义 AndroidEmulator32，它的实现太简陋了，所以需要对它进行个性化定制。AndroidARMEmulator 实现了诸如文件系统 createFileSystem、存储

createMemory、系统调用 createSyscallHandler 等接口。基本的功能要心里有数，哪怕现在不知道是做什么用的，但要知道它有这些 API 存在。

在 MainActivity 中，已经了解到模拟器调用 createDalvikVM 创建了一个 Dalvik VM。接下来，搜索 createDalvikVM(File apkFile)的实现。一步一步点进去看，发现最终在 ApkFactory 做了 ApkFile 的处理。从代码可以看出它是如何对 ApkFile 进行解析的：

```java
class ApkFile implements Apk {

    private final File apkFile;

    ApkFile(File file) {
        this.apkFile = file;
    }

    private ApkMeta apkMeta;

    @Override
    public long getVersionCode() {
        if (apkMeta != null) {
            return apkMeta.getVersionCode();
        }

        try (net.dongliu.apk.parser.ApkFile apkFile = new net.dongliu.apk.parser.ApkFile(this.apkFile)) {
            apkMeta = apkFile.getApkMeta();
            return apkMeta.getVersionCode();
        } catch (IOException e) {
            throw new IllegalStateException(e);
        }
    }


    @Override
    public String getVersionName() {
        if (apkMeta != null) {
            return apkMeta.getVersionName();
        }

        try (net.dongliu.apk.parser.ApkFile apkFile = new net.dongliu.apk.parser.ApkFile(this.apkFile)) {
            apkMeta = apkFile.getApkMeta();
            return apkMeta.getVersionName();
        } catch (IOException e) {
            throw new IllegalStateException(e);
        }
    }
```

```java
    @Override
    public String getManifestXml() {
        try (net.dongliu.apk.parser.ApkFile apkFile = new
net.dongliu.apk.parser.ApkFile(this.apkFile)) {
            return apkFile.getManifestXml();
        } catch (IOException e) {
            throw new IllegalStateException(e);
        }
    }

    @Override
    public byte[] openAsset(String fileName) {
        try (net.dongliu.apk.parser.ApkFile apkFile = new
net.dongliu.apk.parser.ApkFile(this.apkFile)) {
            return apkFile.getFileData("assets/" + fileName);
        } catch (IOException e) {
            throw new IllegalStateException(e);
        }
    }

    private Signature[] signatures;

    @Override
    public Signature[] getSignatures(VM vm) {
        if (signatures != null) {
            return signatures;
        }

        try (net.dongliu.apk.parser.ApkFile apkFile = new
net.dongliu.apk.parser.ApkFile(this.apkFile)) {
            List<Signature> signatures = new ArrayList<>(10);
            for (ApkSigner signer : apkFile.getApkSingers()) {
                for (CertificateMeta meta : signer.getCertificateMetas()) {
                    signatures.add(new Signature(vm, meta));
                }
            }
            this.signatures = signatures.toArray(new Signature[0]);
            return this.signatures;
        } catch (IOException | CertificateException e) {
            throw new IllegalStateException(e);
        }
    }

    @Override
    public String getPackageName() {
```

```
        if (apkMeta != null) {
            return apkMeta.getPackageName();
        }

        try (net.dongliu.apk.parser.ApkFile apkFile = new
net.dongliu.apk.parser.ApkFile(this.apkFile)) {
            apkMeta = apkFile.getApkMeta();
            return apkMeta.getPackageName();
        } catch (IOException e) {
            throw new IllegalStateException(e);
        }
    }

    @Override
    public File getParentFile() {
        return apkFile.getParentFile();
    }

    @Override
    public byte[] getFileData(String path) {
        try (net.dongliu.apk.parser.ApkFile apkFile = new
net.dongliu.apk.parser.ApkFile(this.apkFile)) {
            return apkFile.getFileData(path);
        } catch (IOException e) {
            throw new IllegalStateException(e);
        }
    }
}
```

可以看到 ApkFile 有一些接口，如获取 APK 版本号（getVersionCode）、获取 APK 版本名称（getVersionName）、获取 APK Manifest 文件（getManifestXml）、打开资产文件夹（openAsset）、获取签名（getSignatures）、获取包名（getPackageName）等。

ApkFile 是如何解析的呢？到目前为止，它还只是一个 Java 文件。最终它是如何变成 ApkFile 的？实际上，它是通过 net.dongliu.apk.parser.ApkFile 类的构造函数构造并解析出来的。net.dongliu.apk.parser 是一个外部库，我们可以参考网址 24（参见配书资源文件）查找相关的开源信息。因此，我们需要明白它是如何实现的，它的核心原理是什么，它如何处理这个 APK 文件，以便在我们的项目中轻松调用其中的信息。同时，还需要了解它能够或不能够做到什么地步。

7.3.3 追踪 SO 文件的加载与解析流程

我们继续往下看。"vm.setVerbose(false);"用于设置输出日志级别，改为 true 可以打印详细信息，如图 7-8 所示。

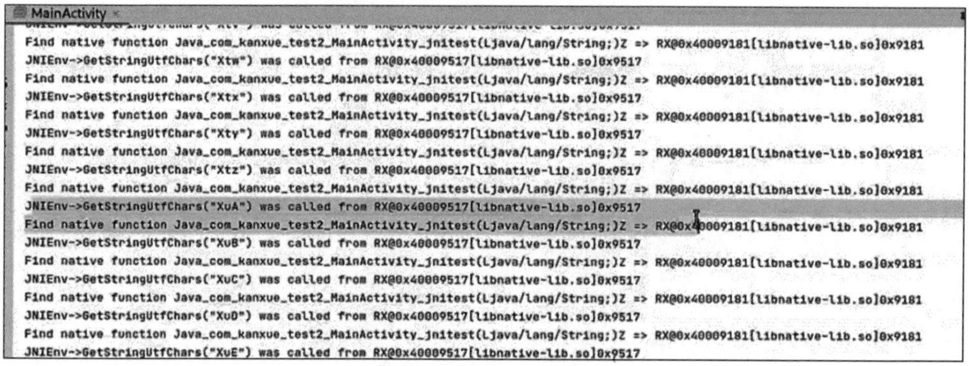

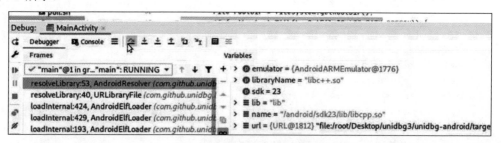

图 7-8　打印日志详细信息

比如，想看它打开的 SO 文件在哪里、何时调用的，就可以加个日志打印或者使用断点调试，如图 7-9 所示。

图 7-9　断点调试 resolveLibrary

可以看到它的调用栈信息。之所以有这么多调用，是因为 SO 文件有很多依赖项。这些依赖项会根据对 SO 文件的依赖进行加载。如果没有这些依赖项，SO 文件在 setLibraryResolver 阶段就已经加载完成。可以在 resolveLibrary 方法中添加日志语句 "System.out.println(name);" 打印加载的 SO 文件的名称，看看加载的 SO 文件有多少个，如图 7-10 所示。

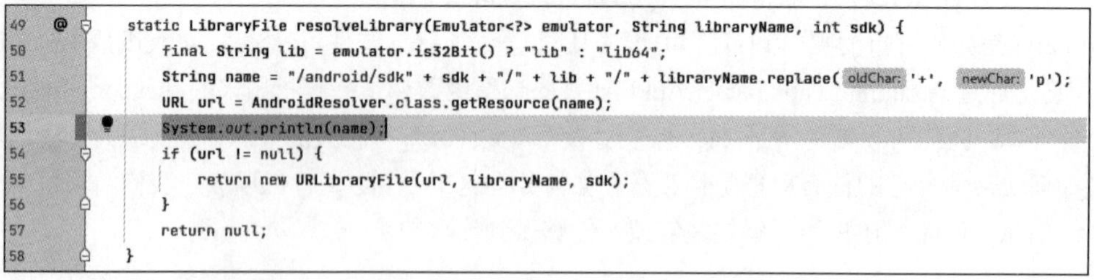

图 7-10　打印加载的 SO 文件

保存并运行，结果如图 7-11 所示。

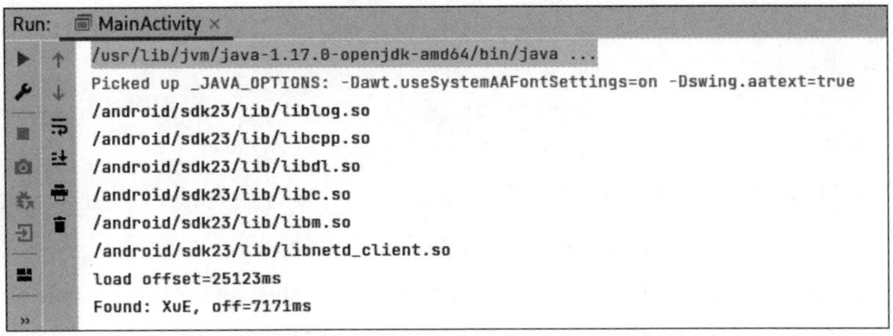

图 7-11 打印加载的 SO 文件名称

也可以看一下它的效率。在使用引擎 DynarmicFactory 的情况下，运行时间如图 7-11 所示，耗时 7171ms。如果使用自带的 Unicorn 引擎，如图 7-12 和图 7-13 所示，耗时为 54368ms。

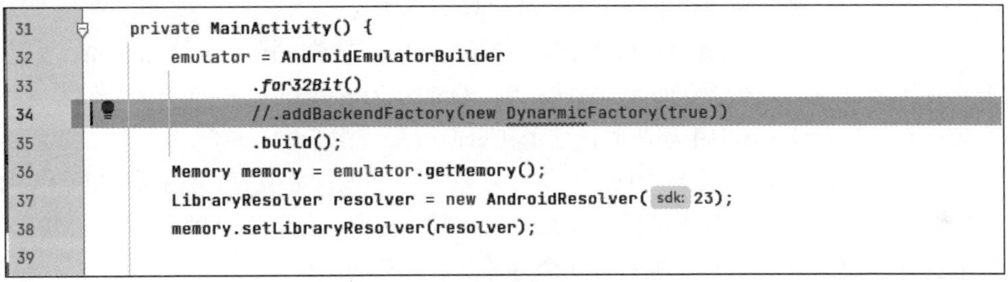

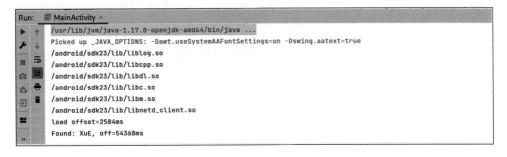

图 7-12 不使用引擎 DynarmicFactory

图 7-13 不使用引擎 DynarmicFactory 的耗时

可以看到，Dynamic 引擎比 Unicorn 引擎效率高得多，因为它是基于动态语言翻译的，跟模拟器不一样。笔者在之前的章节中介绍过，Unicorn 是基于 QEMU 模拟器的模拟执行框架，功能是模拟 CPU。它和 Android Studio 自带的 Monitor 是一样的，因此速度非常慢。

最后，通过 loadLibrary 给 VM 加载 Library。刚刚我们看到，加载 Library 的时候，它对依赖库进行了解析和加载。我们还是进入 loadLibrary 看它的实现，首先在内存加载 ELF 文件，再传入 DalvikModule。

```
@Override
public final DalvikModule loadLibrary(File elfFile, boolean forceCallInit) {
```

```
        Module module = emulator.getMemory().load(new ElfLibraryFile(elfFile),
forceCallInit);
        return new DalvikModule(this, module);
    }
```

再看内存加载 load 做了什么。它也是一个接口,进入实现后发现它调用了 loadInternal 方法。loadInternal 在两个平台 AndroidElfLoader 和 MachOLoader 分别实现。我们关注的是 Android 平台的实现,进入 loadInternal,最终在 private LinuxModule loadInternal(LibraryFile libraryFile)中通过调用类的静态方法 ElfFile.fromBytes(ByteBuffer buffer)对 SO 文件进行了解析。

从 ElfFile 类的定义可以看出它是跟 ELF 文件的结构相匹配。类似地,ELF 文件的具体解析也是通过引入的外部包 net.fornwall.jelf 实现的,网址 25(参见配书资源文件)为 GitHub 项目地址。显然,ElfFile 类就是一个用 Java 写的 ELF 文件的解析器,把 SO 文件解析成 ELF 格式的文件对象并且导出了一些接口。例如,通过 getElfSymbol 可以得到所有的符号表。

回到 loadLibrary,return new DalvikModule(this, module)非常关键。其中,module 就是 SO 文件。在 DalvikModule 中,有一个 public void callJNI_OnLoad(Emulator<?> emulator)函数,很多加固方式都会选择在 JNI_OnLoad 中主动注册,这个在后面的章节中会重点讲解。

题外话:VMP 在 ARM 指令集上的应用并不广泛,它不像 x86 处理器那样性能强大,如果在 ARM 上执行一个 VMP,那么效率降低是非常快的。即使这样,依然有 ARM 上的 VMP。如果是逆向 ARM 上的 VMP,那么将是一项耗时几个月甚至按年算的工程。

7.4 本章小结

本章首先介绍了操作系统 ELF 文件动态加载的概念和理论。然后,通过实战学习了 Unidbg 框架的基本运作原理以及各个组件的基本功能,熟练掌握这个工具对程序的动态分析非常有帮助。

第 8 章

反编译工具 IDA

谈到逆向分析，自然离不开反编译。Java 反编译是指根据 Java 字节码将其"翻译"成源码的过程。本章将介绍一款著名的静态分析工具 IDA。许多读者可能已经对这一工具有所耳闻，但可能还不熟悉它的用法。本章将结合实际案例来学习 IDA 的使用。

8.1 IDA 入门

IDA 在逆向分析界是大名鼎鼎的静态分析工具，它可以分析 x86、x64、ARM、MIPS、Java、.NET 等众多平台的程序代码，功能非常强大。

8.1.1 IDA 的安装与使用

建议在 Windows 10 系统中下载 IDA 7.5，下载完成后，请务必阅读使用说明。注意，不要包含中文路径。运行绿化工具，可以看到，桌面上新增了两个 IDA 的图标：一个用于解析 32 位的 SO 文件，另一个用于解析 64 位的 SO 文件。我们以一个样板 APK 文件为例进行介绍。

首先，使用另一款静态分析工具 JADX 打开样本 APK 进行观察，如图 8-1 所示。onCreate 函数中有个 init 函数加载了一个 native-lib 的 SO 文件，其中进行了 Frida 检测。

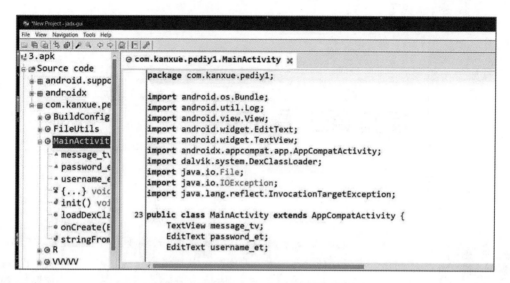

图 8-1　使用 JADX 工具打开样本 APK

在手机上启动 frida-12.8.0，用 64 位 IDA 打开样本 APK。IDA 7.5 相比于 IDA 7.0 是不需要修复头文件的。

在导出（Exports）窗口，SO 文件提供了一些暴露的函数，暴露给 Java 层，以便 Java 层进行调用，如图 8-2 所示。导出窗口列出了文件的入口点。通常用户可在共享库（如 Windows DLL 文件、Linux SO 文件）中找到导出的函数。

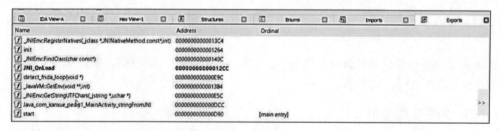

图 8-2　导出窗口

下面来看一下 init 函数。反编译有时会加一些混淆（Ollvm 之类的），看起来特别复杂，这些要从内存中获得。这里我们可以明显地看到调用了一个检测 Frida loop 的函数，如图 8-3 所示。

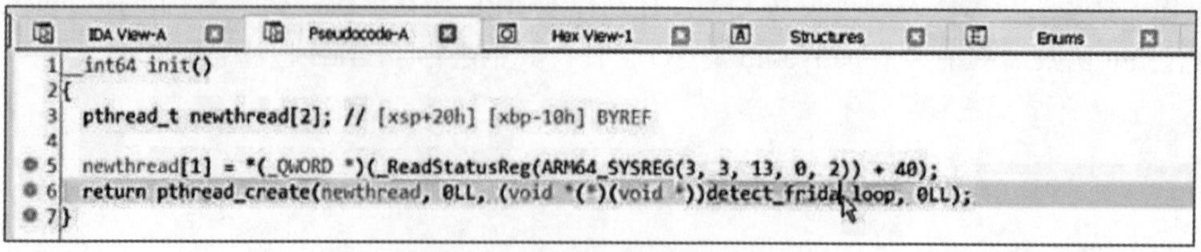

图 8-3　检测 Frida loop 的函数

接下来看这个函数做了什么。右击或者按键盘上的 X 键可以直接看到交叉引用（有哪些函数调用了本函数），如图 8-4 所示。

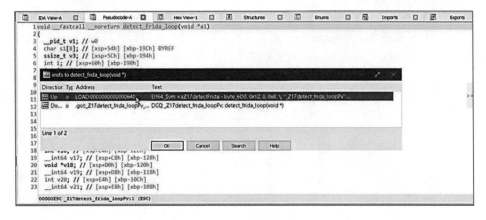

图 8-4　函数的交叉引用

8.1.2　IDA 插件的使用

在逆向分析之前，先介绍一些 IDA 的插件。

IDA 的插件有很多。界面的黑白皮肤效果的使用方式如下：

（1）将安装包解压之后的 Hgl_Dark(_Light)文件夹放入 IDA Pro 7.5\themes 目录下。

（2）依次打开 IDA → Options → Colors... → Current theme，选择 Hgl_Dark(_Light) → ok。

效果如图 8-5 所示。

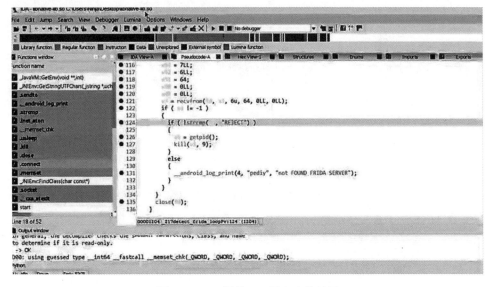

图 8-5　IDA 插件——黑白皮肤效果

GitHub 上也有很多 IDA 的插件，读者可自行尝试。

8.1.3 IDA 反汇编介绍

项目中的 lib 目录下如果是 armeabi，则需要用 IDA32 位来打开。以一个 SO 文件为例，双击进入一个函数可以看到它的流程图，如图 8-6 所示。

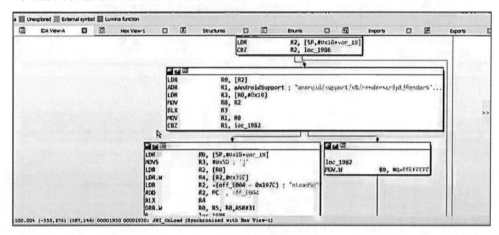

图 8-6　函数流程图

按 F5 键可以看到它注册了哪些函数，如图 8-7 所示。

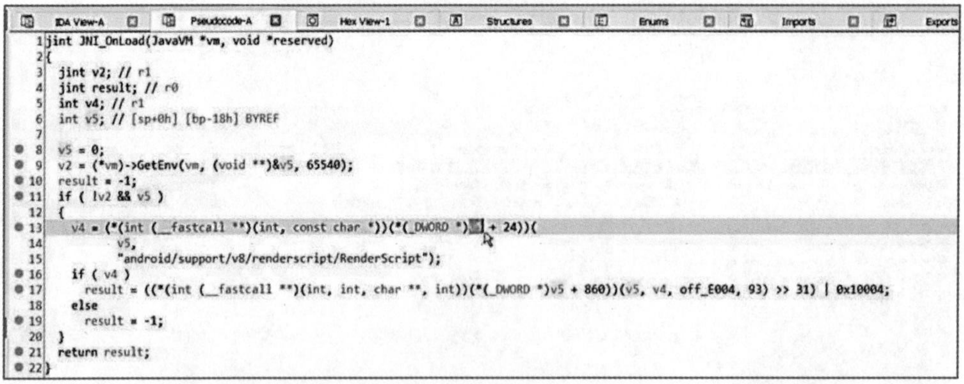

图 8-7　按 F5 键来查看注册的函数

像 v5 这类符号，右击选择 Set→var type，可以改变类型输入，然后输入 JNIEnv *，单击 OK 按钮，如图 8-8 所示。

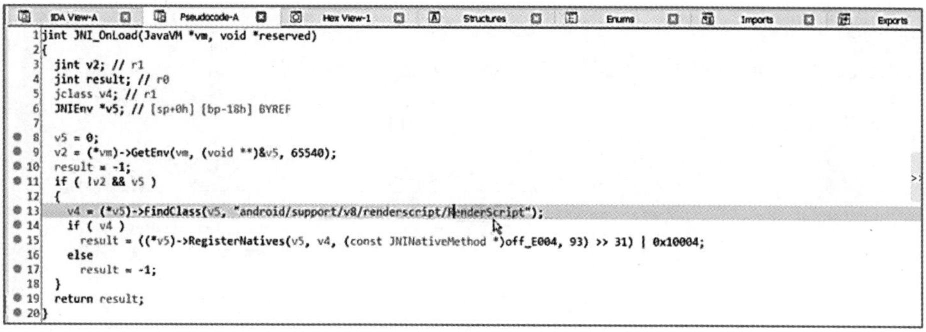

图 8-8 改变类型输入

此处明显调用了 Android 中的库。在 SO 层调用 Java 函数就是用"/"分隔。

再回到 init 函数，可以看到它调用了一个检测 Frida 的函数 detect_frida_loop。在导出表界面，按 Shift+F12 快捷键可以看到 SO 文件的所有字符串都在里面了，如图 8-9 所示。

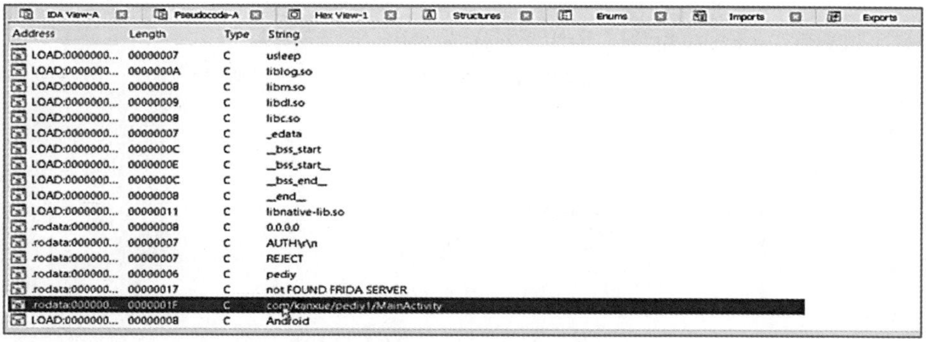

图 8-9 SO 文件的所有字符串

com/kanxue/pediy/MainActivity 明显是一个包名，我们可以进入看一下，如图 8-10 所示。

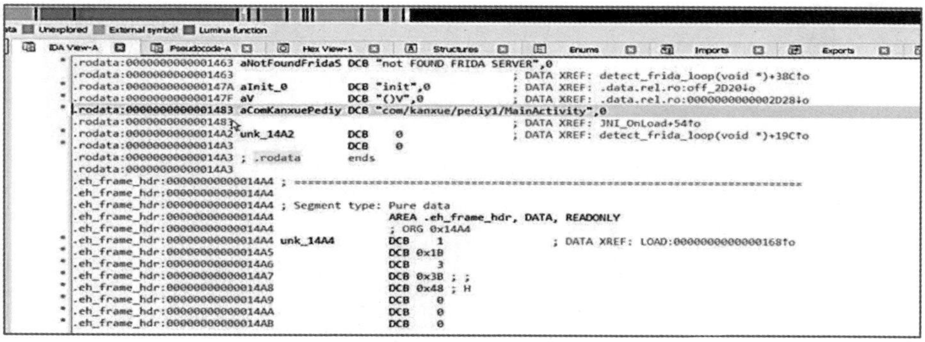

图 8-10 MainActivity

它全部都在 rodata 段里面。rodata 段是只读段，ro 代表 read only。如果这个段里面有改动的话，系统检测到就会报错。

回到 detect_frida_loop 函数，双击进入。可以发现两个入手点：

（1）检测任意地址 inet_aton("0.0.0.0")...，如图 8-11 所示。

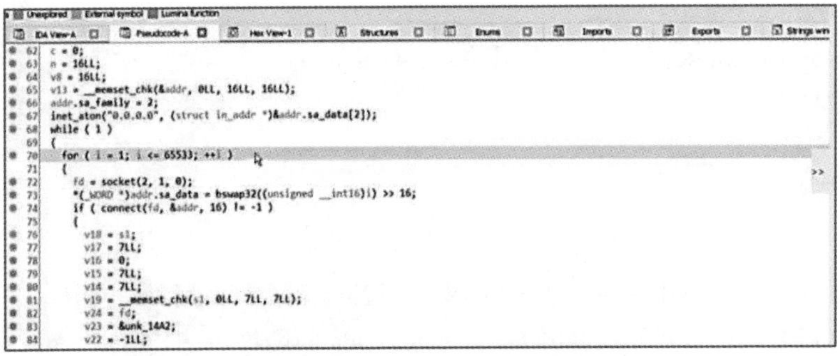

图 8-11　检测任意地址 inet_aton("0.0.0.0")...

（2）检测字符串 if(!strcmp(s1, "REJECT"))...，如图 8-12 所示。

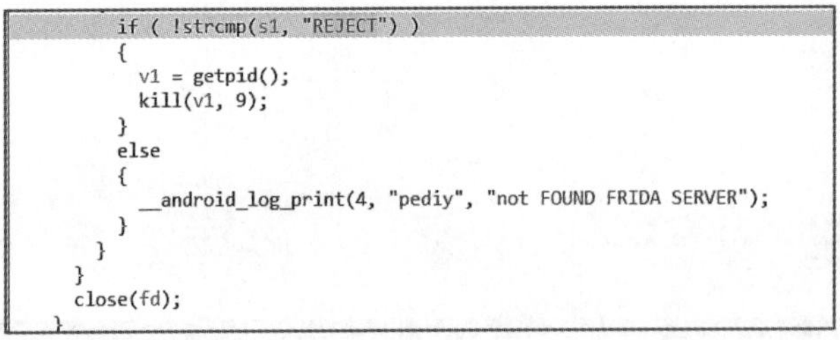

图 8-12　检测字符串 if(!strcmp(s1, "REJECT"))...

先来尝试第一个入手点，检测任意地址 inet_aton("0.0.0.0")。双击地址 0.0.0.0 进入只读段，选择窗口 Hex View-1，如图 8-13 所示。

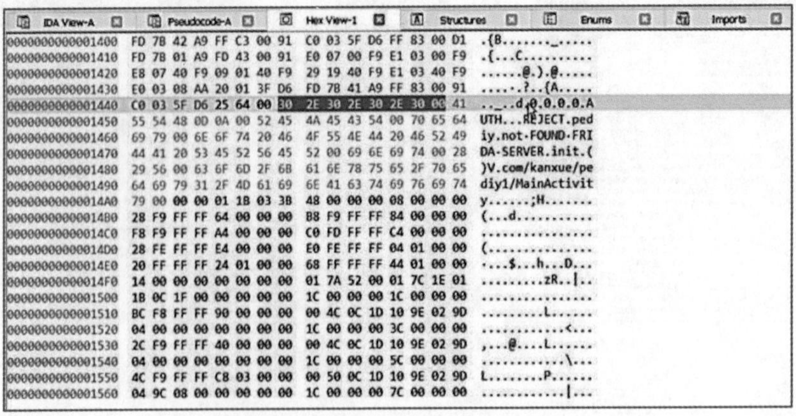

图 8-13　地址 0.0.0.0 十六进制文件的内容

C 语言编码使用的是 ASCII 码表，而 Java 层使用的是 Unicode 码表。我们如何改成 C 语言的十六进制呢？例如，需要将 0.0.0.0 改成 2.2.2.2。由 CyberChef 计算得知，2.2.2.2 对应的 Hex 是 32 2e 32 2e 32 2e 32。将 30 2e 30 2e 30 2e 30 改为 32 2e 32 2e 32 2e 32，更改之后，右击选择 Apply changes，在菜单中选择 Edit→Patch program→Apply patches to input file，如图 8-14 所示。

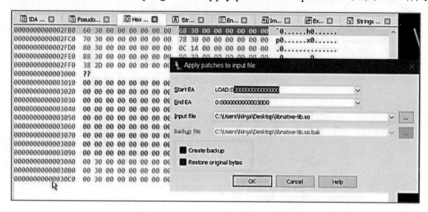

图 8-14　Apply patches to input file

可以看到，地址是从 0000 到 30C0，单击 OK 按钮。回到反编译后的伪代码文件，按 F5 键重新反编译，可以看到 0.0.0.0 变为 2.2.2.2，如图 8-15 所示。

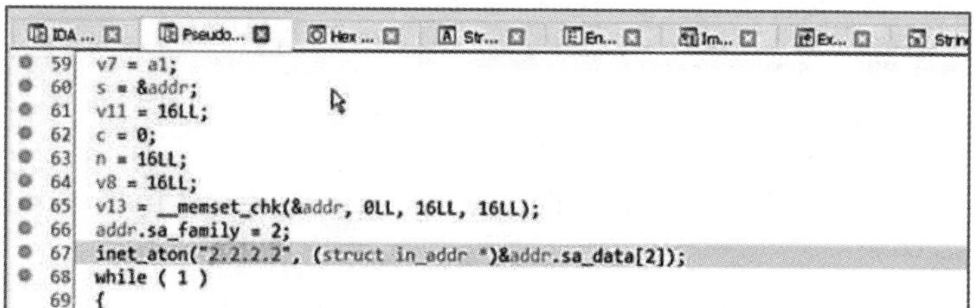

图 8-15　重新反编译

关闭 IDA。硬编码之后，如何将这个文件安装进 App 呢？这时就要用到反编译工具。首先使用 Apktool 反编译 APK，如图 8-16 所示。我们使用改好的 SO 去替换掉反编译生成的文件夹中的原 SO 文件。替换掉后，回编译生成新的 APK，如图 8-17 所示。但是，这样的 APK 直接安装到手机中是有问题的，因为 Android 要求 APK 必须有签名才能安装进去。

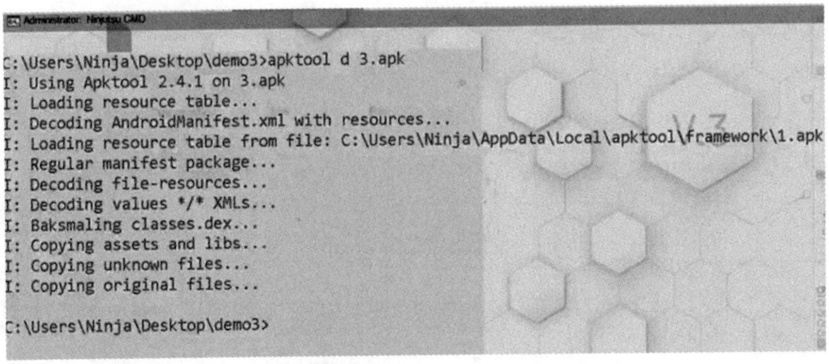

```
C:\Users\Ninja\Desktop\demo3>apktool d 3.apk
I: Using Apktool 2.4.1 on 3.apk
I: Loading resource table...
I: Decoding AndroidManifest.xml with resources...
I: Loading resource table from file: C:\Users\Ninja\AppData\Local\apktool\framework\1.apk
I: Regular manifest package...
I: Decoding file-resources...
I: Decoding values */* XMLs...
I: Baksmaling classes.dex...
I: Copying assets and libs...
I: Copying unknown files...
I: Copying original files...

C:\Users\Ninja\Desktop\demo3>
```

图 8-16　Apktool 反编译

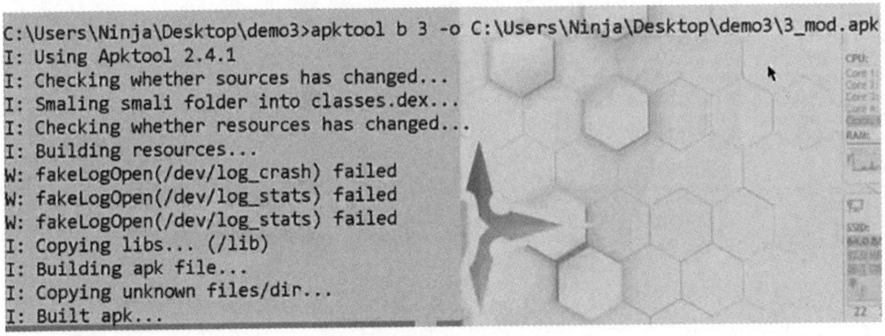

```
C:\Users\Ninja\Desktop\demo3>apktool b 3 -o C:\Users\Ninja\Desktop\demo3\3_mod.apk
I: Using Apktool 2.4.1
I: Checking whether sources has changed...
I: Smaling smali folder into classes.dex...
I: Checking whether resources has changed...
I: Building resources...
W: fakeLogOpen(/dev/log_crash) failed
W: fakeLogOpen(/dev/log_stats) failed
W: fakeLogOpen(/dev/log_stats) failed
I: Copying libs... (/lib)
I: Building apk file...
I: Copying unknown files/dir...
I: Built apk...
```

图 8-17　Apktool 回编译生成新的 APK

- APK 签名教程：参考网址 26（参见配书资源文件）。
- APK 签名工具：参考网址 27（参见配书资源文件）。

从 Releases 下载 uber-apk-signer-1.2.1.jar，使用 uber-apk-signer-1.2.1.jar 对 APK 签名，如图 8-18 所示。

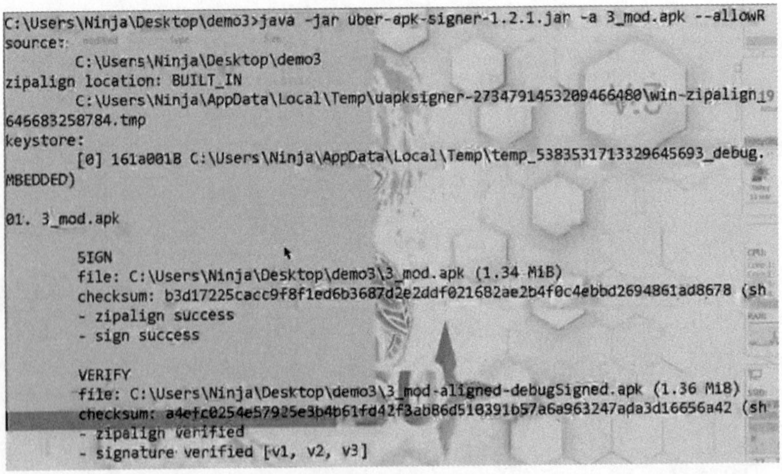

```
C:\Users\Ninja\Desktop\demo3>java -jar uber-apk-signer-1.2.1.jar -a 3_mod.apk --allowR
source:
        C:\Users\Ninja\Desktop\demo3
zipalign location: BUILT_IN
        C:\Users\Ninja\AppData\Local\Temp\uapksigner-2734791453209466480\win-zipalign_19
646683258784.tmp
keystore:
        [0] 161a0018 C:\Users\Ninja\AppData\Local\Temp\temp_5383531713329645693_debug.
MBEDDED)

01. 3_mod.apk

        SIGN
        file: C:\Users\Ninja\Desktop\demo3\3_mod.apk (1.34 MiB)
        checksum: b3d17225cacc9f8f1ed6b3687d2e2ddf021682ae2b4f0c4ebbd2694861ad8678 (sh
        - zipalign success
        - sign success

        VERIFY
        file: C:\Users\Ninja\Desktop\demo3\3_mod-aligned-debugSigned.apk (1.36 MiB)
        checksum: a4efc0254e57925e3b4b61fd42f3ab86d518391b57a6a963247ada3d16656a42 (sh
        - zipalign verified
        - signature verified [v1, v2, v3]
```

图 8-18　使用 uber-apk-signer-1.2.1.jar 对 APK 签名

删除旧的 App，将重新编译并签名的 APK 安装到手机上。运行 App 没有闪退，说明我们成功对抗了 Frida 检测。

接下来再修改一些其他的内容。删除原来反编译、回编译、签名生成的文件和文件夹，重新反编译 3.apk。现在的商业 App 增加了很多检测和对抗，不能随意反编译。这个不在我们讨论的范围内。

另外，IDA 自带一些数据库，保存时选择 DON'T SAVE the database，就不会有这么多数据库出来了。编译不会用到这些数据库文件。

接下来，尝试利用第二个入手点检测字符串 if(!strcmp(s1, "REJECT"))。还是在导出表界面按 Shift+F12 快捷键。找到 REJECT 字符串，如图 8-19 所示。

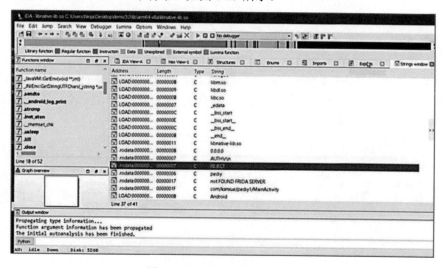

图 8-19　找到 REJECT 字符串

也可以切换到 frida_loop，按一下 F5 键，可以看到这里有一个检测，如果检测匹配，就退出，如图 8-20 所示。

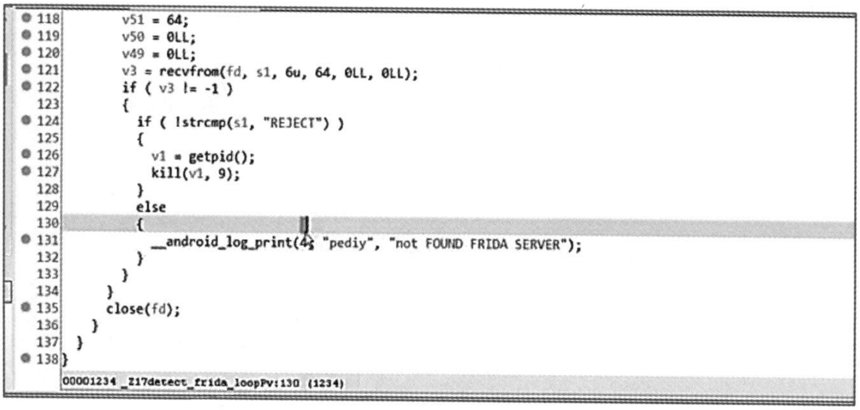

图 8-20　Frida 检测逻辑

双击 REJECT，跳转到只读段。然后选择 Hex View-1，它就自动帮我们定位到十六进制数据了。我们将 REJECT 对应的十六进制数改为 AABBCC 对应的十六进制数 41 41 42 42 43 43。改完之后右击选择 Apply changes 保存。再回到伪代码视图，按 F5 键，REJECT 就变为 AABBCC 了。改好之后，选择 Edit→Patch program→Apply patches to input file，单击 OK 按钮。

再次回编译并签名、安装到手机上。如果启动应用后没有闪退，就说明改成功了。

IDA 数据库文件的用途是 IDA 解析完之后，下一次不用从头去分析，而是直接从数据库读取。

8.1.4　IDA 分析与 Frida Hook 结合

8.1.3 节演示了硬改源程序，本小节将继续介绍 IDA 分析与 Frida Hook 相结合的逆向分析方法。

首先卸载 App，安装回最开始的应用。启动 App，由于 frida-server 还在运行，程序会检测到 Frida 并闪退。

我们之前看到，程序 init 中有 Frida 检测函数。我们将原始 APK 中的 SO 文件拖到 IDA 中分析一下。回到刚才的 strcmp 函数，尝试对其 Hook。由于 strcmp 是 C 的标准库函数，我们来编写 Frida Hook 脚本。

```
function hookstrcmp(){
    Java.perform(function () {

        var addr_strcmp = Module.findExportByName("libc.so","strcmp");
        Interceptor.attach(addr_strcmp, {
            onEnter: function (args) {
                if(ptr(args[1]).readCString().indexOf("REJECT")>=0){
                    console.log("[*] strcmp (" + ptr(args[0]).readCString() + "," + ptr(args[1]).readCString()+")");
                    this.isREJECT = true;
                }
            },onLeave:function(retval){
                if(this.isREJECT){
                    retval.replace(0x1);
                    console.log("the REJECT's result :",retval);
                }
            }
        });
    })
}
setImmediate(hookstrcmp);
```

首先，从 libc.so 的导出表中找到 strcmp 函数地址。然后，attach 附加到 strcmp，进入函数后，在日志中打印 strcmp 的两个参数。然后通过 onLeave 修改 strcmp 函数的返回值。

如果不会编写函数，可以多去看看 Frida 官方文档。

我们使用 Frida 12.8.0 版本，参数 U 代表 USB，参数 a 代表所有 App，如图 8-21 所示。

```
┌──(root💀VXIDr0ysue)-[~/Desktop]
└─# frida-ps -Ua
 PID  Name                       Identifier
----  -----------------------    ------------------------------
1172  Android 系统               android
1377  Android 键盘 (AOSP)        com.android.inputmethod.latin
3663  MTP 主机                   com.android.mtp
2054  NFC服务                    com.android.nfc
2127  Quickstep                  com.android.launcher3
2068  SecureElementApplicati…    com.android.se
3691  Shell                      com.android.shell
2149  com.android.smspush        com.android.smspush
4259  pediyl                     com.kanxue.pediyl
1172  一体化位置信息             com.android.location.fused
2087  内容下载管理器             com.android.providers.downloads
2955  图库                       com.android.gallery3d
3633  外部存储设备               com.android.externalstorage
2087  媒体存储设备               com.android.providers.media
2584  密钥链                     com.android.keychain
3604  文件                       com.android.documentsui
1507  电话和短信存储             com.android.providers.telephony
1507  电话服务                   com.android.phone
1395  系统界面                   com.android.systemui
1172  设置存储                   com.android.providers.settings
3516  软件包安装程序             com.android.packageinstaller
2931  软件包权限帮助程序         com.android.defcontainer
1172  通话管理                   com.android.server.telecom

┌──(root💀VXIDr0ysue)-[~/Desktop]
```

图 8-21　Frida 列出的所有 App 和包名

C 语言提供的标准库也有可能是其他的 App 调用的。启动 Frida 并执行脚本，如图 8-22 所示。

```
function hookstrcmp(){
    Java.perform(function() {
        console.log("I am a Hook function");
        var strcmp = Module.findExportByName("libc.so","strcmp");//这里发现无论"libnative-li
        console.log("find strcmp:",strcmp);
        //Hook strcmp
        Interceptor.attach(strcmp, {
            onEnter: function (args) {
                //hook住后打印strcmp的第一个参数和第二个参数的内容
                if(ptr(args[1]).readCString().indexOf("REJECT")>=0){
                    console.log("[*] strcmp (" + ptr(args[0]).readCString() + "," + pt
                    this.isREJECT = true;
                }
            },onLeave:function(retval){
                if(this.isREJECT){
                    retval.replace(0x1);
                    console.log("the REJECT's result :",retval);
                }
            }
        });
    })
}
setImmediate(hookstrcmp);
```

图 8-22　启动 Frida 并执行脚本

至此，我们成功地通过硬编码和 Frida Hook 绕过了程序中 Frida 的检测。

8.2　动静态 SO 算法还原与脱机

8.2.1　IDA 动态调试 SO 算法

本小节来看“[2021 春节]解题领红包之三”这个 App 是怎么样运行的。在手机中打开 App，启动界面如图 8-23 所示。

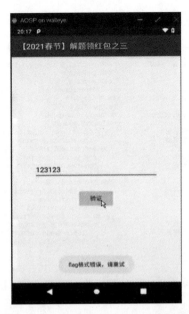

图 8-23 "[2021 春节]解题领红包之三" App 的启动界面

随便输入字符串 123123,单击"验证"按钮会提示"flag 格式错误,请重试"。下载 Jadx 静态分析工具。下载后解压,双击运行 jadx-gui.bat。打开题目,如图 8-24 所示。

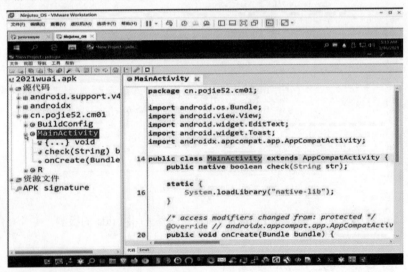

图 8-24 启动 Jadx 并打开题目

从 onCreate 函数中,我们可以看到一些简单的代码运行逻辑。出现以上提示信息是因为输入的字符串长度不等于 30。我们可以再输入 30 个字符测试一下。输入 123456789012345678901234567890,单击"验证"按钮,此时出现的 Toast 提示为"flag 错误,再接再厉"，如图 8-25 所示。

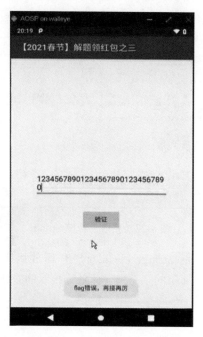

图 8-25　flag 错误，再接再厉

因此，我们推断，这道题目需要逆向分析出真正的 flag。

首先，运行 IDA 64 位打开项目的 SO 文件 libnative-lib.so，在导出表中进入 Java_cn_pojie52_cm01_MainActivity_check 函数，也可以按 Shift+F12 快捷键看 SO 文件中有哪些字符串。从字符串中没有找到有价值的特征。

在流程图界面按 F5 键可以反汇编为伪代码以便阅读。注意，函数的第 3 个参数才是真正传入的参数，前两个参数都是默认的，如图 8-26 所示。

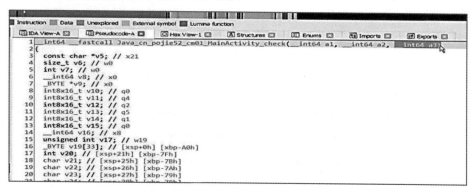

图 8-26　函数传参从第 3 个参数开始

从 Java 层中的 check 函数了解到，此处传递的参数类型是 String。由于 Java 编码和 C 语言编码不同，Java 层传参到 SO 层要有一个 API。前面讲到，伪代码中遇到加号的变量可以右击选择 set →var type 来改变类型输入，如图 8-27 所示。

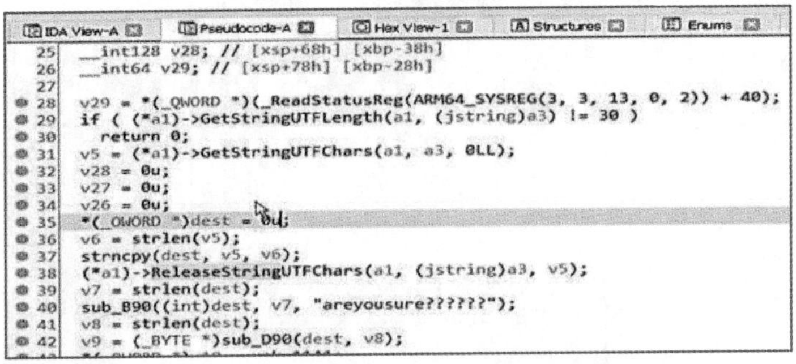

图 8-27　获取、释放 String

我们看到了释放函数 ReleaseStringUTFChars，有释放函数就一定会有对应的获取函数 GetStringUTFChars，用来获取 Java 层的字符串（此处为传入的参数 a3）转成 C 层的字符串。转完之后需要调用 ReleaseStringUTFChars 将字符串释放掉。

用 IDA 调试需要耗费点时间，所以在此之前可以用 Frida 快速获取这个函数传递了哪些参数。

GetStringUTFChar 返回的是 C 语言的字符串。为了便于分析，我们将 Java 的参数 a3 改为 java_str，将返回的 C 的字符串 v5 改为 c_str，将 v6 改为 c_str_len，将 v7 改为 dest_len。strncpy(dest, c_str, c_str_len)将 c_str 复制到 dest，复制的字符串长度为 c_str_len。

C 库函数 strncpy()的声明如下：

```
char *strncpy(char *dest, const char *src, size_t n)
```

带*类型的参数，例如*dest，用 Frida Hook 时要加上 ptr。

接下来调用了 sub_B90 函数。双击进入 sub_B90 函数查看它实现了哪些功能，如图 8-28 所示。

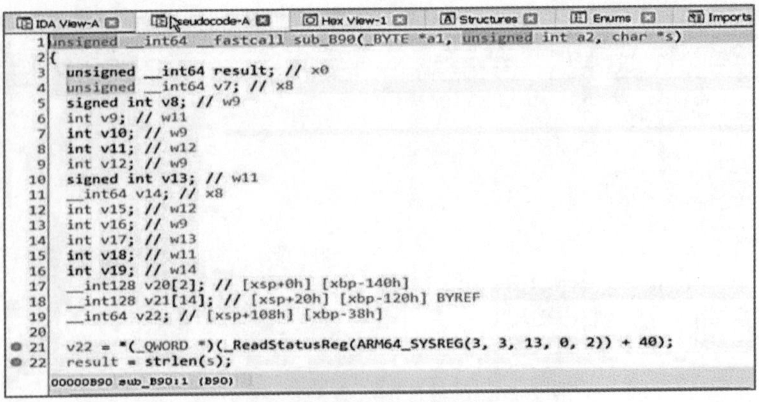

图 8-28　sub_B90 函数

我们要 Hook 函数 sub_B90。每次 App 重启，SO 文件加载到内存中的地址都是不一样的。因此，sub_B90 地址为 SO 文件基址+物理地址（B90），即实际要 Hook 的地址，如图 8-29 所示。

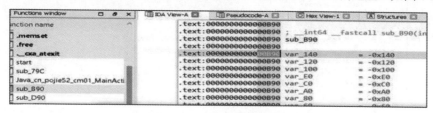

图 8-29　sub_B90 函数地址

sub_B90 函数有 3 个参数：(int)dest、dest_len 和一个自定义字符串，如图 8-28 所示。
Frida Hook 的脚本如下：

```
function inline_hook() {

    Java.perform(function () {

        var addr_libnative = Module.findBaseAddress("libnative-lib.so");

        if (addr_libnative) {
            console.log("SO 基址: ", addr_libnative);
            var sub_B90 = addr_libnative.add(0xB90);
            Interceptor.attach(sub_B90, {
                onEnter: function(args) {
                    console.log("\n",hexdump(args[0]));
                    console.log("\n",args[1]);
                    console.log("\n",hexdump(args[2]));
                },
                onLeave:function(retval){

                }
            });
        }
    });
}

setImmediate(inline_hook);
```

通过 Frida 的 API findBaseAddress 可以直接拿到 SO 文件加载在内存中的基址。拿到基址后，再加上它的物理地址，然后 Hook，否则会报错。

启动 frida-12.8.0，查看进程包名的命令如下：

```
root@VXIDr0ysue:~ # frida-ps -Ua
```

得到当前 App 进程的包名为 cn.pojie52.cm01，运行 Frida Hook 脚本，如图 8-30 所示。

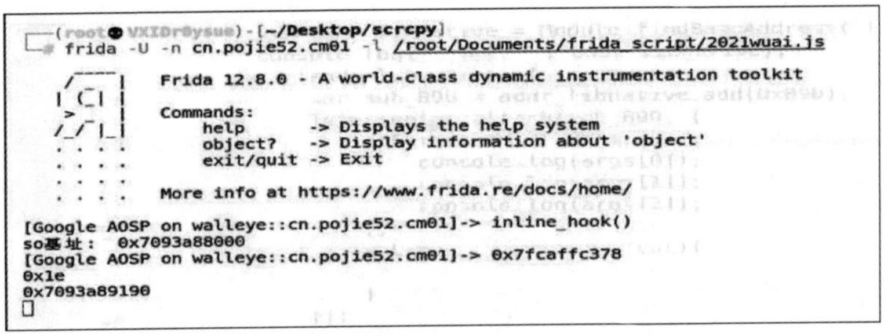

图 8-30 运行 Frida 脚本 Hook sub_B90

结果显示，第一个参数(int)dest 是我们传进来的明文字符串，第二个参数是明文字符串的长度，第三个参数是固定字符串。

我们也可以用 IDA 调试功能。在 IDA 图形视图窗口按空格键可以进入 IDA 文本视图窗口。在 sub_B90 和前一条指令处设置断点，分别为地址 8B0 和地址 8AC 处，右击选择 Add breakpoint，如图 8-31 所示。

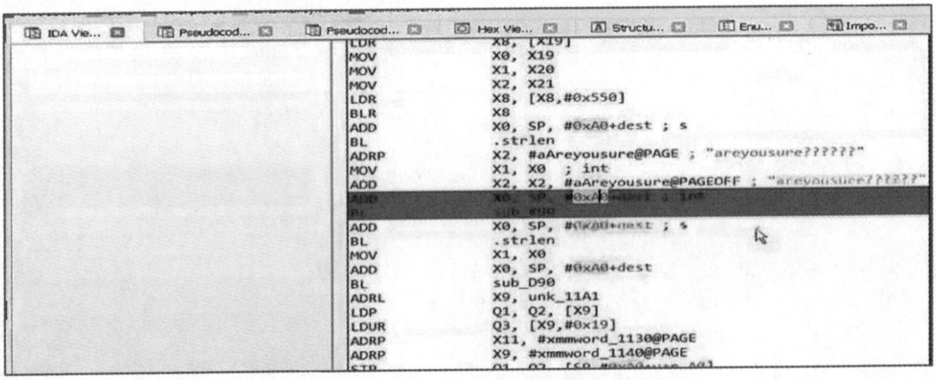

图 8-31 在 sub_B90 处设置断点

然后，将 IDA/dbgsrv 文件夹下的 android-server（32 位）和 android-server64（64 位）push 到手机/data/local/tmp 目录下（frida-server 所在目录），并给它执行权限 chmod 777 android-server64（此处，我们使用的是 android-server64（64 位），对于 android-server（32 位）也是相同的操作）。启动 android-server64，默认端口为 23946，如图 8-32 所示。

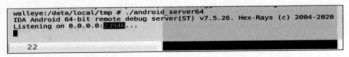

图 8-32 启动 android-server64，默认端口为 23946

依次单击 File→New Instance 重新开启一个空的 IDA，然后选择 GO 就可以了。第一次调试，在菜单栏依次选择 Debugger→Attach→Remote ARM Linux/Android debugger，在弹出的窗口的 Hostname 中填上手机的 IP 地址（可以使用 ifconfig 命令查看手机的 IP 地址），如图 8-33 所示。

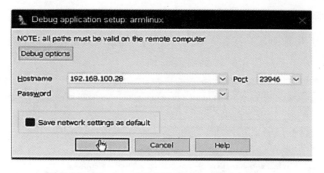

图 8-33　Remote ARM Linux/Android debugger

单击 OK 按钮后可以看到非常多的进程，选择需要 attach 的包名，此处是 cn.pojie52.cm01。如果要调试商业 App，可能需要线程挂起，现在不需要。首先单击"运行"按钮，如图 8-34 所示。

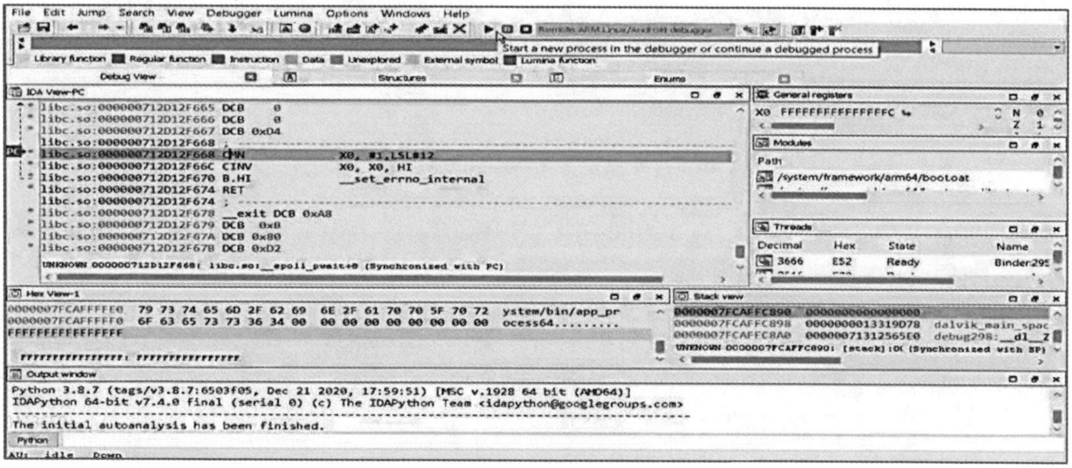

图 8-34　开始调试

现在要 Hook 的函数地址是 8AC，然后拿到 SO 文件的基址。如果需要调试 Frida，同时调试 IDA，必须先把 Frida 运行起来，再使用 IDattach。由于每次运行 App，SO 文件加载到内存中的基址都会改变，因此我们要运行 Frida 脚本来获得 SO 文件的基址，如图 8-35 所示。

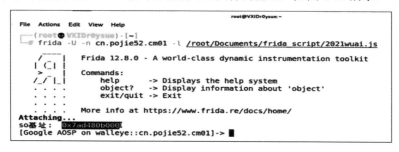

图 8-35　运行 Frida 脚本获得 SO 文件的基址

将断点设置在实际要 Hook 的地址处：SO 基址（0x7ad480b000）+物理地址（B90）=0x7ad480b8AC，如图 8-36 所示。在菜单栏依次选择 Jump→Jump to address...，在弹出的窗口填上地址 0x7ad480b8AC，单击 OK 按钮跳转到对应的地址。

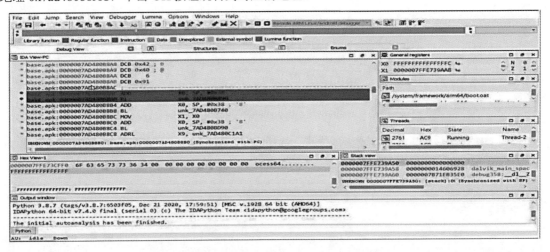

图 8-36　将断点设置在实际要 Hook 的地址处

在 App 运行的界面单击"验证"，实际上已经运行到断点处了。此时，寄存器 x0 还没有赋值，如图 8-37 所示。

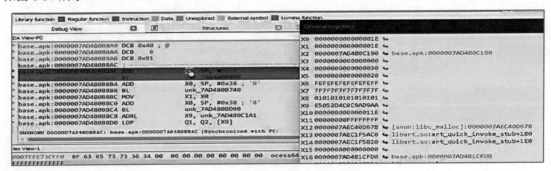

图 8-37　寄存器 x0 还没有赋值

如果单步调试，可以按 F8 键，就会跳到下一行。计算结果存储到寄存器 x0 中了，如图 8-38 所示。单击 x0 后的箭头 Jump in disassembly，可以看到输入的字符串 123456789012345678901234567890，如图 8-39 所示。

第 8 章 反编译工具 IDA   145

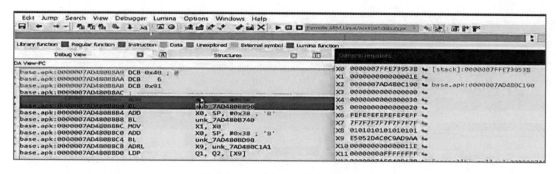

图 8-38 寄存器 x0 的存储结果

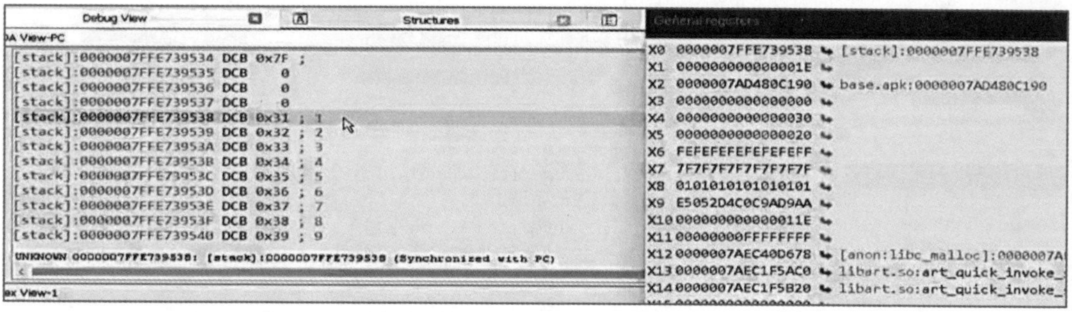

图 8-39 寄存器 x0 存储输入的字符串

这样我们通过 IDA 调试可以获得输入的数据，用 frida-dump 也可以获得，如图 8-40 所示。

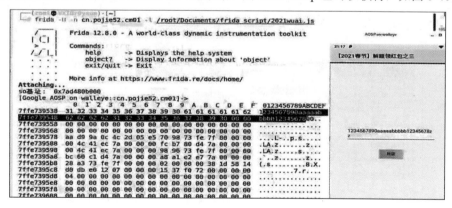

图 8-40 用 frida-dump 获得结果

我们继续研究 sub_B90 函数，可以看到其中有非常多的异或运算。异或运算是可以被还原的。明文 xor 密钥=密文。反之，明文 xor 密文=密钥。我们的目的就是拿到密钥。修改 inline_hook 脚本如下：

```
function inline_hook() {
    Java.perform(function() {
```

```
            var addr_libnative = Module.findBaseAddress("libnative-lib.so");

        if (addr_libnative) {
            console.log("so基址: ", addr_libnative);
            var sub_B90 = addr_libnative.add(0xB90);
            Interceptor.attach(sub_B90, {
                onEnter: function(args) {
                    this.arg0 = args[0];
                    this.arg1 = args[1];
                    this.arg2 = args[2];


                    console.log("\n",hexdump(this.arg0));
                    console.log("\n",this.arg1);
                    console.log("\n",hexdump(this.arg2));
                },
                onLeave:function(retval){

                    console.log("-------sub_B90 retval -------");
                    console.log("\n",hexdump(this.arg0));
                    console.log("\n",this.arg1);
                    console.log("\n",hexdump(this.arg2));
                }
            });
        }
    }

setImmediate(inline_hook);
```

运行脚本，得到 sub_B90 函数运行结束后的第一个参数，如图 8-41 所示。

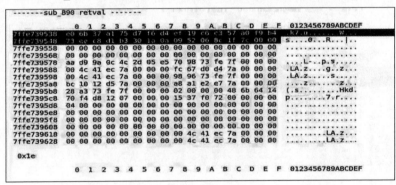

图 8-41　sub_B90 函数运行结束后的第一个参数

我们简单分析一下 sub_B90 函数。它是根据传入的第 3 个参数 s 将 v20 进行了初始化，然后把参数 a1 和 v20 进行了异或运算。参数 a1 原先是我们传入的明文，进行异或操作之后变成了一段密文。由此可以推测，将输入的明文 a1 与密文结果进行异或操作就可以得到密钥。

此时，输入的明文为 30 个 1，也就是 30 个 0x31。然后看密文，第一位是 0xe0，0x31^0xe0 =

D1，第二位是 6b，0x31^0x6b = 5A，按以上方法计算出 30 位密钥（十进制）为[209, 90, 6, 144, 68, 230, 199, 229, 222, 40, 247, 242, 102, 145, 200, 133, 66, 223, 249, 224, 130, 1, 43, 59, 56, 99, 55, 189, 46, 77]。用其他明文验证，得到的结果一致，说明推测正确。

所以，sub_B90 函数的功能主要是传入一段明文，输出一段密文。得到密文后，先通过 strlen() 函数计算其长度，然后将密文作为参数传入 sub_D90 函数中。sub_D90 函数实际上使用的是 Base64 编码。为了便于理解，我们将 sub_D90 函数的返回值名称改为 dest_base64。查看 dest_base64 的交叉引用，发现它在一个循环判断条件中被使用到。根据代码逻辑，我们需要知道 v19 是什么，如图 8-42 所示。

```
82                          v14);
83      v23 = ((1
84          - ((2 * ((((unk_11C8 - 78) ^ 0xB0) - 119) ^ 0xFE) | ((((unsigned __int8)((unk_11C8 - 78) ^ 0xB0) -
85          + 62;
86      v24 = ((1
87          - ((2 * ((((unk_11C9 - 78) ^ 0xBF) - 120) ^ 0xFE) | ((((unsigned __int8)((unk_11C9 - 78) ^ 0xBF) -
88          + 62;
89      while ( dest_base64[v16] == v19[v16] )
90      {
91        if ( dest_base64[v16] )
92        {
93          if ( ++v16 != 41 )
94            continue;
95        }
96        v17 = 1;
97        goto LABEL_9;
98      }
99      v17 = 0;
100 LABEL_9:
101     free(dest_base64);
102     return v17;
103 }
```

图 8-42　dest_base64 与 v19 进行比较

我们的目的是在寄存器中拿到 v19 的值，因为要判断 Base64（密文）是否等于 v19。取得 v19 的值之后，就可以解码 Base64（密文）得到密文，再通过密钥异或从而得到明文。

为了从寄存器中获取 v19 的值，我们再次使用 IDA 调试。运行 App，通过 inline_hook 脚本得到 SO 文件的基址为 0x7ad4d06000。断点处的物理地址为 0xb2c，所以实际地址是 0x7ad4d06000+0xb2c=0x7ad4d06b2c。在菜单栏选择 Jump→Jump to address，填入 0x7ad4d06b2c，单击 OK 按钮，跳转到 0x7ad4d06b2c 地址处，如图 8-43 所示。

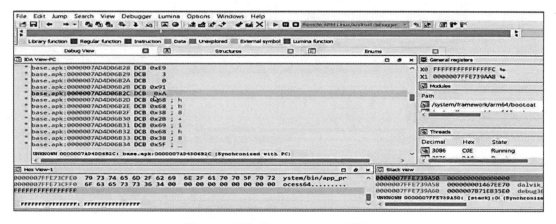

图 8-43　跳转到 0x7ad4d06b2c 地址处

在 0x7ad4d06b2c 地址处设置断点，在 App 界面单击"验证"，运行到断点处。回到反汇编伪代码，当鼠标落在 v19 上的时候，我们可以看到 v19 的值存放在 x8 寄存器中，如图 8-44 所示。

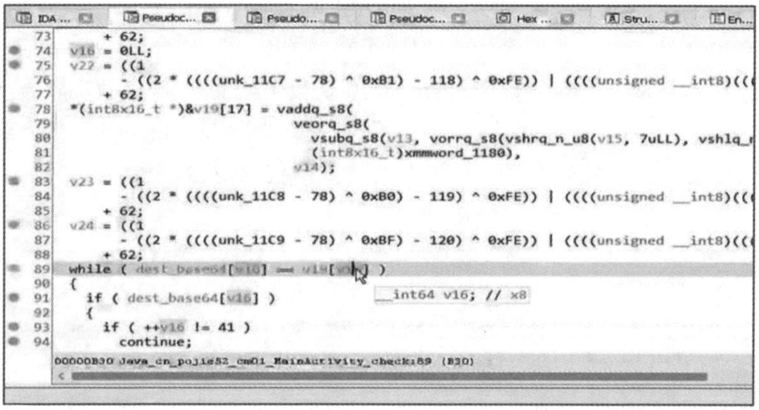

图 8-44　v19 的值存放在 x8 寄存器中

x8 中的这一串密文是 5Gh2/y6Poq2/WIeLJfmh6yesnK7ndnJeWREFjRx8，就是我们要的 v19。注意，这一串并不是真正的密文，而是经过 Base64 编码之后的。我们还需要将它进行 Base64 解码：

```
import base64
data = base64.b64decode('5Gh2/y6Poq2/WIeLJfmh6yesnK7ndnJeWREFjRx8'.encode())
```

data 就是真正的密文。之前我们已经拿到了密钥，再进行一次异或就可以反推出明文。最后推出明文是 52pojieHappyChineseNewYear2021。这里给出 Python 进行异或还原的代码：

```
import base64
xordata = [0xd1,0x5a,0x6,0x90,0x44,0xe6,0xc7,0xe5,0xde,0x28,0xf7,0xf2,0x66,0x91,0xc8,0x85,0x42,0xdf,0xf9,0xe0,0x82,0x1,0x2b,0x3b,0x38,0x63,0x37,0xbd,0x2e,0x4d]
data = base64.b64decode('5Gh2/y6Poq2/WIeLJfmh6yesnK7ndnJeWREFjRx8'.encode())
flag = bytes([xordata[i] ^ data[i] for i in range(len(xordata))]).decode()
print(flag)
```

将结果 52pojieHappyChineseNewYear2021 输入 App 中进行验证，如图 8-45 所示。

可以看到结果正确。

图 8-45　验证结果

8.2.2 Keypatch 原理/实战硬改算法逻辑

可以通过硬改来实现更改算法的逻辑吗？答案是可以的。那就是让返回值 v17 永远等于 1。v17 所在的寄存器是 w19（将鼠标移至 v17 即可见），如图 8-46 所示。

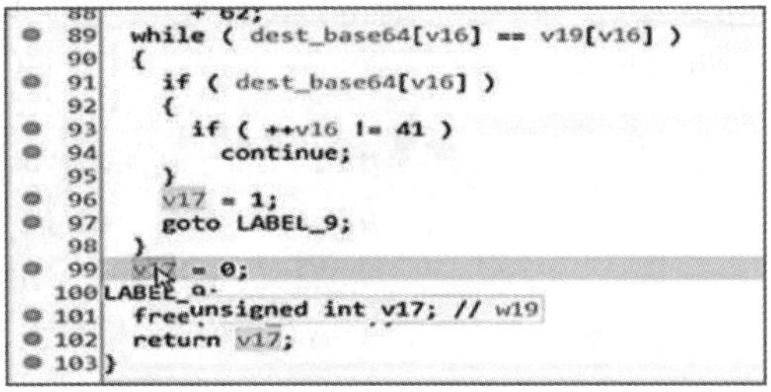

从 52 论坛上下载 Keypatch Patcher 插件，单击 Edit→Plugins→Keypatch Patcher 即可使用，如图 8-47 所示。

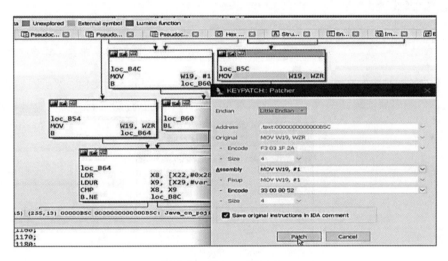

图 8-46 v17 所在的寄存器是 w19

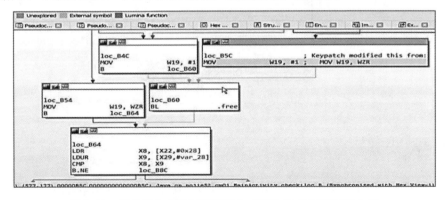

图 8-47 Keypatch Patcher

我们的目的是将右侧分支的 MOV W19，WZR 改成与左侧分支相同的效果，即令 W19 永远等于 1。根据图 8-47 改好之后，单击 Cancel 按钮，结果如图 8-48 所示。

图 8-48 将右侧分支的 MOV W19，WZR 改成 MOV W19, #1

选择 Hex-View，可以看到改了的部分的颜色变化。单击 Edit→Patch program→Apply patches to input file...保存，再单击 OK 按钮就可以写入文件了。按 F5 键重新编译后，可以看到 v17=1，如图 8-49 所示。

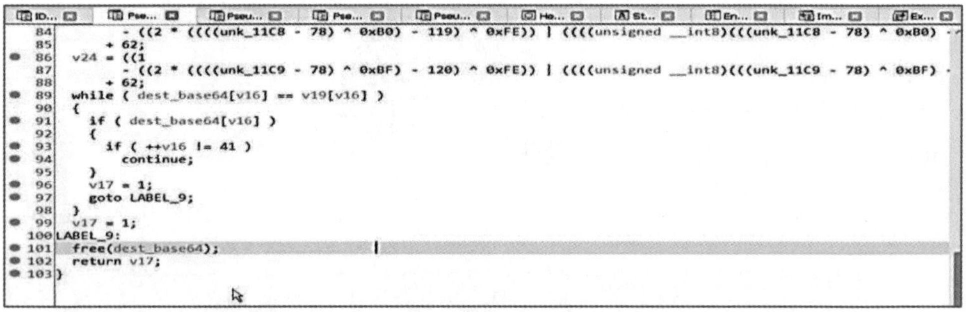

图 8-49　修改 v17=1 成功

反编译并得到对应的文件夹后，用更改后的 SO 文件替换原始 SO 文件。然后回编译并导出新的 APK 文件，签名后安装到手机上。随便输入 30 个字符，可以看到验证成功，如图 8-50 所示。

图 8-50　随便输入 30 个字符，验证成功

8.3　本章小结

由于篇幅有限，本书只对 IDA 的使用进行了简要介绍，但仅用一章学习 IDA 是远远不够的，多加练习才能掌握这一工具。

第 9 章

JNI 接口初识

JNI（Java Native Interface）是连通 Java 层和 Native 层的一个桥梁。通过跟踪 JNI，我们可以分析很多保护手段，比如 Dex2C 和 DexVMP。本章将从 NDK 开始逐步介绍 JNI 的相关知识。还将介绍一些优秀的项目、一些自动追踪的方法以及更深层次的自动实现动态注册追踪的技术。

9.1 JNI 及其工作原理

9.1.1 NDK 简介

在正式介绍 NDK 之前，笔者向大家推荐一系列优秀的文章。通过学习这些文章，我们可以事半功倍。这些文章参考网址 28（参见配书资源文件）。

谷歌提供了两种开发包，供开发人员在 Android 上开发应用程序：一种是 SDK，另一种是 NDK。Android 平台从一开始就支持 C 和 C++语言。在 Android 平台上，越是底层的层次，越是采用 C 和 C++实现，到了 Framework 层或者内核层几乎以 C 语言为主了。实际上，Java 本身就是由 C++开发的，ART（Android Runtime）也是由 C++实现的。越是接近内核的层次，手写汇编的使用就越多。Android 平台的架构如图 9-1 所示。

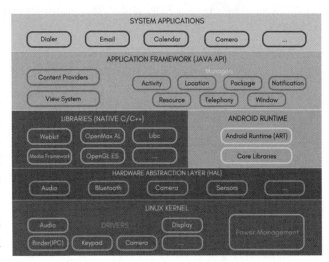

图 9-1　Android 平台架构

如果没有 SDK，整个 Android 项目将无法运行。可以在 Android Studio 的设置中查看 Android

SDK，如图 9-2 所示。

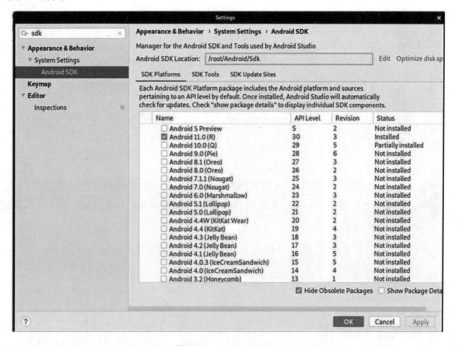

图 9-2 Android SDK

除了 SDK 之外，我们还需要一个 NDK 才能使用 C 和 C++。如果只编写 Android 项目而不涉及 C 和 C++，那么只需要 SDK 就可以编写 Java 项目。但是，如果要编写 Native 项目，则必须要有 NDK。

Android 从一开始就支持 JNI 的编程方式。也就是说，第三方应用可以通过 JNI 调用自己实现的 C 和 C++代码。NDK 通过 NDK Package 将 C 和 C++代码编译成可执行的形式。例如，在 Android/Sdk/ndk/（版本号）目录下，我们常用的调试工具有 ndk-gdb 和 ndk-lldb。

官网对 NDK 的定义如下：

Android NDK 是一个工具集，可以让你使用 C 和 C++等语言以原生代码实现应用的各个部分。对于特定类型的应用，这可以帮助你重复使用以这些语言编写的代码库。

举个例子，我们打开任何一个视频类型的应用。它通常会包含 FFMPEG 这样的一个解码库，这个库可以应用于全平台。因此，我们没有必要用 Java 重新实现一遍解码库。我们只需要用 JNI 调用它的原生（Native）部分，直接运行即可。NDK 本身就是一个交叉的工具链，它包含 Android 的一些库文件。为了方便使用，NDK 提供了一些脚本使得编译 C 和 C++代码更加容易。

为什么要熟练使用 NDK 呢？首先，NDK 适合在平台之间移植应用程序，越是底层越难，因此高级程序员一般都会掌握 NDK。其次，像游戏这种计算密集型应用或者像视频这种对性能要求非常高的应用，一般都是用 NDK 编写的。例如，微信视频/语音电话功能就是用 NDK 编写的，而不是用 Java 编写的。因为 NDK 的性能比 Java 的性能要强得多。此外，NDK 是不依赖 Dalvik 虚拟机的。这一点非常关键。我们在执行 C 算法的时候，它可以不依赖 Dalvik 虚拟机而直接运行在 CPU 上面。因此，测试这些算法的工作情况时，追踪（Trace）虚拟机是没有用的，必须追踪 CPU 才可以。最后，NDK 对于算法的保护效果特别好，这一点也是程序员选择使用 NDK 的原因之一。

从 NDK 到 SO 文件的流程如图 9-3 所示。

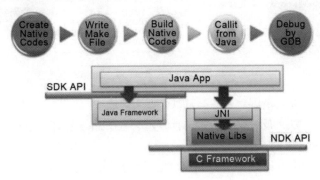

图 9-3　NDK 到 SO 文件

首先编写 C 或 C++代码，然后修改 CMakeLists.txt。有了 Makefile 之后，就可以根据 Makefile 把 NDK 编译成 SO。SO 加载之后，就可以从 Java 层来调用这些 Native 层的算法。

带有 NDK 的 App 是怎么运行的？首先通过 SDK 来运行 Java 应用，然后通过 JNI 来调用 Native 层 SO 文件的一些应用。此外，JNI 和 Java Framework 也可以互相调用。我们用 NDK 进行编译的时候，与 CPU 架构高度相关，编译成的每一个 CPU 架构的版本都是不一样的，而且编译成的每一个 CPU 架构的版本都要在 build.gradle 中单独指定。编译完成后，在 build 文件夹中可以找到对应的 CPU 架构版本的可执行文件。

9.1.2　JNI——NDK 具体的实现接口

JNI 是如何工作的？

在 Java 语言中，Java 本身就可以调用 C++，并不是 Android 自创的。JNI 就是 Java 调用 C++ 的规范，是 Java 调用 Native 方法的一种特性。通过 JNI 能够使 Java 和 C 语言进行交互，既可以在 Java 代码中使用 C 代码，也可以在 C 代码中调用 Java 代码。JNI 其实是 JVM 规范的一部分。因此，我们可以将 JNI 程序在任何实现了 Java 规范的虚拟机中运行。我们可以利用这个特性复用之前用 C 写的代码，并组装成动态库，然后在任意的 Java 程序上运行。一般的 Java 程序使用 JNI 标准可能与 Android 不同。相比之下，Android 的 JNI 用起来更加简单一些。

回顾之前编写的 easyso1 项目中的 stringFromJNI，像这样的一个用 C 编写的函数就是一个 JNI 函数。我们还调用了一个 MD5 算法，这个算法就是用 C 语言实现的。

JNI 函数有两种注册方法，一种是静态注册，另一种是动态注册。JNI 的角色如图 9-4 所示。

JNI 就是一个接口，可以理解为"中间人"。JNI 下涉及 3 个角色，分别是 C/C++代码、Native 方法接口类和 Java 业务类，如图 9-5 所示。举一个不太恰当的例子，JNI 就好比黑白无常，可以穿越"阴阳两界"。

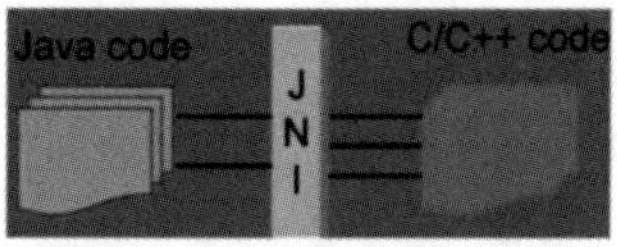

图 9-4 JNI 的角色

图 9-5 C/C++代码、Native 方法接口类和 Java 业务类

我们之前的案例写得非常清晰。业务类可以像调用普通 Java 类一样调用接口类中的方法，从而间接实现调用 Native 方法，反过来也一样。我们先学习在 Java 层调用 Native 层的方法，之后再学习在 Native 层调用 Java 层的方法。

从简单的 Java 层的 stringFromJNI 开始，我们就调用到了 SO 文件中的 Java_com_roysue_easyso1_MainActivity_stringFromJNI 方法，进而调用了 C 语言实现的 MD5 函数。以上就是从 Java 层一步一步调用 Native 层的流程。

所以说，JNI 方法就像是一个桥梁，在 Java 层和 Native 层间起到了承上启下的作用。像 (*env)->GetStringUTFChars 就是 JNI 的函数，把从 Java 层传进来的参数（如 Java String）转换成 Native 层的 char*或 C++ String。由于函数返回值是一个 Java String，因此在 Native 层计算之后，还需要通过 JNI 将 C 语言的 char*转换成 Java String。

我们在 getJString 函数中进行了 JNI 的转换。

```
JNIEXPORT jstring JNICALL
Java_com_roysue_easyso1_MainActivity_method01(JNIEnv *env, jclass clazz, jstring str_) {
    // TODO: implement method01()

    const char *str = (*env)->GetStringUTFChars(env, str_, JNI_FALSE);
    char *result = AES_128_CBC_PKCS5_Encrypt(str);

    (*env)->ReleaseStringUTFChars(env, str_, str);

    jstring jResult = getJString(env, result);
    free(result);

    return jResult;
}

jstring getJString(JNIEnv *env, const char *src) {
    jclass stringClazz = env->FindClass("java/lang/String");
    jmethodID methodID = env->GetMethodID(stringClazz, "<init>", "([BLjava/lang/String;)V");
    jstring encoding = env->NewStringUTF("utf-8");

    jsize length = static_cast<jsize>(strlen(src));
    jbyteArray bytes = env->NewByteArray(length);
    env->SetByteArrayRegion(bytes, 0, length, (jbyte *) src);
```

```
        jstring resultString = (jstring) env->NewObject(stringClazz, methodID, bytes,
encoding);

        env->DeleteLocalRef(stringClazz);
        env->DeleteLocalRef(encoding);
        env->DeleteLocalRef(bytes);

        return resultString;
}
```

getJString 函数通过反射调用 env 中的函数 NewObject 将 C 的 char *转换成 jstring。利用反射机制可以实现任意 Java 层代码的调用。在后续的章节中，我们会具体学习 Java 反射机制。

一个典型的 JNI 函数是如何构成的呢？一开始是系统默认帮我们创建的。新建一个 NDK 项目，系统就会相应地创建一个这样的函数。JNIEXPORT 代表这个函数名不会被 C++作为名称粉碎（Name Mangling），也就是说，经过编译之后名称不会改变。如果是 C 语言，则没有这样的机制，例如 SO 文件中的 MD5。

没有 JNIEXPORT，经过编译之后函数名称会发生变化，这就是 JNIEXPORT 的作用。约定命名格式：Java+包名+类名+方法名，中间用"_"分隔。我们可以看到 Java 层传进来的参数至少有两个。以 stringFromJNI 为例，它在 Java 层是没有传入参数的，而在 Native 层的参数至少有两个：第一个是 JNIEnv，即 JNI 的一个结构体，其中保存了大量的函数，如果需要使用 JNI 的函数，就必须在里面进行调用；第二个参数是 jobject，这个是不固定的，如果声明的是 instance method，那么对应的是 jobject。如果声明的是 static method，那么对应的是 jclass。如果调用 instance method，那么必须先有 instance。而 static method 就不需要有 instance，只需要有 jclass 即可。在这两个参数之后的参数就是从 Java 层正常传入的参数了。

Java_com_roysue_easyso1_MainActivity_method01 中间函数体中的内容主要是通过 env 调用 env 的函数在 Java 和 C 之间进行转换。在 C 运算结束之后，再把结果作为参数传入 env 的函数中进行转换，转换为 Java 的 jstring。

为什么需要通过 env 来调用呢？因为所有的函数都在 env 结构体 JNINativeInterface 的定义中，我们可以从这里调用它的上百种方法，如图 9-6 所示。

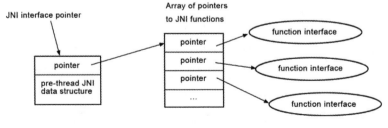

图 9-6 JNI 结构

Java 的参数类型到 Native 层之后会进行怎样的转换呢？举个例子，Java 层中的 String 类型到了 Native 层就变为 jstring。实际上，二者是相等的，只是写法不同。JNI 的原始数据类型及其对应的 Java 类型如图 9-7 所示。

	JNI类型	Java类型	描述
基本数据类型	jboolean	boolean	无符号的char类型
	jbyte	byte	带符号的8位整型
	jchar	char	无符号的16位整型
	jshort	short	带符号的16位整型
	jint	int	带符号的32位整型
	jlong	long	带符号的64位整型
	jfloat	float	32位浮点型
	jdouble	double	64位浮点型
	void	void	无类型
引用类型	jobject	Object	任何Java对象
	jclass	class	Class对象
	jstring	String	字符串对象
	jobjectArray	Object[]	对象数组
	jbooleanArray	boolean[]	布尔类型数组
	jbyteArray	char[]	字符型数组
	jshortArray	short[]	短整型数组
	jintArray	int[]	整型数组
	jlongArray	long[]	长整型数组
	jfloatArray	float[]	浮点型数组
	jdouble	double[]	双浮点型数组
	jthrowable	Throwable	Throwable

图 9-7　JNI 的原始数据类型及其对应的 Java 类型

使用静态注册的话，随便打开一个 SO 文件，都可以直接看到其中的方法体，这样是非常不安全的。用 IDA 导入 SO 文件后，从导出表可以直接看到使用静态注册的函数，如图 9-8 所示。

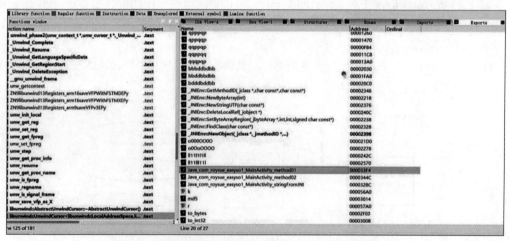

图 9-8　从 IDA 导出表可以直接看到使用静态注册的函数

以上是静态注册的流程。

接下来学习动态注册，改写代码如下：

```cpp
//
// Created by root on 8/22/21
//

#include <jni.h>
#include <string>
#include "tools.h"
#include "aes_utils.h"
#include "junk.h"

#define NELEM(x) ((int) (sizeof(x) / sizeof((x)[0])))

#ifdef __cplusplus
extern "C" {
#endif

JNIEXPORT jstring JNICALL method01(JNIEnv *env, jclass jcls, jstring str_) {
    if (str_ == nullptr) return nullptr;

    const char *str = env->GetStringUTFChars(str_, JNI_FALSE);
    char *result = AES_128_CBC_PKCS5_Encrypt(str);

    env->ReleaseStringUTFChars(str_, str);

    jstring jResult = getJString(env, result);
    free(result);

    return jResult;
}

JNIEXPORT jstring JNICALL method02(JNIEnv *env, jclass jcls, jstring str_) {
    if (str_ == nullptr) return nullptr;

    const char *str = env->GetStringUTFChars(str_, JNI_FALSE);
    char *result = AES_128_CBC_PKCS5_Decrypt(str);

    env->ReleaseStringUTFChars(str_, str);

    jstring jResult = getJString(env, result);
    free(result);

    return jResult;
}

static JNINativeMethod method_table[] = {
    {"method01", "(Ljava/lang/String;)Ljava/lang/String;", (void *)method01},
    {"method02", "(Ljava/lang/String;)Ljava/lang/String;", (void *)method02},
};
```

```cpp
    static int registerMethods(JNIEnv *env, const char *className, JNINativeMethod
*gMethods, int numMethods) {
        jclass clazz = env->FindClass(className);
        if (clazz == nullptr) {
            return JNI_FALSE;
        }
        if (env->RegisterNatives(clazz, gMethods, numMethods) < 0) {
            return JNI_FALSE;
        }
        return JNI_TRUE;
    }

    JNIEXPORT jint JNI_OnLoad(JavaVM *vm, void *reserved) {
        //_JUNK_FUN_0
        _JUNK_FUN_0

        JNIEnv *env = nullptr;

        if (vm->GetEnv((void **) &env, JNI_VERSION_1_6) != JNI_OK) {
            return JNI_ERR;
        }
        assert(env != nullptr);

        // register native method
        if (!registerMethods(env, "com/roysue/easyso1/MainActivity", method_table,
NELEM(method_table))) {
            return JNI_ERR;
        }

        return JNI_VERSION_1_6;
    }

    #ifdef __cplusplus
    }
    #endif
```

上述代码使用了动态注册。在动态注册中，method01 和 method02 是如何运作的呢？从图 9-9 中可以看到，method01 和 method02 的声明已经变红了，并且会报找不到实现的错误。但是，method01 和 method02 却可以调用成功。动态注册的函数的调用名字已经不一样了，不再需要使用约定的命名格式，直接使用方法名就可以。此外，调用 env 中的方法时，可以不使用二级引用 (*env)->GetStringUTFChars(env, str_, JNI_FALSE);，使用一级引用 env->GetStringUTFChars(str_, JNI_FALSE);即可。动态注册代码的逻辑相对于静态代码的逻辑是没有任何变化的。

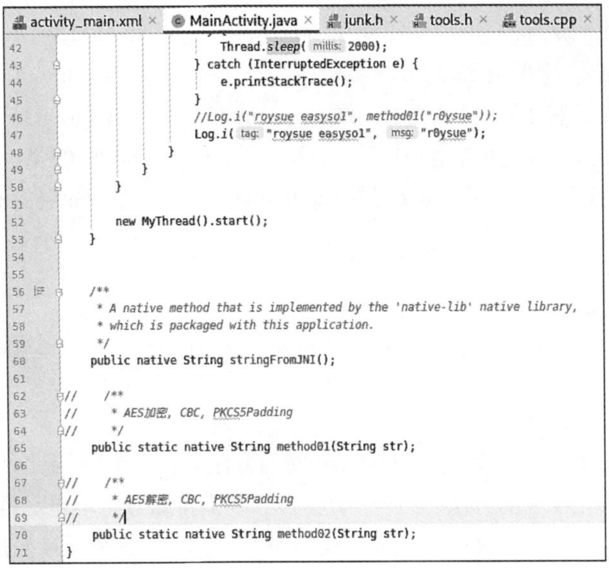

```
🖿 activity_main.xml × | © MainActivity.java × | 🖿 junk.h × | 🖿 tools.h × | 🖿 tools.cpp ×
42                     Thread.sleep( millis: 2000);
43                 } catch (InterruptedException e) {
44                     e.printStackTrace();
45                 }
46                 //Log.i("roysue easyso1", method01("r0ysue"));
47                 Log.i( tag: "roysue easyso1",  msg: "r0ysue");
48             }
49         }
50     }
51
52     new MyThread().start();
53 }
54
55
56 /**
57  * A native method that is implemented by the 'native-lib' native library,
58  * which is packaged with this application.
59  */
60 public native String stringFromJNI();
61
62 //   /**
63 //    * AES加密, CBC, PKCS5Padding
64 //    */
65     public static native String method01(String str);
66
67 //   /**
68 //    * AES解密, CBC, PKCS5Padding
69 //    */
70     public static native String method02(String str);
71 }
```

图 9-9 method01 和 method02 找不到实现

从上述代码中可以了解到如下几点：

（1）SO 文件的特性：在加载完 SO 文件（由 loadLibrary 调用）之后，ELF 首先会运行.init 和.init_array 函数。

（2）JNI 接口的特性：在加载完 SO 文件之后，首先会运行 JNI_Onload 函数。Java 虚拟机会从加载的 SO 文件中寻找 JNI_OnLoad 函数，如果有实现，那么 JNI_Onload 函数会首先运行（自动运行，不需要声明）。如果没有重写该方法，那么系统会自动生成一个。当 SO 文件被卸载时，JNI_OnUnload 会运行。

在 JNI_Onload 函数中，首先得到 env，进而调用注册 Native 方法 registerMethods。registerMethods 关键是要找到 jclass。然后通过传进来的 env 调用接口 RegisterNatives 进行动态注册。RegisterNatives 的第一个参数是 clazz（jclass），第二个参数是 gMethods。gMethods 是一个 method_table，其中对一个个 JNINativeMethod 结构体进行了初始化。在 JNINativeMethod 结构体中，第一个变量是 Java 层的 name，第二个变量是 Java 层的签名，第三个变量是 Native 层的指针，指向方法地址，代码如下：

```
typedef struct {
    const char* name;
    const char* signature;
    void*       fnPtr;
} JNINativeMethod;
```

以上都是固定格式。

这一步成功之后，签名的 method 就与 jclass（此处是 com/roysue/easyso1/MainActivity）进行

了绑定。绑定之后，我们从 Java 层就可以调用 method01 和 method02 方法。Java 层声明的函数名 method01 与 method_table 中的 method01（结构体的第一个变量，即 Java 层的 name）是对应的。method02 同理。我们尝试将 Java 层声明的函数名 method02 改为 method03。method_table 中对应的 method02 改为 method03，以上两者是一致的。此处，method_table 中的第一个变量 method03 与第三个变量 method02 进行了绑定，由 Java 层的 method03 变成了 Native 层的 method02。代码如下：

```
public static native String method03(String str);

static JNINativeMethod method_table[] = {
    {"method01", "(Ljava/lang/String;)Ljava/lang/String;", (void *) method01},
    {"method03", "(Ljava/lang/String;)Ljava/lang/String;", (void *) method02},
};
```

Java 支持重载。也就是说，我们可以定义相同的方法名，带着不同参数的方法。然后 Java 会根据不同的参数寻找对应实现的方法。同样，JNI 必然要支持重载，但是这样的方式到 Native 层要如何实现呢？这就涉及签名的问题。其实 Java 层也有签名，即每一个重载都是一个签名。以 setText 方法为例，如图 9-10 所示。

```
public final void android.widget.TextView.setText(char[],int,int)
public final void android.widget.TextView.setText(int)
public final void android.widget.TextView.setText(int,android.widget.TextView$BufferType)
public final void android.widget.TextView.setText(java.lang.CharSequence)
public final void android.widget.TextView.setTextKeepState(java.lang.CharSequence)
public final void android.widget.TextView.setTextKeepState(java.lang.CharSequence,android.widget.TextView$BufferType)
```

图 9-10　setText 方法重载

每一个重载的方法参数+返回值就是一个签名。根据一个签名信息，我们就可以确定一个类的信息。这个签名的作用就是确定唯一类的方法。

Native 方法签名的组成规则如下：

(参数类型标识 1;参数类型标识 2;...;参数类型标识 n;)返回值类型标识;

这是一个固定格式：(参数)返回值。例如(Ljava/lang/String;)Ljava/lang/String;。Java（基本）类型和类型标识的对应关系如图 9-11 和图 9-12 所示。

类型标识	Java 类型
Z	boolean
B	byte
C	char
S	short
I	int
J	long
F	float
D	double

图 9-11　Java 基本类型和类型标识的对应关系 1

类型标识	Java类型
[签名	数组
[i	int[]
[Ljava/lang/Object	String[]

图 9-12 Java 类型和类型标识的对应关系 2

通过 Java（基本）类型所对应的类型标识构成了 Native 签名，用来确定唯一的方法。再次查看 SO 文件的导出表，我们已经无法从导出表中找到 Java_开头的符号了，如图 9-13 所示。

```
om.roysue.easysol on (Android: 8.1.0) [usb] # memory list exports libroysue.so
ave the output by adding `--json exports.json` to this command
ype      Name                                                          Address
-------  ------------------------------------------------------------  -----------
unction  _ZN7_JNIEnv21ReleaseStringUTFCharsEP8_jstringPKc              0x7664ca3398
unction  qpppqp                                                        0x7664ca4138
unction  qqqpqp                                                        0x7664ca4490
unction  method01                                                      0x7664ca32bc
unction  _ZN7_JNIEnv9NewObjectEP7_jclassP10_jmethodIDz                 0x7664ca6068
unction  qppqqp                                                        0x7664ca46c4
unction  method02                                                      0x7664ca33d8
unction  qqppqp                                                        0x7664ca3c20
unction  _ZN7_JNIEnv12NewStringUTFEPKc                                 0x7664ca6030
unction  qqppqq                                                        0x7664ca3fec
unction  _ZN7_JNIEnv9FindClassEPKc                                     0x7664ca3ba0
unction  JNI_OnLoad                                                    0x7664ca3474
unction  bdddbddbb                                                     0x7664ca5b3c
unction  _ZN7_JNIEnv15RegisterNativesEP7_jclassPK15JNINativeMethodi    0x7664ca3bd8
unction  _ZN7_JNIEnv18SetByteArrayRegionEP11_jbyteArrayiiPKa           0x7664ca5e4c
unction  _ZN7_JNIEnv11GetMethodIDEP7_jclassPKcS3_                      0x7664ca5fe8
unction  _ZN7_JavaVM6GetEnvEPPvi                                       0x7664ca3ad0
unction  _ZN7_JNIEnv17GetStringUTFCharsEP8_jstringPh                   0x7664ca3358
unction  ll11lll11l                                                    0x7664ca6428
unction  ll11ll1lll                                                    0x7664ca61a0
unction  _ZN7_JNIEnv14DeleteLocalRefEP8_jobject                        0x7664ca6168
unction  o0000000                                                      0x7664ca5d6c
unction  o00o0000                                                      0x7664ca5e9c
unction  bbddbbdbb                                                     0x7664ca5910
unction  bbbddbdbb                                                     0x7664ca5a54
unction  _ZN7_JNIEnv12NewByteArrayEi                                   0x7664ca5e14
```

图 9-13 查看 SO 文件的导出表

例如，我们将 Java 层的函数名 method03 改为 decrypt。相应地，method_table 中的 method03（Java 层的 name）也改为 decrypt。运行之后，再次查看 libroysue.so 的导出表，我们发现无法找到 decrypt 方法。最终效果就是，在 Java 层找到了解密函数 decrypt，但是从导出表中无法搜索到 decrypt 函数的实现，只能看到 method02，decrypt 实际上就是 method02。

为了得到动态注册的绑定关系，我们需要逆关键函数 RegisterNatives。由这个方法定位到 gMethods，进而得到 method_table。从 method_table 中，我们可以知道 decrypt 就是 method02。

使用 IDA 反编译 libroysue.so。在导出表中找到 RegisterNatives，单击进入。按 X 键找到调用 RegisterNatives 的地方，如图 9-14 所示。

此处选择第二项进入，直到找到最初调用 RegisterNatives 的地方，如图 9-15 所示。

09

图 9-14　调用 RegisterNatives 的地方

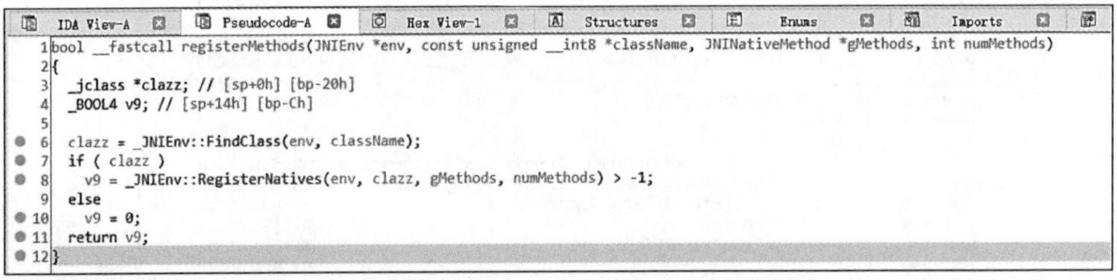

图 9-15　最初调用 RegisterNatives 的地方

假如我们不知道源代码，也可以得知 RegisterNatives 的函数原型：

```
jint RegisterNatives(jclass clazz, const JNINativeMethod* methods, jint nMethods)
```

按 X 键找到调用 registerMethods 的地方，这时我们就可以在调用 registerMethods 的地方看到 method_table 了，如图 9-16 所示。

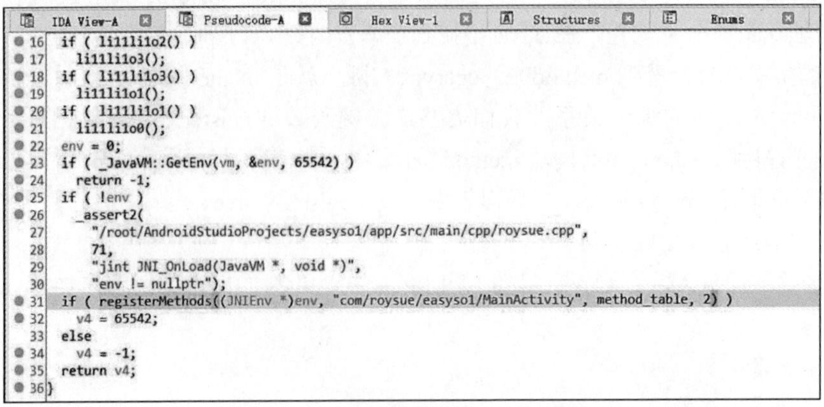

图 9-16　调用 registerMethods 的地方

单击 method_table，如图 9-17 所示。

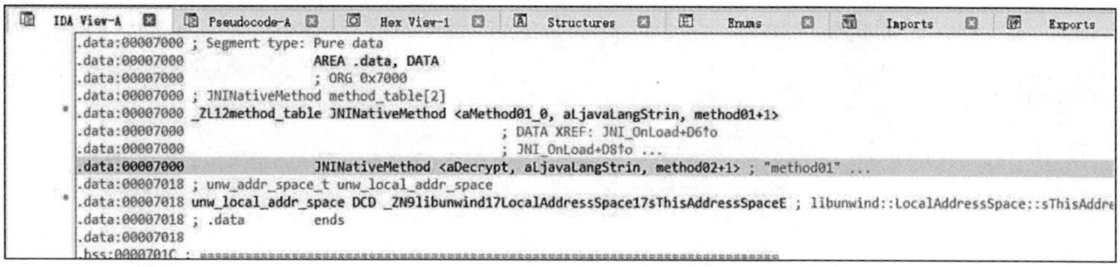

图 9-17　off_8000

我们可以看到 method01 对应的签名(Ljava/lang/String;)Ljava/lang/String;和对应的指针 method01，以及 decrypt 对应的签名(Ljava/lang/String;)Ljava/lang/String;和对应的指针 method02。所以 method02 就是我们要分析的方法，如图 9-18 所示。

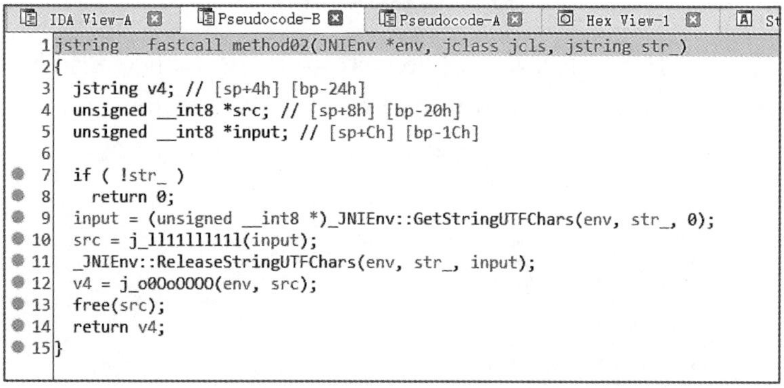

图 9-18　method02 方法体

9.2　Frida 手动追踪 JNI 接口

9.1 节是通过静态分析的方法来发现 decrypt 与 method02 的对应关系。本节将演示动态分析，即 Frida 手动追踪 JNI 接口的方法。我们首先要明确动态注册发生的时机。App 加载后，首先执行的是 MainActivity 的 OnCreate，而 static 早在 MainActivity 触发的时候就已经执行了。动态注册发生得非常早，SO 文件加载的时候首先执行的是.init_array，其次就是 JNI OnLoad。所以，attach 是行不通的。网址 29（参见配书资源文件）提供了已编写好的脚本。

脚本运行效果如图 9-19 所示。

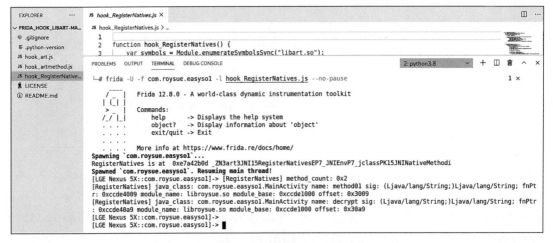

图 9-19 脚本运行效果

我们可以看到，结果打印出了 Java 层的 decrypt，函数主体 Native 层的 libroysue.so 中，偏移是 0x30a9，在 IDA 中按快捷键 G，跳转到 0x30a9 处，如图 9-20 所示。

```
.text:000030A8
.text:000030A8
.text:000030A8
.text:000030A8                 ; jstring method02(JNIEnv *env, jclass jcls, jstring str_)
.text:000030A8                 EXPORT method02
.text:000030A8                 method02
.text:000030A8
.text:000030A8 var_24= -0x24
.text:000030A8 src= -0x20
.text:000030A8 input= -0x1C
.text:000030A8 string= -0x18
.text:000030A8 var_14= -0x14
.text:000030A8 env= -0x10
.text:000030A8 var_C= -0xC
.text:000030A8
.text:000030A8                 PUSH            {R7,LR}
.text:000030AA                 MOV             R7, SP
.text:000030AC                 SUB             SP, SP, #0x20
.text:000030AE                 STR             R0, [SP,#0x28+env]
.text:000030B0                 STR             R1, [SP,#0x28+var_14]
.text:000030B2                 STR             R2, [SP,#0x28+string]
.text:000030B4                 LDR             R0, [SP,#0x28+string]
.text:000030B6                 CMP             R0, #0
.text:000030B8                 BNE             loc_30C2
```

图 9-20 验证 0x30a9 处是 method02

同理，我们也能验证 0x3009 处是 method01，如图 9-21 所示。

```
.text:00003008
.text:00003008
.text:00003008
.text:00003008      ; jstring method01(JNIEnv *env, jclass jcls, jstring str_)
.text:00003008      EXPORT method01
.text:00003008      method01
.text:00003008
.text:00003008      var_24= -0x24
.text:00003008      src= -0x20
.text:00003008      input= -0x1C
.text:00003008      string= -0x18
.text:00003008      var_14= -0x14
.text:00003008      env= -0x10
.text:00003008      var_C= -0xC
.text:00003008
.text:00003008      ; __unwind {
.text:00003008      PUSH              {R7,LR}
.text:0000300A      MOV               R7, SP
.text:0000300C      SUB               SP, SP, #0x20
.text:0000300E      STR               R0, [SP,#0x28+env]
.text:00003010      STR               R1, [SP,#0x28+var_14]
.text:00003012      STR               R2, [SP,#0x28+string]
.text:00003014      LDR               R0, [SP,#0x28+string]
.text:00003016      CMP               R0, #0
.text:00003018      BNE               loc_3022
```

图 9-21　验证 0x3009 处是 method01

也就是说，这个脚本是有效的。接下来，我们来看它是如何实现的。首先，它从 libart.so 中枚举符号并从符号中过滤。过滤后找到真实地址 0xe7a42b0d，并进行 attach。从结构中可以得到 Name Mangling 之后的签名 _ZN3art3JNI15RegisterNativesEP7_JNIEnvP7_jclassPK15JNINativeMethodi。我们可以看到 Java 层的 decrypt，而 Native 层在 libroysue.so 中。

将签名 DeMangling 后可以看到函数原型，说明已经成功 Hook 到了这个函数：

```
root@VXIDr0ysue:~/…/outputs/apk/debug/lib # c++filt
_ZN3art3JNI15RegisterNativesEP7_JNIEnvP7_jclassPK15JNINativeMethodi
   art::JNI::RegisterNatives(_JNIEnv*, _jclass*, JNINativeMethod const*, int)
```

接下来，解析 RegisterNatives 方法。首先逐个取出参数。第一个参数是 env，第二个参数是 jclass。通过调用 Java.vm.tryGetEnv().getClassName(java_class)，传入 jclass，返回 String 类型的 className。类似地，frida-java-bridge 中有很多方法可供使用。第三个参数是 JNINativeMethod，如何解析 JNINativeMethod 是关键。由方法的数量（第四个参数）可知，我们有两个方法需要解析。解析其中的每一个参数，需要通过计算起始地址加上指针的长度才能找到真实数据的地址。

```
typedef struct {
    const char* name;
    const char* signature;
    void*       fnPtr;
} JNINativeMethod;
```

总结一下手动解析的流程：

（1）枚举 ART 的所有符号。

（2）Hook 单个符号。

（3）枚举单个符号的参数并解析。

（4）从参数中找到关键信息。

（5）从内存布局中解析关键信息。

另一个脚本是 hook_art.js，仍然先枚举 ART 的所有符号。从 ART 中找到一些关键的符号，如 GetStringUTFChars、NewStringUTF、FindClass、GetMethodID 等。这也是手动解析的过程。JNI 的函数有上百个，hook_art.js 从中选出 8 个重要的函数进行解析，从而获得了很多关键的内容。以 FindClass 为例，之前在 getJString 中使用的反射可以通过运行这个脚本被追踪到。

JNI 接口不仅是我们自己写的 SO 文件在使用，还包括其他系统的 SO 文件。我们可以从这里获取一些关键的信息。例如，研究加解密算法的主要方法是查看我们传入的数据在哪里被调用。按照一般的逻辑，它会被 GetStringUTFChars 方法调用，如图 9-22 所示。

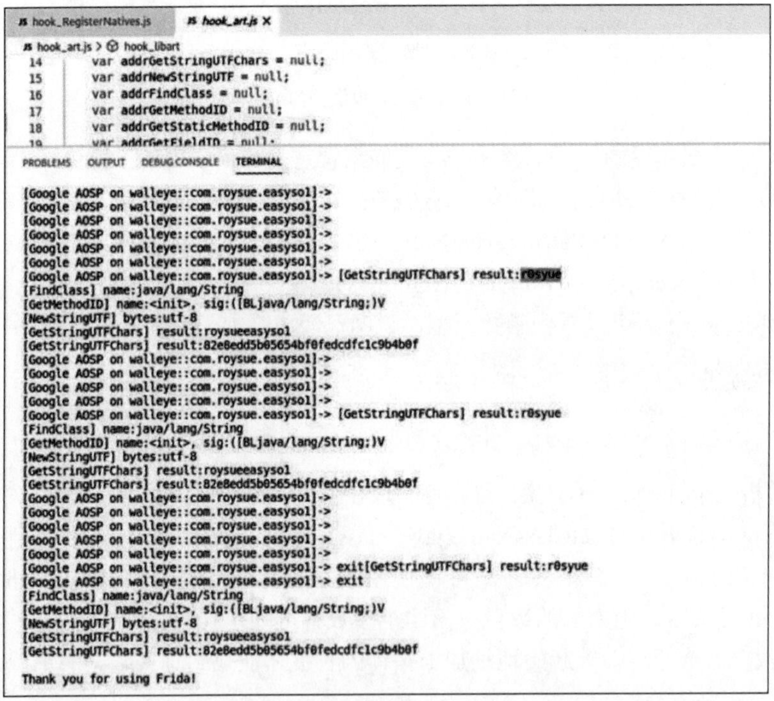

图 9-22　找到 GetStringUTFChars 调用

根据图 9-22，GetStringUTFChars 被调用了。我们可以通过调用栈来看一下是从哪个 SO 文件的哪个位置传过来的。请看以下代码：

```
if (module != null && module.name.indexOf(so_name) == 0) {
    var bytes = Memory.readCString(retval);

    console.log("[GetStringUTFChars] result:" + bytes,
DebugSymbol.fromAddress(this.returnAddress));
```

```
console.log('CCCryptorCreate called from:\n' +
    Thread.backtrace(this.context, Backtracer.ACCURATE)
        .map(DebugSymbol.fromAddress).join('\n') + '\n');
}
```

调用栈打印结果如图 9-23 所示。从调用栈中，我们可以大概追踪到传入的参数是在 method02 处开始执行的。

图 9-23 调用栈打印结果

9.3 jnitrace 自动追踪 JNI

有没有什么方法能够一次性追踪上百个 JNI 函数呢？事实上，借助 jnitrace 工具即可实现这一点。jnitrace 工具可在网址 30 下载（参见配书资源文件）。

如果使用较低版本的 Frida，那么 Frida 和 Objection、jnitrace 版本的对应关系如下：

- Frida: 12.8.0。
- Objection: 1.8.4。
- jnitrace: 3.0.1。

运行命令 pip install jnitrace==v3.0.1 安装 jnitrace。安装成功后，启动 jnitrace：

```
root@VXIDr0ysue:~/Desktop/20210321 # jnitrace -l libroysue.so com.roysue.easyso1
```

效果如图 9-24 所示。

```
┌──(root㉿kali)-[~/Desktop/20210321]
└─# jnitrace -l libroysue.so com.roysue.easysol
Tracing. Press any key to quit...
ERROR: {'type': 'error', 'description': 'Error: access violation accessing 0x17c', 'stack': 'Error: access violation accessin
g 0x17c\n    at NativePointer.readPointer (frida/runtime/core.js:144:17)\n    at JNIEnvInterceptorARM.create (node_modules/jn
itrace-engine/dist/jni/jni_env_interceptor.js:47:13)\n    at InvocationContext.onEnter (node_modules/jnitrace-engine/dist/eng
ine.js:87:21)', 'fileName': 'frida/runtime/core.js', 'lineNumber': 144, 'columnNumber': 17}
Traced library "libroysue.so" loaded from path "/data/app/com.roysue.easysol-LGgZ9ahXr3rsnXF3B-e9vg==/lib/arm".

           /* TID 20423 */
    452 ms [+] JNIEnv->GetEnv
    452 ms |- JavaVM*            : 0xe7c19140
    452 ms |- void**             : 0xfffa32f0
    452 ms |- jint               : 65542
    452 ms |= jint               : 0

    452 ms -----------------------------Backtrace-----------------------------
    452 ms |-> 0xcc95d519: _ZN7_JavaVM6GetEnvEPPvi+0x20 (libroysue.so:0xcc95a000)

           /* TID 20423 */
    643 ms [+] JNIEnv->FindClass
    643 ms |- JNIEnv*            : 0xe7cb12a0
    643 ms |- char*              : 0xcc95f357
    643 ms |:     com/roysue/easysol/MainActivity
    643 ms |= jclass             : 0x85    { com/roysue/easysol/MainActivity }

    643 ms -----------------------------Backtrace-----------------------------
    643 ms |-> 0xcc95c36b: _ZN7_JNIEnv9FindClassEPKc+0x1a (libroysue.so:0xcc95a000)
```

图 9-24　jnitrace 的用法

可以看到这个流程与我们的源码一模一样，依次是 JNIEnv→GetEnv、JNIEnv→FindClass、JNIEnv→RegisterNatives。

最后讲一下，如何实现动态注册的追踪。

一般情况下，程序调用 env->RegisterNatives 来实现动态注册，但是有的应用为了避免逆向研究人员 Hook RegisterNatives，参数不传给 env 的 RegisterNatives，直接传给它的实现，即 ArtMethod 的 RegisterNatives。我们可以自己解析，最终把参数传给 ArtMethod 的 RegisterNatives 进行注册，如图 9-25 所示。

```
Home | History | Annotate | Line# | Navigate | Raw | Download [Search] □ only in /art/runtime/
const void* ArtMethod::RegisterNative(const void* native_method, bool is_fast) {
  CHECK(IsNative()) << PrettyMethod();
  CHECK(!IsFastNative()) << PrettyMethod();
  CHECK(native_method != nullptr) << PrettyMethod();
  if (is_fast) {
    AddAccessFlags(kAccFastNative);
  }
  void* new_native_method = nullptr;
  Runtime::Current()->GetRuntimeCallbacks()->RegisterNativeMethod(this,
                                                                  native_method,
                                                                  /*out*/&new_native_method);
  SetEntryPointFromJni(new_native_method);
  return new_native_method;
}
```

图 9-25　ArtMethod 的 RegisterNatives

9.4　JNI 接口大横评

9.3 节主要介绍了 JNI 的基础知识，包括静态注册和动态注册的方法，以及 Frida 手动追踪 JNI 接口。接下来，我们将学习 jnitrace、Frida、ExAndroidNativeEmu、Unidbg 等工具是如何使用 JNI 的。这些工具的功能总结如图 9-26 所示。

	Frida	jnitrace	ExAndroidNativeEmu	Unidbg
可以跑完整的APK吗？	○	○	×	×
可以Hook Java吗？查看参调返	○	×	×	×
可以Hook JNI接口函数吗？查看参调返	○	○	○	○
可以主动调用JNI接口函数吗？获取返回值	○	×	○	○
如何找到静态绑定的Native函数？	dlopen/dlsym	dlsym	Module	Module
如何找到动态绑定的Native函数？	hookRN	hookRN	hookRN	hookRN
如何Hook(动)静态注册的Native函数？	○	○	×	○
如何主动调用(动)静态注册的Native函数？	○	×	○	○

图 9-26　JNI 接口大横评

很多地方混淆了 Java 层的 Native 函数和 JNI 接口函数，我们需要对它区分一下，否则无法准确地评价这些工具分别实现了哪些功能。

1. Java 层的 Native 函数

以 easyso1 项目为例，在 MainActivity.java 中，public static native String method01(String str)本质上是 Java 函数，它只是通过 JNI 的接口在 Native 层实现。

2. JNI 接口函数

如 RegisterNatives、GetStringUTFChars，从 env 或 VM 中得到的函数被定义为 JNI 接口函数。

3. JNI 静态（注册）函数和 JNI 动态（注册）函数

我们称静态注册的函数为 JNI 静态（注册）函数，类似地，动态注册的函数称为 JNI 动态（注册）函数。为了区分两者，我们将动态注册的函数 method01 修改为静态注册。注意，静态注册的函数名称是高度规范化的，所以需要将 method01 函数名称改为 Java_com_roysue_easyso1_MainActivity_method01，再注释掉 method_table 中的 method01。此外，动态注册的函数可以不放在 extern "C" 中。JNI 动态（注册）函数 method02 与 Java 层的 decrypt 是相互对应的，在 Java 层显示的是 decrypt，而在 Native 层绑定的是 method02。经过 C++编译之后，method02 是看不到的。

4. 纯 Native 函数

本质上跟 Java 层是没有关系的，如 method02。它在行为上与 C++的函数没有区别。纯 Native 函数也可以进行主动调用。

9.4.1　Frida Hook 并主动调用

接下来，我们用 Frida 分别对前面介绍的 4 类函数进行 Hook 和 Invoke 操作。首先对 Java 层的 Native 函数进行 Hook 并主动调用。静态函数可以直接主动调用，动态函数需要一个 object。编写脚本如下：

```javascript
function hookJava(){
    Java.perform(function(){
        Java.use("com.roysue.easyso1.MainActivity").method01.implementation = function(str){
            var result = this.method01(str)
            console.log("str, result is => ", str, result)
            return result;
        }
    })
}

function invokeJava(){
    Java.perform(function(){
        var result = Java.use("com.roysue.easyso1.MainActivity").decrypt("4e8de2f3c674d8157b4862e50954d81c")
        console.log("result is => ", result)
    })

}

function main(){
    hookJava()
}

setImmediate(main)
```

在手机上启动 frida-server-12.8.0 并运行脚本：

```
root@VXIDr0ysue:~/Desktop/20210324 # frida -UF -l h2i.js
```

脚本运行结果如图 9-27 所示。

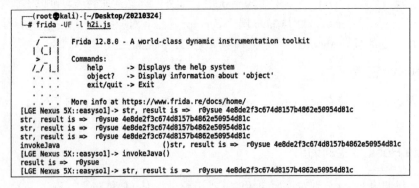

图 9-27　Frida Hook 并主动调用结果

decrypt 函数只在 App 加载时运行了一次，如何把它 Hook 上并打印（参数调用栈的）返回值呢？这在 Java 层很简单，只要将 hookJava()略作修改即可：

```
function hookJava(){
    Java.perform(function(){
        Java.use("com.roysue.easyso1.MainActivity").decrypt.implementation = function(str){
            var result = this.decrypt(str)
            console.log("str, result is => ", str, result)
            return result;
        }
    })
}
```

运行脚本：

```
root@VXIDr0ysue:~/Desktop/20210324 # frida -U -f com.roysue.easyso1 -l h2i.js --no-pause

[LGE Nexus 5X::com.roysue.easyso1]-> str, result is =>
4e8de2f3c674d8157b4862e50954d81c r0ysue
```

Frida 在 Java 层的 Hook 和主动调用没有问题。我们再来看 JNI 接口函数。前面演示了 Hook，接下来尝试主动调用。

主动调用的难点在于如何在 App 启动时 Hook Native 层动态注册的 decrypt。我们在之前的章节中讲过 SO 文件加载和类的绑定流程中的关键函数 dlopen、dlclose、dlsym。

接下来 Hook Native 层的 method02。它本质上是一个动态注册的 Native 函数。为了获取 method02 的地址，可以先 Hook RegisterNatives 函数。如果一开始想要 Hook 静态注册的地址，只需要 Hook dlopen 即可。可以参考网址 31（参见配书资源文件）修改我们的脚本。将原先过滤的 libart.so 改为 libroysue.so，并将出口 onLeave 中的 dump_dex();去掉。在 Android 高版本中，要同时 Hook android_dlopen_ext。

```
var is_hook_libart = false;

function hook_dlopen() {
    Interceptor.attach(Module.findExportByName(null, "dlopen"), {
        onEnter: function(args) {
            var pathptr = args[0];
            if (pathptr !== undefined && pathptr != null) {
                var path = ptr(pathptr).readCString();
                //console.log("dlopen:", path);
                if (path.indexOf("libroysue.so") >= 0) {
                    this.can_hook_libart = true;
                    console.log("[dlopen:]", path);
                }
```

```
            }
        },
        onLeave: function(retval) {
            if (this.can_hook_libart && !is_hook_libart) {
                is_hook_libart = true;
            }
        }
    })

    Interceptor.attach(Module.findExportByName(null, "android_dlopen_ext"), {
        onEnter: function(args) {
            var pathptr = args[0];
            if (pathptr !== undefined && pathptr != null) {
                var path = ptr(pathptr).readCString();
                //console.log("android_dlopen_ext:", path);
                if (path.indexOf("libroysue.so") >= 0) {
                    this.can_hook_libart = true;
                    console.log("[android_dlopen_ext:]", path);
                }
            }
        },
        onLeave: function(retval) {
            if (this.can_hook_libart && !is_hook_libart) {
                is_hook_libart = true;
            }
        }
    });
}
```

运行脚本：

```
root@VXIDr0ysue:~/Desktop/20210324 # frida -U -f com.roysue.easyso1 -l h2i.js
--no-pause

Spawned com.roysue.easyso1. Resuming main thread!
    [LGE Nexus 5X::com.roysue.easyso1]-> [android_dlopen_ext:]
/data/app/com.roysue.easyso1-pcyog1Qt-2FeNY5XgmmfNA==/lib/arm/libroysue.so
```

去掉上述脚本中的两行注释，再次运行脚本。可以看到 dlopen 打开了很多库，包括我们的 SO 库，如图 9-28 所示。

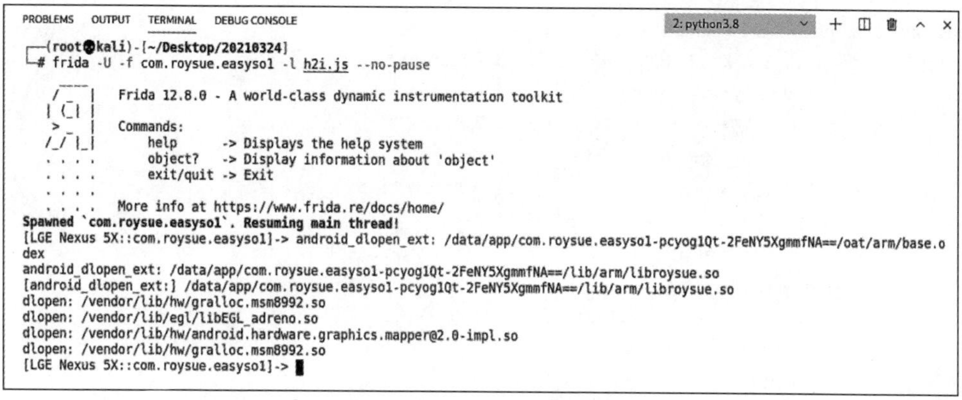

第 9 章　JNI 接口初识

```
PROBLEMS    OUTPUT    TERMINAL    DEBUG CONSOLE                                    2: python3.8    +  □  □  ^  ×
┌──(root㉿kali)-[~/Desktop/20210324]
└─# frida -U -f com.roysue.easyso1 -l h2i.js --no-pause
     ____
    / _  |   Frida 12.8.0 - A world-class dynamic instrumentation toolkit
   | (_| |
    > _  |   Commands:
   /_/ |_|       help      -> Displays the help system
   . . . .       object?   -> Display information about 'object'
   . . . .       exit/quit -> Exit
   . . . .
   . . . .   More info at https://www.frida.re/docs/home/
Spawned `com.roysue.easyso1`. Resuming main thread!
[LGE Nexus 5X::com.roysue.easyso1]-> android_dlopen_ext: /data/app/com.roysue.easyso1-pcyog1Qt-2FeNY5XgmmfNA==/oat/arm/base.o
dex
android_dlopen_ext: /data/app/com.roysue.easyso1-pcyog1Qt-2FeNY5XgmmfNA==/lib/arm/libroysue.so
[android_dlopen_ext]: /data/app/com.roysue.easyso1-pcyog1Qt-2FeNY5XgmmfNA==/lib/arm/libroysue.so
dlopen: /vendor/lib/hw/gralloc.msm8992.so
dlopen: /vendor/lib/egl/libEGL_adreno.so
dlopen: /vendor/lib/hw/android.hardware.graphics.mapper@2.0-impl.so
dlopen: /vendor/lib/hw/gralloc.msm8992.so
[LGE Nexus 5X::com.roysue.easyso1]->
```

图 9-28　dlopen 打开的 SO 库

因此，我们可以在 dlopen 加载一个 SO 文件结束之后对它进行 Hook。无论是动态注册还是静态注册，我们都可以先查看它的函数名是什么。还是用 objection 来查看：

```
root@VXIDr0ysue:~/Desktop/20210324 # objection -g com.roysue.easyso1 explore
com.roysue.easyso1 on (google: 8.1.0) [usb] # memory list exports libroysue.so
```

method01 的符号和方法名是一样的，method02 Name Mangling 之后的方法签名如图 9-29 所示。

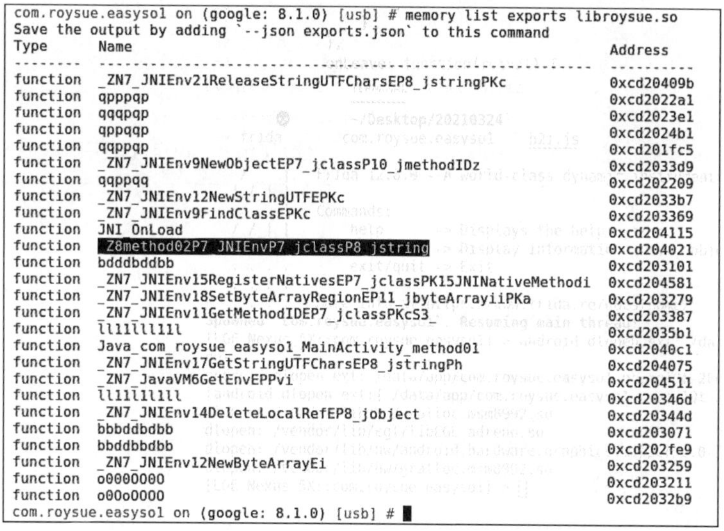

```
com.roysue.easyso1 on (google: 8.1.0) [usb] # memory list exports libroysue.so
Save the output by adding `--json exports.json` to this command
Type       Name                                                              Address
--------   --------------------------------------------------------------    ----------
function   _ZN7_JNIEnv21ReleaseStringUTFCharsEP8_jstringPKc                  0xcd20409b
function   qpppqp                                                            0xcd2022a1
function   qqqpqp                                                            0xcd2023e1
function   qppqqp                                                            0xcd2024b1
function   qqpqpp                                                            0xcd201fc5
function   _ZN7_JNIEnv9NewObjectEP7_jclassP10_jmethodIDz                     0xcd2033d9
function   qqppqq                                                            0xcd202209
function   _ZN7_JNIEnv12NewStringUTFEPKc                                     0xcd2033b7
function   _ZN7_JNIEnv9FindClassEPKc                                         0xcd203369
function   JNI_OnLoad                                                        0xcd204115
function   _Z8method02P7_JNIEnvP7_jclassP8_jstring                           0xcd204021
function   bdddbdddbb                                                        0xcd203101
function   _ZN7_JNIEnv15RegisterNativesEP7_jclassPK15JNINativeMethodi        0xcd204581
function   _ZN7_JNIEnv18SetByteArrayRegionEPll_jbyteArrayiiPKa               0xcd203279
function   _ZN7_JNIEnv11GetMethodIDEP7_jclassPKcS3_                          0xcd203387
function   lllllllll                                                         0xcd2035b1
function   Java_com_roysue_easyso1_MainActivity_method01                     0xcd2040c1
function   _ZN7_JNIEnv17GetStringUTFCharsEP8_jstringPh                       0xcd204075
function   _ZN7_JavaVM6GetEnvEPPvi                                           0xcd204511
function   llllllllll                                                        0xcd20346d
function   _ZN7_JNIEnv14DeleteLocalRefEP8_jobject                            0xcd20344d
function   bbbddbddb                                                         0xcd203071
function   bbddbbddb                                                         0xcd202fe9
function   _ZN7_JNIEnv12NewByteArrayEi                                       0xcd203259
function   o000o000                                                          0xcd203211
function   o0oo0000                                                          0xcd2032b9
com.roysue.easyso1 on (google: 8.1.0) [usb] #
```

图 9-29　method02 Name Mangling 之后的方法签名

通过 c++filt 命令查看 method02 的原型：

```
root@VXIDr0ysue:~/Desktop/20210324 # c++filt _Z8method02P7_JNIEnvP7_jclassP8_jstring
method02(_JNIEnv*, _jclass*, _jstring*)
```

Hook 静态注册的 method01 与动态注册的 method02：

```
        onLeave: function(retval) {
            if (this.can_hook_libart && !is_hook_libart) {
                is_hook_libart = true;
                var method01 = Module.findExportByName("libroysue.so",
"Java_com_roysue_easyso1_MainActivity_method01")
                var method02 = Module.findExportByName("libroysue.so",
"_Z8method02P7_JNIEnvP7_jclassP8_jstring")
                console.log("method01 address is =>", method01)
                console.log("method02 address is =>", method02)
                Interceptor.attach(method01, {
                    onEnter:function(args){
                        console.log("arg[2=>", args[2])
                    },onLeave:function(retval){
                        console.log("retval=>", retval)
                    }
                })
                Interceptor.attach(method02, {
                    onEnter:function(args){
                        console.log("method02 arg[2=>", args[2])
                    },onLeave:function(retval){
                        console.log("method02 retval=>", retval)
                    }
                })
            }
        }
```

运行脚本，Hook 静态注册的 method01 与动态注册的 method02 的结果如图 9-30 所示。

```
┌──(root㉿kali)-[~/Desktop/20210324]
└─# frida -U -f com.roysue.easyso1 -l h2i.js --no-pause
     / _  |   Frida 12.8.0 - A world-class dynamic instrumentation toolkit
    | (_| |
     > _  |   Commands:
    /_/ |_|       help      -> Displays the help system
    . . . .       object?   -> Display information about 'object'
    . . . .       exit/quit -> Exit
    . . . .
    . . . .   More info at https://www.frida.re/docs/home/
Spawned `com.roysue.easyso1`. Resuming main thread!
[LGE Nexus 5X::com.roysue.easyso1]-> android_dlopen_ext: /data/app/com.roysue.easyso1-pcyog1Qt-2FeNY5XgmmfNA==/oat/arm/base.o
dex
android_dlopen_ext: /data/app/com.roysue.easyso1-pcyog1Qt-2FeNY5XgmmfNA==/lib/arm/libroysue.so
[android_dlopen_ext:] /data/app/com.roysue.easyso1-pcyog1Qt-2FeNY5XgmmfNA==/lib/arm/libroysue.so
method01 address is => 0xcd2060c1
method02 address is => 0xcd206021
method02 arg[2=> 0xfffa3b70
method02 retval=> 0xa1
arg[2=> 0xcd17f630
retval=> 0x45
arg[2=> 0xcd17f630
retval=> 0x49
arg[2=> 0xcd17f630
retval=> 0x41
```

图 9-30 打印 Hook 静态注册的 method01 与动态注册的 method02 的结果

然后主动调用 method02 的接口，把 getStringUtfChars 读出来。

```
Interceptor.attach(method02, {
```

```
        onEnter:function(args){
            console.log("method02 arg[2=>", Java.vm.getEnv().getStringUtfChars(args[2],
null).readCString())
        },onLeave:function(retval){
            console.log("method02 retval=>", Java.vm.getEnv().getStringUtfChars(retval,
null).readCString())
        }
    })
```

getStringUtfChars 并不是从 JNI 中获得的,而是从 proxy 中获得的。所以这其实是对 env 的主动调用。我们可以看一下效果,Java String 和 return value 都可以打印出来,而且是单次 Hook 的值,如图 9-31 所示。

```
┌──(root㉿kali)-[~/Desktop/20210324]
└─# frida -U -f com.roysue.easysol -l h2i.js --no-pause
     ____
    / _  |   Frida 12.8.0 - A world-class dynamic instrumentation toolkit
   | (_| |
    > _  |   Commands:
   /_/ |_|       help      -> Displays the help system
                 object?   -> Display information about 'object'
                 exit/quit -> Exit
    . . . .
    . . . .   More info at https://www.frida.re/docs/home/
Spawned `com.roysue.easysol`. Resuming main thread!
[LGE Nexus 5X::com.roysue.easysol]-> android_dlopen_ext: /data/app/com.roysue.easysol-pcyog1Qt-2FeNY5XgmmfNA==/oat/arm/base.o
dex
android_dlopen_ext: /data/app/com.roysue.easysol-pcyog1Qt-2FeNY5XgmmfNA==/lib/arm/libroysue.so
[android_dlopen_ext:] /data/app/com.roysue.easysol-pcyog1Qt-2FeNY5XgmmfNA==/lib/arm/libroysue.so
method01 address is => 0xcd2230c1
method02 address is => 0xcd223021
method02 arg[2=> 4e8de2f3c674d8157b4862e50954d81c
method02 retval=> r0ysue
arg[2=> 0xcd0fe630
retval=> 0x45
```

图 9-31 对 env 的主动调用

最后,如何对 method01 和 method02 的纯 Native 函数进行主动调用?在源代码中,我们可以把 method02 理解为纯 Native 函数。虽然它被绑定到了 Java 层中,但它本身其实是一个纯 Native 函数。

前一节提到的.init 和.init_array 是 SO 文件加载流程中 linker 解析时一定会运行的节区的位置,它们所对应的是 dlopen 和 dlsym。运行顺序是.init 和.init_array、JNI_Onload、其他 SO 文件中的函数。所以说,我们刚刚 Hook 了 dlopen,并从 dlopen 的 onLeave 中 Hook SO 文件里面的静态注册和动态注册函数。其实在 JNI_Onload 中,RegisterNatives 也是一个比较好的时机点。我们可以等 RegisterNatives 结束之后进行赋值,因为这个时候 SO 文件里面已经初始化结束了,Hook 是没有问题的。

```
    var ENV = null;
    var JCLZ = null;

    var method01addr = null;
    var method02addr = null;
    var method02 = null;
```

```
    var addrNewStringUTF = null;
    var NewStringUTF = null;

    function hook_RegisterNatives() {
        var symbols = Module.enumerateSymbolsSync("libart.so");
        var addrRegisterNatives = null;
        for (var i = 0; i < symbols.length; i++) {
            var symbol = symbols[i];

            //_ZN3art3JNI15RegisterNativesEP7_JNIEnvP7_jclassPK15JNINativeMethodi
            if (symbol.name.indexOf("art") >= 0 &&
                    symbol.name.indexOf("JNI") >= 0 &&
                    symbol.name.indexOf("NewStringUTF") >= 0 &&
                    symbol.name.indexOf("CheckJNI") < 0) {
                addrNewStringUTF = symbol.address;
                console.log("NewStringUTF is at ", symbol.address, symbol.name);
                NewStringUTF = new NativeFunction(addrNewStringUTF, 'pointer', ['pointer', 'pointer'])
            }

            if (symbol.name.indexOf("art") >= 0 &&
                    symbol.name.indexOf("JNI") >= 0 &&
                    symbol.name.indexOf("RegisterNatives") >= 0 &&
                    symbol.name.indexOf("CheckJNI") < 0) {
                addrRegisterNatives = symbol.address;
                console.log("RegisterNatives is at ", symbol.address, symbol.name);
            }
        }

        if (addrRegisterNatives != null) {
            Interceptor.attach(addrRegisterNatives, {
                onEnter: function (args) {
                    console.log("[RegisterNatives] method_count:", args[3]);
                    var env = args[0];
                    ENV = args[0];
                    var java_class = args[1];
                    JCLZ = args[1];
                    var class_name = Java.vm.tryGetEnv().getClassName(java_class);
                    //console.log(class_name);

                    var methods_ptr = ptr(args[2]);
                    var method_count = parseInt(args[3]);
                    for (var i = 0; i < method_count; i++) {
                        var name_ptr = Memory.readPointer(methods_ptr.add(i * Process.pointerSize * 3));
                        var sig_ptr = Memory.readPointer(methods_ptr.add(i *
```

```
Process.pointerSize * 3 + Process.pointerSize));
                var fnPtr_ptr = Memory.readPointer(methods_ptr.add(i *
Process.pointerSize * 3 + Process.pointerSize * 2));

                var name = Memory.readCString(name_ptr);
                var sig = Memory.readCString(sig_ptr);
                var find_module = Process.findModuleByAddress(fnPtr_ptr);

                console.log("[RegisterNatives] java_class:", class_name, "name:",
name, "sig:", sig, "fnPtr:", fnPtr_ptr, "module_name:", find_module.name, "module_base:",
find_module.base, "offset:", ptr(fnPtr_ptr).sub(find_module.base));

                if(name.indexOf("method01")>=0){
                    continue;
                }else if(name.indexOf("decrypt")>=0){
                    method02addr = fnPtr_ptr;
                    method02 = new NativeFunction(method02addr, 'pointer', ['pointer',
'pointer', 'pointer'])

                    method01addr = Module.findExportByName("libroysue.so",
"Java_com_roysue_easyso1_MainActivity_method01")
                }
            }
        }
    });
    }
}

function invokemethod01(contents){
    console.log("method01 address is =>", method01addr)
    var method01 = new NativeFunction(method01addr, 'pointer', ['pointer', 'pointer',
'pointer'])
    var NewStringUTF = new NativeFunction(addrNewStringUTF, 'pointer', ['pointer',
'pointer'])
    var result = null;
    Java.perform(function(){

        console.log("Java.vm.getEnv()", Java.vm.getEnv())

        var JSTRING = NewStringUTF(Java.vm.getEnv(), Memory.allocUtf8String(contents))

        result = method01(Java.vm.getEnv(), JSTRING, JSTRING)

        console.log("result is =>", result)

        console.log("result is ", Java.vm.getEnv().getStringUtfChars(result,
```

```
null).readCString()
            result = Java.vm.getEnv().getStringUtfChars(result, null).readCString()
        })
        return result;
    }

    function invokemethod02(contents){
        var result = null;

        Java.perform(function(){

            var JSTRING = NewStringUTF(Java.vm.getEnv(), Memory.allocUtf8String(contents))

            result = method02(Java.vm.getEnv(), JSTRING, JSTRING)

            result = Java.vm.getEnv().getStringUtfChars(result, null).readCString()
        })
        return result;
    }

    rpc.exports = {
        invoke1:invokemethod01,
        invoke2:invokemethod02
    };

    setImmediate(hook_RegisterNatives);
```

创建 Native 层的字符串有两种方法，一种是通过 Java.vm.getEnv().newStringUtf()，另一种是从 libart.so 中解析出 NewStringUTF，将它定义为一个函数，并主动调用它。运行脚本可以获取到 NewStringUTF、RegisterNatives 以及 method02。运行结果如图 9-32 所示。

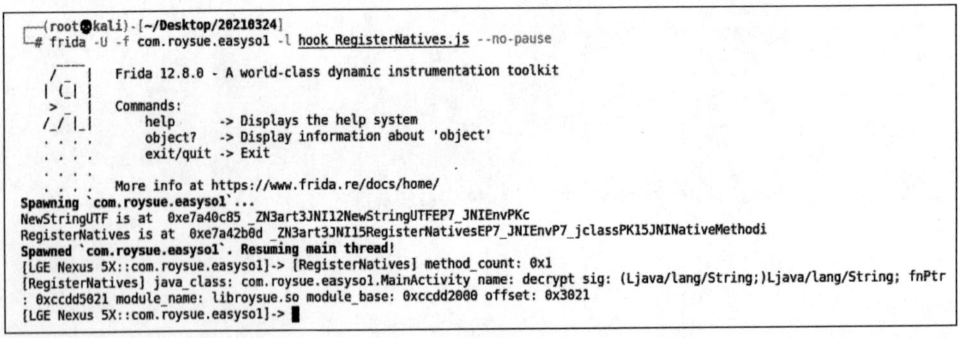

图 9-32　运行 hook_RegisterNatives

主动调用 method01 和 method02，结果如图 9-33 所示。

```
[LGE Nexus 5X::com.roysue.easyso1]-> invokemethod01("r0ysue123")
method01 address is => 0xcd5f90c1
Java.vm.getEnv() [object Object]
result is => 0x55
result is   7251aeae2f6f94de5df58132aa1d2d4b
"7251aeae2f6f94de5df58132aa1d2d4b"
[LGE Nexus 5X::com.roysue.easyso1]-> invokemethod02("7251aeae2f6f94de5df58132aa1d2d4b")
"r0ysue123"
[LGE Nexus 5X::com.roysue.easyso1]->
```

图 9-33　主动调用 method01 和 method02

以上就是纯 Native 的主动调用。

总结一下，Frida 可以做到 Hook 和主动调用 Java 层 Native 实现的函数、env 的函数、JNI 静态函数、JNI 动态函数以及纯 Native 函数。所以说，Frida 是最优秀的工具，其他工具与之相比就相形见绌了。

9.4.2　jnitrace

接下来继续测评 jnitrace 工具。前面讲到，jnitrace 可以把 JNI 的所有函数全部 Hook 出来，但是不能 Hook Java。而且，它是一个自动化的工具，所以不包含主动调用。

jnitrace 找到静态绑定的 Native 函数的原理如下：

首先，既然它主要操作的对象是 FindClass、RegisterNatives 等几百个 JNI 接口函数，那么它是如何 Hook 上这些函数的呢？从源码中，我们可以了解到，只要导入 jnitrace-engine，就可以很轻易地 Hook JNI 函数。也就是说，在 jnitrace 工具中，核心就是 jnitrace-engine。它是如何将这些 JNI 函数解析出来的呢？比如说，我们手动解析 RegisterNatives 的时候，先 Hook libart.so，枚举它所有的符号并解析。然后使用官网提供的 Hook 地址的方法：Interceptor.attach。所以，jnitrace 实际上也是使用 Interceptor.attach。在 jnitrace 的源码 main.ts 中，我们可以找到很多 JNIInterceptor.attach 加方法名，从符号到地址关键就在于这个 JNIInterceptor。根据 JNIInterceptor，我们一步一步了解到，在 jnitrace-engine 的 engine.ts 中进行了 Interceptor.attach，并在 dlsym 中拿到了所有符号，解析出了地址，Hook 了 RegisterNatives，但不能主动调用（动）静态注册的 Native 函数。

9.4.3　ExAndroidNativeEmu

将 easyso1 项目中的 SO 文件移植到 ExAndroidNativeEmu 的 tests/bin 目录下，重命名为 libroysue2.so。相应地，修改 emulator.load_library：

```
lib_module = emulator.load_library("tests/bin/libroysue2.so")
```

method01 不变，method02 符号发生变化：

```
logger.info("Response from JNI call: %s" % emulator.call_symbol(lib_module,
"_Z8method02P7_JNIEnvP7_jclassP8_jstring", emulator.java_vm.jni_env.address_ptr, 0x00,
String('47fcda3822cd10a8e2f667fa49da783f')))
```

ExAndroidNativeEmu 涉及不到 APK 和 Java，可以直接运行：

```
root@VXIDr0ysue:~/Desktop/20210117/ExAndroidNativeEmu # 
/root/.pyenv/versions/3.7.0/bin/python 
/root/Desktop/20210117/ExAndroidNativeEmu/example_jni.py

4847 - 2021-08-30 22:20:01,354 - DEBUG - JNIEnv->DeleteLocalRef(4) was called
4847 - 2021-08-30 22:20:01,355 - INFO - Response from JNI call: JavaString(roysue)
4847 - 2021-08-30 22:20:01,355 - INFO - Exited EMU.
```

注意，我们并没有调用 JNI_OnLoad，更不用谈调用 RegisterNatives，但依然可以调用 JNI 动态注册的函数。这是因为动态注册的目的是将 Native 层的函数与 Java 层的函数绑定起来。实际上，它并不涉及 Java 层，没有在 JNI_Onload 中绑定 RegisterNatives，更不会影响调用 Native 函数。

如果我们取消 call_symbol 的注释，如图 9-34 所示。

```
 93    try:
 94        # Run JNI_OnLoad.
 95        #   JNI_OnLoad will call 'RegisterNatives'.
 96        emulator.call_symbol(lib_module, 'JNI_OnLoad', emulator.java_vm.address_ptr, 0x00)
 97
 98        # Do native stuff.
 99        # main_activity = MainActivity()
100        # logger.info("Response from JNI call: %s" % main_activity.string_from_jni(emulator))
101        logger.info("Response from JNI call: %s" % emulator.call_symbol(lib_module,"Java_com_roysue_easyso1_MainAct
102        logger.info("Response from JNI call: %s" % emulator.call_symbol(lib_module,"_Z8method02P7_JNIEnvP7_jclassP8
103
```

图 9-34　打开 call_symbol

则会运行报错：

```
root@VXIDr0ysue:~/Desktop/20210117/ExAndroidNativeEmu # /root/.pyenv/versions/3.7.0/bin/python /root/Desktop/20210117/ExAndroidNativeEmu/example_jni.py

Could not find class 'com/roysue/easyso1/MainActivity' for JNIEnv.
```

因为运行 JNIEnv 的时候就会调用 registerMethods 函数，进而找 Java 层的 class，返回一个 jclass。显然，这时找不到 Java 层的 class。因此，我们需要构造一个 Java 层的 class 给它。案例中 example_jni.py 提供了构造方法，根据我们的项目稍作修改：

```
class MainActivity(metaclass=JavaClassDef,
jvm_name='com/roysue/easyso1/MainActivity'):

    def __init__(self):
        pass

    @java_method_def(name='decrypt',
signature='(Ljava/lang/String;)Ljava/lang/String;', native=True)
    def string_from_jni(self, mu):
        pass

    def test(self):
        pass
```

再运行就不会报错了。也就是调用了 RegisterNatives,打印出的日志如图 9-35 所示。它本质上什么也没有做,而是直接把参数传给 ArtMethod 的 RegisterNatives 进行解析。

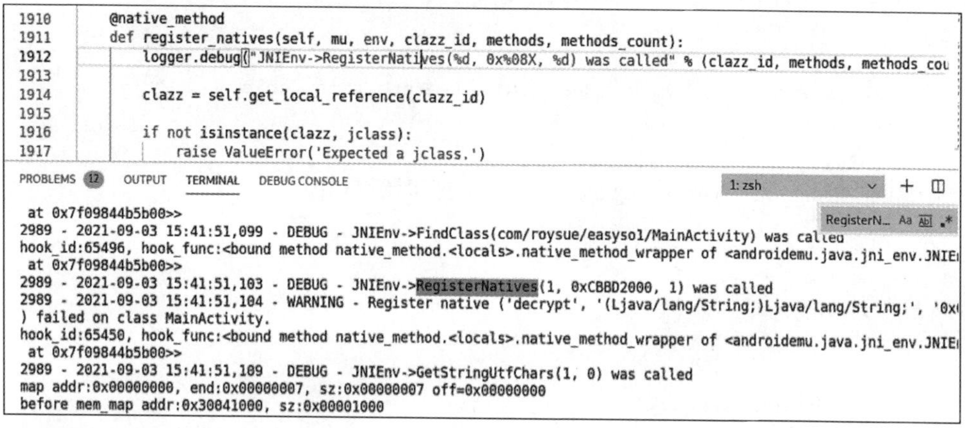

```
1910        @native_method
1911        def register_natives(self, mu, env, clazz_id, methods, methods_count):
1912            logger.debug("JNIEnv->RegisterNatives(%d, 0x%08X, %d) was called" % (clazz_id, methods, methods_cou
1913
1914            clazz = self.get_local_reference(clazz_id)
1915
1916            if not isinstance(clazz, jclass):
1917                raise ValueError('Expected a jclass.')
```

```
PROBLEMS 12    OUTPUT    TERMINAL    DEBUG CONSOLE                                                        1: zsh
at 0x7f09844b5b00>>                                                                              RegisterN... Aa Abl .*
2989 - 2021-09-03 15:41:51,099 - DEBUG - JNIEnv->FindClass(com/roysue/easysol/MainActivity) was called
hook_id:65496, hook_func:<bound method native_method.<locals>.native_method_wrapper of <androidemu.java.jni_env.JNIE
at 0x7f09844b5b00>>
2989 - 2021-09-03 15:41:51,103 - DEBUG - JNIEnv->RegisterNatives(1, 0xCBBD2000, 1) was called
2989 - 2021-09-03 15:41:51,104 - WARNING - Register native ('decrypt', '(Ljava/lang/String;)Ljava/lang/String;', '0x
) failed on class MainActivity.
hook_id:65450, hook_func:<bound method native_method.<locals>.native_method_wrapper of <androidemu.java.jni_env.JNIE
at 0x7f09844b5b00>>
2989 - 2021-09-03 15:41:51,109 - DEBUG - JNIEnv->GetStringUtfChars(1, 0) was called
map addr:0x00000000, end:0x00000007, sz:0x00000007 off=0x00000000
before mem_map addr:0x30041000, sz:0x00001000
```

图 9-35 打印出 RegisterNatives 日志

以上就是补全 Java 类的过程。另外,在 Android 手机上,一个 SO 文件的地址会不断地发生变化,而在 ExAndroidNativeEmu 就不会变。因为它并没有实现地址随机化(Address Space Layout Randomization,ASLR),所有的组件都是固定的地址,也即函数都在固定的地址之上,所有的地址都是自定义硬编码的,如图 9-36 所示。

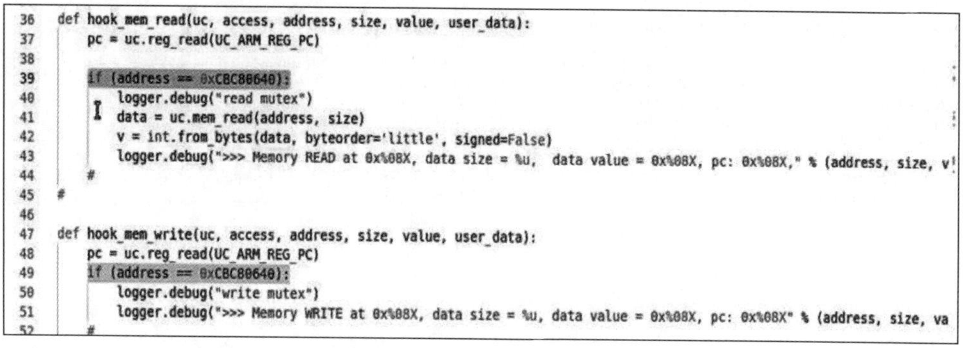

```
36     def hook_mem_read(uc, access, address, size, value, user_data):
37         pc = uc.reg_read(UC_ARM_REG_PC)
38
39         if (address == 0xCBC80640):
40             logger.debug("read mutex")
41             data = uc.mem_read(address, size)
42             v = int.from_bytes(data, byteorder='little', signed=False)
43             logger.debug(">>> Memory READ at 0x%08X, data size = %u, data value = 0x%08X, pc: 0x%08X," % (address, size, v
44         #
45         #
46
47     def hook_mem_write(uc, access, address, size, value, user_data):
48         pc = uc.reg_read(UC_ARM_REG_PC)
49         if (address == 0xCBC80640):
50             logger.debug("write mutex")
51             logger.debug(">>> Memory WRITE at 0x%08X, data size = %u, data value = 0x%08X, pc: 0x%08X" % (address, size, va
52
```

图 9-36 自定义硬编码地址

可以 Hook JNI 接口函数并查看参数调用返回值,可以主动调用 JNI 接口函数。我们刚刚观察 register_native 的过程其实就是参数调用返回值的过程。我们在这里再加一个调用栈的日志就行了。在 jni_env.py 中定义一个参数调用栈 TraceStack:

```
logger = logging.getLogger(__name__)

def TraceStack():
    print ("--------------------")
    frame=sys._getframe(1)
```

```
        while frame:
            print (frame.f_code.co_filename + " " + frame.f_code.co_name + " " +
str(frame.f_lineno))
            frame=frame.f_back

    # This class attempts to mimic the JNINativeInterface table.
    class JNIEnv:
```

分别在 get_string_utf_chars 和 register_native 中打印调用栈：

```
    @native_method
    def get_string_utf_chars(self, mu, env, string, is_copy_ptr):
        logger.debug("JNIEnv->GetStringUtfChars(%u, %x) was called" % (string,
is_copy_ptr))
        logger.debug(TraceStack())

    @native_method
    def register_natives(self, mu, env, clazz_id, methods, methods_count):
        logger.debug("JNIEnv->RegisterNatives(%d, 0x%08X, %d) was called" % (clazz_id,
methods, methods_count))
        logger.debug(TraceStack())
```

运行结果如图 9-37 所示。

```
2689 - 2021-09-10 10:10:11,633 - DEBUG - JNIEnv->GetStringUtfChars(1, 0) was called
-------------------
/root/Desktop/20210117/ExAndroidNativeEmu/androidemu/java/jni_env.py get_string_utf_chars 1598
/root/Desktop/20210117/ExAndroidNativeEmu/androidemu/java/helpers/native_method.py native_method_wrapper 109
/root/Desktop/20210117/ExAndroidNativeEmu/androidemu/hooker.py _hook 100
/root/.pyenv/versions/3.7.0/lib/python3.7/site-packages/unicorn/unicorn.py _hookcode_cb 438
/root/.pyenv/versions/3.7.0/lib/python3.7/site-packages/unicorn/unicorn.py emu_start 286
/root/Desktop/20210117/ExAndroidNativeEmu/androidemu/emulator.py call_native 206
/root/Desktop/20210117/ExAndroidNativeEmu/androidemu/emulator.py call_symbol 188
/root/Desktop/20210117/ExAndroidNativeEmu/example_jni.py <module> 102
```

图 9-37　参数调用栈打印结果

如何找到静态绑定的 Native 函数？在加载 SO 文件时，它会对符号表进行完整地解析，解析完之后存放在符号表中。所以说，流程是一样的。如何找到动态绑定的 Native 函数？也是通过 Hook RegisterNatives。

为什么不能 Hook 静态注册的 Native 函数呢？比如说，decrypt 函数在 App 刚加载时运行过一次，那么我们能够在 SO 文件中直接 Hook method02 或 method01 吗？答案是不能。作为 Unicorn，它可以 Hook 每一个地址。但它不可以 Hook method01 或 method02。这一点 Unidbg 就实现了，而且它不是自己实现的，而是通过第三方库来实现的。

最后，Unidbg 和 ExAndroidNativeEmu 基本上是一样的。唯一不同的地方在于，Unidbg 可以 Hook。inline Hook 可以 Hook 每一个地址。运行 IDEA，我们从官方案例中了解了符号 Hook 和 inline Hook。首先从 Module 中寻找符号，之后回调，这个流程和 Xpose 类似。inline Hook 代码如下：

```
    hookZz.instrument(module.base + 0x00000F5C + 1, new
InstrumentCallback<Arm32RegisterContext>() {
        @Override
        public void dbiCall(Emulator<?> emulator, Arm32RegisterContext ctx, HookEntryInfo
info) { // 通过 base+offset inline wrap 内部函数，在 IDA 看到的是 sub_xxx
            System.out.println("R3=" + ctx.getLongArg(3) + ", R10=0x" +
Long.toHexString(ctx.getR10Long()));
        }
    });
```

Dobby.replace 可以类比 Interceptor.replace。ExAndroidNativeEmu 和 Unidbg 模拟实现了 dlsym 以找到静态绑定的 Native 函数。

9.5　本章小结

本章介绍了 JNI 流程的原理和手动追踪调用的方法，还对 JNI 函数进行了分类，并测评和对比了几种 JNI 接口分析工具。我们发现，Frida 是功能最强大的工具，而其他几种工具则各有其优劣之处。

第 10 章

JNI 的特性：Java/Native 互相调用、反射/全局/局部引用

本章主要学习反射的思想，这一思想在逆向工作中非常重要，几乎在方方面面都会用到，尤其是像 Hook 工具、动静态分析等方面。本章主要讲解反射对于类的属性的影响以及如何获取和使用对象的基本属性，还介绍了如何设置、获取、调用类和对象的域与方法。

推荐阅读网址 32~网址 35 的文章。

10.1 反射"滥用"类和对象的基本属性

10.1.1 反射的概念与相关的 Java 类

在日常的第三方应用开发过程中，经常会遇到某个类的某个成员变量、方法或属性是私有的或者只对系统应用开放，这时就可以利用 Java 的反射机制，通过反射来获取所需的私有成员或方法。

在运行状态中，对于任意一个类都能够知道这个类的所有属性和方法，对于任意一个对象都能够调用它的任意方法和属性，这种动态获取信息以及动态调用对象方法的功能称为 Java 语言的反射机制。

反射主要是指程序可以访问、检测和修改它本身状态或行为的一种能力。所有的 Hook 工具都会注入目标进程中，从而可以动态地获取甚至修改类和对象的各种属性和方法。

除了解反射的功能外，我们还需要知道类和对象有哪些基本属性，以及操作对象都有哪些。简单来说，类是一个模板，对象是类的一个实例，有状态和行为。读者可以参考网址 36（参见配书资源文件）的教程来了解类和对象。

通过反射可以访问和动态地修改域和方法，也可以对构造函数进行主动调用来实例化一个类。

而且，只要把 Xposed 模块注入对方的进程中，就拥有了和对方代码相同的权限。

与 Java 反射相关的类如下。

- Class 类：代表类的实体，在运行的 Java 应用程序中表示类和接口。
- Field 类：代表类的成员变量（成员变量也称为类的属性/域）。
- Method 类：代表类的方法。
- Constructor 类：代表类的构造方法。

获取类相关的方法主要有以下几种（常用）。

- getClassLoader()：获取类的加载器。
- getClasses()：返回一个数组，数组中包含该类中所有公共类和接口类的对象。
- getDeclaredClasses()：返回一个数组，数组中包含该类中所有类和接口类的对象。
- forName(String className)：根据类名返回类的对象。
- getName()：获取类的完整路径名字。
- newInstance()：创建类的实例。
- getPackage()：获取类的包。
- getSuperclass()：获取当前类继承的父类的名字。
- getInterfaces()：获取当前类实现的类或接口。

获取类中的属性相关的方法有以下几种。

- - getField(String name)：获取某个公有的属性对象。
- - getFields()：获取所有公有的属性对象。
- - getDeclaredField(String name)：获取某个属性对象。
- - getDeclaredFields()：获取所有属性对象。

一般情况下，getDeclaredField(String name)和 getDeclaredFields()使用的比较多。先获取所有属性，再从其中过滤。

获取类中的方法有以下几种方法。

- getMethod(String name, Class...<?> parameterTypes)：获取该类某个公有的方法。
- getMethods()：获取该类所有公有的方法。
- getDeclaredMethod(String name, Class...<?> parameterTypes)：获取该类某个方法。
- getDeclaredMethods()：获取该类所有方法。

getDeclaredMethod(String name, Class...<?> parameterTypes)和 getDeclaredMethods()不分公有、私有，所以我们一般使用这两种方法。先用 getDeclaredMethods()获取该类的所有方法，再进行过滤，可以更快一些。

注解和构造器用得都比较少，构造器是一种特殊的方法，可以和方法一起学习。

10.1.2　实例：Xposed 刷机和编译使用的插件

本小节将结合 GravityBox 工具来学习如何使用类。GravityBox 是一款依赖 Xposed 框架支持的全能系统设置 DIY 工具，它是 Xposed 史上最强大的系统修改插件。

首先，下载 GravityBox 项目到本地：

```
root@VXIDr0ysue:~/Desktop/20210328 # git clone https://github.com/GravityBox/GravityBox.git -b nougat-as
```

我们需要将手机刷机，选择 7.1.2 (N2G47O, May 2017)，如图 10-1 所示。

图 10-1　bullhead 工厂镜像

```
root@VXIDr0ysue:~/Desktop # wget https://dl.google.com/dl/android/aosp/bullhead-n2g47o-factory-3ff32f55.zip?hl=zh-CN
```

```
root@VXIDr0ysue:~/Desktop # 7z x bullhead-n2g47o-factory-3ff32f55.zip\?hl=zh-CN
```

```
root@VXIDr0ysue:~/Desktop # cd bullhead-n2g47o
```

手机关机后，按住音量下键和开关键启动，插入 USB 线。

```
root@VXIDr0ysue:~/Desktop # ./flash-all.sh
```

flash-all 结束后，经过一系列设置，打开开发者选项，在开发者选项中设置不锁定屏幕，关闭自动系统更新、打开 USB 调试。将其他工具使用 ADB 推送（push）到相应路径下：

```
root@VXIDr0ysue:~/Desktop/20210328 # adb push SR5-SuperSU-v2.82-SR5-20171001224502.zip /sdcard/
SR5-SuperSU-v2.82-SR5-20171001224502.zip: 1 file pushed, 0 skipped. 8.4 MB/s (6882992 bytes in 0.785s)
```

```
root@VXIDr0ysue:~/Desktop/20210328 # adb push userinit.sh /sdcard/
userinit.sh: 1 file pushed, 0 skipped. 0.8 MB/s (209 bytes in 0.000s)
```

```
root@VXIDr0ysue:~/Desktop/20210328 # adb push frida-server-12.8.0-android-arm64 /data/local/tmp
```

第 10 章　JNI 的特性：Java/Native 互相调用、反射/全局/局部引用

```
    frida-server-12.8.0-android-arm64: 1 file pushed, 0 skipped. 10.4 MB/s (38545744 bytes
in 3.543s)

    root@VXIDr0ysue:~/Desktop/20210328 # adb install
de.robv.android.xposed.installer_3.1.5-43_minAPI15\(nodpi\)_apkmirror.com.apk
    Performing Streamed Install
    Success

    root@VXIDr0ysue:~/Desktop/20210328 # adb reboot bootloader

    root@VXIDr0ysue:~/Desktop/20210328 # fastboot flash recovery
twrp-3.3.0-0-bullhead.img
    Sending 'recovery' (16317 KB)                OKAY [  2.115s]
    Writing 'recovery'                           OKAY [  0.306s]
    Finished. Total time: 2.463s
```

拔掉 USB 线，在手机上按两次音量下键，再按电源键进入 recovery mode。稍等片刻之后，就会进入 twrp 3.3.0-0 系统。直接滑开即可解锁，表示允许修改系统，我们将可以看到非常丰富的功能。

在手机上启动 Xposed Installer，点击"安装/更新"下的 Version 89，如图 10-2 所示。

授权超级用户，安装完成后选择重启设备。开启 Xposed Installer，可以看到 Xposed Status 显示已激活，如图 10-3 所示。

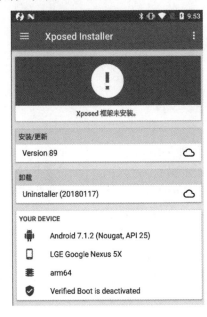

图 10-2　"安装/更新"Version 89　　　图 10-3　Xposed Status 显示已激活

点击左上角的菜单选择"模块"，然后就可以在 Android Studio 中点击"运行"安装 App，如图 10-4 所示。

再次重启设备，GravityBox 就可以使用了。例如，可以改变状态栏的图标颜色，如图 10-5 所

示，说明刚刚的源码是生效的。

图 10-4　GravityBox 安装完毕

图 10-5　改变状态栏的图标颜色

如果开机后 WiFi 有感叹号，显示"已连接，但无法访问互联网"，且时间无法同步，解决办法是在手机的 Shell 中以 root 用户执行：

```
# settings put global captive_portal_http_url https://www.google.cn/generate_204
# settings put global captive_portal_https_url https://www.google.cn/generate_204
# settings put global ntp_server 1.hk.pool.ntp.org
# reboot
```

后续只要把时区调整正确，时间就会自动同步。

接下来打开 GravityBox 项目。首先 Android Studio 会同步下载相应版本的 Gradle。

Xposed 项目入口类为 com.ceco.nougat.gravitybox.GravityBox。从入口来看，它提供了 Hook 孵化器，实现了孵化器的接口。首先来看 GravityBox.class.getPackage().getName()，这其实是一个反射，GravityBox.class 获取一个 Class 实例，并调用 getPackage().getName() 方法。

GravityBox.class.getPackage().getName 也存在于 Objection 中。也就是说，不仅可以用 Xposed 进行反射，在 Frida 和 Objection 中也大量使用了反射。反射是如何找到并修改类和对象的基本属性的呢？让我们来看一下 Objection 中的反射是如何使用的。其中，最经典的是 android.os.Build：

```
export const androidPackage = (): Promise<IAndroidPackage> => {
return wrapJavaPerform(() => {

    // https://developer.android.com/reference/android/os/Build.html
    const Build: any = Java.use("android.os.Build");

    return {
      application_name: getApplicationContext().getPackageName(),
      board: Build.BOARD.value.toString(),
```

```
        brand: Build.BRAND.value.toString(),
        device: Build.DEVICE.value.toString(),
        host: Build.HOST.value.toString(),
        id: Build.ID.value.toString(),
        model: Build.MODEL.value.toString(),
        product: Build.PRODUCT.value.toString(),
        user: Build.USER.value.toString(),
        version: Java.androidVersion,
    };
});
};
```

Build.BOARD.value.toString()就是一个反射。

例如，在 GravityBox 的案例中，反射的经典用法如下：

```
// Firstly, make sure we are able to get to pid field of ProcessImpl class
final Class<?> classProcImpl = Class.forName("java.lang.UNIXProcess");
final Field fieldPid = classProcImpl.getDeclaredField("pid");
fieldPid.setAccessible(true);
```

首先获取 java.lang.UNIXProcess 类，然后通过 getDeclaredField 获取这个类的域 pid，因为 pid 是私有属性，最后使用 setAccessible 修改权限，将 pid 设为允许访问，否则后续使用中会报错。如果想查看 field 有哪些方法，可以按住 Ctrl 键，点进 getInt()方法中，我们就可以进入 Field.class，查看其中的方法，如图 10-6 所示。我们不仅可以获取它的值，还可以设置它的值。

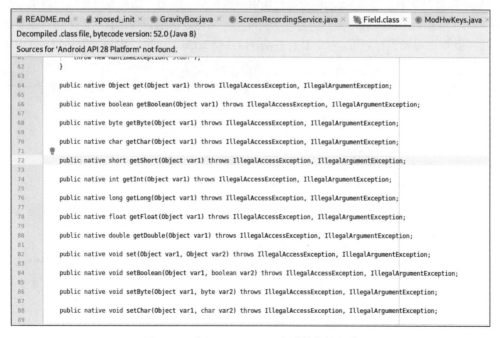

图 10-6　进入 Field.class，查看其中的方法

10.1.3 反射设置/调用类和对象的域和方法

以上是反射获取类和对象的域和方法，然后获取其中的值。10.1.2 节的案例是静态且私有的（static），接下来我们来看实例（instance）的 Field 类型如何操作。

- equals(Object obj)：若属性与 obj 相等，则返回 true。
- get(Object obj)：获取 obj 中对应的属性值。
- set(Object obj, Object value)：设置 obj 中对应的属性值。

```
Field fMeasuredWidth = View.class.getDeclaredField("mMeasuredWidth");
fMeasuredWidth.setAccessible(true);

fMeasuredWidth.setInt(param.thisObject, newWidth);
```

我们可以看到获取实例的域后，设置了它的值。第一个参数必须是一个对象。因为设置对象的域首先需要一个对象，然后是一个值。类似地，getDeclaredFields()返回一个 field 数组，然后进行单个过滤。再比如 getDeclaredMethod(String name, Class...<?> parameterTypes)：

```
Method m = mPhoneWindowManagerClass.getDeclaredMethod("takeScreenshot", int.class);
m.setAccessible(true);
return m;
```

首先，指定方法名字 takeScreenshot，再指定参数类型，因为方法可能有重载。如果方法有重载，就必须通过参数类型来确定唯一的方法。实际上，无论方法是否有重载，都必须明确写出它的参数类型。获取之后要进行调用，对于私有方法，仍然要 setAccessible，取消权限检查。调用如下：

```
private static void takeScreenshot(final long delayMs) {
    final Handler handler = (Handler) XposedHelpers.getObjectField(mPhoneWindowManager, "mHandler");
    if (handler == null) return;

    final Method m = getNativeScreenshotMethod();
    if (m != null) {
        handler.postDelayed(new Runnable() {
            @Override
            public void run() {
                try {
                    m.invoke(mPhoneWindowManager, 1);
                } catch (Throwable t) {
                    GravityBox.log(TAG, t);
                }
            }
        }, delayMs);
        return;
    }
```

由于 takeScreenshot 是 instance 方法，因此 invoke 调用时必须传入对象 mPhoneWindowManager。final Handler handler = (Handler) XposedHelpers.getObjectField(mPhoneWindowManager, "mHandler"); 可以追溯到 getDeclaredField。之后，再调用 get 方法返回一个 Object，最后进行类型强制转换，转换成 handler。也就是说，XposedHelpers.getObjectField = getDeclaredField + setAccessible + get。由此可见，Hook 框架对于反射的使用是普遍存在的。

类似的使用方法还有：

```
- XposedHelpers.setObjectField = getDeclaredField + setAccessible + set
- XposedHelpers.callMethod(mTelephonyManager, "setDataEnabled", enabled); = Method m
= mPhoneWindowManagerClass getDeclaredMethod("takeScreenshot", int.class); +
m.setAccessible(true); + m.invoke(mPhoneWindowManager, 1);
```

还有很多 API 跟反射是高度相关的，可以在 Frida 中直接使用。

Method 类主要有一个方法是主动调用的：

- invoke(Object obj, Object... args)传递 object 对象及参数，调用该对象对应的方法。如果调用一个 static method，那么第一个参数可以为空。例如，Xposed 中的 callStaticMethod 最终还是会用到 invoke。
- Constructor 类是一个特殊的 Method，代表类的构造方法。Constructor 调用 newInstance(Object... initargs)，根据传递的参数创建类的对象。

接下来介绍 Frida 中反射的写法。例如，通过域.value 直接获取域的值：

```
const Build: any = Java.use("android.os.Build");

board: Build.BOARD.value.toString()
```

value=设置域的值。类似地，获取一个对象的方法也是通过"对象.方法"，例如：

```
this.socket.value.getLocalAddress().toString().split(":")[0].split("/").pop()
```

获取了方法之后，我们可以把值传进去，并调用这个方法，例如：

```
Java.use("android.util.Log").getStackTraceString(Java.use("java.lang.Throwable").$
new()).toString()
```

这是一个典型的主动调用。我们可以看到，Frida 的主动调用比 Xposed 简单很多，直接构造对象作为参数传递。其中，$new()表示 newInstance。

构造一个 String 的方法如下：

```
Java.use("java.security.KeyStore").getInstance("PKCS12", "BC");
ks.load(null, null);
ks.setKeyEntry("client", pri,
Java.use('java.lang.String').$new(p12Password).toCharArray(), chain);
try {
    var out = Java.use("java.io.FileOutputStream").$new(p12Path);
```

```
      ks.store(out, Java.use('java.lang.String').$new(p12Password).toCharArray())
    } catch (exp) {
      console.log(exp)
    }
```

首先，通过 Java.use 找到类，再主动调用 getInstance 构造类的一个实例 ks，传进去的参数是两个 String，ks 再主动调用 setKeyEntry 方法。所有的主动调用都是 invoke，所有的$new 都是 newInstance。

Xposed 项目是注入对方的进程中的，然后也能访问对方的类和方法。

10.1.4 来自 Native 层的反射调用追踪

SO 文件本身就在 App 进程中，它是否可以直接访问 Java 层的类和方法？

Xposed 和 Frida 需要注入进程中，才能够找到 Java 的类和方法。但 SO 文件本身就在 App 进程中，并且还跟 MainActivity 进行了绑定。绑定之后，它就隶属于 MainActivity 类，那么它是否可以直接寻找到其他的类呢？

更改 easyso1 项目中的 MainActivity.java 代码：

```
import android.os.Build;

  tv.setText(decrypt("4e8de2f3c674d8157b4862e50954d81c") + " " + Build.BRAND + " " + Build.DEVICE);
```

运行结果如图 10-7 所示。

图 10-7　改变状态栏图标的颜色

不用反射,在进程中可以直接使用 Build。因此,我们从 Native 层就可以通过反射获取到 Build，然后从中得到一些信息。我们之前用过反射的一个案例：getJString。

```cpp
jstring getJString(JNIEnv *env, const char *src) {
    jclass stringClazz = env->FindClass("java/lang/String");
    jmethodID methodID = env->GetMethodID(stringClazz, "<init>",
"([BLjava/lang/String;)V");
    jstring encoding = env->NewStringUTF("utf-8");

    jsize length = static_cast<jsize>(strlen(src));
    jbyteArray bytes = env->NewByteArray(length);
    env->SetByteArrayRegion(bytes, 0, length, (jbyte *) src);
    jstring resultString = (jstring) env->NewObject(stringClazz, methodID, bytes,
encoding);

    env->DeleteLocalRef(stringClazz);
    env->DeleteLocalRef(encoding);
    env->DeleteLocalRef(bytes);

    return resultString;
}
```

首先，获取线程中的 env，从 env 中选择 FindClass，这是 env 接口的固定写法。jclass 等同于 Java 中的 Class。JNIEnv 中有 3 个函数可以获取 jclass：

（1）FindClass(const char* clsName)：通过类的名称（类的全名，这时包名不是用 "."，而是用 "/" 来区分的）来获取 jclass。

（2）GetObjectClass(jobject obj)：通过对象实例来获取 jclass，相当于 Java 反射中的 getClass() 函数。

（3）getSuperClass(jclass obj)：通过 jclass 获取其父类的 jclass 对象。

获取对象之后，再获取它的属性和方法。

在 Native 本地代码中访问 Java 层的代码，一个常见的场景是获取 Java 类的属性和方法。为了在 C/C++ 获取 Java 层的属性和方法，JNI 在 jni.h 头文件中定义了 jfieldID 和 jmethodID 这两种类型来分别代表 Java 端的属性和方法。在访问或者设置 Java 某个属性时，首先要在本地代码中取得代表该 Java 类的属性的 jfieldID，然后才能在本地代码中进行 Java 属性的操作，同样，在需要调用 Java 类的某个方法时，也需要取得代表该方法的 jmethodID 才能进行 Java 方法的操作。

常见的调用 Java 层的方法如下（一般使用 JNIEnv 进行操作）：

- GetFieldID：获取某个 instance 属性。
- GetMethodID：获取某个 instance 方法。
- GetStaticFieldID：获取某个静态属性。
- GetStaticMethodID：获取某个静态方法。

JNI 中常用 NewObject(jclass clazz, jmethodID methodID,...)创建对象，对应的是 Java 层的 newInstance(Object... initargs)。如果有多个构造函数，则可以通过签名来判断。

JNINativeInterface 中的 call、get、set 方法分别对应 Java 层的 invoke、get、set 方法。

回到刚刚的 getJString，通过 env->FindClass 找到类之后，获取 jmethodID，<init>表示构造函数，([BLjava/lang/String;)V 是签名。最后，调用 env->NewObject 传入 jclass、jmethodID、jbyteArray 和 UTF-8 编码，相当于 Java 的 new String(bytes, "UTF-8");。以上既然调用了 JNI 的 API，我们是否可以进行一些简单的追踪（Trace）呢？

启动 frida-server：

```
bullhead:/ $ su
bullhead:/ # cd /data/local/tmp
bullhead:/data/local/tmp # chmod 777 *
bullhead:/data/local/tmp # ./frida-server-12.8.0-android-arm64
```

执行 hook_art.js 脚本：

```
root@VXIDr0ysue:~/Desktop/20210328 # pyenv local 3.8.0
root@VXIDr0ysue:~/Desktop/20210328 # frida -U -f com.roysue.easyso1 -l hook_art.js
--no-pause

[FindClass] name:java/lang/String
[GetMethodID] name:<init>, sig:([BLjava/lang/String;)V
[NewStringUTF] bytes:utf-8
[GetStringUTFChars] result:roysue easyso1
CCCryptorCreate called from:
0xe9052960 libart.so!0xc2960
0xe904bc24 libart.so!0xbbc24
```

结果显示，我们可以手动追踪到 FindClass。

10.2 设计简单风控 SDK 并主动调用观察效果

10.1 节主要讲解了 Java 层的反射。Java 层的反射是基础，只有学会了 Java 层的反射，才能开始学习 JNI 层的反射。JNI 写在 Native 层主要是为了提高性能。从安全的角度来说，在 Native 层开发风控是因为 Native 层更难逆向、更难分析。

本节将设计一个简单的风控 SDK 并主动调用观察效果。风控是安全开发的一个应用，我们将通过这个案例学习 Native 层对于反射的开发。Native 层与 Java 层开发唯一的区别就在于 JNI，通过 JNI 可以进入 Java 的世界。如果是纯 Native 开发，则不需要 JNI 接口，可直接在 Native 层进行开发。

10.2.1 收集设备关键信息的常见方向和思路

风控的目的和方向是什么，我们可以参考比较大的厂商，看看它们做了哪些事情。以数字联盟（网址 37，参见配书资源文件）为例。

我们可以看到首页的 Banner 写得非常全面，其产品的核心目的是保证用户真实、推广有效、数据准确，这 3 点非常关键。

- 对 Android、iOS 平台提供 App 内部识别设备唯一、有效的数据基准。

这是风控在客户端上的一个根本目的，即识别某个手机是唯一的手机。一些非法行为最核心的目标就是更改这些内容，将一台手机设备识别成数台手机设备。所以说，如何确定一台手机设备是唯一的手机设备非常关键。一些"黑产"利用改进软件把一台手机虚拟成数台手机，然后借此领取优惠券，领取一些 App 接管的费用。针对以上非法行为，唯一设备识别技术通过识别硬件和模拟器，从而防止一台设备被虚拟成多台设备。

- 不依赖操作系统反馈的方式，独立获取设备信息，不受任何劫持程序的影响，保证设备信息准确有效。

说明它可能没有使用到 Java 层的任何代码，也没有使用到 C 库中的任何代码，而是通过手写汇编进行开发，直接与内核进行交互。当然，我们依旧有办法对它进行劫持。

其他的一些终端防伪技术包括模拟器甄别、代理服务器甄别、调试状态甄别、传感器状态评估（生物指纹识别）等。

这些都可以当作判断唯一、真实设备的依据。根据这些依据，综合各个维度，给这台设备评分。当然，也可以不在设备上进行任何操作，而是将收集的数据上传到服务器上，然后根据服务器积累的数据，来判断它是否为一台"作弊"的手机。目前，大部分风控都是这种运作模式。它知道你在做什么，但是它不会阻止你，不会崩溃，也不会退出。你可以正常地使用设备，过一段时间，你就会发现调用接口传回来的数据逐渐就不对了。

近些年风控的改变并不大，只是开发的技术在不断迭代。最开始写在 Java 层，之后写在 Native 层，用一些 C 的 API 来做。到现在为止，许多安全厂商都是通过手写汇编代码来获取各种信息的。

所以说，如果要 bypass 客户端的风控，我们只能逆向 SDK。SDK 又加上了一些防护，比如 SO 文件的整体加密、SO 文件的方法体加密，还有整个 SO 文件的落地加载——直接打开一个文件并解压，用 linker 加载一个 SO 文件。最强的还是 OLLVM 和 ARM 的 VMP。如果用上这两个防护，想要逆向是一件非常困难的事。所以说，单纯看它的开发，如果没有加上任何防护，逆风控是非常简单的。反之，如果加上了多种防护，则逆风控是非常困难的，只能做一些粒度非常细的沙箱。

我们可以上报设备风险环境评估结果，也可以在设计"错误 SIGN"的时候加入评估结果，让一些主动调用工具在调用时直接生成一个"错误 SIGN"。"错误 SIGN"上传服务器并解析之后，如果发现是用机器主动调用的，而不是正常环境下生成的"SIGN"，则直接丢弃请求或者返回一个错误的结果。

10.2.2　Native 层使用反射调用 Java 层 API 获取设备信息

如何让我们的产品有一个最简单的风控？

比如说，我们要进行一个查询。在查询之前要对这个查询的内容加一个 SIGN。将这个查询的内容和 SIGN 的结果一起发送给服务器。这是一个很常见的场景。首先,在 Java 层声明一个 Native 方法 SIGN:

```
package com.roysue.easyso1;
public class MainActivity extends AppCompatActivity {

    protected void onCreate(Bundle savedInstanceState) {

        class MyThread extends Thread {
            @Override
            public void run () {
                while (true) {

                    Log.i("roysue easyso1", method01("r0ysue"));
                    Log.i("roysueSIGN", Sign("requestUserInfo"));
                }
            }
        }
    }

    public static native String Sign(String str);
}
```

当然,这个 Native 方法现在还没有实现。我们选择动态注册来实现:

```
JNIEXPORT jstring JNICALL ngis(JNIEnv *env, jclass jcls, jstring str_) {
    if (str_ == nullptr) return nullptr;

    const char *str = env->GetStringUTFChars(str_, JNI_FALSE);
    char *result = AES_128_CBC_PKCS5_Encrypt(str);

    env->ReleaseStringUTFChars(str_, str);

    jstring jResult = getJString(env, result);
    free(result);

    return jResult;
}

static JNINativeMethod method_table[] = {
    //{"method01", "(Ljava/lang/String;)Ljava/lang/String;", (void *) method01},
    {"decrypt", "(Ljava/lang/String;)Ljava/lang/String;", (void *) method02},
    {"Sign", "(Ljava/lang/String;)Ljava/lang/String;", (void *) ngis},
};
```

现在,我们有了一个基本的 nativeSign,并且无法从方法名称猜测出 SIGN 和 native Sign(ngis)之间的联系。连上手机并运行,可以看到结果是没有问题的,在发送的时候可以把 requestUserInfo

和 43f5 一起发送出去，如图 10-8 所示。

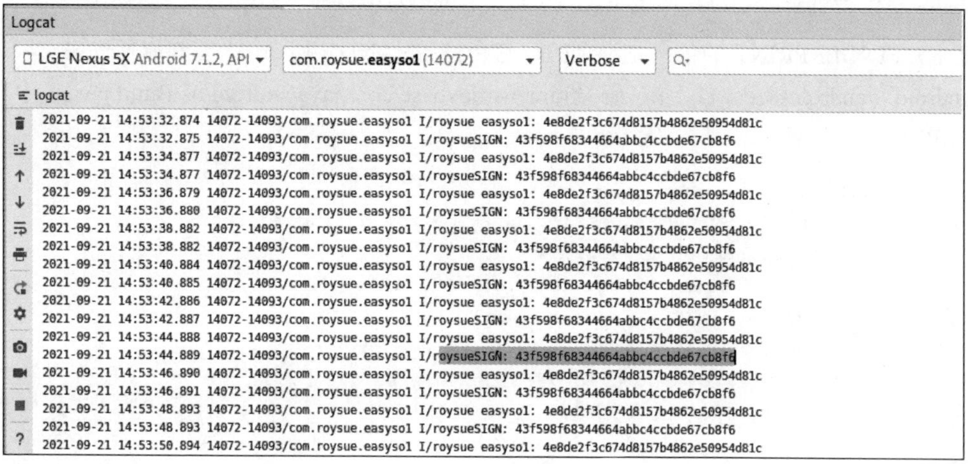

图 10-8　requestUserInfo 和 43f5 一起发出去

当然，这个里面还没有任何设备风险评估的信息。下面来设计一些简单的设备风险评估信息。由于 SIGN 是在 JNI_OnLoad 动态注册的，因此在 Java 层无法找到它的实现。如果要查找 Java 层 SIGN 与 Native 层的哪个函数进行了绑定，就要追踪 RegisterNatives。

接下来，添加一些环境监控。这里最基本的是取得 android.os.Build 里面的信息。首先，在手机上启动 frida-server，并运行 Objection：

```
root@VXIDr0ysue:~/Desktop/20210402 # pyenv local 3.8.0

root@VXIDr0ysue:~/Desktop/20210402 # objection -g com.roysue.easyso1 explore
com.roysue.easyso1 on (google: 7.1.2) [usb] # plugin load
/root/Desktop/Wallbreaker-master
    Loaded plugin: wallbreaker
    com.roysue.easyso1 on (google: 7.1.2) [usb] # plugin wallbreaker classdump
android.os.Build

    package android.os

    class Build {

        /* static fields */
        static String BOARD; => bullhead
        static String BOOTLOADER; => BHZ11m
        static String BRAND; => google
        static String CPU_ABI; => armeabi-v7a
        static String CPU_ABI2; => armeabi
        static String DEVICE; => bullhead
        static String DISPLAY; => N2G47O
```

```
            static String FINGERPRINT; =>
google/bullhead/bullhead:7.1.2/N2G47O/3852959:user/release-keys
```

其中，FINGERPRINT 的信息是最全的，它会把 BRAND、用户 ID 等都囊括进去。我们可以在 Android android-8.1.0_r1 源码 /frameworks/base/core/java/android/os/Build.java 中查看 FINGERPRINT 的信息：

```
    private static String deriveFingerprint() {
        String finger = SystemProperties.get("ro.build.fingerprint");
        if (TextUtils.isEmpty(finger)) {
            finger = getString("ro.product.brand") + '/' +
                getString("ro.product.name") + '/' +
                getString("ro.product.device") + ':' +
                getString("ro.build.version.release") + '/' +
                getString("ro.build.id") + '/' +
                getString("ro.build.version.incremental") + ':' +
                getString("ro.build.type") + '/' +
                getString("ro.build.tags");
        }
        return finger;
    }
```

我们从 Native 层该如何取得 android.os.Build 中的信息呢？这时就要用到反射。

首先，Java 层的 Class 对应 Native 层的 jclass，需要调用 env->FindClass：

```
jclass buildClazz = env->FindClass("android/os/Build");
```

然后，通过 env->GetStaticFieldID 获取 FINGERPRINT，需要 3 个参数，分别是 jclass buildClazz、FieldName "FINGERPRINT"和签名"Ljava/lang/String; "：

```
jfieldID FINGERPRINT =
env->GetStaticFieldID(buildClazz,"FINGERPRINT","Ljava/lang/String;");
```

我们还要取出 FINGERPRINT 的值，它是 String 类型的，到 Native 层就会变成 jstring，调用 env->GetStaticObjectField 时，需要的两个参数分别是 buildClazz 和 FINGERPRINT。由于它返回的是 Object 类型，因此需要进行强制转换：

```
jstring fingerprint = static_cast<jstring>(env->GetStaticObjectField(buildClazz,
FINGERPRINT));
```

接下来就可以判断 fingerprint 是否包含 AOSP 了。因为没有一个真正的用户会将 AOSP 用于日常工作，但逆向分析人员往往喜欢使用它。基于这一经验，我们设计了一个简单的 SIGN：如果 fingerprint 中包含 AOSP，我们将不把它当作真实用户使用的设备：

```
    char* sign = "REAL";

    if(strstr( env->GetStringUTFChars(fingerprint, JNI_FALSE),"aosp")){
```

```
        sign = "FAKE";
    }
```

最后，上传 SIGN，加到 request 的后面。完整代码如下：

```
#include<android/log.h>

#define TAG "roysuejni" // 这个是自定义的 LOG 的标识
#define LOGD(...) __android_log_print(ANDROID_LOG_DEBUG,TAG,__VA_ARGS__) // 定义 LOGD 类型
#define LOGI(...) __android_log_print(ANDROID_LOG_INFO,TAG,__VA_ARGS__) // 定义 LOGI 类型
#define LOGW(...) __android_log_print(ANDROID_LOG_WARN,TAG,__VA_ARGS__) // 定义 LOGW 类型
#define LOGE(...) __android_log_print(ANDROID_LOG_ERROR,TAG,__VA_ARGS__) // 定义 LOGE 类型
#define LOGF(...) __android_log_print(ANDROID_LOG_FATAL,TAG,__VA_ARGS__) // 定义 LOGF 类型
...
JNIEXPORT jstring JNICALL ngis(JNIEnv *env, jclass jcls, jstring str_) {
    if (str_ == nullptr) return nullptr;

    char *str = const_cast<char *>(env->GetStringUTFChars(str_, JNI_FALSE));

    char* sign = "REAL";

    jclass buildClazz = env->FindClass("android/os/Build");
    jfieldID FINGERPRINT = env->GetStaticFieldID(buildClazz,"FINGERPRINT","Ljava/lang/String;");
    jstring fingerprint = static_cast<jstring>(env->GetStaticObjectField(buildClazz, FINGERPRINT));

    if(strstr( env->GetStringUTFChars(fingerprint, JNI_FALSE),"aosp")){
        sign = "FAKE";
    }

    strcat(str,sign);

    LOGI("before entering aes => %s",str);

    char *result = AES_128_CBC_PKCS5_Encrypt(str);

    env->ReleaseStringUTFChars(str_, str);

    jstring jResult = getJString(env, result);
    free(result);
```

```
        return jResult;
    }
```

此时，如果是一个 AOSP 手机，就能检测到 SIGN 是 FAKE，如图 10-9 所示。

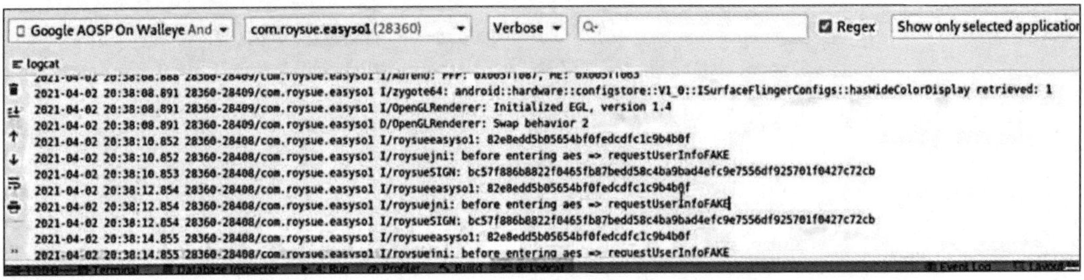

图 10-9 检测到 AOSP 手机，SIGN 是 FAKE

以上就是 Native 层使用反射调用 Java 层 API 的方法。设计一个 SIGN 不能只使用 AES，我们再设计一个 MD5 与 AES 进行拼接，防止 SIGN 被篡改。添加的 MD5 代码如下：

```
JNIEXPORT jstring JNICALL ngis(JNIEnv *env, jclass jcls, jstring str_) {
        strcat(str,sign);
    jstring jstr = getJString(env, str);


    LOGI("before entering aes => %s",str);

    ////    byte[] bytes = MessageDigest.getInstance("MD5").digest(input.getBytes());

    jclass MessageGigest = env->FindClass("java/security/MessageDigest");
    jmethodID getInstance = env->GetStaticMethodID(MessageGigest, "getInstance",
"(Ljava/lang/String;)Ljava/security/MessageDigest;");
    jobject md5 = env->CallStaticObjectMethod(MessageGigest, getInstance,
getJString(env, "MD5"));
    jmethodID digest = env->GetMethodID(MessageGigest, "digest", "([B)[B");

    jclass stringClazz = env->FindClass("java/lang/String");
    jmethodID getbytes = env->GetMethodID(stringClazz, "getBytes", "()[B");
    jbyteArray jba = static_cast<jbyteArray>(env->CallObjectMethod(jstr, getbytes));
    jbyteArray md5result = static_cast<jbyteArray>(env->CallObjectMethod(md5, digest,
jba));

    char *cmd5 = reinterpret_cast<char *>(env->GetByteArrayElements(md5result, 0));

    char md5string[33];
    for(int i = 0; i < 16; ++i)
        sprintf(&md5string[i*2], "%02x", (unsigned int)cmd5[i]);

    char *result = AES_128_CBC_PKCS5_Encrypt(str);

    strcat(result,md5string);
```

```
    jstring finalResult = getJString(env, result);
    env->ReleaseStringUTFChars(str_, str);
    //jstring jResult = getJString(env, result);
    free(result);

    return finalResult;
}
```

运行结果如图 10-10 所示。最终传进来的数据是 requestUserInfo，也就是请求的用户名，后面跟着一个 SIGN。这个 SIGN 的后几位是 MD5，去掉后剩下的部分是 AES，使用 Key 和 IV 解密后得到的值，计算其 MD5 与后面的值进行比较。如果匹配，则该请求有效；否则，该请求被视为无效，即被篡改了。

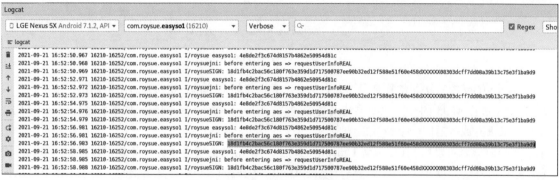

图 10-10　AES 与 MD5 拼接的结果

结果的正确性可以通过 CyberChef 进行验证，如图 10-11 所示。服务器对 SIGN 进行验证时，先把 AES 加密后的数据还原回去，再计算 MD5，看与后面的数据是否一致。

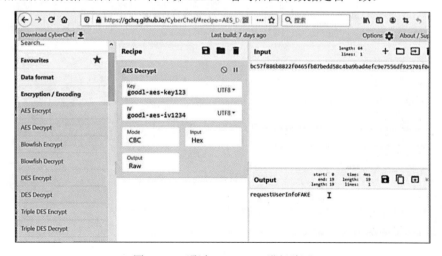

图 10-11　通过 CyberChef 进行验证

计算 MD5，结果与第二段数据一致，如图 10-12 所示。

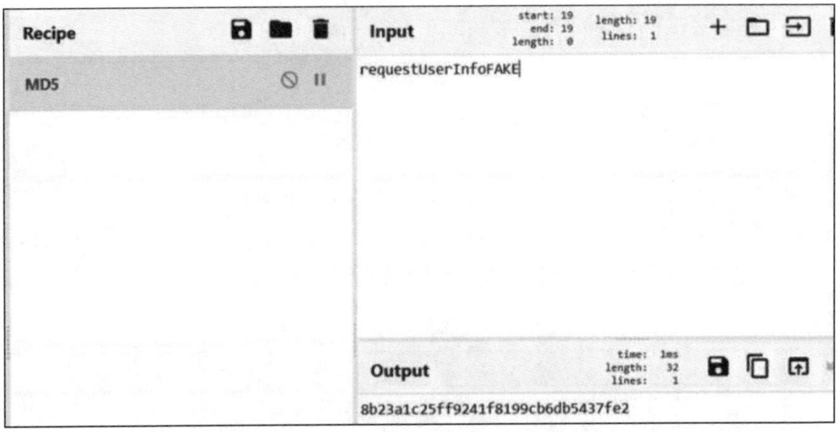

图 10-12　计算 MD5 的结果

还原 SIGN 的算法非常复杂。因为它是两个算法拼接而成的。拼接后，服务器可以解密，并取得 request 进行 MD5 的计算。如果结果是正确的，则说明请求没有问题。如果解密后出现 FAKE，就将请求丢弃或者返回一个错误的结果。

这个时候就可以用 Frida 或者 Unicorn 进行主动调用来生成这样的一个 SIGN。但生成 SIGN 的过程跟设备环境高度相关。如果设备环境是 AOSP，生成的 SIGN 就是错误的。设备环境也可以进行其他的检测。AOSP 是用 Java 层的 FINGERPRINT 来检测的。这其实很好绕过，示例代码如下：

```
    setTimeout(function () {
        Java.perform(function () {
            Java.use("android.os.Build").FINGERPRINT.value = "12345"
        })
    }, 0);

    function invoke() {
        Java.perform(function () {
            var result =
Java.use("com.roysue.easyso1.MainActivity").Sign("requestUserInfo");
            console.log("result is => ", result);

        })
    }
```

主动调用结果如图 10-13 所示。

第 10 章　JNI 的特性：Java/Native 互相调用、反射/全局/局部引用

```
PROBLEMS   OUTPUT   TERMINAL   DEBUG CONSOLE                           2: python3.8

┌─(root㉿kali)-[~/Desktop/20210402]
└─# frida -UF -l invoke.js
     /_  |    Frida 12.8.0 - A world-class dynamic instrumentation toolkit
    | (_ |
    > _  |   Commands:
   /_/ |_|       help      -> Displays the help system
    . . . .      object?   -> Display information about 'object'
    . . . .      exit/quit -> Exit
    . . . .
    . . . .    More info at https://www.frida.re/docs/home/
[LGE Nexus 5X::easysol]-> invoke()
result is =>  18d1fb4c2bac56c180f763e359d1d7175ef88da8ba6e6d2cb0c6c777c4af627eXXXXXX08303dcff7dd08a39b13c75e3f1ba9d9
[LGE Nexus 5X::easysol]-> []
```

图 10-13　主动调用结果得到正确的 SIGN

我们再添加 Native 层的检测代码，直接调用 C 库的 __system_property_get 函数获取 manufacturer 和 model：

```cpp
bool system_getproperty_check() {
    char man[256], mod[156];
    /* A length 0 value indicates that the property is not defined */
    int lman = __system_property_get("ro.product.manufacturer", man);
    int lmod = __system_property_get("ro.product.model", mod);
    int len = lman + lmod;
    char *pname = NULL;
    if (len > 0) {
        pname = static_cast<char *>(malloc(len + 2));
        snprintf(pname, len + 2, "%s/%s", lman > 0 ? man : "", lmod > 0 ? mod : "");
    }

    bool b = false;
    if(strstr(pname,"Google"))b=true;
    LOGE("[roysue device]: [%s] result is => %d\n", pname ? pname : "N/A",b);
    return b;
}

JNIEXPORT jstring JNICALL ngis(JNIEnv *env, jclass jcls, jstring str_) {
        if(system_getproperty_check() || strstr( env->GetStringUTFChars(fingerprint,JNI_FALSE),"aosp")){
            sign = "FAKE";
        }
    }
```

此时，只是主动调用 android.os.Build 无法影响 SIGN 的值，还需要 Hook Native 层，让 __system_property_get 返回其他的结果。

另外，添加检查 tracerPID 的代码如下：

```cpp
#include <bitset>
#include <fstream>
```

```cpp
    bool function_check_tracerPID() {
        bool b = false ;
        int pid = getpid();
        std::string file_name = "/proc/pid/status";
        std::string line;
        file_name.replace(file_name.find("pid"), 3, std::to_string(pid));
        LOGE("replace file name => %s", file_name.c_str());
        std::ifstream myfile(file_name, std::ios::in);
        if (myfile.is_open()) {
            while (getline(myfile, line)) {
                size_t TracerPid_pos = line.find("TracerPid");
                if (TracerPid_pos == 0) {
                    line = line.substr(line.find(":") + 1);
                    LOGE("file line => %s", line.c_str());
                    if (std::stoi(line.c_str()) != 0) {
                        LOGE("trace pid => %s, i want to exit.", line.c_str());
                        b = true ;
//                        kill(pid, 9);
                        break;
                    }
                }
            }
            myfile.close();
        }
        return b ;
    }

    JNIEXPORT jstring JNICALL ngis(JNIEnv *env, jclass jcls, jstring str_) {
            if(function_check_tracerPID() || system_getproperty_check() ||
strstr( env->GetStringUTFChars(fingerprint, JNI_FALSE),"aosp")){
            sign = "FAKE";
        }
    }
```

为了验证判断条件是否有效，在条件语句if(function_check_tracerPID() || system_getproperty_check() || strstr(env->GetStringUTFChars(fingerprint, JNI_FALSE),"aosp"))处设置断点，调试得到的tracerPID 如图10-14 所示，返回值为true。

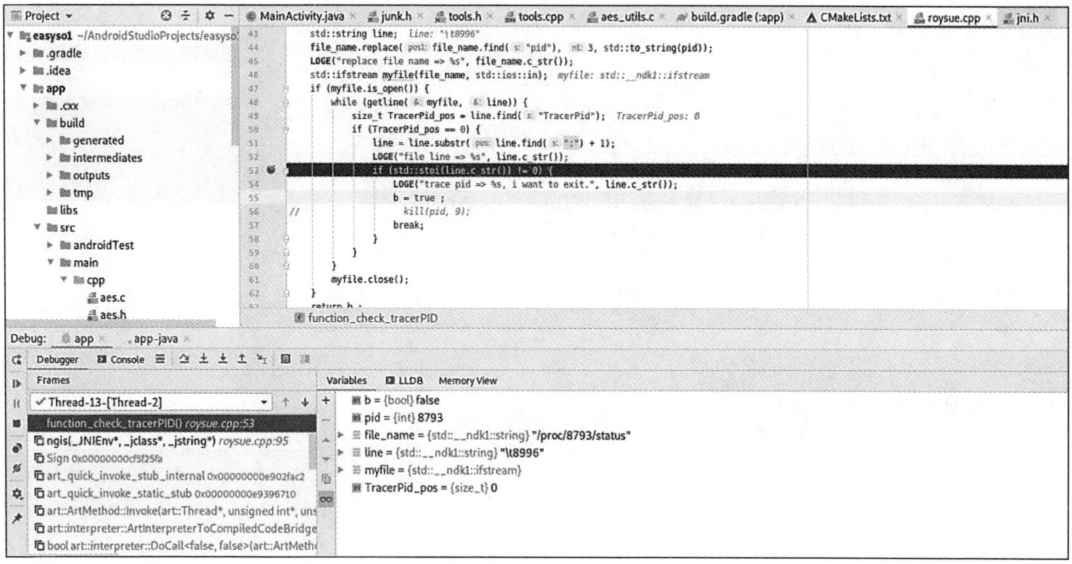

图 10-14　tracerPID

```
bullhead:/ # ps |grep easy
u0_a96    8793   601   1031020  59584 ptrace_sto 00e96d9528 t com.roysue.easyso1
u0_a96    8993  8987    6744    1268 __skb_recv 0000301f2c S
/data/data/com.roysue.easyso1/lldb/bin/lldb-server
u0_a96    8994  8993    8312    1908 poll_sched 00003172c0 S
/data/data/com.roysue.easyso1/lldb/bin/lldb-server
u0_a96    8996  8994    9304    3716 poll_sched 0000301f2c S
/data/data/com.roysue.easyso1/lldb/bin/lldb-server
bullhead:/ # cat /proc/7840/status
Name:    .roysue.easyso1
State:   t (tracing stop)
Tgid:    8793
Pid:     8793
PPid:    601
TracerPid:       8996
```

结果中 TracerPid（lldb-server）的进程号为 8996，与函数体中得到的 TracerPid 一致。条件判断语句第一个条件返回值为 true 后就不用再判断后面的条件了。用类似的方法可以验证第二个条件。

经过一系列的分析，我们可以得出一个结论：生成一个正确的 SIGN，关键在于 strcat(str,sign); 这个函数。

启动 Desktop/idea-IC-211.7628.21/bin./idea.sh，运行 kanxue.test2 目录下的 MainActivity，可以看到它会把所有解析出来的符号都 Trace 一遍。

我们在 java 目录下新建一个包 com.r0ysue.easyso，在包下创建一个 class MainActivity。

编写 MainActivity 代码如下：

```java
    package com.r0ysue.easyso;

    import com.github.unidbg.AndroidEmulator;
    import com.github.unidbg.LibraryResolver;
    import com.github.unidbg.arm.backend.DynarmicFactory;
    import com.github.unidbg.linux.android.AndroidEmulatorBuilder;
    import com.github.unidbg.linux.android.AndroidResolver;
    import com.github.unidbg.linux.android.dvm.*;
    import com.github.unidbg.linux.android.dvm.jni.ProxyDvmObject;
    import com.github.unidbg.memory.Memory;

    import java.io.File;

    /**
     * https://bbs.pediy.com/thread-263345.htm
     */
    public class MainActivity extends AbstractJni {

        public static void main(String[] args) {
            long start = System.currentTimeMillis();
            com.r0ysue.easyso.MainActivity mainActivity = new com.r0ysue.easyso.MainActivity();
            System.out.println("load offset=" + (System.currentTimeMillis() - start) + "ms");
            mainActivity.crack();
        }

        private final AndroidEmulator emulator;
        private final VM vm;
        private final DvmClass dvmClass;
        private String className = "com/roysue/easyso1/MainActivity";

        private MainActivity() {
            emulator = AndroidEmulatorBuilder
                    .for32Bit()
//                    .addBackendFactory(new DynarmicFactory(true))
                    .build();
            Memory memory = emulator.getMemory();
            LibraryResolver resolver = new AndroidResolver(23);
            memory.setLibraryResolver(resolver);

            vm = emulator.createDalvikVM();
            System.out.println(vm);
            vm.setVerbose(true);
            vm.setJni(this);
            DalvikModule dm = vm.loadLibrary(new File("unidbg-android/src/test/resources/example_binaries/armeabi-v7a/libroysue.so"), false);
            dm.callJNI_OnLoad(emulator);
            dvmClass = vm.resolveClass(className);
```

第 10 章　JNI 的特性：Java/Native 互相调用、反射/全局/局部引用

```
    }
    private void crack() {
        DvmObject result =
dvmClass.callStaticJniMethodObject(emulator,"Sign(Ljava/lang/String;)Ljava/lang/String;","requestUserInfo");
        long start = System.currentTimeMillis();
    }
}
```

构造了一个 MainActivity 对象，有了 MainActivity 对象之后，调用 crack()。在 MainActivity 构造函数中，进行模拟器的生成，取得 memory，解析并 set SDK 的版本号。创建 DalvikVM，set 输出日志为 true。然后创建 DalvikModule。最后调用 callJNI_OnLoad。

我们将 libroysue.so 复制到对应目录下 /root/Desktop/unidbg3/unidbg-android/src/test/resources/example_binaries/armeabi-v7a/，记得检查一下权限：

```
root@VXIDr0ysue:~/.../test/resources/example_binaries/armeabi-v7a # ls -alit
total 11000
4206594 drwxr-xr-x 2 root root    4096 Sep 23 09:55 .
4206589 drwxr-xr-x 6 root root    4096 Jul 13 10:21 ..
4206598 -rw-r--r-- 1 root root 7055180 Jul 13 10:21 libtmessages.29.so
4206597 -rw-r--r-- 1 root root  522128 Jul 13 10:21 libnative-lib.so
4206596 -rw-r--r-- 1 root root   87432 Jul 13 10:21 libjnidispatch.so
4206595 -rw-r--r-- 1 root root    9700 Jul 13 10:21 libd-lib.so
4207585 -rw-r--r-- 1 root root 3569004 Jan  1  1981 libroysue.so

root@VXIDr0ysue:~/.../test/resources/example_binaries/armeabi-v7a # chmod 777 *
```

运行结果如图 10-15 所示。

```
/android/sdk23/lib/libnetd_client.so
JNIEnv->FindClass(com/roysue/easysol/MainActivity) was called from RX@0x40037f03[libroysue.so]0x37f03
JNIEnv->RegisterNatives(com/roysue/easysol/MainActivity, RW@0x40080000[libroysue.so]0x80000, 2) was called from RX@0x4003b81f[libroysue.so]0x3b81f
RegisterNative(com/roysue/easysol/MainActivity, decrypt(Ljava/lang/String;)Ljava/lang/String;, RX@0x40038ba1[libroysue.so]0x38ba1)
RegisterNative(com/roysue/easysol/MainActivity, Sign(Ljava/lang/String;)Ljava/lang/String;, RX@0x400392e1[libroysue.so]0x392e1)
load offset=1207ms
Find native function Java_com_roysue_easysol_MainActivity_Sign(Ljava/lang/String;)Ljava/lang/String; => RX@0x400392e1[libroysue.so]0x392e1
JNIEnv->GetStringUtfChars("requestUserInfo") was called from RX@0x40038c17[libroysue.so]0x38c17
JNIEnv->FindClass(android/os/Build) was called from RX@0x40037f03[libroysue.so]0x37f03
JNIEnv->GetStaticFieldID(android/os/Build.FINGERPRINTLjava/lang/String;) => 0xe947c974 was called from RX@0x40039599[libroysue.so]0x39599
[10:21:47 455] WARN [com.github.unidbg.linux.ARM32SyscallHandler] (ARM32SyscallHandler:467) - handleInterrupt intno=2, NR=-1873744224, svcNumber=0x143, PC=unidbg@0xfffe04
java.lang.UnsupportedOperationException Create breakpoint : android/os/Build->FINGERPRINT:Ljava/lang/String;
    at com.github.unidbg.linux.android.dvm.AbstractJni.getStaticObjectField(AbstractJni.java:83)
    at com.github.unidbg.linux.android.dvm.AbstractJni.getStaticObjectField(AbstractJni.java:45)
    at com.github.unidbg.linux.android.dvm.DvmField.getStaticObjectField(DvmField.java:39)
    at com.github.unidbg.linux.android.dvm.DalvikVM$68.handle(DalvikVM.java:1613)
    at com.github.unidbg.linux.ARM32SyscallHandler.hook(ARM32SyscallHandler.java:103)
    at com.github.unidbg.linux.arm.backend.UnicornBackend$6.hook(UnicornBackend.java:299)
    at unicorn.Unicorn$NewHook.onInterrupt(Unicorn.java:128)
    at unicorn.Unicorn.emu_start(Native Method)
```

图 10-15　MainActivity 运行报错

在 /root/Desktop/unidbg3/unidbg-android/src/main/java/com/github/unidbg/linux/android/dvm/api/ 目录下新建 class Build，编写代码如下：

```
package com.github.unidbg.linux.android.dvm.api;
```

```java
import com.github.unidbg.linux.android.dvm.DvmClass;
import com.github.unidbg.linux.android.dvm.DvmObject;

public class Build extends DvmObject<String> {

    public static final String FINGERPRINT = "12345input_method";

    protected Build(DvmClass objectType, String value) {
        super(objectType, value);
    }
}
```

然后在 /root/Desktop/unidbg3/unidbg-android/src/main/java/com/github/unidbg/linux/android/dvm/ AbstractJni 中根据报错添加 case：

```java
case "android/os/Build->FINGERPRINT:Ljava/lang/String;":
    return new StringObject(vm, Build.FINGERPRINT);
```

再次运行后可以看到打印结果，如图 10-16 所示。

```
final result is =>
18d1fb4c2bac56c180f763e359d1d717500787ee90b32ed12f588e51f60e458dXXXXXX08303dcff7dd08a39
b13c75e3f1ba9d9
```

图 10-16 打印出 final result

10.3 本章小结

在本章中，我们学习了 Java 和 Native 层反射，以及在 Xposed 和 Frida 中如何使用反射。通过反射的使用，我们设计了一个简单的风控软件开发工具包 SDK，并主动调用观察效果。

第 11 章 onCreate 进行 Native 化和引用

在一个加固的应用中，我们经常看到应用的 onCreate 函数被隐藏了，这是如何实现的呢？本章将对 onCreate 的抽取方法进行讲解并学习一些应用加固技术。

11.1 将 onCreate 函数 Native 化

首先，启动 Android Studio，打开 easyso1 项目。为了简单起见，将 onCreate 函数删减至以下内容：

```java
@Override
protected void onCreate(Bundle savedInstanceState) {
    super.onCreate(savedInstanceState);
    setContentView(R.layout.activity_main);

    // Example of a call to a native method
    TextView tv = findViewById(R.id.sample_text);
    tv.setText("r0ysue");
}
```

接下来，为了将 onCreate 函数 Native 化，我们的任务就是"翻译" onCreate 函数中的 4 条语句。首先注释掉 Java 层初始的 onCreate 函数，并声明一个 Java 层的 Native 函数 onCreate，单击选择 Create JNI function for onCreate，如图 11-1 所示。

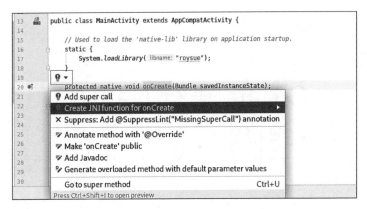

图 11-1　声明一个 Java 层的 Native 函数 onCreate

首先，找到 onCreate 的父类 AppCompatActivity。获取 AppCompatActivity 有以下两种方法：

（1）使用 Objection 在内存中找，再调用 env->FindClass：

```
root@VXIDr0ysue:~/Desktop/20210409 # objection -g com.roysue.easyso1 explore

com.roysue.easyso1 on (google: 7.1.2) [usb] # android hooking search classes AppCompatActivity
androidx.appcompat.app.AppCompatActivity

Found 1 classes

jclass AppCompatActivity_class1 = env->FindClass("androidx/appcompat/app/AppCompatActivity");
jclass MainActivity_class1 = env->FindClass("com/roysue/easyso1/MainActivity");
```

（2）GetObjectClass：由对象获得它的类，再由类获得其父类：

```
jclass MainActivity_class2 = env->GetObjectClass(thiz);
jclass AppCompatActivity_class2 = env->GetSuperclass(MainActivity_class2);
```

然后，再找它的方法（即函数），也有两种途径：

```
jmethodID onCreate = env->GetMethodID(AppCompatActivity_class2, "onCreate", "(Landroid/os/Bundle;)V");
```

调用父类的 onCreate 方法时，要加上 CallNonvirtualVoidMethod：

```
env->CallNonvirtualVoidMethod(thiz, AppCompatActivity_class2, onCreate, saved_instance_state);
```

第一条语句 super.onCreate(savedInstanceState);完整翻译如下：

```
jclass MainActivity_class2 = env->GetObjectClass(thiz);
jclass AppCompatActivity_class2 = env->GetSuperclass(MainActivity_class2);

jmethodID onCreate = env->GetMethodID(AppCompatActivity_class2, "onCreate", "(Landroid/os/Bundle;)V");
```

```
env->CallNonvirtualVoidMethod(thiz, AppCompatActivity_class2, onCreate,
saved_instance_state);
```

另一种方法，使用 JADX 直接看 APK 中的方法签名，如图 11-2 所示。

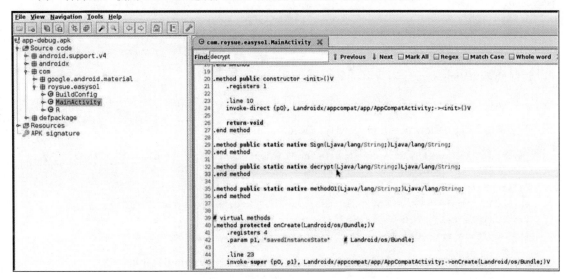

图 11-2　使用 JADX 查看方法签名

接下来，我们来看第二句话在 Native 层如何实现。由于 setContentView 在 AppCompatActivity 中，因此可以直接找 jmethodID：

```
jmethodID setContentView = env->GetMethodID(AppCompatActivity_class2,
"setContentView", "(I)V");
```

之后，R.layout.activity_main 参数需要通过 env->FindClass 和 GetStaticFieldID 来获取。有了 jfieldID，我们就可以获取它的值：

```
jclass Rlayout = env->FindClass("com/roysue/easyso1/R$layout");
jfieldID activity_main = env->GetStaticFieldID(Rlayout, "activity_main", "I");
jint activity_main_value = env->GetStaticIntField(Rlayout, activity_main);
```

由于 setContentView 是一个 non-static method，因此我们通过 CallVoidMethod 来主动调用它：

```
env->CallVoidMethod(thiz, setContentView, activity_main_value);
```

第二条语句 setContentView(R.layout.activity_main);完整翻译如下：

```
jmethodID setContentView = env->GetMethodID(AppCompatActivity_class2,
"setContentView", "(I)V");
jclass Rlayout = env->FindClass("com/roysue/easyso1/R$layout");
jfieldID activity_main = env->GetStaticFieldID(Rlayout, "activity_main", "I");
jint activity_main_value = env->GetStaticIntField(Rlayout, activity_main);
env->CallVoidMethod(thiz, setContentView, activity_main_value);
```

在第三条语句中,我们可以用相似的方法获取 findViewById(jmethodID)和 R.id.sample_text:

```
    jmethodID findViewById = env->GetMethodID(AppCompatActivity_class2, "findViewById",
"(I)Landroid/view/View;");
    jclass Rid = env->FindClass("com/roysue/easyso1/R$id");
    jfieldID sample_text = env->GetStaticFieldID(Rid,"sample_text","I");
    jint sample_text_value = env->GetStaticIntField(Rid,sample_text);
```

返回值 TextView 是一个对象(object):

```
    jobject tv = env->CallObjectMethod(thiz, findViewById, sample_text_value);
```

第三条语句 TextView tv = findViewById(R.id.sample_text);完整翻译如下:

```
    jmethodID findViewById = env->GetMethodID(AppCompatActivity_class2, "findViewById",
"(I)Landroid/view/View;");
    jclass Rid = env->FindClass("com/roysue/easyso1/R$id");
    jfieldID sample_text = env->GetStaticFieldID(Rid,"sample_text","I");
    jint sample_text_value = env->GetStaticIntField(Rid,sample_text);
    jobject tv = env->CallObjectMethod(thiz, findViewById, sample_text_value);
```

最后一条语句的参数是一个 String "r0ysue",可用 env->NewStringUTF 来构造:

```
    jstring r0ysue = env->NewStringUTF("r0ysue");
```

获取 jclass textView,再获取 jmethodID setText:

```
    jclass textView = env->FindClass("android/widget/TextView");
    jmethodID setText =
env->GetMethodID(textView,"setText","(Ljava/lang/CharSequence;)V");
```

最后调用 setText:

```
    env->CallVoidMethod(tv,setText,r0ysue);
```

最后一句 TextView tv = findViewById(R.id.sample_text);完整翻译如下:

```
    jstring r0ysue = env->NewStringUTF("r0ysue");
    jclass textView = env->FindClass("android/widget/TextView");
    jmethodID setText =
env->GetMethodID(textView,"setText","(Ljava/lang/CharSequence;)V");
    env->CallVoidMethod(tv,setText,r0ysue);
```

至此,我们已经翻译完 onCreate 的 4 条语句,运行结果如图 11-3 所示,说明我们翻译得没有问题。

第 11 章 onCreate 进行 Native 化和引用　213

图 11-3　翻译完 onCreate 的 4 条语句后的运行结果

通过这种方式，我们将 onCreate 进行了完整的 Native 化。将 APK 反编译后，我们只能在 Java 层看到一个空的 onCreate 函数。

下面将 onCreate 改为动态注册。此时借助 IDA 仍然可以找到 Native 化的 onCreate，也可以看到我们写的内容都在，跟明文没有什么区别，而且可以快速定位，如图 11-4 和图 11-5 所示。

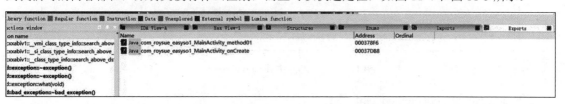

图 11-4　借助 IDA 仍然可以找到 Native 化的 onCreate

图 11-5　Native 层实现的 onCreate

这是我们希望避免的。最基础的保护方法是把 Native 层的 onCreate 名称换掉。我们需要从 JADX 中找到 onCreate 签名，如图 11-6 所示。

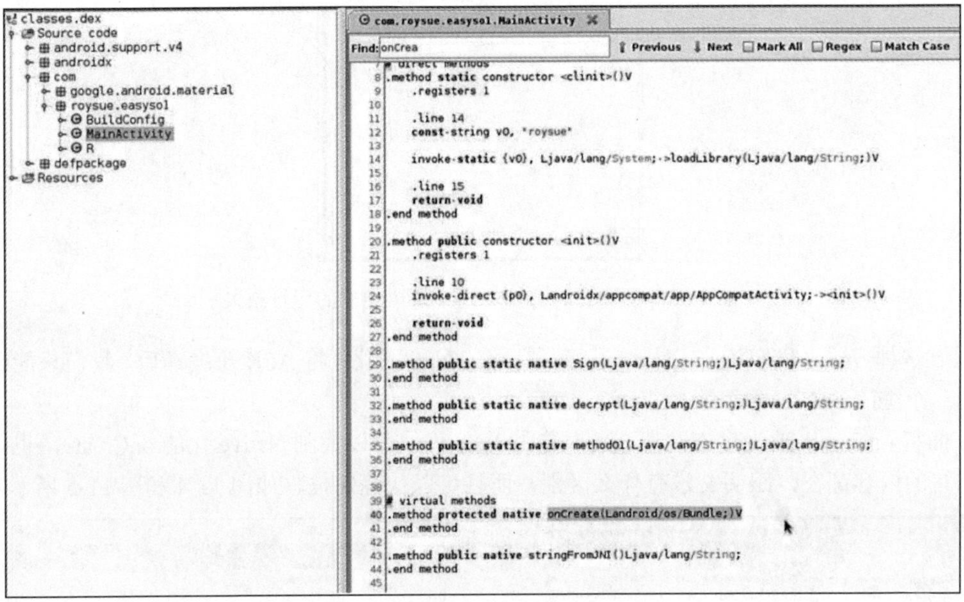

图 11-6　从 JADX 中找到 onCreate 签名

将 onCreate 添加到 method_table 中：

```
static JNINativeMethod method_table[] = {
    //{"method01", "(Ljava/lang/String;)Ljava/lang/String;", (void *) method01},
    {"decrypt", "(Ljava/lang/String;)Ljava/lang/String;", (void *) method02},
    {"Sign", "(Ljava/lang/String;)Ljava/lang/String;", (void *) ngis},
    {"onCreate", "(Landroid/os/Bundle;)V", (void *) ononon},
};
```

同样，将 Native 层的静态注册命名改为动态注册命名，如图 11-7 所示，并将 ononon 函数体移到 method_table 前面，否则可能会找不到。

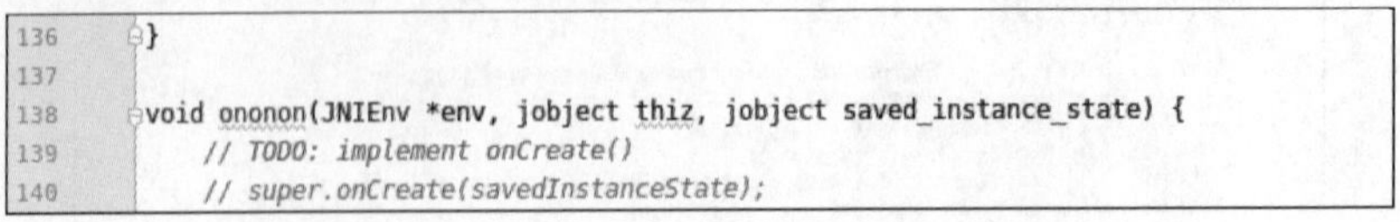

图 11-7　将 Native 层的静态注册命名改为动态注册命名

这样就将 onCreate 改为动态注册了。动态注册的好处是不能直接找到 Java 层和 Native 层的对应关系，如图 11-8 所示，但是运行时看不出任何区别。

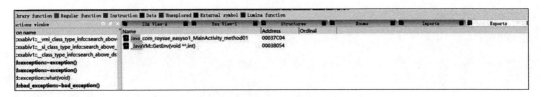

图 11-8　搜索不到 Java 层和 Native 层的对应关系了

尽管我们对 onCreate 做了 Native 化和动态注册的保护，但仍然可以通过追踪找到。如果使用 IDA 静态分析，可以通过 JNI_OnLoad 去找，method_table 是关键，里面有对应关系。

如果使用 Frida 动态查看，可以直接运行 hook_RegisterNatives.js：

```
root@VXIDr0ysue:~/Desktop/20210321/frida_hook_libart-master # frida -U -f com.roysue.easyso1 -l hook_RegisterNatives.js --no-pause

RegisterNatives is at 0xe92077b1
_ZN3art3JNI15RegisterNativesEP7_JNIEnvP7_jclassPK15JNINativeMethodi
Spawned com.roysue.easyso1. Resuming main thread!
[LGE Nexus 5X::com.roysue.easyso1]-> [RegisterNatives] method_count: 0x3
[RegisterNatives] java_class: com.roysue.easyso1.MainActivity name: decrypt sig: (Ljava/lang/String;)Ljava/lang/String; fnPtr: 0xce35dde9 module_name: libroysue.so module_base: 0xce325000 offset: 0x38de9
[RegisterNatives] java_class: com.roysue.easyso1.MainActivity name: Sign sig: (Ljava/lang/String;)Ljava/lang/String; fnPtr: 0xce35e529 module_name: libroysue.so module_base: 0xce325000 offset: 0x39529
[RegisterNatives] java_class: com.roysue.easyso1.MainActivity name: onCreate sig: (Landroid/os/Bundle;)V fnPtr: 0xce35e94d module_name: libroysue.so module_base: 0xce325000 offset: 0x3994d
```

这样就有了 onCreate、签名和 Native 的对应关系，offset 是 0x3994d。另一个工具 Hook 的是 ArtMethod 的 RegisterNatives：参考网址 38（参见配书资源文件）。这个工具可以将动态注册和静态注册的函数都打印出来。我们将代码复制到本地并运行（需要适配系统 Android 8.1）。

```
const void* ArtMethod::RegisterNative(const void* native_method, bool is_fast) {
    CHECK(IsNative()) << PrettyMethod();
    CHECK(!IsFastNative()) << PrettyMethod();
    CHECK(native_method != nullptr) << PrettyMethod();
    if (is_fast) {
        AddAccessFlags(kAccFastNative);
    }
    void* new_native_method = nullptr;
    Runtime::Current()->GetRuntimeCallbacks()->RegisterNativeMethod(this, native_method,
        /*out*/&new_native_method);
    SetEntryPointFromJni(new_native_method);
    return new_native_method;
```

```
    }

    /*
    仅在 Android 8.1 下测试成功,其他版本可能需要重新修改适配
    */

    var ArtMethod_PrettyMethod
    function readStdString(str) {
        if ((str.readU8() & 1) === 1) { // size LSB (=1) indicates if it's a long string
            return str.add(2 * Process.pointerSize).readPointer().readUtf8String();
        }
        return str.add(1).readUtf8String();
    }
    function attach(addr) {
        Interceptor.attach(addr, {
            onEnter: function (args) {
                this.arg0 = args[0]
            },
            onLeave: function (retval) {
                var modulemap = new ModuleMap()
                modulemap.update()
                var module = modulemap.find(retval)
                var string = Memory.alloc(0x100)
                ArtMethod_PrettyMethod(string, this.arg0, 1)
                if (module != null) {
                    console.log('<' + module.name + '> method_name =>', readStdString(string), ',offset=>', ptr(retval).sub(module.base), ',module_name=>', module.name)
                }else{
                    console.log('<anonymous> method_name =>', readStdString(string), ', addr =>', ptr(retval))
                }
            }
        });
    }

    function hook_RegisterNative() {
        var libart = Process.findModuleByName('libart.so')
        var symbols = libart.enumerateSymbols()
        for (var i = 0; i < symbols.length; i++) {
            if (symbols[i].name.indexOf('PrettyMethod') > -1 && symbols[i].name.indexOf('ArtMethod') > -1 && symbols[i].name.indexOf("Eb") >= 0) {
                ArtMethod_PrettyMethod = new NativeFunction(symbols[i].address, "void", ['pointer', "pointer", "bool"])
            }
            if (symbols[i].name.indexOf('RegisterNative') > -1 && symbols[i].name.indexOf('ArtMethod') > -1 && symbols[i].name.indexOf('RuntimeCallbacks')
```

```
< 0) {
            attach(symbols[i].address)
        }
    
    }
}
function main() {
    hook_RegisterNative()
}
setImmediate(main)
```

从运行结果可以获取动态注册的函数，如图 11-9 所示。

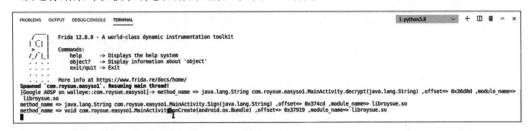

图 11-9　运行 hook_ArtMethodRegisterNatives.js 获取所有动态注册的函数

为了体现 method01，我们将 Java_com_roysue_easyso1_MainActivity_method01 放到前面，修改代码如下，实现 method01 对 r0ysue 加密，并用反射调用 method01：

```
    // tv.setText("r0ysue");
    jstring r0ysue = env->NewStringUTF("r0ysue");
    jclass textView = env->FindClass("android/widget/TextView");
    jmethodID setText = env->GetMethodID(textView,"setText","(Ljava/lang/CharSequence;)V");
    jstring r0ysueEn = Java_com_roysue_easyso1_MainActivity_method01(env, MainActivity_class2, r0ysue);
    jmethodID Javamethod01 = env->GetStaticMethodID(MainActivity_class2,"method01","(Ljava/lang/String;)Ljava/lang/String;");
    jstring r0ysueEn2 = static_cast<jstring>(env->CallStaticObjectMethod(MainActivity_class2, Javamethod01, r0ysue));
    env->CallVoidMethod(tv,setText,r0ysueEn);
```

运行结果如图 11-10 所示。

图 11-10　method01 对 r0ysue 加密

再次运行 hook_ArtMethodRegisterNatives.js，我们就可以看到 method01 了，如图 11-11 所示。

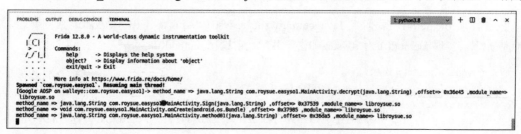

图 11-11　Hook 得到 method01

对于这个工具来说，通过 Hook RegisterNatives 能否得到方法信息的区别如下：

- 动态注册的方法：都会显示。
- 静态注册的方法：只有 Java 调用过才会显示。

hook_ArtMethodRegisterNatives.js 逻辑分析：在 hook_RegisterNative 函数中，首先找到虚拟机并枚举所有符号。之后将所有符号拿出来，满足条件之后，定义 NativeFunction，即 Native 的主动调用。

我们只要抽取 Android 8 的进程出来查看即可，例如 32 位进程 2129，如图 11-12 所示。根据进程号搜索 libart.so，并将它保存到 sdcard 中，如图 11-13 所示。

```
walleye:/ # cat /proc/2129/maps|grep libart
```

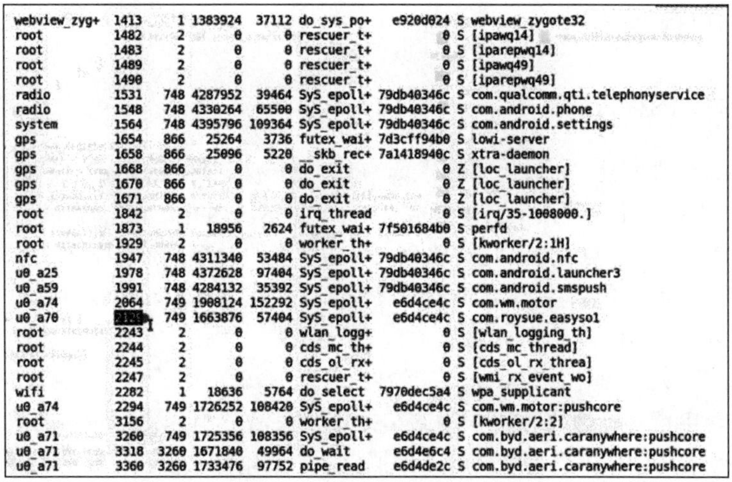

```
webview_zyg+  1413     1 1383924  37112 do_sys_po+   e920d024 S webview_zygote32
root          1482     2       0      0 rescuer_t+          0 S [ipawq14]
root          1483     2       0      0 rescuer_t+          0 S [iparepwq14]
root          1489     2       0      0 rescuer_t+          0 S [ipawq49]
root          1490     2       0      0 rescuer_t+          0 S [iparepwq49]
radio         1531   748 4287952  39464 SyS_epoll+ 79db40346c S com.qualcomm.qti.telephonyservice
radio         1548   748 4330264  65500 SyS_epoll+ 79db40346c S com.android.phone
system        1564   748 4395796 109364 SyS_epoll+ 79db40346c S com.android.settings
gps           1654   866   25264   3736 futex_wai+ 7d3cff94b0 S lowi-server
gps           1658   866   25096   5220 __skb_rec+ 7a1418940c S xtra-daemon
gps           1668   866       0      0 do_exit             0 Z [loc_launcher]
gps           1670   866       0      0 do_exit             0 Z [loc_launcher]
gps           1671   866       0      0 do_exit             0 Z [loc_launcher]
root          1842     2       0      0 irq_thread          0 S [irq/35-1008000.]
root          1873     1   18956   2624 futex_wai+ 7f501684b0 S perfd
root          1929     2       0      0 worker_th+          0 S [kworker/2:1H]
nfc           1947   748 4311340  53484 SyS_epoll+ 79db40346c S com.android.nfc
u0_a25        1978   748 4372628  97404 SyS_epoll+ 79db40346c S com.android.launcher3
u0_a59        1991   748 4284132  35392 SyS_epoll+ 79db40346c S com.android.smspush
u0_a74        2064   749 1908124 152292 SyS_epoll+   e6d4ce4c S com.wm.motor
u0_a70        2129   749 1663876  57404 SyS_epoll+   e6d4ce4c S com.roysue.easyso1
root          2243     2       0      0 wlan_logg+          0 S [wlan_logging_th]
root          2244     2       0      0 cds_mc_th+          0 S [cds_mc_thread]
root          2245     2       0      0 cds_ol_rx+          0 S [cds_ol_rx_threa]
root          2247     2       0      0 rescuer_t+          0 S [wmi_rx_event_wo]
wifi          2282     1   18636   5764 do_select  7970dec5a4 S wpa_supplicant
u0_a74        2294   749 1726252 108420 SyS_epoll+   e6d4ce4c S com.wm.motor:pushcore
root          3156     2       0      0 worker_th+          0 S [kworker/2:2]
u0_a71        3260   749 1725356 108356 SyS_epoll+   e6d4ce4c S com.byd.aeri.caranywhere:pushcore
u0_a71        3318  3260 1671840  49964 do_wait      e6d4e6c4 S com.byd.aeri.caranywhere:pushcore
u0_a71        3360  3260 1733476  97752 pipe_read    e6d4de2c S com.byd.aeri.caranywhere:pushcore
```

图 11-12　选取 32 位进程 2129

```
walleye:/ # cp /system/lib/libart.so /sdcard/

root@VXIDr0ysue:~/Desktop/20210409 # adb pull /sdcard/libart.so
```

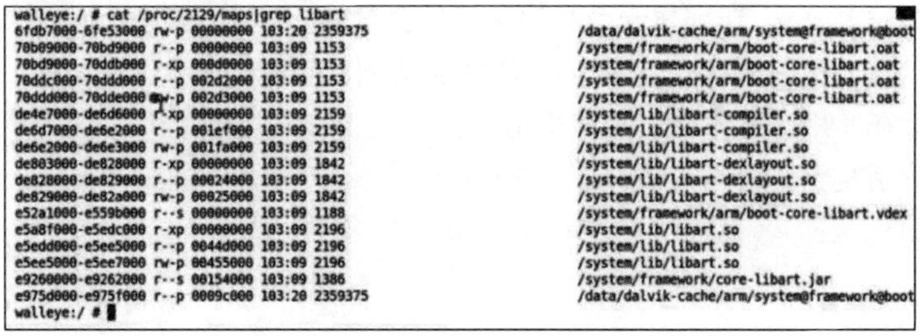

```
walleye:/ # cat /proc/2129/maps|grep libart
6fdb7000-6fe53000 rw-p 00000000 103:20 2359375     /data/dalvik-cache/arm/system@framework@boot
70b09000-70bd9000 r--p 00000000 103:09 1153        /system/framework/arm/boot-core-libart.oat
70bd9000-70ddb000 r-xp 000d0000 103:09 1153        /system/framework/arm/boot-core-libart.oat
70ddc000-70ddd000 r--p 002d2000 103:09 1153        /system/framework/arm/boot-core-libart.oat
70ddd000-70dde000 rw-p 002d3000 103:09 1153        /system/framework/arm/boot-core-libart.oat
de4e7000-de6d6000 r-xp 00000000 103:09 2159        /system/lib/libart-compiler.so
de6d7000-de6e2000 r--p 001ef000 103:09 2159        /system/lib/libart-compiler.so
de6e2000-de6e3000 rw-p 001fa000 103:09 2159        /system/lib/libart-compiler.so
de803000-de828000 r-xp 00000000 103:09 1842        /system/lib/libart-dexlayout.so
de828000-de829000 r--p 00024000 103:09 1842        /system/lib/libart-dexlayout.so
de829000-de82a000 rw-p 00025000 103:09 1842        /system/lib/libart-dexlayout.so
e52a1000-e559b000 r--s 00000000 103:09 1188        /system/framework/arm/boot-core-libart.vdex
e5a8f000-e5edc000 r-xp 00000000 103:09 2196        /system/lib/libart.so
e5edd000-e5ee5000 r--p 0044d000 103:09 2196        /system/lib/libart.so
e5ee5000-e5ee7000 rw-p 00455000 103:09 2196        /system/lib/libart.so
e9260000-e9262000 r--s 00154000 103:09 1386        /system/framework/core-libart.jar
e975d000-e975f000 r--p 0009c000 103:20 2359375     /data/dalvik-cache/arm/system@framework@boot
walleye:/ #
```

图 11-13　根据进程号搜索 libart.so

用 IDA 打开分析 libart.so，就可以找出它所有的符号。回到 hook_ArtMethodRegisterNatives.js，它寻找的主要是两个函数：PrettyMethod 和 RegisterNative。其中，重点还是 RegisterNative 函数。在 attach 中，把参数传递给返回值，返回值是一个 new_native_method 指针。然后就可以用 ModuleMap 判断 new_native_method 属于哪个 Module。ModuleMap 可以缓存一份当前内存中所有 SO 文件的模块表。ArtMethod_PrettyMethod(string, this.arg0,1)第一个参数是一段内存空间，第二个参数是 ArtMethod 指针，第三个参数表示是否带有签名。所以说，它的功能就是给它一个 ArtMethod 方法类型指针，返回一个方法的签名、类的信息、返回值，方法名写入第一个参数中。最后用 readStdString 读取出来并打印。

11.2 Java 内存管理

11.2.1 C 和 Java 内存管理的差异

在 C 语言中，只要新建（new）一个对象，就对应一次删除（delete）操作。同样地，只要申请（malloc）一段内存，不再使用后必须释放（free）掉这段内存，否则程序就会内存泄露。如果内存中的对象越来越多，却不回收，这时就会占用非常多的内存。而且在 C 中，内存还分栈空间和堆空间。函数的形参、（临时）变量等放在栈空间，而用 new、malloc 申请的内存放在堆空间。以上就是 C 语言最基本的内存管理方式。

对于 Java 而言，这些都不用管，直接交给内存管理器就可以了。手动管理内存非常麻烦，而且需要细心。手动申请和释放内存对程序的掌握比较灵活，不会受到其他的限制。而我们编写 Android 程序的时候会受到 ART 或者 Dalvik 虚拟机版本的限制。它们的版本一开始只有 16~24MB 或者 32~64MB，后来 DVM 不断变大。相比之下，对于 C 来讲，内存管理完全不受虚拟机限制，可以用到完整的 4GB。所以在 Native 层，我们需要手动实现一部分内存管理。

11.2.2 JNI 的三种引用

使用 Java 编写程序时，在创建一个类的实例后会得到一个实例的引用，通过这个引用可以访问类的成员变量以及成员函数。从 JVM 创建的对象传到 C/C++代码会产生引用，根据 Java 的垃圾回收机制，只要存在引用，就不会触发该引用所指向的 Java 对象的垃圾回收。因此，在 Native 中存在 3 种引用，分别是全局引用（Global Reference）、局部引用（Local Reference）、弱全局引用（Weak Global Reference），从而实现在 C 中对 Java 对象的一些管理。

1. 局部引用

局部引用也称作本地引用，通常在函数中创建并使用，会阻止垃圾回收（Garbage Collector，GC）。局部引用是最常见的引用类型，基本上通过 JNI 返回的引用都属于局部引用，例如调用 NewObject 函数创建出来的实例将返回一个局部引用。局部引用值在当前的 Native 函数中有效，所有在该函数中产生的局部引用都会在函数返回时自动释放，也可以调用 DeleteLocalRef 函数手动释放该引用。实际上，当存在局部引用时，它会防止其指向的对象被垃圾回收器回收，尤其是当局部变量引用指向一个很庞大的对象，或者在循环中生成局部引用时。调用 DeleteLocalRef 函数手动释放引用是为了确保在垃圾回收器触发时局部变量引用被回收。在局部引用的有效期内，它可以传递到其他本地函数中。需要强调的是，它的有效期仅限于第一次的 Java 本地函数调用中，因此千万不能用 C++全局变量保存它或把它定义为 C++静态局部变量。

通过 NewLocalRef 和各种 JNI 接口创建（FindClass、NewObject、GetObjectClass 和 NewCharArray 等），会阻止垃圾回收所引用的对象。函数返回后，局部引用所引用的对象会被 JVM 自动释放，或调用 DeleteLocalRef 释放。

2. 全局引用

全局引用可以跨方法、跨线程使用，直到被开发人员显式释放。类似于局部引用，一个全局引用在被释放前可以保证引用对象不被垃圾回收。和局部引用不同的是，没有那么多函数能够创建全局引用。能创建全局引用的函数只有 NewGlobalRef，而释放它需要调用 ReleaseGlobalRef 函数。

调用 NewGlobalRef 函数基于局部引用创建全局引用，全局引用会阻止垃圾回收器回收所引用的对象，并且可以跨方法、跨线程使用。JVM 不会自动释放全局引用，必须调用 DeleteGlobalRef 函数手动释放。

3. 弱全局引用

调用 NewWeakGlobalRef 函数基于局部引用或全局引用创建弱全局引用，弱全局引用不会阻止垃圾回收所引用的对象，可以跨方法、跨线程使用。弱全局引用不会自动释放，在 JVM 认为应该回收它的时候（比如内存紧张的时候）才会进行回收，或调用 DeleteWeakGlobalRef 函数手动释放。

关于 JNI 的 3 种引用更详细的介绍，可以参考网址 39（参见配书资源文件）的文章。

11.3 本章小结

本章介绍了对 Java 层的 onCreate 函数进行 Native 化的加固，以增加逆向分析工作的难度。此外，我们还学习了 JNI 的 3 种引用方式，即局部引用、全局引用和弱全局引用，并进行了对比。

第 12 章

JNI 动静态绑定和追踪

本章将讲解 ROM,在源码中添加日志能够实现不依赖于 Frida 的追踪。虽然 Frida 的用处非常广泛,但只要应用程序中有检测 Frida 的代码,那么依赖于 Frida 的逆向分析就无法进展下去。

回顾第 11 章,我们使用反射将 onCreate 函数 Native 化。尽管使用反射实现比较复杂,但其原理是一模一样的,这也是 dextoC 加固的基础。本质上可以把 Java 层所有函数都 Native 化,将它们转换为 Native 层的函数,并在 Native 层中使用反射调用 Java 层的各种类库。比如 super.onCreate,需要先找到它的父类,再获取 jmethodID,并使用 env 中 Call 开头的函数进行调用。重点在于观察 env 的 RegisterNatives,其中 RegisterNatives 的第二个参数 gMethods 指向了 method_table 的起始位置。

12.1 Dalvik 下动静态注册流程追踪

在之前的章节中,我们讲解了通过 Hook RegisterNatives 来对 method_table 中的函数绑定关系进行解析。那么 RegisterNatives 是如何进行动态注册的呢?首先来看一下 Dalvik 中的 RegisterNatives 函数。

Android android-4.0.1_r1 源码参考网址 40(参见配书资源文件)。

在 Definition 中查找 RegisterNatives,在右侧的 Project 中选择 dalvik,发现它在 Jni.cpp 中:

```
static jint RegisterNatives(JNIEnv* env, jclass jclazz,
    const JNINativeMethod* methods, jint nMethods)
{
    ScopedJniThreadState ts(env);

    ClassObject* clazz = (ClassObject*) dvmDecodeIndirectRef(env, jclazz);
```

```
    if (gDvm.verboseJni) {
        LOGI("[Registering JNI native methods for class %s]",
            clazz->descriptor);
    }

    for (int i = 0; i < nMethods; i++) {
        if (!dvmRegisterJNIMethod(clazz, methods[i].name,
                methods[i].signature, methods[i].fnPtr))
        {
            return JNI_ERR;
        }
    }
    return JNI_OK;
}
```

从注释中，我们了解到，RegisterNatives 的功能是在一个 Java 类中注册一个或多个 Native 的函数，可以在一个方法中调用多次，并且允许调用者重新定义方法的实现。在 RegisterNatives 函数的参数中，最重要的是第 3 个参数 methods。我们可以看到函数体中遍历了所有的 methods，并调用了 dvmRegisterJNIMethod。下面继续查看 dvmRegisterJNIMethod：

```
/*
 * Register a method that uses JNI calling conventions
 */
static bool dvmRegisterJNIMethod(ClassObject* clazz, const char* methodName,
...
    Method* method = dvmFindDirectMethodByDescriptor(clazz, methodName, signature);
    if (method == NULL) {
        method = dvmFindVirtualMethodByDescriptor(clazz, methodName, signature);
    }
    if (method == NULL) {
        dumpCandidateMethods(clazz, methodName, signature);
        return false;
    }

    if (!dvmIsNativeMethod(method)) {
        LOGW("Unable to register: not native: %s.%s:%s", clazz->descriptor, methodName,
signature);
        return false;
    }

    if (fastJni) {
        // In this case, we have extra constraints to check...
        if (dvmIsSynchronizedMethod(method)) {
            // Synchronization is usually provided by the JNI bridge,
            // but we won't have one
```

```
            LOGE("fast JNI method %s.%s:%s cannot be synchronized",
                    clazz->descriptor, methodName, signature);
            return false;
        }
        if (!dvmIsStaticMethod(method)) {
            // There's no real reason for this constraint, but since we won't
            // be supplying a JNIEnv* or a jobject 'this', you're effectively
            // static anyway, so it seems clearer to say so
            LOGE("fast JNI method %s.%s:%s cannot be non-static",
                    clazz->descriptor, methodName, signature);
            return false;
        }
    }

    if (method->nativeFunc != dvmResolveNativeMethod) {
        /* this is allowed, but unusual */
        LOGV("Note: %s.%s:%s was already registered", clazz->descriptor, methodName,
signature);
    }

    method->fastJni = fastJni;
    dvmUseJNIBridge(method, fnPtr);

    LOGV("JNI-registered %s.%s:%s", clazz->descriptor, methodName, signature);
    return true;
}
```

Java 层的 Method 实际上是 Native 层的一个结构体。从 Method 结构体中输出 name 就能够得到方法名。在获取一个结构体对象之后，从里面索引 nativeFunc，如果它不等于 dvmResolveNativeMethod，就说明绑定过了。

接下来，调用 dvmUseJNIBridge，传入 method 结构体和 fnPtr 对象（指向 Native 层函数的指针）作为参数。

```
    /*
     * Point "method->nativeFunc" at the JNI bridge, and overload "method->insns"
     * to point at the actual function
     */
    void dvmUseJNIBridge(Method* method, void* func) {
        method->shouldTrace = shouldTrace(method);
        ...
        DalvikBridgeFunc bridge = gDvmJni.useCheckJni ? dvmCheckCallJNIMethod :
dvmCallJNIMethod;
        dvmSetNativeFunc(method, bridge, (const u2*) func);
    }
```

dvmUseJNIBridge 把 Native 函数指到 JNI bridge 上，并重载 method->insns 指向一个实际的函

数地址。进入 dvmSetNativeFunc：

```
void dvmSetNativeFunc(Method* method, DalvikBridgeFunc func,
    const u2* insns)
{
    ClassObject* clazz = method->clazz;

    ...

    if (insns != NULL) {
        /* update both, ensuring that "insns" is observed first */
        method->insns = insns;
        android_atomic_release_store((int32_t) func,
            (volatile int32_t*)(void*) &method->nativeFunc);
    } else {
        /* only update nativeFunc */
        method->nativeFunc = func;
    }

    dvmLinearReadOnly(clazz->classLoader, clazz->virtualMethods);
    dvmLinearReadOnly(clazz->classLoader, clazz->directMethods);
}
```

最核心的一行代码是 method->nativeFunc = func;，表示动态注册的本质就是把 Method 结构体中的 nativeFunc 指针指向 func。

我们在源码中添加一些日志：

```
static jint RegisterNatives(JNIEnv* env, jclass jclazz,
    const JNINativeMethod* methods, jint nMethods)
{
    ScopedJniThreadState ts(env);

    ClassObject* clazz = (ClassObject*) dvmDecodeIndirectRef(env, jclazz);

    if (gDvm.verboseJni) {
        LOGI("[Registering JNI native methods for class %s]", clazz->descriptor);
    }

    for (int i = 0; i < nMethods; i++) {
        LOGE("[RegisterNatives]Class:%s, methodname:%s, sig:%s, addr:%p",
clazz->descriptor, methods[i].name, methods[i].signature, methods[i].fnPtr);
        if (!dvmRegisterJNIMethod(clazz, methods[i].name,
                methods[i].signature, methods[i].fnPtr))
        {
            return JNI_ERR;
        }
```

```
        }
        return JNI_OK;
    }

    /*
     * Register a method that uses JNI calling conventions
     */
    static bool dvmRegisterJNIMethod(ClassObject* clazz, const char* methodName,
        const char* signature, void* fnPtr)
    {
        if (fnPtr == NULL) {
            return false;
        }
        LOGE("[dvmRegisterJNIMethod]Class:%s, methodname:%s, sig:%s, addr:%p",
clazz->descriptor, methods[i].name, methods[i].signature, methods[i].fnPtr);
        // If a signature starts with a '!', we take that as a sign that the native code doesn't
        // need the extra JNI arguments (the JNIEnv* and the jclass)
        bool fastJni = false;
        if (*signature == '!') {
            fastJni = true;
            ++signature;
            LOGV("fast JNI method %s.%s:%s detected", clazz->descriptor, methodName,
signature);
        }

        if (method->nativeFunc != dvmResolveNativeMethod) {
            /* this is allowed, but unusual */
            LOGE("Note: %s.%s:%s was already registered", clazz->descriptor, methodName,
signature);
        }

        method->fastJni = fastJni;
        dvmUseJNIBridge(method, fnPtr);

        LOGV("JNI-registered %s.%s:%s", clazz->descriptor, methodName, signature);
        return true;
    }

    /*
     * Point "method->nativeFunc" at the JNI bridge, and overload "method->insns"
     * to point at the actual function.
     */
    void dvmUseJNIBridge(Method* method, void* func) {
        method->shouldTrace = shouldTrace(method);
```

```c
        LOGE("[dvmUseJNIBridge]Class:%s, methodname:%s, addr:%p",
method->clazz->descriptor, method->name, func);
        // Does the method take any reference arguments?
        method->noRef = true;
        const char* cp = method->shorty;
        while (*++cp != '\0') { // Pre-increment to skip return type.
            if (*cp == 'L') {
                method->noRef = false;
                break;
            }
        }

        DalvikBridgeFunc bridge = gDvmJni.useCheckJni ? dvmCheckCallJNIMethod :
dvmCallJNIMethod;
        dvmSetNativeFunc(method, bridge, (const u2*) func);
    }

    void dvmSetNativeFunc(Method* method, DalvikBridgeFunc func,
        const u2* insns)
    {
        ClassObject* clazz = method->clazz;

        assert(func != NULL);

        /* just open up both; easier that way */
        dvmLinearReadWrite(clazz->classLoader, clazz->virtualMethods);
        dvmLinearReadWrite(clazz->classLoader, clazz->directMethods);

        if (insns != NULL) {
            /* update both, ensuring that "insns" is observed first */
            method->insns = insns;
            android_atomic_release_store((int32_t) func,
                (volatile int32_t*)(void*) &method->nativeFunc);
        } else {
            LOGE("[dvmSetNativeFunc]Class:%s, methodname:%s, addr:%p",
method->clazz->descriptor, method->name, func);
            /* only update nativeFunc */
            method->nativeFunc = func;
        }

        dvmLinearReadOnly(clazz->classLoader, clazz->virtualMethods);
        dvmLinearReadOnly(clazz->classLoader, clazz->directMethods);
    }
```

用 Nexus 5 刷机，将 easyso1 项目中调用的 method01 相关代码注释掉。

在之前的章节中，我们提到过 SO 文件的加载流程，最先运行的是 linker 的 .init 和 .init_array，它们运行结束后各种 env 环境就生成了，然后，就可以由 JNI_OnLoad 进行加载。在 JNI 环境执行 JNI_OnLoad 函数，在 JNI_OnLoad 函数中，我们调用了 registerMethods 函数进行动态注册。所以可以看到日志输出。

如果要验证 onCreate 是否可以进行二次绑定，以及最终执行的是哪个绑定的函数。可复制 ononon 函数，并将其函数名改为 ononon2。然后在 method_table 中与 onCreate 函数进行绑定。运行后打印出的日志中会显示二次注册的信息。

两次注册的地址不同。第一次是 0x755e89d5，第二次是 0x755e8cc1。也就是说，对应的是两个不同的 JNI 函数。此时，手机上显示的是 ononon2。

12.2　ART 下动静态注册流程追踪

接下来，我们对 ART 源码进行修改。参考源码 Android android-8.0.0_r1（网址 41，参见配书资源文件）。

在 Definition 中查找 RegisterNatives，在右侧的 Project 中选择 art，发现它有两个 RegisterNatives，分别在 check_jni.cc 和 jni_internal.cc 中，选择 jni_internal.cc 中的 RegisterNatives，如图 12-1 所示。

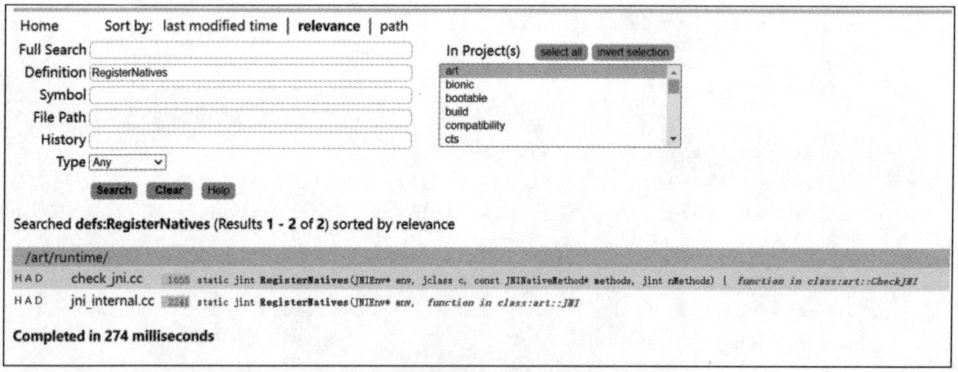

图 12-1　选择 jni_internal.cc 中的 RegisterNatives

通常，我们会在 Hook 时把 check_jni 过滤掉。

```
static jint RegisterNatives(JNIEnv* env, jclass java_class, const JNINativeMethod* methods, jint method_count) {
    return RegisterNativeMethods(env, java_class, methods, method_count, true);
}
```

我们发现 RegisterNatives 只是调用了 RegisterNativeMethods，单击 RegisterNativeMethods，发现结果有两个，但是参数不同。根据 RegisterNatives 调用传参，我们得知应选择第二个参数是 jclass 的 RegisterNativeMethods，如图 12-2 所示。

第 12 章　JNI 动静态绑定和追踪

图 12-2　选择第二个参数是 jclass 的 RegisterNativeMethods

```
    static jint RegisterNativeMethods(JNIEnv* env, jclass java_class, const
JNINativeMethod* methods, jint method_count, bool return_errors) {
        if (UNLIKELY(method_count < 0)) {
            JavaVmExtFromEnv(env)->JniAbortF("RegisterNatives", "negative method
count: %d", method_count);
    // Not reached except in unit tests
            return JNI_ERR;
        }
        CHECK_NON_NULL_ARGUMENT_FN_NAME("RegisterNatives", java_class, JNI_ERR);
        ScopedObjectAccess soa(env);
        StackHandleScope<1> hs(soa.Self());
        Handle<mirror::Class> c = hs.NewHandle(soa.Decode<mirror::Class>(java_class));
        if (UNLIKELY(method_count == 0)) {
            LOG(WARNING) << "JNI RegisterNativeMethods: attempt to register 0 native methods
for " << c->PrettyDescriptor();
            return JNI_OK;
        }
        CHECK_NON_NULL_ARGUMENT_FN_NAME("RegisterNatives", methods, JNI_ERR);
        for (jint i = 0; i < method_count; ++i) {
            const char* name = methods[i].name;
            const char* sig = methods[i].signature;
            const void* fnPtr = methods[i].fnPtr;
            if (UNLIKELY(name == nullptr)) {
                ReportInvalidJNINativeMethod(soa, c.Get(), "method name", i,
return_errors);
                return JNI_ERR;
            } else if (UNLIKELY(sig == nullptr)) {
                ReportInvalidJNINativeMethod(soa, c.Get(), "method signature", i,
return_errors);
                return JNI_ERR;
            } else if (UNLIKELY(fnPtr == nullptr)) {
                ReportInvalidJNINativeMethod(soa, c.Get(), "native function", i,
return_errors);
                return JNI_ERR;
```

```cpp
        }
        bool is_fast = false;
        // Notes about fast JNI calls:
        //
        // On a normal JNI call, the calling thread usually transitions
        // from the kRunnable state to the kNative state. But if the
        // called native function needs to access any Java object, it
        // will have to transition back to the kRunnable state
        //
        // There is a cost to this double transition. For a JNI call
        // that should be quick, this cost may dominate the call cost
        //
        // On a fast JNI call, the calling thread avoids this double
        // transition by not transitioning from kRunnable to kNative and
        // stays in the kRunnable state
        //
        // There are risks to using a fast JNI call because it can delay
        // a response to a thread suspension request which is typically
        // used for a GC root scanning, etc. If a fast JNI call takes a
        // long time, it could cause longer thread suspension latency
        // and GC pauses
        //
        // Thus, fast JNI should be used with care. It should be used
        // for a JNI call that takes a short amount of time (eg. no
        // long-running loop) and does not block (eg. no locks, I/O,
        // etc.)
        //
        // A '!' prefix in the signature in the JNINativeMethod
        // indicates that it's a fast JNI call and the runtime omits the
        // thread state transition from kRunnable to kNative at the
        // entry.
        if (*sig == '!') {
            is_fast = true;
            ++sig;
        }

        // Note: the right order is to try to find the method locally
        // first, either as a direct or a virtual method. Then move to
        // the parent
        ArtMethod* m = nullptr;
        bool warn_on_going_to_parent =
down_cast<JNIEnvExt*>(env)->vm->IsCheckJniEnabled();
        for (ObjPtr<mirror::Class> current_class = c.Get();
             current_class != nullptr;
             current_class = current_class->GetSuperClass()) {
            // Search first only comparing methods which are native
```

```cpp
            m = FindMethod<true>(current_class.Ptr(), name, sig);
            if (m != nullptr) {
                break;
            }

            // Search again comparing to all methods, to find non-native methods that match
            m = FindMethod<false>(current_class.Ptr(), name, sig);
            if (m != nullptr) {
                break;
            }

            if (warn_on_going_to_parent) {
                LOG(WARNING) << "CheckJNI: method to register \"" << name << "\" not in the given class. " << "This is slow, consider changing your RegisterNatives calls.";
                warn_on_going_to_parent = false;
            }
        }

        if (m == nullptr) {
            c->DumpClass(
                LOG_STREAM(return_errors
                    ? ::android::base::ERROR
                    : ::android::base::FATAL_WITHOUT_ABORT),
                mirror::Class::kDumpClassFullDetail);
            LOG(return_errors ? ::android::base::ERROR : ::android::base::FATAL)
                << "Failed to register native method "
                << c->PrettyDescriptor() << "." << name << sig << " in "
                << c->GetDexCache()->GetLocation()->ToModifiedUtf8();
            ThrowNoSuchMethodError(soa, c.Get(), name, sig, "static or non-static");
            return JNI_ERR;
        } else if (!m->IsNative()) {
            LOG(return_errors ? ::android::base::ERROR : ::android::base::FATAL)
                << "Failed to register non-native method "
                << c->PrettyDescriptor() << "." << name << sig
                << " as native";
            ThrowNoSuchMethodError(soa, c.Get(), name, sig, "native");
            return JNI_ERR;
        }

        VLOG(jni) << "[Registering JNI native method " << m->PrettyMethod() << "]";

        if (UNLIKELY(is_fast)) {
            // There are a few reasons to switch:
            // 1) We don't support !bang JNI anymore, it will turn to a hard error later
            // 2) @FastNative is actually faster. At least 1.5x faster than !bang JNI
```

```
                //    and switching is super easy, remove ! in C code, add annotation in .java
code
                // 3) Good chance of hitting DCHECK failures in ScopedFastNativeObjectAccess
                //    since that checks for presence of @FastNative and not for ! in the
descriptor
                LOG(WARNING) << "!bang JNI is deprecated. Switch to @FastNative for " <<
m->PrettyMethod();
                is_fast = false;
            // TODO: make this a hard register error in the future
            }

            const void* final_function_ptr = m->RegisterNative(fnPtr, is_fast);
            UNUSED(final_function_ptr);
        }
    return JNI_OK;
}
```

我们首先关心的是 java_class，它在 Handle<mirror::Class> c = hs.NewHandle(soa.Decode<mirror::Class>(java_class));先转化成了一个 Handle c。c 在 ObjPtr<mirror::Class> current_class = c.Get();获取了 current_class 后，通过 FindMethod<true>(current_class.Ptr(), name, sig);将返回值赋给 Artmethod* m。我们继续查看 ArtMethod，发现它是一个 class。其中有一些域和方法：

```
    protected:
    // Field order required by test "ValidateFieldOrderOfJavaCppUnionClasses"
    // The class we are a part of
    GcRoot<mirror::Class> declaring_class_;

    // Access flags; low 16 bits are defined by spec
    // Getting and setting this flag needs to be atomic when concurrency is
    // possible, e.g. after this method's class is linked. Such as when setting
    // verifier flags and single-implementation flag
    std::atomic<std::uint32_t> access_flags_;

    /* Dex file fields. The defining dex file is available via declaring_class_->dex_cache_
*/

    // Offset to the CodeItem
    uint32_t dex_code_item_offset_;

    // Index into method_ids of the dex file associated with this method.
    uint32_t dex_method_index_;

    /* End of dex file fields */

    // Entry within a dispatch table for this method. For static/direct methods the index
is into
    // the declaringClass.directMethods, for virtual methods the vtable and for interface
methods the
```

```
    // ifTable
    uint16_t method_index_;

    // The hotness we measure for this method. Managed by the interpreter. Not atomic,
as we allow
    // missing increments: if the method is hot, we will see it eventually
    uint16_t hotness_count_;

    // Fake padding field gets inserted here

    // Must be the last fields in the method
    struct PtrSizedFields {
      // Short cuts to declaring_class_->dex_cache_ member for fast compiled code access.
      ArtMethod** dex_cache_resolved_methods_;

      // Pointer to JNI function registered to this method, or a function to resolve the
JNI function
      // or the profiling data for non-native methods, or an ImtConflictTable, or
      //  the single-implementation of an abstract/interface method
      void* data_;

      // Method dispatch from quick compiled code invokes this pointer which may cause
bridging into
      // the interpreter
      void* entry_point_from_quick_compiled_code_;
    } ptr_sized_fields_;
```

declaring_class_用于找到方法对应的 class；access_flags_用于表示访问权限，例如 public、private；dex_code_item_offset_ 是方法体对应的偏移；dex_method_index_ 是对应的方法索引。PtrSizedFields 中的 data_ 是指向方法注册的 JNI function，我们拿到每一个 ArtMethod 都可以输出对应域的值，就可以知道方法对应的 JNI function 的地址。这与 Hook 没有什么区别。

回到 RegisterNativeMethods，有了 ArtMethod 之后，输出 PrettyMethod() 方法。最终调用了 m->RegisterNative(fnPtr, is_fast);：

```cpp
const void* ArtMethod::RegisterNative(const void* native_method, bool is_fast) {
  CHECK(IsNative()) << PrettyMethod();
  CHECK(!IsFastNative()) << PrettyMethod();
  CHECK(native_method != nullptr) << PrettyMethod();
  if (is_fast) {
    AddAccessFlags(kAccFastNative);
  }
  void* new_native_method = nullptr;
  Runtime::Current()->GetRuntimeCallbacks()->RegisterNativeMethod(this,
                                                                  native_method,
                                                                  /*out*/&new_native_method);
  SetEntryPointFromJni(new_native_method);
  return new_native_method;
}
```

RegisterNative 中只是把简单地把 native_method 赋给 new_native_method，并传入 SetEntryPointFromJni，经过一系列调用，最终设置好 data。

综上所述，绑定关系的本质就是把 ArtMethod 里面的结构体中的一个域的指针设置成 native 函数指针。

我们在源码中添加一些日志：

```
    static jint RegisterNatives(JNIEnv* env, jclass java_class, const JNINativeMethod* methods, jint method_count) {
        LOG(WARNING) << "[jni_internal.cc->RegisterNatives]" << method_count;
        return RegisterNativeMethods(env, java_class, methods, method_count, true);
    }


    static jint RegisterNativeMethods(JNIEnv* env, jclass java_class, const JNINativeMethod* methods, jint method_count, bool return_errors) {

        for (jint i = 0; i < method_count; ++i) {
          const char* name = methods[i].name;
          const char* sig = methods[i].signature;
          const void* fnPtr = methods[i].fnPtr;
          LOG(WARNING) <<"[jni_internal.cc->RegisterNativeMethods]" << i << ":methodname:" << name << ".sig:" << sig << ", addr:" << fnPtr;
          if (UNLIKELY(name == nullptr)) {
            ReportInvalidJNINativeMethod(soa, c.Get(), "method name", i, return_errors);
            return JNI_ERR;
          } else if (UNLIKELY(sig == nullptr)) {
            ReportInvalidJNINativeMethod(soa, c.Get(), "method signature", i, return_errors);
            return JNI_ERR;
          } else if (UNLIKELY(fnPtr == nullptr)) {
            ReportInvalidJNINativeMethod(soa, c.Get(), "native function", i, return_errors);
            return JNI_ERR;
          }
          bool is_fast = false;
          // Notes about fast JNI calls
          //
          // On a normal JNI call, the calling thread usually transitions


    const void* ArtMethod::RegisterNative(const void* native_method, bool is_fast) {
        CHECK(IsNative()) << PrettyMethod();
        CHECK(!IsFastNative()) << PrettyMethod();
        CHECK(native_method != nullptr) << PrettyMethod();
```

```
        if (is_fast) {
            AddAccessFlags(kAccFastNative);
        }
        void* new_native_method = nullptr;
        Runtime::Current()->GetRuntimeCallbacks()->RegisterNativeMethod(this,
native_method, /*out*/&new_native_method);
        LOG(WARNING) << "[art_method.cc->ArtMethod::RegisterNative]methodname:" <<
PrettyMethod() << ", addr:" << native_method;
        SetEntryPointFromJni(new_native_method);
        return new_native_method;
    }

    void SetEntryPointFromJni(const void* entrypoint) {
        DCHECK(IsNative());
        LOG(WARNING) << "[art_method.h->ArtMethod::SetEntryPointFromJni]methodname:" <<
PrettyMethod() << ", addr:" << entrypoint;
        SetEntryPointFromJniPtrSize(entrypoint, kRuntimePointerSize);
    }
```

再次刷机。修改 roysue.cpp 的代码，加入 fingerprint 和反调试：

```
    void onomon2(JNIEnv *env, jobject thiz, jobject saved_instance_state) {
        ...
        env->CallVoidMethod(tv,setText,r0ysue);

        jclass buildClazz = env->FindClass("android/os/Build");
        jfieldID FINGERPRINT =
env->GetStaticFieldID(buildClazz,"FINGERPRINT","Ljava/lang/String;");
        jstring fingerprint = static_cast<jstring>(env->GetStaticObjectField(buildClazz,
FINGERPRINT));

        if(function_check_tracerPID() || system_getproperty_check() ||
strstr( env->GetStringUTFChars(fingerprint, JNI_FALSE),"aosp")){
            int a, b, c;
            a = 1;
            b = 0;
            c = a/b;
        }
    }
```

之所以用除零错误使进程结束，是因为 kill 函数容易被 Hook 住，而用除零错误可以保证程序崩溃退出。

先注释掉反调试的内容，安装并运行。查看日志，如图 12-3 所示。

```
lanager: Start proc 3026:com.roysue.easysol/u0a65 for activity com.roysue.easysol/.MainActivity
 Late-enabling -Xcheck:jni
hread: go into handleBindApplication
hread: go into performLaunchActivity
 [art_method.h->ArtMethod::SetEntryPointFromJni]java.lang.String com.roysue.easysol.MainActivity.Sign(java.lang.String),addr0xe80a0f11
 [art_method.h->ArtMethod::SetEntryPointFromJni]java.lang.String com.roysue.easysol.MainActivity.decrypt(java.lang.String),addr0xe80a0f11
 [art_method.h->ArtMethod::SetEntryPointFromJni]java.lang.String com.roysue.easysol.MainActivity.method01(java.lang.String),addr0xe80a0f1
 [art_method.h->ArtMethod::SetEntryPointFromJni]void com.roysue.easysol.MainActivity.onCreate(android.os.Bundle),addr0xe80a0f11
 [art_method.h->ArtMethod::SetEntryPointFromJni]java.lang.String com.roysue.easysol.MainActivity.stringFromJNI(),addr0xe80a0f11
 [jni_internal.cc->RegisterNatives]4
 [jni_internal.cc->RegisterNativeMethods]0:methodname:decrypt,sig:(Ljava/lang/String;)Ljava/lang/String;,addr:0xcc57ae95
 [art_method.cc->ArtMethod::RegisterNative]java.lang.String com.roysue.easysol.MainActivity.decrypt(java.lang.String),addr0xcc57ae95
 [art_method.h->ArtMethod::SetEntryPointFromJni]java.lang.String com.roysue.easysol.MainActivity.decrypt(java.lang.String),addr0xcc57ae95
 [jni_internal.cc->RegisterNativeMethods]1:methodname:Sign,sig:(Ljava/lang/String;)Ljava/lang/String;,addr:0xcc57b589
 [art_method.cc->ArtMethod::RegisterNative]java.lang.String com.roysue.easysol.MainActivity.Sign(java.lang.String),addr0xcc57b589
 [art_method.h->ArtMethod::SetEntryPointFromJni]java.lang.String com.roysue.easysol.MainActivity.Sign(java.lang.String),addr0xcc57b589
 [jni_internal.cc->RegisterNativeMethods]2:methodname:onCreate,sig:(Landroid/os/Bundle;)V,addr:0xcc57b9d5
 [art_method.cc->ArtMethod::RegisterNative]void com.roysue.easysol.MainActivity.onCreate(android.os.Bundle),addr0xcc57b9d5
 [art_method.h->ArtMethod::SetEntryPointFromJni]void com.roysue.easysol.MainActivity.onCreate(android.os.Bundle),addr0xcc57b9d5
 [jni_internal.cc->RegisterNativeMethods]3:methodname:onCreate,sig:(Landroid/os/Bundle;)V,addr:0xcc57bcc1
 [art_method.cc->ArtMethod::RegisterNative]void com.roysue.easysol.MainActivity.onCreate(android.os.Bundle),addr0xcc57bcc1
 [art_method.h->ArtMethod::SetEntryPointFromJni]void com.roysue.easysol.MainActivity.onCreate(android.os.Bundle),addr0xcc57bcc1
 Rejecting re-init on previously-failed class java.lang.Class<androidx.core.view.ViewCompat$2>: java.lang.NoClassDefFoundError: Failed re
tListener;
  at void androidx.core.view.ViewCompat.setBackground(android.view.View, android.graphics.drawable.Drawable) (ViewCompat.java:2559)
```

图 12-3　查看动态注册流程日志

首先进行 SetEntryPointFromjni，依次打印出 Sign、decrypt、method01、onCreate、stringFromJNI。其中，method01 是唯一一个静态注册的。其他几个都是动态注册的。但是，它们都显示为同一个地址 0f11，这是为什么呢？这其实是实际运行之前预先设置的值。这是一种“懒绑定”的思想，函数在运行之前先绑定在一个 proxy method 上，后面再对这个地址进行修改、查找并调用。然后打印出动态注册函数的数量为 4，以及它们的调用流程。

我们从日志中没有看到 method01。因为没有调用就没有绑定。打开 ononon2 的 method01 的静态绑定并打开反调试。运行后再次查看日志，如图 12-4 所示。

```
adb shell
I zygote  :   at void android.app.ActivityThread.handleLaunchActivity(android.app.ActivityThread$ActivityClientRecord, android.content.I
)
I zygote  :   at void android.app.ActivityThread.-wrap11(android.app.ActivityThread, android.app.ActivityThread$ActivityClientRecord, an
read.java:-1)
I zygote  :   at void android.app.ActivityThread$H.handleMessage(android.os.Message) (ActivityThread.java:1624)
I zygote  :   at void android.os.Handler.dispatchMessage(android.os.Message) (Handler.java:105)
I zygote  :   at void android.os.Looper.loop() (Looper.java:164)
I zygote  :   at void android.app.ActivityThread.main(java.lang.String[]) (ActivityThread.java:7119)
I zygote  :   at java.lang.Object java.lang.reflect.Method.invoke(java.lang.Object, java.lang.Object[]) (Method.java:-2)
I zygote  :   at void com.android.internal.os.Zygote$MethodAndArgsCaller.run() (Zygote.java:240)
I zygote  :   at void com.android.internal.os.ZygoteInit.main(java.lang.String[]) (ZygoteInit.java:767)
I zygote  :
W zygote  : [art_method.cc->ArtMethod::RegisterNative]java.lang.String com.roysue.easysol.MainActivity.method01(java.lang.String),addr0x
W zygote  : [art_method.h->ArtMethod::SetEntryPointFromJni]java.lang.String com.roysue.easysol.MainActivity.method01(java.lang.String).a
E roysuejni: replace file name => /proc/4009/status
E roysuejni: file line =>       0
E roysuejni: [roysue device]: [google/AOSP on msm8996] result is => 0
D OpenGLRenderer: HWUI GL Pipeline
I Adreno  : QUALCOMM build                   : 7142022, Ib5823dd10c
I Adreno  : Build Date                       : 06/23/17
I Adreno  : OpenGL ES Shader Compiler Version: EV031.18.00.00
I Adreno  : Local Branch                     : 011A
I Adreno  : Remote Branch                    :
I Adreno  : Remote Branch                    :
I Adreno  : Reconstruct Branch               :
I Adreno  : PFP: 0x005ff087, ME: 0x005ff063
I OpenGLRenderer: Initialized EGL, version 1.4
D OpenGLRenderer: Swap behavior 2
```

图 12-4　method01 调用后生效

我们发现 method01 调用后生效了，而在调用之前，method01 一直是一个虚假的地址。

为了进一步增强混淆效果，我们可以再对 onCreate 函数进行静态注册。这样就有了 3 个 onCreate

函数。静态注册只是个幌子，连动态注册都可以注册很多次，可以在一定的条件触发之后，再进行动态注册，从而把原来的方法替换掉。

如果再对 onCreate 函数进行静态注册，最后会执行哪一个呢？我们来尝试一下。代码的修改非常简单，此处就不展示了。修改代码后运行，发现实际执行的是动态注册的第二个。也就是说，我们可以无限覆盖 Java 层的 ArtMethod。

12.3 本章小结

在本章中，我们主要通过阅读源码学习了 Dalvik 和 ART 下动静态注册流程追踪，并通过修改源码添加日志输出绑定关系。此外，我们通过二次修改绑定关系打造"真假"onCreate 达到混淆的目的。

第 13 章

MD5 算法分析和魔改

本章主要讲解如何实现一个 MD5（消息摘要）算法。MD5 是一个经典的哈希算法。

哈希（Hash）也称为散列，就是把任意长度的输入通过散列算法变换成固定长度的输出，这个输出值就是散列值。

作为 App 的开发人员，在学习本章之后，我们就能够知道应该怎么更改算法才会让算法不易被人破解。

为什么要学习加解密算法？主要出于以下 3 个原因。

1. SO 层的加解密算法识别难度大

在 Java 层逆向过程中，我们不需要了解哈希算法实现上的细节，Java 实现的 MD5 加密过程如图 13-1 所示。而到了 SO 层，我们必须要跟算法的实现细节打交道，这就要求我们对加解密的原理和常见的实现有更深的理解，SO 层实现的 MD5 加密过程如图 13-2 所示。

```
public static final String getMessageDigestUP(byte[] bArr) {
    char[] cArr = {'0', '1', '2', TopUpUtil.CG_CHARACTERISTIC, '4', '5', '6', '7', '8', '9', 'A', 'B', 'C', 'D', 'E', 'F'};
    try {
        MessageDigest instance = MessageDigest.getInstance("MD5");
        instance.update(bArr);
        byte[] digest = instance.digest();
        int length = digest.length;
        char[] cArr2 = new char[(length * 2)];
        int i = 0;
        for (byte b : digest) {
            int i2 = i + 1;
            cArr2[i] = cArr[(b >>> 4) & 15];
            i = i2 + 1;
            cArr2[i2] = cArr[b & TType.f40176m];
        }
        return new String(cArr2);
    } catch (Exception unused) {
        return null;
    }
}
```

图 13-1　Java 实现的 MD5 加密过程

```
v222 = (v6 << 16) | (*(unsigned __int8 *)(v5 - 127) << 8) | *(unsigned __int8 *)(v5 - 128) | (*(unsigned __int8 *)(v5 - 125) << 24);
v8 = __ROR4__(((v209 ^ v4) & v3 ^ v209) - 680876936 + v222 + v221, 25);
v7 = v3 + v8;
v9 = v3
   + v8
   + __ROR4__(
       v209
     - 389564586
     + ((*(unsigned __int8 *)(v5 - 122) << 16) | (*(unsigned __int8 *)(v5 - 123) << 8) | *(unsigned __int8 *)(v5 - 124) | (*(unsigned __int8 *)(v5 - 121) <
     + ((v4 ^ v3) & (v3 + v8) ^ v4),
       20);
v210 = (*(unsigned __int8 *)(v5 - 122) << 16) | (*(unsigned __int8 *)(v5 - 123) << 8) | *(unsigned __int8 *)(v5 - 124) | (*(unsigned __int8 *)(v5 - 121) <<
v11 = __ROR4__(
        v4
      + 606105819
      + ((*(unsigned __int8 *)(v5 - 118) << 16) | (*(unsigned __int8 *)(v5 - 119) << 8) | *(unsigned __int8 *)(v5 - 120) | (*(unsigned __int8 *)(v5 - 117) ·
      + (((v3 + v8) ^ v3) & v9 ^ v3),
        15);
v10 = v9 + v11;
v212 = (*(unsigned __int8 *)(v5 - 114) << 16) | (*(unsigned __int8 *)(v5 - 115) << 8) | *(unsigned __int8 *)(v5 - 116) | (*(unsigned __int8 *)(v5 - 113) <<
v12 = (v9 + v11) ^ v9;
v14 = __ROR4__(v3 - 1044525330 + v212 + ((v9 ^ v7) & (v9 + v11) ^ v7), 10);
v13 = v10 + v14;
v213 = (*(unsigned __int8 *)(v5 - 110) << 16) | (*(unsigned __int8 *)(v5 - 111) << 8) | *(unsigned __int8 *)(v5 - 112) | (*(unsigned __int8 *)(v5 - 109) <<
v15 = v10 + v14 + __ROR4__(v213 - 176418897 + v7 + (v12 & (v10 + v14) ^ v9), 25);
v211 = (*(unsigned __int8 *)(v5 - 118) << 16) | (*(unsigned __int8 *)(v5 - 119) << 8) | *(unsigned __int8 *)(v5 - 120) | (*(unsigned __int8 *)(v5 - 117) <<
v214 = (*(unsigned __int8 *)(v5 - 106) << 16) | (*(unsigned __int8 *)(v5 - 107) << 8) | *(unsigned __int8 *)(v5 - 108) | (*(unsigned __int8 *)(v5 - 105) <<
v17 = __ROR4__(v214 + 1200080426 + v9 + (((v10 + v14) ^ v10) & v15 ^ v10), 20);
v16 = v15 + v17;
v19 = __ROR4__(
```

图 13-2　SO 层实现的 MD5 加密过程

2. 在 Native 中存在普遍对算法魔改的情况

MD5 魔改常数如图 13-3 所示。

```
int __fastcall sub_65540(int a1, int a2, int *a3)
{
  char v5[24]; // [sp+4h] [bp-7Ch] BYREF
  int v6[25]; // [sp+1Ch] [bp-64h] BYREF

  v6[0] = 0x67552301;          MD5 魔改常数
  v6[1] = 0xEDCDAB89;
  v6[2] = 0x98BADEFE;
  v6[3] = 0x16325476;
  v6[4] = 0;
  v6[5] = 0;

_DWORD *__fastcall CYP_Init(_DWORD *result)
{
  *result = 0;
  result[1] = 0;
  result[5] = 0x10325476;          MD5 魔改常数
  result[4] = 0x9812343E;
  result[3] = 0xEABCDB89;
  result[2] = 0x666454BA;
```

图 13-3　MD5 魔改常数

另外，如果要更改加密算法，要考虑以下几个问题：

（1）为什么要改？

（2）一般怎么改？

（3）我们想魔改算法增强公司 App 的安全性，会不会适得其反，降低了安全性给公司造成损失？

（4）在逆向分析算法的时候，样本看着像 MD5，但验证后结果对不上，这是怎么回事呢？我们该如何排查问题？

13

3. 减少对 SO 的恐惧感

这一点非常关键。作为 SO 逆向的新手，看到复杂的 C 语言代码，很容易被吓退。

执行一个常见的 SO 函数可能涉及 JNI 调用、系统库函数、系统调用、加密算法和自定义算法，这些都不容易理解。

13.1　MD5 算法的描述

MD5 算法的输入和输出如下。

- 算法的输入：任意长度的消息。
- 算法的输出：128bit 长（32 个十六进制数）的消息摘要。

我们在 CyberChef 网站上用 MD5 计算 123456 得到的结果为一个 32 位的十六进制数：e10adc3949ba59abbe56e057f20f883e，如图 13-4 所示。

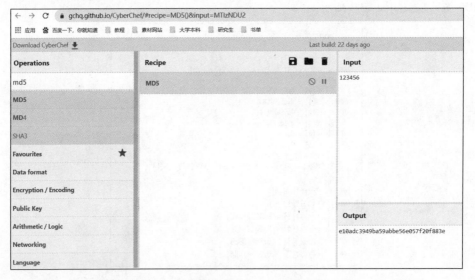

图 13-4　用 MD5 计算 123456 得到的结果为 e10adc3949ba59abbe56e057f20f883e

如果清空输入，那么会发现在输出中也会有一个 32 位的十六进制数，这是怎么做到的呢？我们将在后续的章节中进行讲解。

MD5 算法的工作原理流程图如图 13-5 所示。

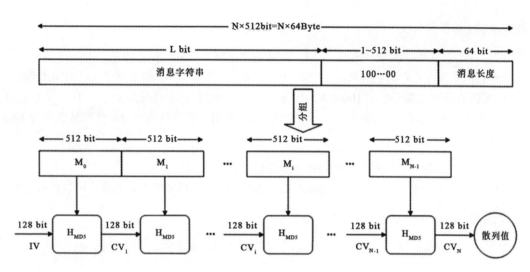

图 13-5 MD5 算法的工作原理流程图

1. 对消息进行填充

填充规则：使得其 bit 长度除以 512 余 448。

明文：123456。

首先，对明文进行编码，因为计算机只能对二进制进行读写。123456 对应的 ASCII 编码为 31 32 33 34 35 36，共 48bit。我们需要将它填充到 448bit 长。填充方法是（二进制）先填充一个 1，后面跟上足够多的 0，直到符合要求。

2. 附加长度信息

余下的 64（512-448）bit 长度，用于填充真实数据的长度信息，长度单位为 bit。我们要填充的数据是 48，由于 MD5 是小端序，因此填充的内容是 300...0，而不是 000...030。我们在之后也会面临很多端序转化的问题。

> **注意** 假如明文编码后本身就是 448bit，这时就需要填充 512bit。也就是说，无论明文是什么样的，都必须填充，填充的长度为 8~512bit。

MD5 算法支持无限长度的输出，但是附加长度只有 64bit，最多表示 $2^{64}-1$ 位。如果实际长度超过它可以表示的长度，怎么办？遇到这种情况就取低 64 位。其他的哈希算法都是有长度限制的，因为像 MD5 算法这样的设计可能存在安全隐患。

请思考以下问题：

（1）为什么要填充到 512 的倍数？

因为我们要做一个分组的数据。假如输入是无限长度的或者说是可变的，而输出是固定的 128bit。只能通过分组迭代来对输入的数据进行处理。MD5 算法选择的是 512 位的分组，所以我们才需要凑 512 这样的长度。

（2）为什么填充方式是先填充一个 1，后面跟上足够多的 0？

考虑到以下情况：有两个明文，一个是 1000，另一个是 10000。假如直接在后面填上 0，填充到 8 位，那么两个明文都变成了 10000000，这显然是不合理的。而如果先填上一个 1，再填 0 补充，1000 被填充为 10001000，10000 被填充为 10000100。我们可以看到两个明文分别被填充为不同的待处理的值。

（3）填充完 100… 后，为什么还要加上长度信息？

这是为了应对长度扩展攻击（Length Extension Attack）。读者感兴趣的话，可以自行了解。

3. 运算

将明文分组为 M1~M16，每组 64bit，接下来都用十六进制表示：

M1 = 31 32 33 34

M2 = 35 36 80 00

M3~M14 = 0

M15 = 30 00 00 00

M16 = 0

端序转化后的结果如下：

M1 = 34 33 32 31

M2 = 00 80 36 35

M3~M14 = 0

M15 = 30

M16 = 0

接下来有 64 轮的步骤，每一轮都会对 M1~M16 中的某一个分组进行处理。在经过 64 轮之后，每一个分组都会被处理 4 次。MD5 算法的运算结构如图 13-6 所示。

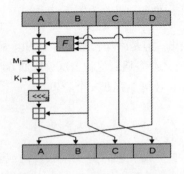

图 13-6　MD5 算法的运算结构

在图 13-6 中，A、B、C、D、F 函数和 K_i 都只是算法的一部分，实际上在操作变化的只有 M_i。A、B、C、D 称为初始化链接变量，假设：

A：01 23 45 67

B：89 ab cd ef

C：fe dc ba 98

D：76 54 32 10

但是，由于 MD5 算法采用小端序，因此实际上 MD5 算法的 A、B、C、D 是：

A：67 45 23 01

B：ef cd ab 89

C：98 ba dc fe

D：10 32 54 76

在图 13-6 中，最直观的是 B→C，C→D，D→A，这 3 个变化的过程是简单的平移，我们不用考虑。

A1：10 32 54 76

B1：?

C1：ef cd ab 89

D1：98 ba dc fe

计算 B1：

① A+F(B,C,D)

F 称为非线性函数，或者逻辑函数，它由 4 个函数组成：f1、f2、f3 和 f4。每 16 轮会发生变化，f1 用于第 1~16 轮，f2 用于第 17~32 轮，f3 用于第 33~48 轮，f4 用于第 49~64 轮。每个函数的运算都不同，如图 13-7 所示。

> 图中的F函数并不是一个函数，而是由4个函数组成的
> ✓ F(X,Y,Z)=(X&Y)|((~X)&Z)
> ✓ G(X,Y,Z)=(X&Z)|(Y&(~Z))
> ✓ H(X,Y,Z)=X^Y^Z
> ✓ I(X,Y,Z)=Y^(X|(~Z))

图 13-7 4 个函数使用的位运算

计算第一轮，过程如下：

将十六进制表示的 B（ef cd ab 89）、C（98 ba dc fe）、D（10 32 54 76）分别转成二进制：

```
1110 1111 1100 1101 1010 1011 1000 1001
1001 1000 1011 1010 1101 1100 1111 1110
0001 0000 0011 0010 0101 0100 0111 0110
```

为了计算简便，f1 可等价于 if x then y else z，将 B、C、D 分别代入公式中的 x、y、z，按位运算得到结果 F(B,C,D)：

1001 1000 1011 1010 1101 1100 1111 1110 => 98 ba dc fe （此处计算出的结果与 C 相同，纯属巧合）

A + F(B,C,D) => 67 45 23 01 + 98 ba dc fe = ff ff ff ff

手算 MD5 有另一个好处：基于 Unicorn 的动态 Trace，可以追踪和分析寄存器的变化，动态地判断和分析它使用了哪种加密算法。

② 将①中的结果与 Mi 相加

每轮对于 Mi 的选择：第 1~16 轮对应 M1~M16，第 17~64 轮变为随机化的顺序，这是为了抵御密码学上的攻击。

M1 + ff ff ff ff => 34 33 32 31 + ff ff ff ff = 1 34 33 32 30

③ 将②中的结果与 Ki 相加

与 Mi 的选取不同，每轮的 Ki 都是不同的，固定为 K1~K64。这也是算法的识别或者自动化检测中很重要的一环。

1 34 33 32 30 + K1 => 1 34 33 32 30 + d7 6a a4 78 = 2 0b 9d d6 a8

MD5 只处理 32 位二进制数。将取后 8 位（对 2^3^{2}取模）得 0b 9d d6 a8。

④ 循环左移

将移出的低位放到高位，如图 13-8 所示。

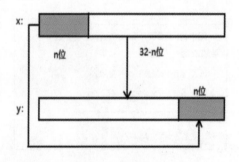

图 13-8　第一轮计算

循环左移的位数同样是规定好的。第一轮循环左移 7 位，0b 9d d6 a8 转成二进制为：

0000 1011 1001 1101 1101 0110 1010 1000

循环左移后的结果为：

1100 1110 1110 1011 0101 0100 0000 0101 => ce eb 54 05

⑤ 将④中的结果与 B 相加

ef cd ab 89 + ce eb 54 05 = 1 be b8 ff 8e

所有这些完成之后，将得到的 A、B、C、D 分别在其基础上加上初始 A、B、C、D，进行下一轮运算。

- Mi 以及循环左移位数是怎么选的？

随机选择，没有规律。

Ki = 2^3^{2} * |sini|，sini 的 i 是弧度。例如，K1=3614090360（十进制），转化成十六进制为 d76aa478。为什么要这样构造 K 表呢？实际上，算法中的任何步骤都是为了尽可能扩散和混淆，sin 函数生成的 64 个数相对来说是无迹可寻的。用户也可以使用自己生成的随机数，或者像 SHA 256，取自然数中前面 64 个素数的立方根的小数部分的前 32 位，避开一些明显有缺陷的值就好。例如全 0、全 1、强自相关等。

- 为什么是这些非线性函数？

这没什么最优解，也可以使用别的非线性函数，或者自己实现时把顺序颠倒。

但这可能会造成算法安全性下降，而且这是用户无法把控的部分，因为这部分内容的专业性非常强，属于密码学的内容。

接下来，我们计算 MD5 64 轮中的最后一轮。我们需要先得到 64 轮初始的 A、B、C、D（可以通过断点调试 MD5 算法的实现来获取，为了演示方便，这里笔者直接给出结果）。

A：b7 e8 71 af

B：bf 25 79 c0

C：2e 55 bb 7c

D：d2 96 e7 e0

这个时候，F 函数是 f4。类似于 f1，f2、f3 也有更简便的等价表示形式：

f2 = if z then x else y

f3 = if x==y then z else ~z

f4(B,C,D) = C^(B|(~D))

~D：0010 1101 0110 1001 0001 1000 0001 1111

B：1011 1111 0010 0101 0111 1001 1100 0000

B|(~D)：1011 1111 0110 1101 0111 1001 1101 1111

C：0010 1110 0101 0101 1011 1011 0111 1100

C^(B|(~D))：1001 0001 0011 1000 1100 0010 1010 0011 => 91 38 c2 a3

f4(B,C,D) + A = 91 38 c2 a3 + b7 e8 71 af = 1 49 21 34 52

1 49 21 34 52 + M16 = 1 49 21 34 52 + 0 = 1 49 21 34 52

1 49 21 34 52 + K64 = 1 49 21 34 52 + eb 86 d3 91 = 2 34 a8 07 e3

去掉最高位后，34 a8 07 e3 => 0011 0100 1010 1000 0000 0111 1110 0011，循环左移 21 位后，得 1111 1100 0110 0110 1001 0101 0000 0000 => fc 66 94 00

fc 66 94 00 + B = fc 66 95 00 + bf 25 79 c0 = bb 8c 0e c0

新的 A、B、C、D 如下：

A：d2 96 e7 e0

B：bb 8c 0e c0

C：bf 25 79 c0

D：2e 55 bb 7c

将结果加上初始的 A、B、C、D，得：

A：39 dc 0a e1

B：ab 59 ba 49

C：57 e0 56 be

D：3e 88 0f f2

将 A、B、C、D 转换大小端序并级联后，可以得到最终结果：e10adc3949ba59abbe56e057f20f883e。

13.2　MD5 工程实现

本节介绍 MD5 工程实现中的哈希算法。

1. 实现方法及特点

1）查表法

例如，在实际项目中，我们不会在运行时计算 K 表。这样很浪费资源，而且没有必要。我们会预定义好 K 表，用空间换时间。但这也引发了一个问题：我们相对容易利用这些特征在 SO 文件中确定它是否使用了这些算法。

2）支持流加密

几乎所有的加解密都支持这样的一种方式：三段式。

（1）init 初始化。
（2）update 把明文不断填进去。
（3）final 最后计算出一个结果。

这样做有一个好处，就是处理文件的时候，可以分块放入数据，而不必一次性放入所有数据。

3）封装

以 OpenSSL 为例，由于哈希算法大都采用 MD 结构，或者可以直接称为哈希结构，因此 OpenSSL 在其中进行了封装，这样的好处是不需要为每个算法都实现一整套流程，只需要在一些宏定义的常数或运算轮次上进行变化。

2. OpenSSL 实战案例

首先，从 Java 层传入明文，初始化 MD5 结构体：

```
typedef struct MD5state_st1{
    MD5_LONG A,B,C,D;  // ABCD
    MD5_LONG Nl,Nh;          // 数据的 bit 数计数器(对2^64取余),Nh存储高32位,Nl存储低32位。
这种设计服务于32位处理器，MD5的设计就是为了服务于32位处理器的
    MD5_LONG data[MD5_LBLOCK];//数据缓冲区,用于存储加密结果
    unsigned int num;
}MD5_CTX; // 存放 MD5 算法相关信息的结构体定义
```

若传入长度超过分组长度，则会把用户第一个分组的内容进行一轮计算，并返回对应的 A、B、C、D。若长度不超过 512，则赋值即可。

按照 OpenSSL 的规定，运算结束之后，我们需要调用函数清除上下文中的结构体，随机化或者赋 0，以防止攻击。最后，将结果转成十六进制的输出。

3. MD5 安全性讨论

我们可以将 MD5 理解为一种超损压缩，只能从明文计算出哈希值，而无法逆向这个过程。如果将 MD5 和 AES 相比，AES 是加密算法，MD5 是哈希算法，两者是没有可比性的。

哈希算法不可逆，只能找到哈希值的多个原文，MD5 的输入可以无限大，输出总是 128bit 长度。因此，每一个哈希值都对应着无数个明文。

逆向 MD5 的感觉是哪来的呢？可以简单理解成字典穷举，存储了较短长度或者固定字符范围的明文的哈希值。所以只要 MD5 稍作修改、MD5 仅作为加密的中间一环、使用 HMAC-MD5 等办法，就可以避免被逆向的风险。

13.3 哈希算法逆向分析

13.3.1 Findcrypt/Signsrch 源码剖析

接下来，我们测试 Findcrypt/Signsrch 的一系列 IDA 相关的算法检测插件，看它是否能正确地将哈希算法检测出来，以及它是用什么方法检测出来的。

仍然使用之前的 MD5 样本。在 SO 中计算 123456 的哈希值。代码是从 OpenSSL 中截取出来的。先来看它是否能被 IDA 的算法识别插件检测出来。我们可以清楚地看到 4 个初始化链接变量，如图 13-9 所示。

图 13-9　4 个初始化链接变量

我们可以看到 MD5 的初始化，然后以明文传进 Update，并在 Final 中得出结果，如图 13-10 所示。最后将结果返回给 Java。

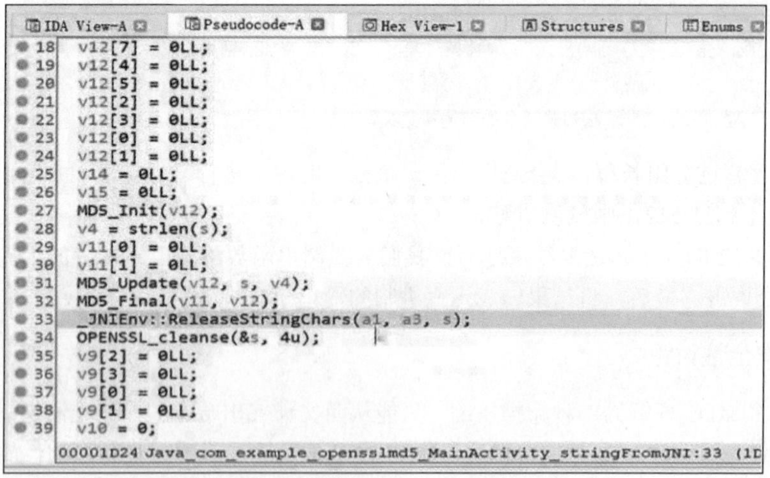

图 13-10　MD5 的关键函数

我们使用 IDA 算法插件检测，此处使用两个插件：Findcrypt 和 Signsrch，这两个插件都可以在 GitHub 中下载。

Findcrypt 插件可在网址 42 下载（参见配书资源文件）。

Signsrch 插件可在网址 43 下载（参见配书资源文件）。

下载后，解压安装包并将对应的 Python 脚本和 XML 规则文件复制到 IDA\plugins 目录下，重启 IDA。在菜单栏中选择 Edit→Plugins 就能找到对应的插件，如图 13-11 所示。首先运行 Findcrypt，发现没有检测到任何结果，如图 13-12 所示。

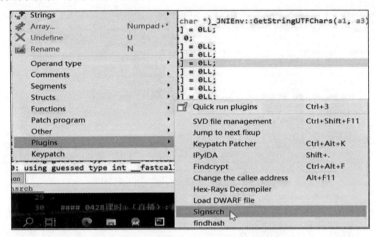

图 13-11　Findcrypt 和 Signsrch 插件

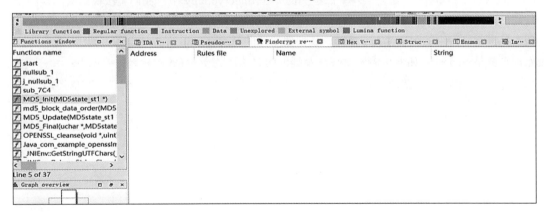

图 13-12　Findcrypt 没有检测到任何结果

Signsrch 没有独立的结果显示窗口，我们可以在输出中查看。结果显示窗口可以看到 Signsrch 找到了 0 个匹配结果，如图 13-13 所示。

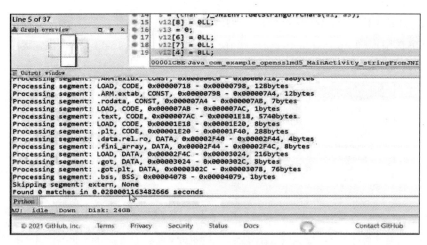

图 13-13 Signsrch 找到了 0 个匹配结果

这两个算法识别工具在这里都失效了，这是为什么呢？我们从两方面分析。第一，查看源码；第二，针对当前 SO 分析它为什么会失效。

首先来看 Findcrypt 是如何实现算法检测的？在 findcrypt3.py 中，我们无法找到实际检测相关的代码。这是因为它采用了这样一种设计方式：使用 YARA 工具。基于它可以方便地编写规则，实现对二进制文件中的内容匹配。所以，我们要关注的是 findcrypt3.rules 文件，其中有我们想要的规则，也就是 findcrypt 如何检测哈希算法及其他的算法。关于 MD5 的检测主要如下：

```
rule MD5_Constants {
    meta:
        author = "phoul (@phoul)"
        description = "Look for MD5 constants"
        date = "2014-01"
        version = "0.2"
    strings:
        // Init constants
        $c0 = { 67452301 }
        $c1 = { efcdab89 }
        $c2 = { 98badcfe }
        $c3 = { 10325476 }
        $c4 = { 01234567 }
        $c5 = { 89ABCDEF }
        $c6 = { FEDCBA98 }
        $c7 = { 76543210 }
        // Round 2 K
        $c8 = { F4D50d87 }
        $c9 = { 78A46AD7 }
    condition:
        5 of them
}
```

它检测了 MD5 大小端序的初始化链接变量，以及 K 表中的两个值（大端序和小端序各找了一个数）。对于 ARM SO 只存在小端序，但由于 Findcrypt 旨在面向各种各样逆向分析任务的插件，它还要考虑其他的二进制文件，所以它需要兼容大小端序。

在这 10 个规则中，如果匹配到 5 个，就认为它在 SO 中检测到 MD5 的常量。大端序和小端序是不可能同时出现在一个文件中的。所以必须有 4 个初始化常量，以及一个 K 表中的常量，我们才认为它包含在 MD5 中。

另一部分是 MD5 的 API。

```
rule MD5_API {
    meta:
        author = "_pusher_"
        description = "Looks for MD5 API"
        date = "2016-07"
    strings:
        $advapi32 = "advapi32.dll" wide ascii nocase
        $cryptdll = "cryptdll.dll" wide ascii nocase
        $MD5Init = "MD5Init" wide ascii
        $MD5Update = "MD5Update" wide ascii
        $MD5Final = "MD5Final" wide ascii
    condition:
        ($advapi32 or $cryptdll) and ($MD5Init and $MD5Update and $MD5Final)
}
```

MD5 的 API 检测了一些字符串常量。例如，advapi32.dll 和 cryptdll.dll 是 Windows 上的加解密动态库。此外，还检测了一些 MD5 常见的函数名，分别是 MD5Init、MD5Update、MD5Final。

为什么它无法在 SO 中检测到 MD5 算法呢？

在 IDA 的汇编代码中，我们能够查找到 K 表中的数，但是 f5 反汇编插件错误地将这个数识别成了它的补码。Invert sign 之后的值才是正确的。

这是因为 Findcrypt、Signsrch 这样的静态检测工具是在 HEX 中从上至下扫描的。IDA 帮我们将两条指令识别成了一条汇编指令，如图 13-14 所示。MOV R1,#0xF4D50D87 将 R1 的低位设置成 0x0D87，再通过另一条指令将 R1 的高位设置成 0xF4D5。所以说，在它的 HEX 中并不存在我们想要的特征码。

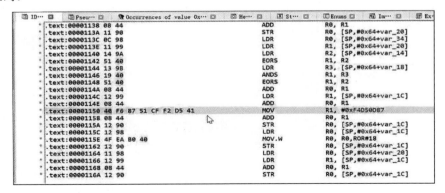

图 13-14　IDA 将两条指令识别成了一条汇编指令

所以 Findcrypt 失效了。我们再看一下 Signsrch。与 Findcrypt 不同的是，Signsrch 没有用到 YARA 库，它自己实现了一个简单的遍历。

```
        <p t="MD5 digest [32.le.272&amp;]">0123456789ABCDEFFEDCBA987654321078A46AD7...
91D386EB</p>
        <p t="MD5 digest [32.be.272&amp;]">67452301EFCDAB8998BADCFE10325476D76AA478...
EB86D391</p>
        <p t="MD5_DS [32.le.36]">78A46AD756B7C7E8DB702024EECEBDC1AF0F7CF52AC68747134630
A8019546FD00000000</p>
        <p t="MD5_DS [32.be.36]">D76AA478E8C7B756242070DBC1BDCEEEF57C0FAF4787C62AA8304613
FD46950100000000</p>
        <p t="Misty md5const
[32.le.256]">78A46AD756B7C7E8DB702024EECEBDC1AF0F7CF5...91D386EB</p>
        <p t="Misty md5const
[32.be.256]">D76AA478E8C7B756242070DBC1BDCEEEF57C0FAF...EB86D391</p>
        <p t="MD5 constants [32.le.16&amp;]">0123456777543210022345677654321O</p>
        <p t="MD5 constants [32.be.16&amp;]">6745230110325477674523021O325476</p>
```

前两条规则要匹配初始化链接变量以及它的 K 表。第 3~6 条规则只匹配 K 表，最后两条规则只匹配初始化链接变量。我们发现这些规则之间是有重复的，这是因为不同的密码实现库，或者说不同时期有不同的人对这些规则进行维护，尽可能避免遗漏。

前两条规则比较苛刻。实际上，初始化链接变量与 K 表连在一起在 SO 中是不常见的，一般是分成几块的。

出于这些原因，Findcrypt、Signsrch 都没有在 HEX 文件中找到这样的特征，所以它们失败了。那么怎么样才能检测出来呢？

在 IDA 中直接查找（Search）0x67452301 这样一个立即数，并勾选 Find all occurences 复选框，如图 13-15 所示。

图 13-15　在 IDA 中直接查找 0x67452301 立即数

这样我们就可以在 SO 中找到它了。对于我们当前分析的样本来说，这样的方法是奏效的。我们再来看下一个样本：OpenSSL 的标准库，其中包含 OpenSSL 的所有加解密算法。

首先用 Findcrypt 检测，结果如图 13-16 所示。

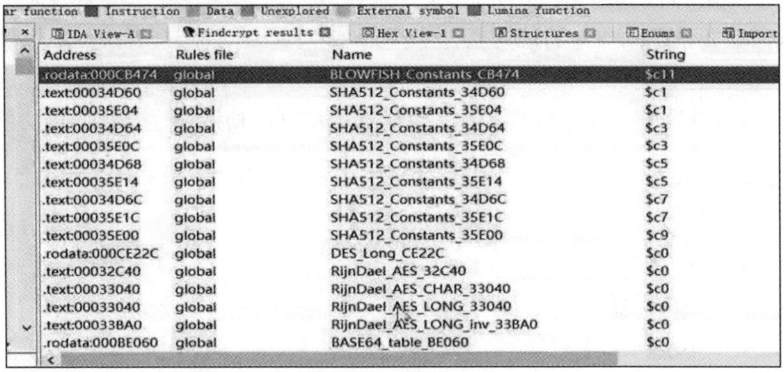

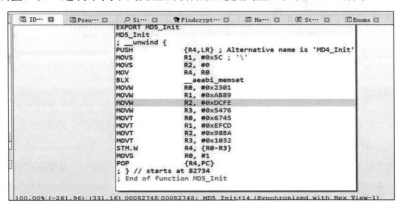

图 13-16　Findcrypt 只检测出了 SHA512

我们发现，Findcrypt 只检测出了 SHA512，别的哈希算法都没有。这种静态分析的插件弱点很明显。Signsrch 也只检测出了 SHA256 的 K 表。

为什么这两个插件如此流行呢？这是因为它们主要用于检测其他加密算法。哈希算法的常量比较少，容易漏掉。而对称加密算法或者非对称加密算法都有非常多的常量表。在这种情况下，静态检测工具能比较好地检测到这些算法。

从汇编代码中，我们看到 IDA 并没有进行优化。MOVW 和 MOVT 分别对寄存器中的高 16 位以及低 16 位赋值，在二进制中找不到完整的初始化链接变量，如图 13-17 所示。

图 13-17　MOVW 和 MOVT 分别对寄存器中的高 16 位以及低 16 位赋值

而且，在这样的一个样本中，我们也无法通过直接搜索 0x67452301 来检测算法。

此外，Signsrch 还会检测 padding 表。

 采用静态的、二进制的匹配识别，哈希算法的准确率非常低。

13.3.2 算法识别插件的核心原理与改进方向

目前，算法识别插件以静态分析为主，动态分析是未来趋势。静态检测的优点是效率高、扩展性强，我们只需要在里面添加新的规则就行了，而且编写起来简单。它的缺点是容易误报和漏报。

更好的检测方法是，将算法检测看成一个自然语言处理（Natural Language Processing，NLP）的任务，常数检测和符号检测就是情感词典。但显然有更好的检测办法，比如算法的结构、动态分析、旁路攻击等。

算法动态监测理论：

- 进行大量的循环运算。
- 依赖运算相关的汇编指令（70%）。
- 加解密中信息熵的变化。

在以往的研究中，一个比较成熟的框架如图 13-18 所示。

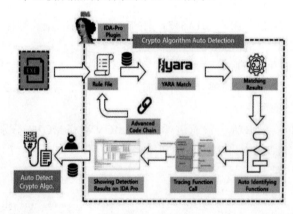

图 13-18　检测框架

输入一个 SO 文件，首先通过 IDA 插件进行解析，类似于通过 Findcrypt 的 RULES 文件去做一个常量的识别。然后根据代码中的逻辑关系构建一个高级的代码链。它是将一些常见的加解密算法库的汇编实现中的一些关键点进行匹配。接着进行符号的识别。关键在于动态追踪（Trace）。它采用某一个动态追踪工具去得到寄存器的值的变化或内存的相关变化，从而分辨出是哪一个算法。最后把结果展示在 IDA 插件中。

10 年前使用的是动态插桩工具 PIN，但是它在 Android 上支持得较差。这部分工作我们可以采用 Unicorn 或者 Unidbg 来完成。后续章节会详细讲解这部分内容。

综上所述，有以下改进方向：

（1）引用动态分析（Unicorn/Unidbg）。
（2）使用更多"巧妙的间接方法"。

13.3.3 使用 findhash 插件检测哈希算法

我们再来尝试一款检测哈希算法的插件：findhash。在哈希算法上，findhash 是比 Findcrypt 更好的检测工具。从网址 44（参见配书资源文件）下载并安装 findhash 插件，运行该插件，结果如图 13-19 所示。

图 13-19　findhash 插件运行结果

首先在二进制文件中检索哈希算法常量，找到两个可疑的函数。然后，它生成了一个 Frida Hook 脚本 libnative-lib_findhash_1635146525.js：

```
function monitor_constants(targetSo) {
    let const_array = [];
    let const_name = [];
    let const_addr = [];

    for (var i = 0; i < const_addr.length; i++) {
        const_array.push({base:targetSo.add(const_addr[i][1]),size:0x1});
        const_name.push(const_addr[i][0]);
    }

    MemoryAccessMonitor.enable(const_array, {
        onAccess: function (details) {
            console.log("\n");
            console.log("监控到疑似加密常量的内存访问\n");
            console.log(const_name[details.rangeIndex]);
            console.log("访问来自:"+details.from.sub(targetSo)+"(可能有误差)");
        }
    });
}

function hook_suspected_function(targetSo) {
    const funcs = [['MD5_Init(MD5state_st1 *)', '函数 MD5_Init(MD5state_st1 *)疑似哈希函数，包含初始化魔数的代码。', '0x7f1'], ['md5_block_data_order(MD5state_st1 *,void const*,uint)', '函数 md5_block_data_order(MD5state_st1 *,void const*,uint)疑似哈希函数运算部分。', '0x84d']];
    for (var i in funcs) {
```

```
                let relativePtr = funcs[i][2];
                let funcPtr = targetSo.add(relativePtr);
                let describe = funcs[i][1];
                let handler = (function() {
                    return function(args) {
                        console.log("\n");
                        console.log(describe);
console.log(Thread.backtrace(this.context,Backtracer.ACCURATE).map(DebugSymbol.fromAddress).join("\n"));
                    };
                })();
            Interceptor.attach(funcPtr, {onEnter: handler});
    }
}

function main() {
    var targetSo = Module.findBaseAddress('libnative-lib.so');
    // 对疑似哈希算法常量的地址进行监控,使用 Frida MemoryAccessMonitor API,有几个缺陷,在这里比较鸡肋
    // 1.只监控第一次访问,如果此区域被多次访问,那么后续访问无法获取。可以根据网址 45(参见配书资源文件)这篇文章进行改良和扩展
    // 2.ARM 64 无法使用
    // 3.无法查看调用栈
    // 在这里用于验证这些常量是否被访问,访问了就说明可能使用该哈希算法
    // MemoryAccessMonitor 在别处可能有较大用处,比如 OLLVM 过的 SO,或者 ida xref 失效/过多等情况
    // hook 和 monitor 这两个函数,只能分别注入和测试,两个同时会出错,这可能涉及 frida inline hook 的原理
    // 除非 hook_suspected_function 没结果,否则不建议使用 monitor_constants
    // monitor_constants(targetSo);

    hook_suspected_function(targetSo);
}

setImmediate(main);
```

为什么 findhash 能在哈希算法上取得比较好的效果呢?我们可以这样想,在开发时,首先需要做初始化工作 init。而在初始化过程中,我们会给结构体的链接变量赋值。它的代码结构是有特点的:在很短的区域里会出现 3~4 个 8 位十六进制的数。所以,我们可以直接使用正则表达式匹配具有特殊结构的结构体。

```
Suspected_magic_num = [i[1] for i in re.findall(
"(]|\+ \d{1,3}\)) = -?(0?x?[0-9A-FL]{8,20});", decompilerStr)]
```

这可能会带来一些误报。为了应对这些误报,生成了一个 Frida 脚本,只需运行该脚本进行检测即可。但是,我们更担心的是漏报。所以,可以采用一种宽松的策略,尽可能筛选出可能的函数,并将它们全都交由 Frida 进行检测。在 Frida 运行后,我们会对可疑的函数进行 Hook,并打印出可疑的函数名以及对应的调用栈。除了被调用的函数外,其他的函数都被认为是误报的。

输出的结果中还会包含哪些是可疑的运算部分。运算部分是怎么确定的呢?在开发过程中,

我们的每次运算都会大量地调用一些函数。而在样本的反编译结果中，我们也可以看到大量的函数被调用。所以，我们可以通过检测代码中是否有大量的函数调用来确定可疑的运算部分：

```
Suspected_transform_funcs = re.findall(" ([^ (*]{2,}?)\(", decompilerStr)[1:]
funcs_count = list(Counter(Suspected_transform_funcs).values())
max_func_num = max(funcs_count) if funcs_count else 0

if max_func_num > 60:
```

findhash.py 脚本中检测的是 transform 函数（诸如 OpenSSL 等库会把这样的运算过程称作 transform）。此处设置成 60 轮。我们来看一下哈希算法主要的"SHA 家族"的数据资料，如图 13-20 所示。

算法和变体		输出散列值长度 (bits)	中继散列值长度 (bits)	资料区块长度 (bits)	最大输入消息长度 (bits)	循环次数	使用到的运算符	碰撞攻击 (bits)	性能示例[3] (MiB/s)
MD5（作为参考）		128	128 (4 × 32)	512	无限[4]	64	And, Xor, Rot, Add (mod 2^{32}), Or	≤18 (发现碰撞)	335
SHA-0		160	160 (5 × 32)	512	$2^{64} − 1$	80	And, Xor, Rot, Add (mod 2^{32}), Or	<34 (发现碰撞)	-
SHA-1		160	160 (5 × 32)	512	$2^{64} − 1$	80		<63[5] (发现碰撞[6])	192
SHA-2	SHA-224 SHA-256	224 256	256 (8 × 32)	512	$2^{64} − 1$	64	And, Xor, Rot, Add (mod 2^{32}), Or, Shr	112 128	139
	SHA-384 SHA-512 SHA-512/224 SHA-512/256	384 512 224 256	512 (8 × 64)	1024	$2^{128} − 1$	80	And, Xor, Rot, Add (mod 2^{64}), Or, Shr	192 256 112 128	154
SHA-3	SHA3-224 SHA3-256 SHA3-384 SHA3-512	224 256 384 512	1600 (5 × 5 × 64)	1152 1088 832 576	无限[7]	24[8]	And, Xor, Rot, Not	112 128 192 256	-
	SHAKE128 SHAKE256	d (arbitrary) d (arbitrary)		1344 1088				min(d/2, 128) min(d/2, 256)	-

图 13-20 "SHA 家族"的数据资料

除了 SHA3 外，其他的循环次数都在 60 轮以上。SHA3 是一个新的标准，目前使用的还比较少。

另外，笔者推荐一个插件：IPyIDA。IPyIDA 是一个编写插件的插件。它使用 IPython 的交互方式，可以让我们快速地对 IDA 的 API 进行测试，如图 13-21 所示。

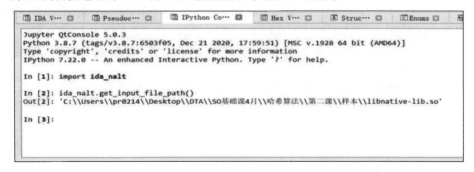

图 13-21 IPyIDA 插件

网址 46（参见配书资源文件）为 IPyIDA 项目。

13.3.4　SHA1 算法逆向分析实战

我们先来做一个 SHA1 算法的测试。在 CyberChef 中输入 123456，选择 SHA1 算法并将轮次设置为 0，如图 13-22 所示。

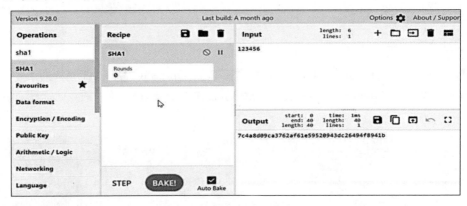

图 13-22　0 轮的 SHA1 算法

实际上，在工具中，0 轮表示默认的轮数。SHA1 算法默认为 80 轮。因此，它是一个 80 轮的 SHA1 算法。

接下来介绍 SHA0 和 SHA1 的区别。

安全哈希算法（Secure Hash Algorithm，SHA）是一个密码哈希函数家族。其中，SHA0 和 SHA1 在进行明文分组时，SHA1 会循环左移 1 位，而 SHA0 不会，所以在现实中我们遇到的都是 SHA1 算法。SHA1 比 SHA0 安全性增强了，但运算量没有任何变化。示例如下：

```
for i in range(16, 80):
    # SHA1
    w[i] = leftRotate((w[i - 3] ^ w[i - 8] ^ w[i - 14] ^ w[i - 16]), 1)
    # SHA0
    w[i] = (w[i - 3] ^ w[i - 8] ^ w[i - 14] ^ w[i - 16])
```

回顾 MD5，它将明文分成了 M1~M16，共 16 块。在之后的 64 轮运算中，每一块都运算了 4 次。而 SHA1 和 SHA0 不是这么做的。除了将明文分成 16 块外，又扩展了 64 块，即 M17~M80。之后，它进行了 80 轮运算，每一轮都使用了不同的明文块，而不像 MD5，每个明文块都用了 4 次。

完整的 SHA1 和 SHA0 算法如下：

```
# 0xffffffff is used to make sure numbers dont go over 32

def chunks(messageLength, chunkSize):
    chunkValues = []
    for i in range(0, len(messageLength), chunkSize):
        chunkValues.append(messageLength[i:i + chunkSize])

    return chunkValues
```

```python
def leftRotate(chunk, rotateLength):
    return ((chunk << rotateLength) | (chunk >> (32 - rotateLength))) & 0xffffffff

def sha1Function(message):
    # initial hash values
    h0 = 0x67452301
    h1 = 0xEFCDAB89
    h2 = 0x98BADCFE
    h3 = 0x10325476
    h4 = 0xC3D2E1F0

    messageLength = ""

    # preprocessing
    for char in range(len(message)):
        messageLength += '{0:08b}'.format(ord(message[char]))

    temp = messageLength
    messageLength += '1'

    while (len(messageLength) % 512 != 448):
        messageLength += '0'

    messageLength += '{0:064b}'.format(len(temp))
    chunk = chunks(messageLength, 512)

    for eachChunk in chunk:
        words = chunks(eachChunk, 32)
        w = [0] * 80
        for n in range(0, 16):
            w[n] = int(words[n], 2)

        for i in range(16, 80):
            # sha1
            # w[i] = leftRotate((w[i - 3] ^ w[i - 8] ^ w[i - 14] ^ w[i - 16]), 1)
            # sha0
            w[i] = (w[i - 3] ^ w[i - 8] ^ w[i - 14] ^ w[i - 16])

            # Initialize hash value for this chunk:
        a = h0
        b = h1
        c = h2
        d = h3
        e = h4

        # main loop:
        for i in range(0, 80):
            if 0 <= i <= 19:
                f = (b & c) | ((~b) & d)
                k = 0x5A827999
```

```
            elif 20 <= i <= 39:
                f = b ^ c ^ d
                k = 0x6ED9EBA1

            elif 40 <= i <= 59:
                f = (b & c) | (b & d) | (c & d)
                k = 0x8F1BBCDC

            elif 60 <= i <= 79:
                f = b ^ c ^ d
                k = 0xCA62C1D6

            a, b, c, d, e = ((leftRotate(a, 5) + f + e + k + w[i]) & 0xffffffff, a,
leftRotate(b, 30), c, d)

        h0 = h0 + a & 0xffffffff
        h1 = h1 + b & 0xffffffff
        h2 = h2 + c & 0xffffffff
        h3 = h3 + d & 0xffffffff
        h4 = h4 + e & 0xffffffff

    return '%08x%08x%08x%08x%08x' % (h0, h1, h2, h3, h4)

plainText = "123456"
sha1Hash = sha1Function(plainText)
print(sha1Hash)
```

在 SHA1 样本中，初始化过程如图 13-23 所示。

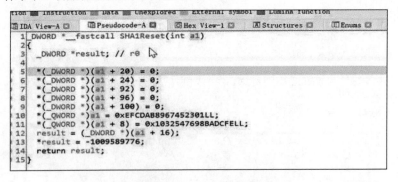

图 13-23 SHA1 初始化过程

类似于(a1 + xx)这种形式表示的是结构体。SHA 家族的哈希算法都需要维护一个结构体，以保证它的内容不断地进行循环迭代。幸运的是，笔者找到了这一样本的源代码实现，参考网址 47（参见配书资源文件）。其中，结构体定义如下：

```
typedef struct SHA1Context
{
    unsigned Message_Digest[5];        /* Message Digest (output)         */
```

```
    unsigned Length_Low;              /* Message length in bits      */
    unsigned Length_High;             /* Message length in bits      */

    unsigned char Message_Block[64];/* 512-bit message blocks        */
    int Message_Block_Index;          /* Index into message block array */

    int Computed;                     /* Is the digest computed?     */
    int Corrupted;                    /* Is the message digest corruped? */
} SHA1Context;
```

我们将以上结构体定义导入 IDA。在菜单栏选择 View→Open subviews→Local types，然后在打开的界面右击 Insert，把 SHA1Context 结构体复制到文本编辑框中，如图 13-24 所示。

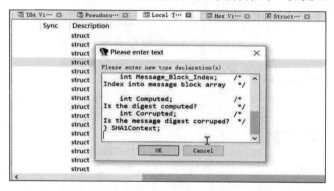

图 13-24　把 SHA1Context 结构体复制到文本编辑框中

此时，Local Types 窗口中就多了一个我们自己定义的结构体，如图 13-25 所示。

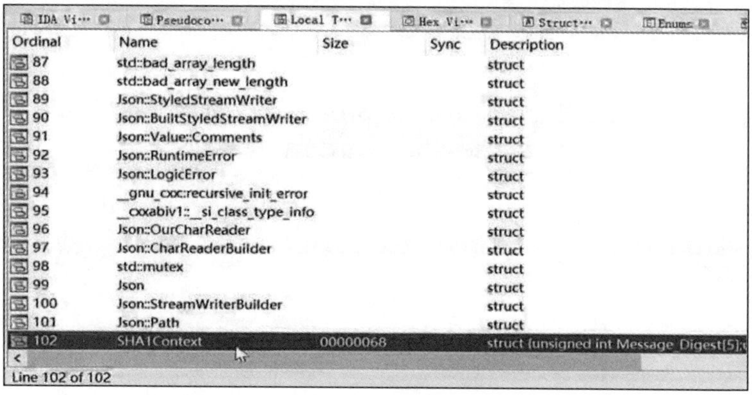

图 13-25　Local Types 中插入的 SHA1Context 结构体

在 SHA1 初始化过程 SHA1Reset 的函数定义中，右击参数类型 int，选择 Convert to struct *...，如图 13-26 所示。

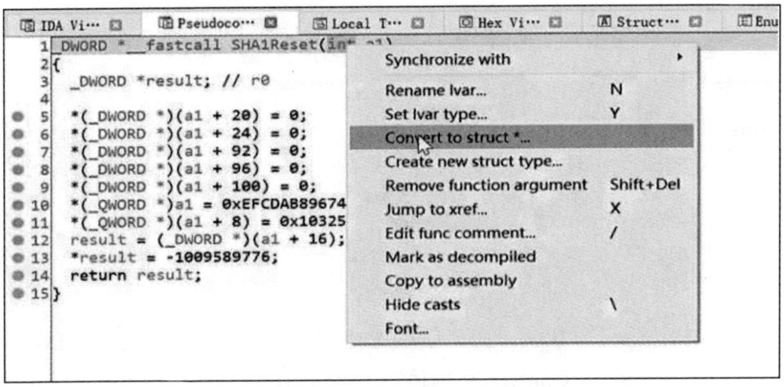

图 13-26 将 int 类型转换成 SHA1Context 结构体指针类型

结构体导入后代码如下：

```
unsigned int *__fastcall SHA1Reset(SHA1Context *a1)
{
  unsigned int *result; // r0

  a1->Length_Low = 0;
  a1->Length_High = 0;
  a1->Message_Block_Index = 0;
  a1->Computed = 0;
  a1->Corrupted = 0;
  *(_QWORD *)a1->Message_Digest = 0xEFCDAB8967452301LL;
  *(_QWORD *)&a1->Message_Digest[2] = 0x1032547698BADCFELL;
  result = &a1->Message_Digest[4];
  *result = -1009589776;
  return result;
}
```

对比结构体导入后的反汇编代码和源程序的 SHA1Reset，如图 13-27 所示。

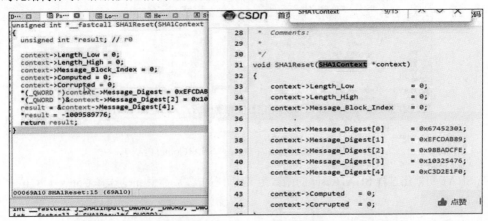

图 13-27 结构体导入后的反汇编代码和源程序的 SHA1Reset 对比

首先将记录长度的成员变量 Length_Low 和 Length_High 赋值为 0，然后将消息分组长度 Message_Block_Index 赋值为 0，同时需要赋值的还有 5 个初始化链接变量等。我们可以看到，在导入结构体之后，SHA1 初始化函数 SHA1Reset 实现了比较完整的还原，几乎和源码没有什么区别。

我们再用相同的方法还原 SHA1Input，代码如下：

```c
SHA1Context *__fastcall SHA1Input(SHA1Context *context, const unsigned __int8 
*message_array, unsigned int length)
    {
      unsigned int v3;         // r5
      SHA1Context *v5;         // r4
      int v6;                  // r0
      unsigned __int8 v7;      // r1
      int v8;                  // r0
      int v9;                  // r0

      v3 = length;
      v5 = context;
      if ( length )
      {
        if ( context->Computed || context->Corrupted )
        {
          context = (SHA1Context *)(&dword_0 + 1);
          v5->Corrupted = 1;
        }
        else
        {
          do
          {
            v6 = v5->Message_Block_Index;
            v7 = *message_array;
            v5->Message_Block_Index = v6 + 1;
            v5->Message_Block[v6] = v7;
            v8 = v5->Length_Low + 8;
            v5->Length_Low = v8;
            if ( !v8 )
            {
              v9 = v5->Length_High + 1;
              v5->Length_High = v9;
              if ( !v9 )
                v5->Corrupted = 1;
            }
            context = (SHA1Context *)v5->Message_Block_Index;
            --v3;
            if ( context == (SHA1Context *)&dword_40 )
              context = (SHA1Context *)j_SHA1ProcessMessageBlock(v5);
```

```
      if ( !v3 )
        break;
      context = (SHA1Context *)v5->Corrupted;
      ++message_array;
    }
    while ( !context );
  }
}
return context;
}
```

经过对比发现结果和源码差距非常小。以上过程就是面向谷歌编程,面向 CSDN 逆向。在一些 SO 比较复杂的情况下,这种符号还原的操作可以使反汇编代码更容易理解。

13.4 哈希算法的扩展延伸

本节将介绍一些关于哈希算法的扩展、延伸和补充。

13.4.1 哈希算法的特征

1. MD5

MD5 算法的特征及说明如表 13-1 所示。

表13-1 MD5算法的特征及说明

特 征	说 明
入参	不限长度
输出	16 字节(32 个十六进制数);4 个初始化常数,每个长 4 字节,结果是运算之后的 4 字节拼接
K 表	来自 sin 函数;有 64 个不同的 K,可以对比运算中的前 3 个 K 值是不是和标准的一样,如果前 3 个 K 值和标准的一样,基本可以断定它没有经过魔改
初始化魔数	A = 67 45 23 01,B = ef cd ab 89,C = 98 ba dc fe,D = 10 32 54 76

我们在之后的章节中会讲解对称加密算法和非对称加密算法,它们会用到一个称为初始化链接变量的参数。本质上,哈希算法的常数就是一种初始化链接变量 IV,只是哈希算法只保存了最后一块链接变量。

对于哈希算法来说,输出都来自初始化魔数运算后的拼接。

2. MD5-half

MD5-half 即 8 字节 MD5,将正常的 MD5 结果截取中间的一半。例如,MD5(123456)=e10abc3949ba59abbe56e057f20f883e,MD5-half(123456)=49ba59abbe56e057。这种算法在中国是比较常见的,如图 13-28 所示。

图 13-28　MD5-half

3. 加盐 MD5

加盐 MD5 传入的明文经过加密后的输出结果不是标准的 MD5，但长度又与标准的 MD5 相同。

- 逆向技巧：传入空值，将输出值上传到 MD5 在线破解网站碰运气。如果成功了，则可以节省半天甚至更久的时间。对于这些网站，盐在 9 位以内，往往可以成功破解。
- 原理：彩虹表攻击，可以将较短的哈希明文与密文的对应关系做成字典，以便于查询，并不是真正的逆向。

 盐（Salt）：在密码学中，指通过在密码任意固定位置插入特定的字符串，让哈希运算后的结果和使用原始密码的哈希结果不相符，这种过程称为"加盐"。

4. SHA1

SHA1 算法的特征及说明如表 13-2 所示。

表13-2　SHA1算法的特征及说明

特　　征	说　　明
入参	不限长度
输出	20 字节（40 个十六进制数）；5 个初始化常数，每个长 4 字节，结果是运算之后的 4 字节拼接
K 表	只有 4 个 K 值，每 20 轮用同一个 K 进行变换。K1 = 5a 82 79 99，K2 = 6e d9 eb a1，K3 = 8f 1b bcdc，K4 = ca 62 c1 d6。可以对比运算中的前 3 个 K 值是不是和标准的一样，如果前 3 个 K 值和标准的一样，基本可以断定它没有经过魔改
初始化魔数 IV	A = 67 45 23 01，B = ef cd ab 89，C = 98 ba dc fe，D = 10 32 54 76，E = c3 d2 e1 f0

和 MD5 相比，SHA1 有 5 个初始化链接变量，而且前 4 个链接变量与 MD5 完全相同。一般来说，对 SHA1 进行魔改只会改变其初始化常数，而不会对 K 表进行修改。因此，可以通过这 4 个 K 来判断是不是 SHA1。

5. SHA256

SHA256 算法的特征及说明如表 13-3 所示。

表13-3　SHA256算法的特征及说明

特　　征	说　　明
入参	不限长度
输出	32 字节（64 个十六进制数）；8 个初始化常数，每个长 4 字节，结果是运算之后的 4 字节拼接
K 表	有 64 个 K 值，每轮 1 个 K 值，K 值来自素数（质数）。K1 = 42 8a 2f 98，K2 = 71 37 44 91，K3 = b5 c0 fb cf，K4 = e9 b5 db a5
初始化魔数 IV	8 个初始化链接变量：A = 6a 09 e6 67，B = bb 67 ae 85，C = 3c 6e f3 72，D = a5 4f f5 3a，E = 51 0e 52 7f，F = 9b 05 68 8c，G = 1f 83 d9 ab，H = 5b e0 cd 19

6. SHA512

SHA512 算法的特征及说明如表 13-4 所示。

表13-4　SHA512算法的特征及说明

特　　征	说　　明
入参	不限长度
输出	64 字节（128 个十六进制数）；8 个初始化常数，每个长 8 字节，结果是运算之后的 8 字节拼接
K 表	有 80 个 K 值，每轮 1 个 K 值，K 值来自素数，且和 SHA256 的 K 有关联。前半部分 K 值与 SHA256 一致，看是否有最后 16 个 K
初始化魔数 IV	8 个初始化链接变量，IDA 反编译也时常显示为 16 个：A=6a 09 e6 67 f3 bc c9 08，B=bb 67 ae 85 84 ca a7 3b，C = 3c 6e f3 72 fe 94 f8 2b，D = a5 4f f5 3a 5f 1d 36 f1，E = 51 0e 52 7f ad e6 82 d1，F = 9b 05 68 8c 2b 3e 6c 1f，G = 1f 83 d9 ab fb 41 bd 6b，H = 5b e0 cd 19 13 7e 21 79

13.4.2　大厂最爱：HMAC-MD5/SHA1 详解

HMAC-XXX：加盐+双重哈希方案。

如何从明文和密钥得到结果？中间进行了哪些操作？

HMAC 定义如图 13-29 所示。

Definition [edit]

This definition is taken from RFC 2104:

$$\text{HMAC}(K, m) = H\big((K' \oplus \text{opad}) \parallel H((K' \oplus \text{ipad}) \parallel m)\big)$$

$$K' = \begin{cases} H(K) & K \text{ is larger than block size} \\ K & \text{otherwise} \end{cases}$$

图 13-29　HMAC 定义

1. HMAC-MD5

明文：123456（UTF-8）。

密钥：12345（UTF-8）。

结果：b8e91a37b5a5b21ea7aec961b3eea43b。

第1步：密钥扩展。

MD5：512bit——64字节——128hex 分组长度。
K=31 32 33 34 35。
31 32 33 34 35（填充118个0）：
3132333435000000000000000000000000000000000000000000000000000000000000000000000000000000000000000000000000000000000000000000000000

如果密钥长度大于64字节，先对其进行一次MD5，再补0。

第2步：K' xor ipad。

K'=3132333435000000000000000000000000000000000000000000000000000000000000000000000000000000000000000000000000000000000000000000000000

ipad=0x36

K' xor ipad = 070405020336363636363636363636363636363636363636363636363636363636363636363636363636363636363636363636363636363636363636363636

可以通过CyberChef来计算两者的异或值，结果如图13-30所示。

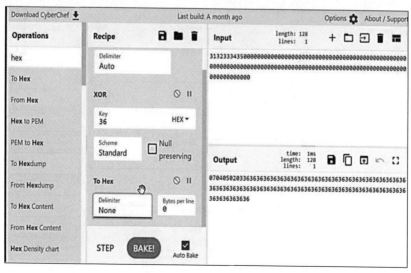

图13-30　K'异或ipad的结果

第 3 步：级联||。

将 K'异或 ipad 的结果与明文级联。级联就是拼起来：0704050203363636363636363636363636363636363636363636363636363636363636363636363636363636363636363636363636363636363636363636363631323334 3536。

第 4 步：MD5。

H 函数表示采用的哈希方案。我们对上一步的结果做 MD5。使用 CyberChef 时注意输入的转换，CyberChef 的输入默认是字符串，此处我们需要的是 hex 的输入。

MD5 结果：1147f3110ffb80b11ecd00d040d2ed5e。

第 5 步：opad=0x5C。

K'异或 opad 的结果如图 13-31 所示。

K' xor opad = 6d6e6f68695c5c5c5c5c5c5c5c5c5c5c5c5c5c5c5c5c5c5c5c5c5c5c5c5c5c5c5c5c5c5c5c5c5c5c5c5c5c5c5c5c5c5c5c5c5c5c5c5c5c5c5c5c5c5c5c5c5c5c

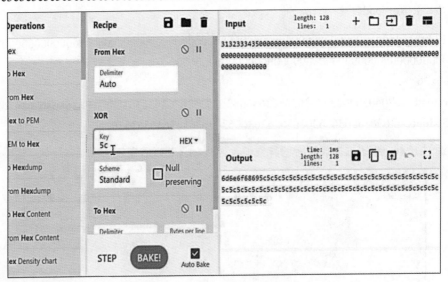

图 13-31　K'异或 opad 的结果

第 6 步：级联。

将第 5 步的结果与第 4 步的结果级联的结果：6d6e6f68695c5c5c5c5c5c5c5c5c5c5c5c5c5c5c5c5c5c5c5c5c5c5c5c5c5c5c5c5c5c5c5c5c5c5c5c5c5c5c5c5c5c5c5c5c5c5c5c5c5c5c5c5c5c5c5c5c5c5c1147f3110ffb80b11ecd00d040d2ed5e。

第 7 步：MD5。

将第 6 步的结果求哈希值，结果为 b8e91a37b5a5b21ea7aec961b3eea43b。

经比较，我们计算出来的结果与用 CyberChef 直接 HMAC-MD5 计算出来的结果一致。

回顾整个流程，HMAC-MD5 进行了两次加盐，两次 MD5。如果在反汇编或者分析 SO 时发现里面有一轮运算包含 0x36（十进制 54）或者 0x5C（十进制 92），那么应意识到这可能是一个 HMAC 方案。

OpenSSL 反汇编中的 HMAC 方案实现如图 13-32 所示。

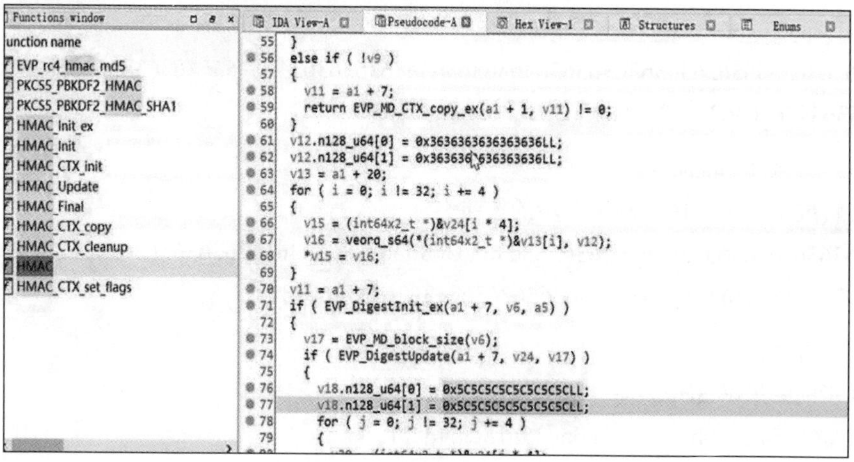

图 13-32　OpenSSL 反汇编中的 HMAC 方案实现

2. HMAC-SHA1

明文：test（UTF-8）。

密钥：2111（hex）。

结果：6e3a2a24b997b7138fcc48a56427d196a056bf8b。

HMAC-SHA1 方案如图 13-33 所示。

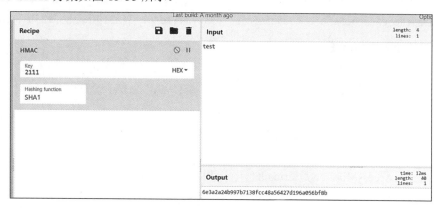

图 13-33　HMAC-SHA1 方案

第 1 步：密钥扩展。

将密钥扩展到 SHA1 分组大小（64 字节/128 个十六进制数）：

21110000000000000000000000000000000000000000000000000000000000000000000000000000000000000000000000000000000000000000000000000000

第 2 步：ipad 0x36。

17273636363636363636363636363636363636363636363636363636363636363636363636363636363636363636363636363636363636363636363636363636

第 3 步：拼接（append）。

将明文转成十六进制拼接到第 2 步的结果之后：

172736363636363636363636363636363636363636363636363636363636363636363636363636363636363636363636363636363636363636363636363636363674657374

第 4 步：SHA1。

将第 3 步的结果求 SHA1：

c765a87a411d40e734a219f863b65e25b18cfd4d

第 5 步：

将扩展之后的密钥 21110000000000000000000000000000000000000000000000000000000000000000000000000000000000000000000000000000000000000000000000000000 与 0x5c 做异或：

7d4d5c5c5c5c5c5c5c5c5c5c5c5c5c5c5c5c5c5c5c5c5c5c5c5c5c5c5c5c5c5c5c5c5c5c5c5c5c5c5c5c5c5c5c5c5c5c5c5c5c5c5c5c5c5c5c5c5c5c5c5c5c5c

第 6 步：将第 5 步的结果与第 4 步的结果拼在一起：

7d4d5c5c5c5c5c5c5c5c5c5c5c5c5c5c5c5c5c5c5c5c5c5c5c5c5c5c5c5c5c5c5c5c5c5c5c5c5c5c5c5c5c5c5c5c5c5c5c5c5c5c5c5c5c5c5c5c5c5c5c5c5c5cc765a87a411d40e734a219f863b65e25b18cfd4d

第 7 步：将第 6 步的结果求 SHA1：

6e3a2a24b997b7138fcc48a56427d196a056bf8b

经比较，我们计算出来的结果与用 CyberChef 直接 HMAC-SHA1 计算出来的结果一致。

13.5 Frida MemoryAccessMonitor 的使用场景

Frida MemoryAccessMonitor 是 Frida 的一个 API，它可以监控一个或者多个内存区块。笔者认为它"食之无味，弃之可惜"，而且 findhash 已经将其涵盖。

在 IDA 中运行 findhash 插件，它会对每个函数进行反汇编（批量执行 f5），然后在它的结果中匹配初始化魔数以及运算部分。这个过程可能会比较久。

结果如图 13-34 所示。

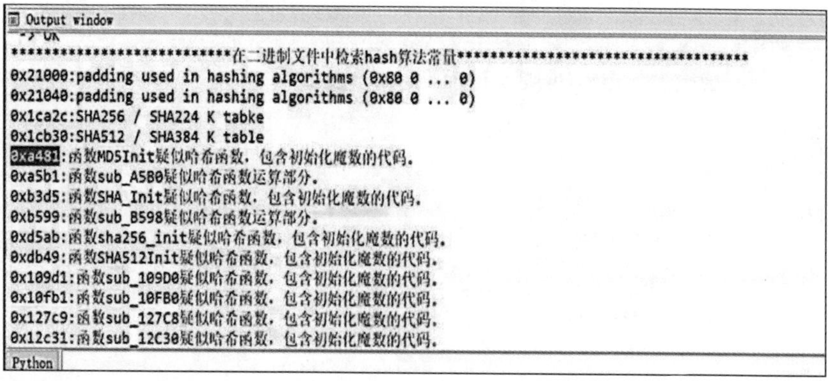

第 13 章　MD5 算法分析和魔改

图 13-34　findhash 结果

从结果中可以看出，findhash 正确地识别出了 MD5。findhash 的结果中也存在一些误报，但是没关系，我们可以用最后生成的 Frida Hook 脚本去验证它，如图 13-35 所示。

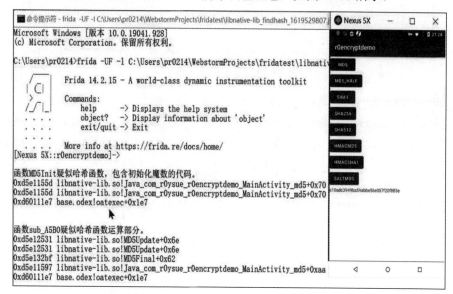

图 13-35　Frida Hook 脚本验证

查看生成的脚本，可以发现脚本非常短，如图 13-36 所示。

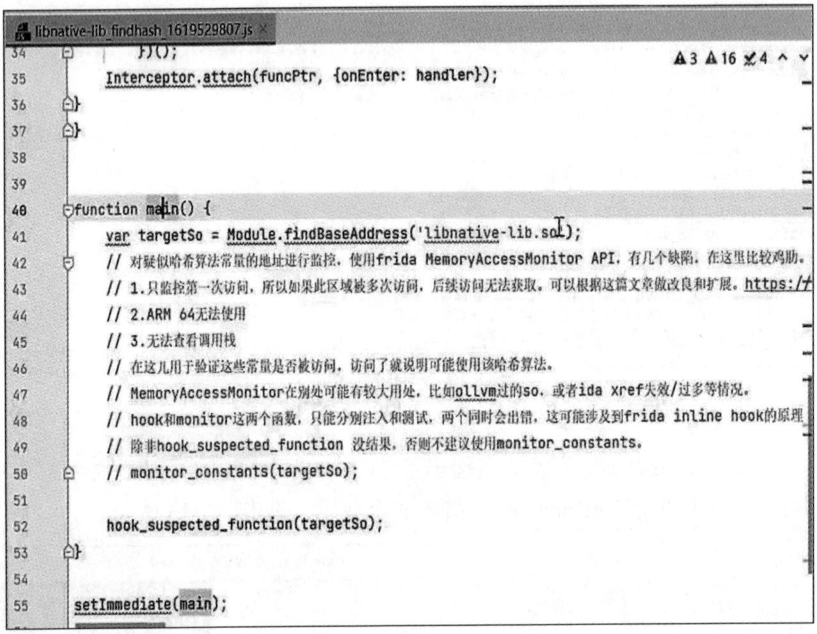

图 13-36 查看生成的脚本

在 main 函数中，第一条语句的功能是找 SO 基址。余下的部分一共有两个函数，一个是监控常量的函数 monitor_constants(targetSo)，另一个是 Hook 可疑函数 hook_suspected_function(targetSo)。监控常量的函数被默认注释掉是因为它的作用不大，如果 hook_suspected_function(targetSo)没有结果，再考虑调用 monitor_constants(targetSo)函数。

这两个函数只能启用一个，否则会相互干扰。运行脚本测试 monitor_constants(targetSo)函数。在手机的测试 App 中点击 MD5，发现返回结果显示"监控到疑似加密常量的内存访问""访问来自：0xa59a（可能有误差）"，如图 13-37 所示。

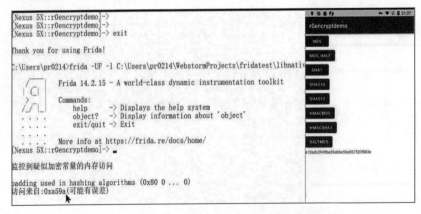

图 13-37 测试 monitor_constants(targetSo)函数 MD5

在我们单击 MD5 时，地址 0xa59a 处的语句发生了对 MD5 常数的访问。我们查看此函数的交叉引用，如图 13-38 所示。

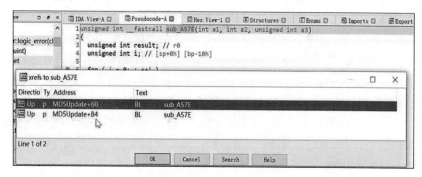

图 13-38　查看疑似访问 MD5 常数的函数的交叉引用

可以发现，此处就是 MD5 的 update。也就是说，monitor_constants(targetSo)函数确实帮助我们找到了 MD5 发生的地方。

我们再尝试在手机的测试 App 上点击 SHA1，出现了和 MD5 相似的结果，如图 13-39 所示。

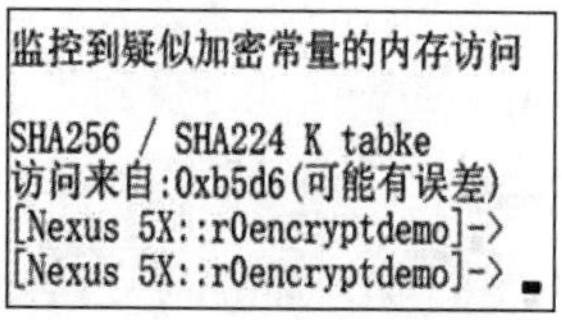

图 13-39　测试 monitor_constants(targetSo)函数 SHA1

检测到疑似访问来自地址 0xb5d6。在 IDA 中找到该地址，发现 0x5A827999，这正是我们说的 SHA1 算法的运算部分，如图 13-40 所示。

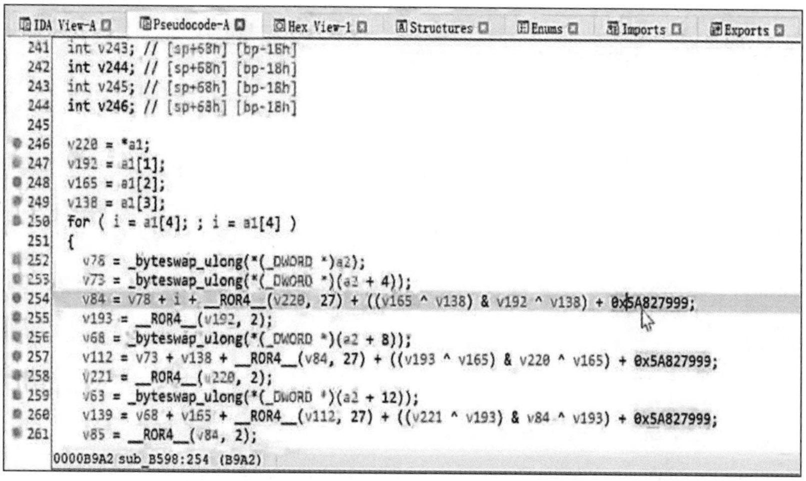

图 13-40　查看疑似访问来自地址 0xb5d6，发现 0x5A827999

通过以上测试结果，这个 API Frida MemoryAccessMonitor 看似有用，能够帮我们检测到一些常量访问，并且告诉我们是哪里访问了它。但问题是它无法处理对内存的多次访问，在这种情况下是无法用于监控工作的。我们希望的是，只要访问流经目标函数，我们就能把它拦下来。但 MemoryAccessMonitor 只会监控第一次访问/读写，这可能会对我们的逆向分析造成误导。这是第一个问题。第二个问题是，它无法用于 ARM64 的 SO 文件，只能用于 ARM32。我们知道 ARM64 是当下的主流，但 MemoryAccessMonitor 尚未"与时俱进"，从而进一步减少了它的用处。第三个问题是它无法查看调用栈。读者可能会觉得，MemoryAccessMonitor 已经有比较好地显示访问来自哪处的地址，与调用栈没有什么区别。其实并非如此。我们不能忽略结果提示"（可能有误差）"。因为，有的时候访问并不是来自这个 SO 文件，访问地点可能来自 libc 或者 libart。基于以上三个原因，MemoryAccessMonitor 只能作为一个补充检测的 API 来使用。

MemoryAccessMonitor 的使用场景：假设我们通过 findhash 或者 Findcrypt 发现了某个常量，但是没法进行下一步的工作。为什么没法进行下一步的工作呢？有两种可能：

（1）发现了多条交叉引用，我们就无法确定分析的这个函数是否用到了这个常量或在什么地方用到了这个常量。所以可以用这个 API 进行尝试，看能否得知哪个地方访问/读写了这个常量。如果得到错误的结果，则不会产生什么影响，否则会大大加快逆向分析的速度。

（2）没有交叉引用。如果一个 App 的一些地方受到了保护或者针对 IDA 进行了对抗（这种情况不在少数），我们就可以通过 MemoryAccessMonitor 去看哪个地方对它进行了调用。

在以上两种情况下，我们就可以尝试使用 MemoryAccessMonitor，看它能否给我们一些结果。我们尝试对 Demo App 使用 trace_natives 脚本，参考网址 48（参见配书资源文件）。

经过测试，这个脚本在一些没有遇到加固或者没有强保护的 SO 文件上是可以正常运行的，这得益于 Frida 本身的能力。脚本的用法：直接把 Python 脚本复制到 IDA 7.5 目录下的 plugins 中即可。

```python
# -*- coding:UTF-8 -*-
import os
from idaapi import plugin_t
from idaapi import PLUGIN_PROC
from idaapi import PLUGIN_OK
import ida_nalt
import idaapi
import idautils
import idc
import time

# 获取 SO 文件名和路径
def getSoPathAndName():
    fullpath = ida_nalt.get_input_file_path()
    filepath,filename = os.path.split(fullpath)
    return filepath,filename
```

```python
# 获取代码段的范围
def getSegAddr():
    textStart = []
    textEnd = []

    for seg in idautils.Segments():
        if (idc.get_segm_name(seg)).lower() == '.text' or (
        idc.get_segm_name(seg)).lower() == 'text':
            tempStart = idc.get_segm_start(seg)
            tempEnd = idc.get_segm_end(seg)

            textStart.append(tempStart)
            textEnd.append(tempEnd)

    return min(textStart), max(textEnd)


class traceNatives(plugin_t):
    flags = PLUGIN_PROC
    comment = "traceNatives"
    help = ""
    wanted_name = "traceNatives"
    wanted_hotkey = ""

    def init(self):
        print("traceNatives(v0.1) plugin has been loaded.")
        return PLUGIN_OK

    def run(self, arg):
        # 查找需要的函数
        ea, ed = getSegAddr()
        search_result = []
        for func in idautils.Functions(ea, ed):
            try:
                functionName = str(idaapi.ida_funcs.get_func_name(func))
                if len(list(idautils.FuncItems(func))) > 10:
                    # 如果是thumb模式,地址+1
                    arm_or_thumb = idc.get_sreg(func, "T")
                    if arm_or_thumb:
                        func += 1
                    search_result.append(hex(func))
            except:
                pass

        so_path, so_name = getSoPathAndName()
        search_result = [f"-a '{so_name}!{offset}'" for offset in search_result]
        search_result = " ".join(search_result)
```

```
            script_name = so_name.split(".")[0] + "_" + str(int(time.time())) +".txt"
            save_path = os.path.join(so_path, script_name)
            with open(save_path, "w", encoding="utf-8")as F:
                F.write(search_result)

            print("使用方法如下: ")
            print(f"frida-trace -UF -O {save_path}")

    def term(self):
        pass


def PLUGIN_ENTRY():
    return traceNatives()
```

用 IDA 打开目标 SO 文件,从菜单栏中选择 plugins→traceNatives 即可运行脚本,待运行结束,输出窗口中又生成对应的脚本,运行生成的脚本。我们发现,它做了大量的追踪(Trace)。因为这个 IDA 插件利用了 Frida 自带的 Frida Trace,所以效果比较好。我们再在手机 Demo App 中点击 MD5,打印出的结果如图 13-41 所示。

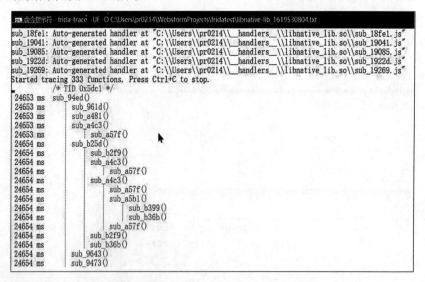

图 13-41　traceNatives 对 MD5 的结果

通过在 IDA 中验证,函数在 SO 文件中是否流经这样一个途径:

```
sub_94ed() -> Java_com_r0ysue_r0encryptdemo_MainActivity_md5
    sub_961d() -> GetStringUTFChars(a1, a3)
    sub_a481() -> j_MD5Init()
    sub_a4c3() -> j_MD5Update()
```

我们可以根据追踪的结果图静态地分析所要还原的算法,而绕过一些不会执行的逻辑。但是在使用中也存在一些令人困扰的问题：一是追踪的结果给出的地址与 IDA 反编译后显示的地址差 1；二是我们不能确定地址所对应的函数。

我们可以修改脚本，汇编代码超过 10 行，就需要进行追踪。假如我们的 App 中的函数比较多，逻辑比较复杂，可以将 10 改成 20、30，这样 App 不容易崩溃。如果是小 App，想要事无巨细地追踪，那么可以把这个数字改成 5、6 等。我们可以根据自己的需要来更改这个范围。

trace_natives 的优化方案有两种：

（1）修改 Frida Trace 的原脚本，缺点是操作复杂，不便于使用。

① 将 traceNatives.py 放进 IDA plugins 目录中。
② 替换本地 frida-tools 包中的 tracer.py 和 tracer_agent.js。
③ 在 IDA 中，单击 Edit→Plugins→traceNatives。

运行效果如图 13-42 所示。

```
Started tracing 333 functions. Press Ctrl+C to stop.
         /* TID 0x5dc1 */
12240 ms  Java_com_r0ysue_r0encryptdemo_MainActivity_md5()
12240 ms  │  _JNIEnv::GetStringUTFChars(_jstring *,uchar *)()
12240 ms  │  MD5Init()
12240 ms  │  MD5Update()
12240 ms  │   │  sub_A57E()
12240 ms  │  MD5Final()
12240 ms  │   │  sub_B2F8()
12240 ms  │   │  MD5Update()
12240 ms  │   │   │  sub_A57E()
12241 ms  │   │  MD5Update()
12241 ms  │   │   │  sub_A57E()
12241 ms  │   │   │  sub_A5B0()
12241 ms  │   │   │   │  sub_B398()
12241 ms  │   │   │   │  sub_B36A()
12241 ms  │   │   │  sub_A57E()
12241 ms  │   │  sub_B2F8()
12241 ms  │   │  sub_B36A()
12241 ms  │  _JNIEnv::ReleaseStringUTFChars(_jstring *,char const*)()
12241 ms  │  _JNIEnv::NewStringUTF(char const*)()
```

图 13-42　优化后的 traceNatives 效果

结果正确地打印出了以 IDA 中函数命名显示的调用流程。

（2）很多时候，我们会分析一些商用 App，它们的函数名不会有明显的函数功能提示信息。此时，我们只能慢慢分析出函数的功能，在 IDA 中为函数重命名后，再运行 traceNatives。

13.6　本章小结

在本章中，我们主要学习了哈希算法中的 MD5 算法，并且演示了两轮计算流程。另外，还介绍了两款算法检测插件，以及它们的检测原理。最后，介绍了 Frida MemoryAccessMonitor 插件，它的用处不大，读者简单了解即可。

第 14 章

对称加密算法逆向分析

加解密算法在 SO 文件的分析中是很重要的内容。读者可能会有疑问：

（1）是否可以在不了解加密算法细节的情况下，对 SO 文件中的函数进行分析和还原？答案是可以的，但是要花费更多的时间和精力，甚至可能劳而无功。

（2）是否可以对加密算法的主函数进行 Frida Hook？答案也是可以的，但不一定总是行得通。

本章将学习另一种重要的加密算法——对称加密算法及其逆向分析。对称加密算法指加密和解密使用相同密钥的加密算法。经典的对称加密算法包括 DES、AES、IDEA、RC 等。

14.1 DES 详解

数据加密标准（Data Encrypt Standard，DES）是一种经典的加密算法，从研发至今应用了 20 多年，直到更加强大的高级加密标准（Advanced Encrypt Standard，AES）诞生之后，DES 仍然被广泛应用。作为安全技术工程师，有必要了解 DES 的特点和实现，有助于我们认识变体以及其他分组加密算法，并很快地复现。

标准的 DES 包括入参和出参，其中：

入参：

明文：64 比特（16 个十六进制数，8 字节）。
密钥：64 比特（16 个十六进制数，8 字节）。
出参：64 比特（16 个十六进制数，8 字节）。

DES 的实现过程分为两部分，一是明文处理，二是密钥编排。下面我们根据给出的明文和密钥使用 DES 加密来具体实现。

给出的明文和密钥如下：

明文：0123456789abcdef（HEX）。
密钥：133457799bbcdff1（HEX）。

我们还是用 CyberChef 来对上述明文和密钥进行 DES 加密计算，结果如图 14-1 所示。

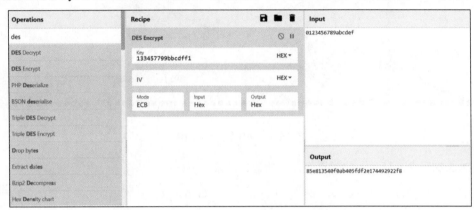

图 14-1　CyberChef 计算 DES 加密结果

读者看到这里可能会发现，CyberChef 计算 DES 加密的结果是 128 比特而不是 64 比特。这一原因我们稍后揭晓。

这里的输出为：85e813540f0ab405fdf2e174492922f8（HEX）。

下面来看具体的实现步骤。

第 1 步：明文处理。

对明文进行预处理（初始置换），首先将明文和密钥转化成二进制分别填入代码中。

明文：0123456789abcdef（HEX）。
-> 00000001 00100011 01000101 01100111 10001001 10101011 11001101 11101111（binary）

根据 IP 表（Initial Permutation）对明文进行置换，即重新排列。IP 表如图 14-2 所示。

		IP					
58	50	42	34	26	18	10	2
60	52	44	36	28	20	12	4
62	54	46	38	30	22	14	6
64	56	48	40	32	24	16	8
57	49	41	33	25	17	9	1
59	51	43	35	27	19	11	3
61	53	45	37	29	21	13	5
63	55	47	39	31	23	15	7

图 14-2　IP 表

IP 表从左至右、从上到下置换数据。例如，IP 表的第 1 个数字是 58，我们就要去原明文分组中找对应的第 58 位，即 1，放到新数据的第 1 位，第 50 位放到新数据的第 2 位，以此类推。初始置换示意图如图 14-3 所示。

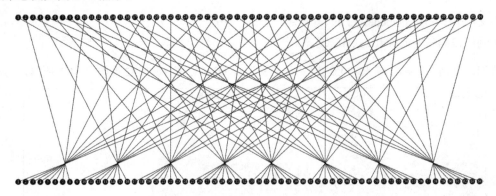

图 14-3　初始置换示意图

在 DES 加密中，充满了此类置换。初始置换结果如下：

> 1, 1, 0, 0, 1, 1, 0, 0, 0, 0, 0, 0, 0, 0, 0, 0, 1, 1, 0, 0, 1, 1, 0, 0, 1, 1, 1, 1, 1, 1, 1, 1, 1, 1, 1, 1, 0, 0, 0, 0, 1, 0, 1, 0, 1, 0, 1, 0, 1, 1, 1, 1, 0, 0, 0, 0, 1, 0, 1, 0, 1, 0, 1, 0

以上对明文进行了预处理，接下来我们来看如何对密钥进行编排。

第 2 步：密钥的编排。

初始置换做好之后，接下来要进行的工作是密钥的编排——生成 16 轮次的子密钥。对于 DES 来说，它的密钥有 64 位，但只有 56 位被使用。

密钥：00010011 00110100 01010111 01111001 10011011 10111100 11011111 11110001。

我们在预处理中，首先将明文根据 IP 表进行了初始置换。在密钥的编排中，也需要将其进行初始置换，根据 PC-1 表（置换选择表 1）重排密钥。PC-1 表如图 14-4 所示。

PC-1						
57	49	41	33	25	17	9
1	58	50	42	34	26	18
10	2	59	51	43	35	27
19	11	3	60	52	44	36
63	55	47	39	31	23	15
7	62	54	46	38	30	22
14	6	61	53	45	37	29
21	13	5	28	20	12	4

图 14-4　PC-1 表

PC-1 表中只包含 56 个索引（7 列 8 行）。所以，密钥经过编排后只留下了 56 位，每字节的最后一位都没有采用。所以我们说，DES 的密钥的有效位只有 56 位。重新编排后的密钥如下：

1, 1, 1, 1, 0, 0, 0, 0, 1, 1, 0, 0, 1, 1, 0, 0, 1, 0, 1, 0, 1, 0, 1, 0, 1, 1, 1, 1,
0, 1, 0, 1, 0, 1, 0, 1, 0, 1, 1, 0, 0, 1, 1, 0, 0, 1, 1, 1, 1, 0, 0, 0, 1, 1, 1, 1

接下来将这 56 比特长的密钥分成左右两部分，并命名为 L0 和 R0。

L0 = 1111000 0110011 0010101 0101111

R0 = 0101010 1011001 1001111 0001111

接下来循环左移规定的位数。密钥的编排这一步骤的目的是根据输入的密钥生成 16 个不同的、对应运算部分的、每一轮的密钥。DES 一共有 16 轮运算，每轮需要一个子密钥。子密钥通过循环左移生成，第 i 轮左移位数对应 SHIFT 中的第 i 个元素，得到 C1、D1 到 C16、D16 的 16 个子密钥。

SHIFT = [1,1,2,2,2,2,2,2,1,2,2,2,2,2,2,1]

左移 1：

L1 = 1110000 1100110 0101010 1011111

R1 = 1010101 0110011 0011110 0011110

L1R1 = 1110000 1100110 0101010 1011111 1010101 0110011 0011110 0011110

根据 PC-2（置换选择表 2）重新编排 L1R1，输入为 56 位，输出进一步缩小至 48 位：

K1 = 000110 110000 001011 101111 111111 000111 000001 110010

K2 的运算同理。先计算 L2、R2。

查阅 SHIFT 得知，L2、R2 分别在 L1、R1 的基础上循环左移 1 位。然后将 L2、R2 拼接得到 L2R2，L2R2 再经过 PC-2 置换，输出的结果即为 K2。

K2 = 011110 011010 111011 011001 110110 111100 100111 100101

以此类推，可以得到 K3~K16，如下：

K2 = 011110 011010 111011 011001 110110 111100 100111 100101
K3 = 010101 011111 110010 001010 010000 101100 111110 011001
K4 = 011100 101010 110111 010110 110110 110011 010100 011101
K5 = 011111 001110 110000 000111 111010 110101 001110 101000
K6 = 011000 111010 010100 111110 010100 000111 101100 101111
K7 = 111011 001000 010010 110110 111101 100010 100010 111100
K8 = 111101 111000 101000 111010 110000 010011 101111 111011
K9 = 111000 001101 101111 101011 111011 011110 011110 000001
K10 = 101100 011111 001101 000111 101110 100100 011001 001111

K11 = 001000 010101 111111 010011 110111 101101 001110 000110
K12 = 011101 010111 000111 110101 100101 000110 011111 101001
K13 = 100101 111100 010111 010001 111110 101011 101001 000001
K14 = 010111 110110 000110 110111 111100 101110 011100 111010
K15 = 101111 111001 000110 001101 001111 010011 111100 001010
K16 = 110010 110011 110110 001011 000011 100001 011111 110101

子密钥的生成到此结束，在后续 DES 的 16 轮运算中，第 n 轮就是用 Kn。

第 3 步：明文的运算。

到目前为止，我们已经对明文进行了重排，并得到了 16 个子密钥。接下来就是明文的 16 轮运算。首先将明文分成左右两半，像上面的密钥一样。

初始置换后的明文：11001100 00000000 11001100 11111111 11110000 10101010 11110000 10101010。

L0 = 11001100 00000000 11001100 11111111
R0 = 11110000 10101010 11110000 10101010

Ln 和 Rn 的计算公式如下：

Ln = **Rn-1**
Rn = **Ln-1** + **f**(**Rn-1**,**Kn**)

不知道读者会不会想到 MD5，MD5 中的每一轮只改变 A、B、C、D 中的一个，其余 3 个进行平移赋值，Ln 和 Rn 也是同样的道理。需要注意的是，MD5 中凡是 "+" 运算，指的就是简单的加法，如果超过 0xFFFFFFFF 大小，则取低 32 位，而 DES 中 "+" 指的是按位异或运算。

以第一轮为例：

L1 = R0 = 11110000 10101010 11110000 10101010
R1 = L0 xor f(R0, K1)

接下来的关注点就在 F 函数上了。在 MD5 中也有一个 F 函数，但它只是运算的一部分，而 DES 中的 F 函数代表了 16 轮运算中一轮的全部运算。以 n = 1 为例，F 函数传入了 R0 与 K1，即第一个子密钥。

由于 R0 是 32 比特长，K1 是 48 比特长，长度上就不相同，因此没法按比特进行运算。第一步是对 R0 进行扩展，同样是使用一个表，这个表叫 Expand 表，即扩展表。在扩展表中，有 16 个比特位被使用了两次。因此，R0 长度从 32 比特被扩展成了 48 比特。扩展表如图 14-5 所示。

```
E BIT-SELECTION TABLE
32   1   2   3   4   5
 4   5   6   7   8   9
 8   9  10  11  12  13
12  13  14  15  16  17
16  17  18  19  20  21
20  21  22  23  24  25
24  25  26  27  28  29
28  29  30  31  32   1
```

图 14-5　扩展表

R0 = 11110000 10101010 11110000 10101010
E(R0) = 011110 100001 010101 010101 011110 100001 010101 010101

接下来用 E(R0) 和 K1 异或，这一步在 AES 中同样存在，专业说法叫密钥混合。

E(R0) = 011110 100001 010101 010101 011110 100001 010101 010101
K1 = 000110 110000 001011 101111 111111 000111 000001 110010
K1+**E**(**R0**) = 011000 010001 011110 111010 100001 100110 010100 100111

最后我们得到了 48 比特的结果。从上面数据的表示上可以看出，我们把它视为 8 个 6 比特的数据，这是因为我们后续将按这种方式使用它。

接下来是 DES 加密的核心——S 盒。首先，将上面一步的结果当成 8 块 6 比特长的输入，用 Bn 来表示每一块的数据。

K1+**E**(**R0**) = **B1B2B3B4B5B6B7B8**

那么：

B1 = 011000

B2 = 010001

B3 = 011110

B4 = 111010

B5 = 100001

B6 = 100110

B7 = 010100

B8 = 100111

之后，我们把 B1~B8 的值当成地址（或者说索引），去 S1~S8 盒中找对应的值。

规则如下：

以 B1 为例，值为 011000，一个 6 比特长的二进制数，我们将它分成两部分，第一部分 i 为首位与末位的拼接 00，第二部分 j 为中间的 4 位 1100。

  i = 00 即 0

  j = 1100 即 12

  =>(0, 12)

将 i、j 组成坐标(i,j)，即(0, 12)，代表 S1 盒(0, 12)处的值，S1 盒如图 14-6 所示。

S1																
	Column Number															
Row No.	0	1	2	3	4	5	6	7	8	9	10	11	12	13	14	15
0	14	4	13	1	2	15	11	8	3	10	6	12	5	9	0	7
1	0	15	7	4	14	2	13	1	10	6	12	11	9	5	3	8
2	4	1	14	8	13	6	2	11	15	12	9	7	3	10	5	0
3	15	12	8	2	4	9	1	7	5	11	3	14	10	0	6	13

图 14-6 S1 盒

B1：(0, 12) = 5。

因为 B 是 6 比特长，所以用 b1b2b3b4b5b6 来表示 6 比特的数据。b1b6 的范围是 0~3，b2b3b4b5 的范围是 0~15，即与表实现了完全的对应，不会有 B 的"索引"超出表的大小。B1 在 S1 中找，B2 在 S2 中找，以此类推，最后 8 个 6 比特的值在 S 盒作用下变成 8 个 4 比特的值。

S(**K1**+**E**(**R0**)) = 0101 1100 1000 0010 1011 0101 1001 0111

实际上，S 盒替换是分组加密算法的常见处理方式，是 DES 安全性的核心，没有 S 盒，DES 脆弱不堪。之后，在逆向分析的加密算法检测中，我们主要关注这个 S 盒。如果想检测 DES 或者 AES，就会通过 S 盒下手。

接下来进行 P 盒置换，就是简单地将 32 比特进行重新排列置换，P 盒如图 14-7 所示。

P			
16	7	20	21
29	12	28	17
1	15	23	26
5	18	31	10
2	8	24	14
32	27	3	9
19	13	30	6
22	11	4	25

图 14-7 P 盒

P(S(**K1**+**E**(**R0**))) = 0010 0011 0100 1010 1010 1001 1011 1011

F 函数大功告成。

F = P(S(**K1**+**E**(**R0**)))

最后将 F 的结果与 L0 异或：

L1 = R0 = 11110000 10101010 11110000 10101010
R1 = L0 xor f(R0, K1) = 11101111 01001010 01100101 01000100

第一轮彻底结束，接下来进行 16 轮完全一致的操作：

L1：11110000 10101010 11110000 10101010
R1：11101111 01001010 01100101 01000100
L2：11101111 01001010 01100101 01000100
R2：11001100 00000001 01110111 00001001
L3：11001100 00000001 01110111 00001001
R3：10100010 01011100 00001011 11110100
L4：10100010 01011100 00001011 11110100
R4：01110111 00100010 00000000 01000101
L5：01110111 00100010 00000000 01000101
R5：10001010 01001111 10100110 00110111
L6：10001010 01001111 10100110 00110111
R6：11101001 01100111 11001101 01101001
L7：11101001 01100111 11001101 01101001
R7：00000110 01001010 10111010 00010000
L8：00000110 01001010 10111010 00010000
R8：11010101 01101001 01001011 10010000
L9：11010101 01101001 01001011 10010000
R9：00100100 01111100 11000110 01111010
L10：00100100 01111100 11000110 01111010
R10：10110111 11010101 11010111 10110010
L11：10110111 11010101 11010111 10110010
R11：11000101 01111000 00111100 01111000
L12：11000101 01111000 00111100 01111000
R12：01110101 10111101 00011000 01011000
L13：01110101 10111101 00011000 01011000
R13：00011000 11000011 00010101 01011010
L14：00011000 11000011 00010101 01011010
R14：11000010 10001100 10010110 00001101
L15：11000010 10001100 10010110 00001101
R15：01000011 01000010 00110010 00110100

L16：01000011 01000010 00110010 00110100
R16：00001010 01001100 11011001 10010101

得到 L16 和 R16 后，左右调一下顺序，结果为 R16L16。

R16L16 = 00001010 01001100 11011001 10010101 01000011 01000010 00110010 00110100

第 4 步：末置换。

接下来还差最后一步，末置换，末置换当然也要用表，表叫 FP（Final Permutation）表。

从图 14-8 可以很清晰地发现末置换就是初始置换的逆运算。

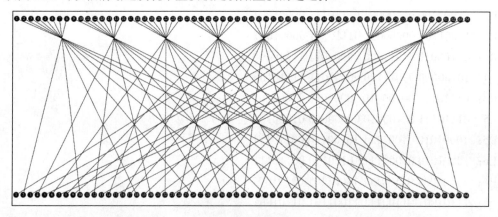

图 14-8　末置换

这里与初始置换互相呼应。因此，如果一个 64 比特的明文首先经过初始置换，不做任何处理再进行一次末置换，则值等于它自身。R16L16 经过末置换得到最终结果：

10000101 11101000 00010011 01010100 00001111 00001010 10110100 00000101

转换成 HEX 为 85e813540f0ab405。

经过 CyberChef 验证，结果完全正确。

经过以上演示，总体来看，DES 算法只是比较烦琐，但是并不难。上述的一些置换表以及运算细节都可以当作算法分析的特征。

14.1.1　分组密码的填充与工作模式

对于 DES、AES 等算法来说，其输入的明文要求都是固定的分组长度，称为分组密码。例如 DES，其输入、密钥、输出的分组长度必须是 64 比特或其整数倍。

问题来了，如果输入不是 64 比特或其整数倍，怎么办？答案是明文不足 64 比特时，就需要填充。填充方式考虑以下 3 种情况：

（1）明文分组长度与 64 位相差整数字节。

这种情况的解决办法是，假设明文为 56 位，与 64 位相差一字节，则在原明文末尾填充 01。

（2）明文分组长度与 64 位相差非整数字节。

这种情况的解决办法是，假设明文为（HEX）011，与 64 位相差 52 位（6.5 字节），则将原明文填充为 0101060606060606。

（3）明文分组长度为 64 位。

解决办法是，假设明文为（HEX）0123456789abcdef，则将原明文填充为 0123456789abcdef0808080808080808。

在 CBC 和 ECB 模式下，采用 PKCS#7 的填充方式。需要注意的是，密钥没有相应的填充规则，但如果没有输入符合要求的密钥，那么程序如何执行是无法预料的。

例如在 ECB 工作模式下，如果填充后的原文长度大于分组长度（64 比特），则将原文分组并分别进行 DES 加密。缺点是：相同的明文分组输入计算出来的密文结果是完全一致的。不建议在任何开发中采用 ECB 模式。我们可以通过图 14-9 和图 14-10 来感受一下其危害。

图 14-9　ECB 模式加密前　　　　图 14-10　ECB 模式加密后

可以看到，采用 ECB 模式加密后，企鹅的特征仍然存在，从肉眼就能分辨出图像中是一只企鹅。或者说，我们能够从密文中得到一些信息。

使用除了 ECB 之外的模式，即使是相同的明文和密钥，不同分组的加密结果也是不一样的。在 ECB 模式下，可以不用输入 IV，而其他模式必须输入 IV，且输入和输出都不一定是 8 字节的倍数。

14.1.2　三重 DES

DES 的密钥只有短短的 8 字节，很难给现代化的应用提供安全保障，这种密钥 20 世纪 90 年代就已经被破解了。因此，DES 的设计公司 IBM 很快就设计了 3DES。3DES 并不是简单地将 DES 做三遍。三重 DES 加密原理如图 14-11 所示。

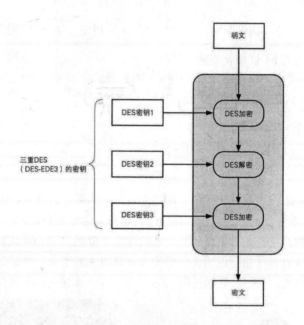

图 14-11　三重 DES 加密原理示意图

三重 DES 为什么要这样设计呢？让我们来做个实验。
请看下面的示例。

明文：12345。
密钥：0123456789abcdef0123456789abcdef0123456789abcdef（HEX）。

K1、K2、K3：
K1 = 0123456789abcdef（HEX）
K2 = 0123456789abcdef（HEX）
K3 = 0123456789abcdef（HEX）
3DES 结果：a304f525d2d52cbf。
DES 结果：a304f525d2d52cbf。

我们发现，3DES 的结果和 DES 的结果是一模一样的。这是为什么呢？我们来看一下 3DES 的加密流程。

12345→DES 加密（0123456789abcdef）→DES 解密（0123456789abcdef）→12345→DES 加密（0123456789abcdef）

实际上等同于：12345→DES 加密（0123456789abcdef）。

这就是 3DES 的结果和 DES 的结果相同的原因。那么问题来了，既然 3DES 的结果和 DES 的结果相同，为什么还要用 3DES 呢？这其实是一个设计上的问题。我们设置的是 K1、K2、K3 相等，在这种情况下，3DES 的结果和 DES 的结果是一致的。IBM 公司考虑到，DES 被广泛应用于银行、企业等各个领域的服务中。现在要推广 3DES，就存在一个兼容性的问题。一些使用 DES 的通信协议，如果想换到 3DES，一时换不过来，所以要保持兼容性。这个时候，只要把密钥设置成和原来一样的重复三个，结果就和原来一样了。这样可以带来向下兼容。

3DES 的安全性如何保证呢？如果 K1、K2、K3 不相等（K1、K2 不相等即可），就是真正的三重 DES：

12345→DES 加密（K1）→DES 解密（K2）

K1、K2 不相等会带来一个问题，即解密时用了错误的密钥是解不出来的，并且会导致更严重的混淆。因此，DES 解密相当于又一次的加密。

我们可以这样来理解 3DES 加密：当 K1、K2 相等的时候，它就是简单的 DES 加密；当 K1、K2 不相等的时候，它就是复杂的 3DES 加密。

14.2 AES

14.2.1 AES 初识

作为替代 DES 的加密算法 AES，又称为 Rijndael 加密法，目前已得到广泛使用。AES 已然成为对称密钥加密中最流行的算法之一。在学习 AES 的过程中，我们将结合一些样本，从样本中发现问题并解决问题。逆向分析一个真实的 SO 文件，往往会经历"踩坑、翻车、耗时间"的过程，这是非常正常的，也是非常有意义的。如果分析一个简单的 SO 样本需要两小时，那么分析一个正常的 SO 样本则需要半天到一天，而分析商用的 SO 样本甚至可能需要几天甚至一周的时间。更不用说复杂的、带有强保护的 SO 样本，花上几个月的时间来分析它也是正常的。无论是简单还是复杂的 SO 样本，我们都不可避免地会遇到加密算法，而其中 AES 算法占有十分重要的地位。

在正式学习 AES 之前，我们先来回顾一下 SO 中的密码学算法，主要分为 3 类：哈希算法（主要用于签名校验）、对称加密算法（DES、AES 等）、非对称加密算法（RSA、ECC 等）。

1. 哈希算法

哈希算法比较简单，只要传入一个明文，就能得到一个哈希后的结果。它更多的时候用于进行签名校验。我们不需要了解其原理和底层，这对于分析程序有影响，但影响不大。

2. 对称加密算法

对称加密算法的重点是 AES，它的工作模式包括 ECB、CBC 等，KEY 是密钥，IV 是初始化链接变量，Padding 是填充规则。

3. 非对称加密算法

非对称加密算法包括 RSA、ECC 等，它的优点是安全性好，缺点是运算复杂且慢。RSA+AES

常用作通信中的混合加密，其中 RSA 只用于密钥处理（即密钥的加解密）。

接下来，我们先来看样本 1——MT（代指美团 App，我们一般用代号）。在手机上运行样本 1 并启动 frida-server，之后运行 callJava.js 脚本。

```javascript
function testsiua(plainText){

    Java.perform(function () {

        var datacollect = Java.use('com.meituan.android.common.datacollection.DataCollectionJni');

        var ByteString = Java.use("com.android.okhttp.okio.ByteString");

        var Base64 = Java.use("java.util.Base64");

        // 参数 1 是上下文

        var currentApplication= Java.use("android.app.ActivityThread").currentApplication();

        var context = currentApplication.getApplicationContext();

        // 参数 2 是明文字节数组

        // 使用 String 类将我们传入的字符串转成数组

        var JavaString = Java.use("java.lang.String");

        var bytesStr = JavaString.$new(plainText).getBytes();

        // 参数 3 是明文的长度

        var textLength = plainText.length;

        // 主动调用

        var result = datacollect.packData(context, bytesStr, textLength)

        // 返回 Base64 编码后的结果，使用系统类转换成十六进制
```

```
        var base64result = ByteString.of(result).utf8()

        var base64before = Base64.getDecoder().decode(base64result);

        // console.log("base64 result:"+base64result);

        console.log("result:"+ByteString.of(base64before).hex());
    });
}
```

callJava.js 的参数 1 是上下文，参数 2 是传入的字节数组，参数 3 是明文的长度。我们直接构造一个主动调用的脚本，这里不去分析具体的业务逻辑，运行脚本，结果如图 14-12 所示。

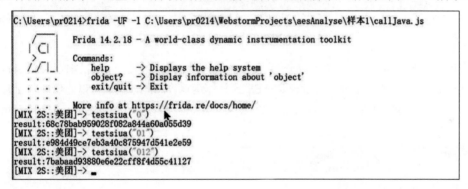

图 14-12　callJava.js 运行结果

假如我们对 SO 层一无所知，怎么猜测它用了什么算法？从结果来看，我们发现，无论传入的明文是什么样的，它都会返回一个固定长度的数——32 个十六进制数，即 128 比特，刚好是 AES 的分组长度。而且，当明文长度等于或超过 16 字节时，结果变成了两个分组大小。这就很符合 AES 的特征。因此，我们猜测它可能用到了 AES。

接下来，我们分析一个疑似包含 AES 的样本 SO。既然我们怀疑这里面有 AES，那么可以通过 Findscript 查找其是否有 AES 的相关特征。发现确实有包含 AES 的字符串，如图 14-13 所示。

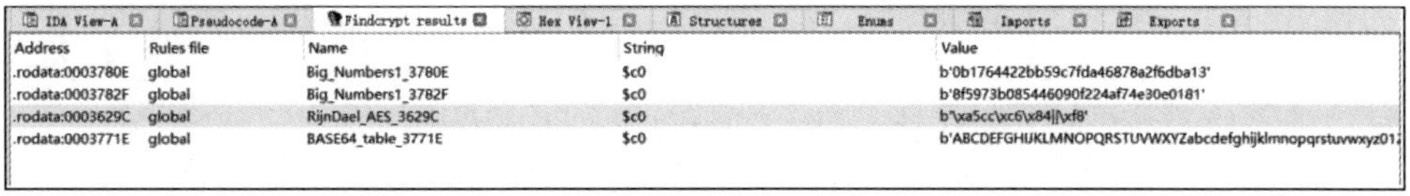

图 14-13　通过 Findscript 查找其是否有 AES 的相关特征

按 x 查看交叉引用，我们发现只有一个函数 sub_2F88 调用了 RijnDael_AES_3629C，如图 14-14 所示。

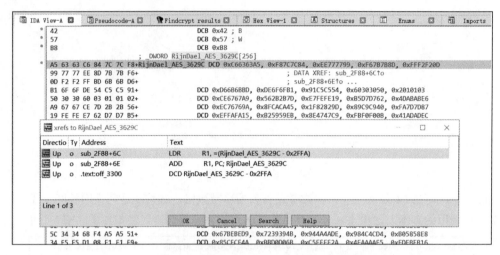

图 14-14 sub_2F88 调用了 RijnDael_AES_3629C

双击进入 sub_2F88 函数，并按 Tab 键查看伪代码。我们发现看不懂这些代码，但可以猜想，它可能用到了 AES 中的一个表，或者说是 AES 的具体运算。因此，我们可以先给它加上注释：疑似 AES 具体运算步骤。再查看 sub_2F88 函数的交叉引用,也只看到一个函数 sub_36AC 引用了它,如图 14-15 所示，双击进入。

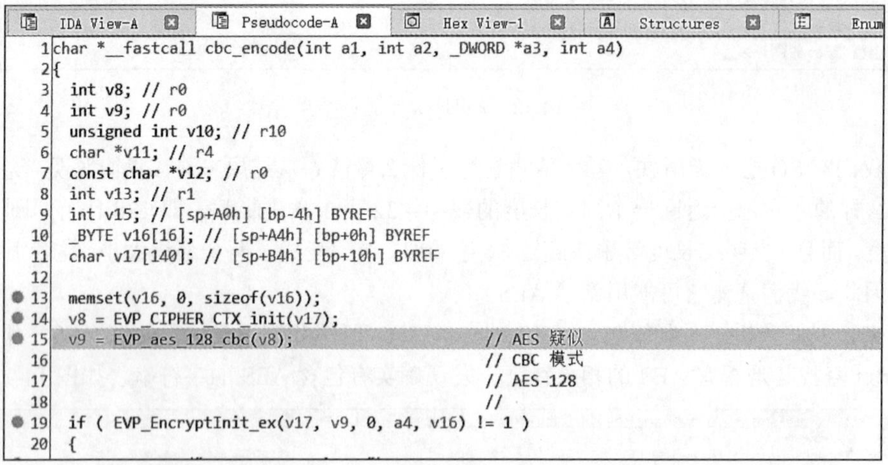

图 14-15 查看 sub_2F88 函数的交叉引用

如何分析 sub_2F88 函数呢？可以 Hook 它的 3 个参数。有一些经验的逆向人员可能会发现，此时已经具备一些 AES 算法的特征了。

请注意，AES 算法的相关知识：

（1）AES 分组长度为 128 比特，DES 分组长度为 64 比特。

（2）AES 密钥长度：

- 128 比特长 16 字节。
- 192 比特长 24 字节。
- 256 比特长 32 字节。

密钥越长，安全性越高。但无论 AES 密钥是 128 比特、192 比特还是 256 比特，分组大小以及密文输出大小都是不变的。

（3）在 CyberChef 中，由于密钥长度是按字节计算的，如果输入的密钥不是整字节数的长度，例如，Key=1111111（3.5 字节），那么 CyberChef 会给它补 0，变为 Key=11111101。

了解了一些 AES 相关知识以后，我们回到样本上来，编写脚本 Hook sub_2F88 函数，代码如下：

```
// 2F88 函数
// 确定 SO 的基地址
var base_addr = Module.findBaseAddress("libmtguard.so");

Interceptor.attach(real_addr, {
    onEnter: function (args) {

        this.arg0 = args[0];
        this.arg1 = args[1];
        this.arg2 = args[2];

        console.log("\n");
        console.log("Before 2f88");
        console.log(hexdump(this.arg0));
        console.log(hexdump(this.arg1));
        console.log(hexdump(this.arg2));

    }, onLeave: function (retval) {

        console.log("After 2f88");
        console.log(hexdump(this.arg0));
        console.log(hexdump(this.arg1));
        console.log(hexdump(this.arg2));
    }
});

function main() {
    // 确定 SO 的基地址
    var base_addr = Module.findBaseAddress("libmtguard.so");
    hook_2F88(base_addr, 0x2F88+1); // thumb
}
```

主动调用结果如图 14-16 所示。

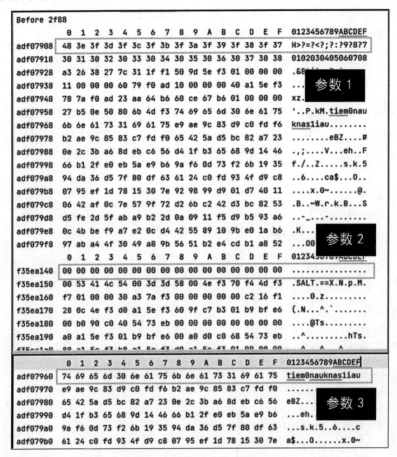

图 14-16　Hook sub_2F88 脚本结果

从结果中可以发现，输入的参数 1、参数 3 在输出时，结果并未改变。而参数 2 在输入时全 0，这意味着参数 2 很有可能是一个 buffer，用来存放最终输出的结果。经验证，输出的参数 2 为 68c7…5d39，确实为最终结果。

到现在为止，由于我们不了解 AES 的具体流程和实现细节，因此无法从样本 1 的反汇编伪代码中获取更多的信息。

接下来，我们要做的是暂时放下样本 1 的分析，探索样本 2——librong360.so，从中了解更多的 AES 的工作模式、具体流程和实现细节。这个样本同样用到了 AES，如图 14-17 所示。

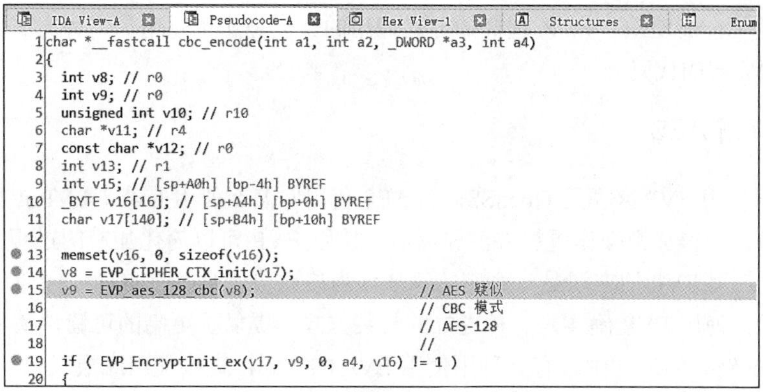

```
  1  char *__fastcall cbc_encode(int a1, int a2, _DWORD *a3, int a4)
  2  {
  3    int v8; // r0
  4    int v9; // r0
  5    unsigned int v10; // r10
  6    char *v11; // r4
  7    const char *v12; // r0
  8    int v13; // r1
  9    int v15; // [sp+A0h] [bp-4h] BYREF
 10    _BYTE v16[16]; // [sp+A4h] [bp+0h] BYREF
 11    char v17[140]; // [sp+B4h] [bp+10h] BYREF
 12
 13    memset(v16, 0, sizeof(v16));
 14    v8 = EVP_CIPHER_CTX_init(v17);
 15    v9 = EVP_aes_128_cbc(v8);             // AES 疑似
 16                                          // CBC 模式
 17                                          // AES-128
 18                                          //
 19    if ( EVP_EncryptInit_ex(v17, v9, 0, a4, v16) != 1 )
 20    {
```

图 14-17 样本 2 用到了 AES

从函数名可以得知它用到了 AES 加密算法，工作模式为 CBC，且密钥长度为 128 位。还缺什么信息去还原它呢？我们不知道明文在哪里，也不知道它的密钥和 IV 在哪里。

其实，从函数名中还可以得到一个关键信息——EVP。EVP 是 OpenSSL 一个标志性的特征。OpenSSL 实现了很多加密算法，调用这些加密算法的方法有两种：一种是从方法底层调用，找到 AES/DES 算法的实现，缺点是比较烦琐；另一种是用 EVP，它是 OpenSSL 封装的 API。下面我们来看 EVP 的用法。

初始化结构体：

```
int EVP_CIPHER_CTX_init(EVP_CIPHER_CTX *a);
```

初始化加密算法：

```
int EVP_EncryptInit_ex(EVP_CIPHER_CTX *ctx, const EVP_CIPHER *type, ENGINE *impl, unsigned char *key, unsigned char *iv);
```

第一个参数是结构体，第二个参数表示的是使用的算法，参数 4 是密钥，参数 5 是 IV，IV 是在算法中生成的。我们看到 memset 将 IV 设置成 16 字节的 0x0（参数 1 是被操作的内存块 IV，参数 2 是填充的值 0，参数 3 是填充长度）。

数据加密：

```
int EVP_EncryptUpdate(EVP_CIPHER_CTX *ctx, unsigned char *out, int *outl, unsigned char *in, int inl);
```

我们发现 cbc_encode 函数的第一个参数就是明文。

清除上下文：

```
int EVP_CIPHER_CTX_cleanup(EVP_CIPHER_CTX *a);
```

为了防止在内存中可以直接搜索到加密相关信息，清除上下文，将 EVP_CIPHER_CTX 结构体进行随机化或全部赋 0。

可以发现，我们完全没有经过动态调试，单凭百度搜索就得到了样本中的一些 AES 的信息。

它实际上用的是 OpenSSL 的 EVP 这样的方式调用的。但是还可以用另一种方式调用——不使用 EVP，直接使用底层的代码。

14.2.2 深入了解 AES

为什么需要使用 EVP 来调用 OpenSSL 的 AES 代码？怎么使用 EVP？EVP 提供了一种更高层次的抽象和封装，使得加密操作更加方便和灵活。使用 EVP 可以简化加密的过程，并提供了更多的功能和选项。使用 EVP 的方法是将对应的算法作为参数传递给 EVP 的第二个参数。相比直接调用底层算法代码，使用 EVP 确实是一种更好的封装方式，提供了更高的可读性和易用性。

在基于 OpenSSL 的加密中，有一种不使用 EVP 的方式，直接使用底层代码进行操作。然而，在实际的逆向分析中，这种情况相对较少，通常会使用 EVP 进行加密操作。在不使用 EVP 的情况下，使用 Hook 技术可以比较快速地获取相关信息；而使用 EVP 则比较麻烦，因为它传入的是结构体，需要结合经验和对 OpenSSL 的深入了解才能开始逆向分析。

对于上面简单的样本来说，不了解 AES 算法的原理，对逆向分析的影响不大，因为它做了比较好的封装。

样本 3——testtinyaesdemo1 的目标是还原每个 sub_xxxx 函数符号的含义，如图 14-18 所示。

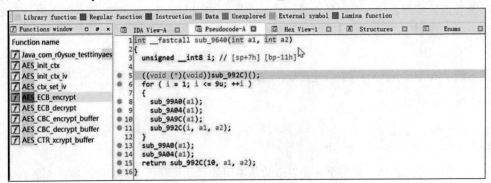

图 14-18　还原每个 sub_xxxx 函数符号的含义

AES 的原理涉及两大核心知识点：

- 密钥的编排。
- 明文的运算。

DES 是用比特流的方式操作的，而在 AES 中，操作都是基于字节的。假如明文是一个字符串：0123456789abcdef，则 DES 是先把它变成二进制比特流的形式：00110000　00110001　00110010　00110011　00110100　00110101　00110110　00110111　00111000　00111001　01100001　01100010　01100011　01100100　01100101　01100110，然后分成左右两部分进行操作。AES 则是将这 16 个字符转成 16 个十六进制数，放到 4×4 的表格中，如下所示（注意是按从上至下、从左至右的顺序依次放置的）。

```
|------|------|------|------|
| 30   | 34   | 38   | 63   |
| 31   | 35   | 39   | 64   |
| 32   | 36   | 61   | 65   |
| 33   | 37   | 62   | 66   |
```

密钥同样是一个 4×4 的表，如图 14-19 所示。表中的每个元素都是一字节大小。

之后，明文和密钥分别进入两个流程，明文进入加密流程，密钥进入密钥编排流程，如图 14-20 所示。一般会先做密钥的编排，因为没有密钥是没办法进行明文加密的。

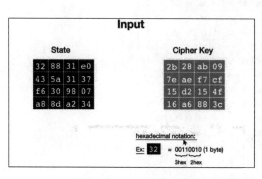

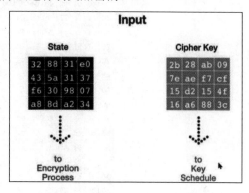

图 14-19　AES 分组输入　　　　　　图 14-20　加密流程与密钥编排流程

明文操作基于字节，即：

AES-128：10 轮运算。

AES-192：12 轮运算。

AES-256：14 轮运算。

以 AES-128 10 轮运算为例，要经过初始运算、9 轮完全相同的运算、位运算，流程如图 14-21 所示。

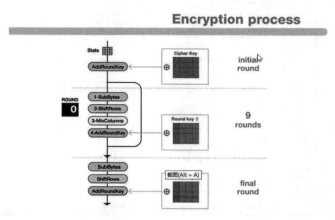

图 14-21　初始运算、9 轮完全相同的运算、位运算

下面对图 14-21 进行具体的说明。

其中，初始运算（addRoundKey）将明文和密钥 K 进行异或。

在 9 轮完全相同的运算中，包含以下 4 个子步骤。

1. S 盒替换

这一步非常简单，同时也是逆向分析的突破口。S 盒如图 14-22 所示。

	00	01	02	03	04	05	06	07	08	09	0a	0b	0c	0d	0e	0f
00	63	7c	77	7b	f2	6b	6f	c5	30	01	67	2b	fe	d7	ab	76
10	ca	82	c9	7d	fa	59	47	f0	ad	d4	a2	af	9c	a4	72	c0
20	b7	fd	93	26	36	3f	f7	cc	34	a5	e5	f1	71	d8	31	15
30	04	c7	23	c3	18	96	05	9a	07	12	80	e2	eb	27	b2	75
40	09	83	2c	1a	1b	6e	5a	a0	52	3b	d6	b3	29	e3	2f	84
50	53	d1	00	ed	20	fc	b1	5b	6a	cb	be	39	4a	4c	58	cf
60	d0	ef	aa	fb	43	4d	33	85	45	f9	02	7f	50	3c	9f	a8
70	51	a3	40	8f	92	9d	38	f5	bc	b6	da	21	10	ff	f3	d2
80	cd	0c	13	ec	5f	97	44	17	c4	a7	7e	3d	64	5d	19	73
90	60	81	4f	dc	22	2a	90	88	46	ee	b8	14	de	5e	0b	db
a0	e0	32	3a	0a	49	06	24	5c	c2	d3	ac	62	91	95	e4	79
b0	e7	c8	37	6d	8d	d5	4e	a9	6c	56	f4	ea	65	7a	ae	08
c0	ba	78	25	2e	1c	a6	b4	c6	e8	dd	74	1f	4b	bd	8b	8a
d0	70	3e	b5	66	48	03	f6	0e	61	35	57	b9	86	c1	1d	9e
e0	e1	f8	98	11	69	d9	8e	94	9b	1e	87	e9	ce	55	28	df
f0	8c	a1	89	0d	bf	e6	42	68	41	99	2d	0f	b0	54	bb	16

图 14-22　Rijndael S 盒

明文的一字节由 8 位组成，以这 8 位的高 4 位作为 S 盒表的行值，低 4 位作为 S 盒表的列值，在其中寻找对应的代替结果。例如，十六进制数{19}表示代替结果在第 1 行第 9 列，因此结果为{d4}。

在 IDA 插件中可以通过查找 S 盒来确定 SO 中是否有 AES 算法，如图 14-23 所示。

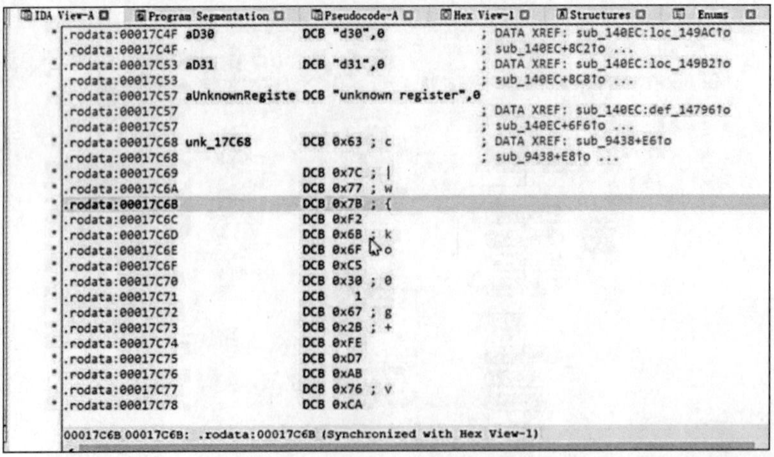

图 14-23　样本中的 S 盒

2. 循环左移

与 DES 不同的是，AES 算法中运算最小的粒度是字节。

实现方法非常简单。方阵的第一行不变，方阵的第二行循环左移一字节，方阵的第三行循环左移两字节，方阵的第四行循环左移三字节，如图 14-24 所示。

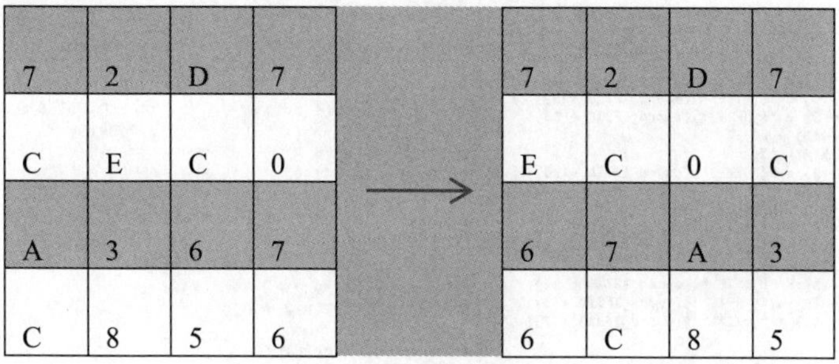

图 14-24　循环左移图示

在 IDA 中，循环左移的一种实现方式如图 14-25 所示。

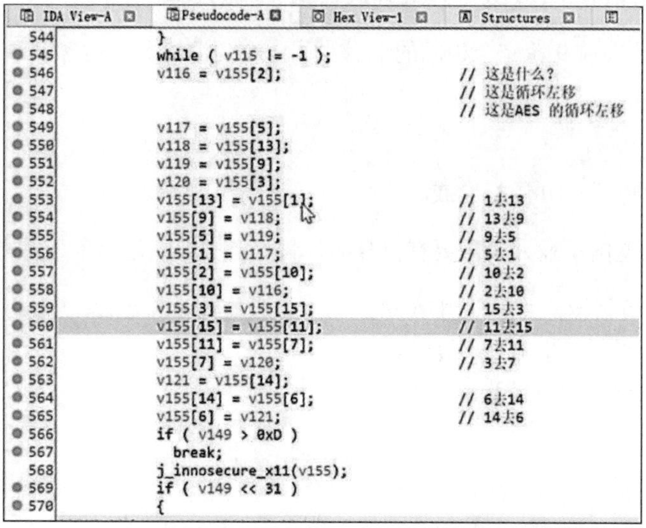


```
}
while ( v115 != -1 );
v116 = v155[2];                         // 这是什么？
                                        // 这是循环左移
                                        // 这是AES 的循环左移
v117 = v155[5];
v118 = v155[13];
v119 = v155[9];
v120 = v155[3];
v155[13] = v155[1];                     // 1去13
v155[9] = v118;                         // 13去9
v155[5] = v119;                         // 9去5
v155[1] = v117;                         // 5去1
v155[2] = v155[10];                     // 10去2
v155[10] = v116;                        // 2去10
v155[3] = v155[15];                     // 15去3
v155[15] = v155[11];                    // 11去15
v155[11] = v155[7];                     // 7去11
v155[7] = v120;                         // 3去7
v121 = v155[14];
v155[14] = v155[6];                     // 6去14
v155[6] = v121;                         // 14去6
if ( v149 > 0xD )
  break;
j_innosecure_x11(v155);
if ( v149 << 31 )
{
```

图 14-25　在 IDA 中，循环左移的实现方式 1

在 IDA 中，循环左移的另一种实现方式如图 14-26 所示。

识别出 SO 中循环左移步骤的代码，然后以此为基点，上面和下面全部打通。

3. 列混淆

列混淆的原理涉及很多近世代数（抽象代数）的知识。由于篇幅有限，这里就不一一讲解了，读者感兴趣的话可以自行了解。列混淆有两个关键数字：0x1B 和 0x80，可以把它们看作是列混淆的一个特征，如图 14-27 所示。

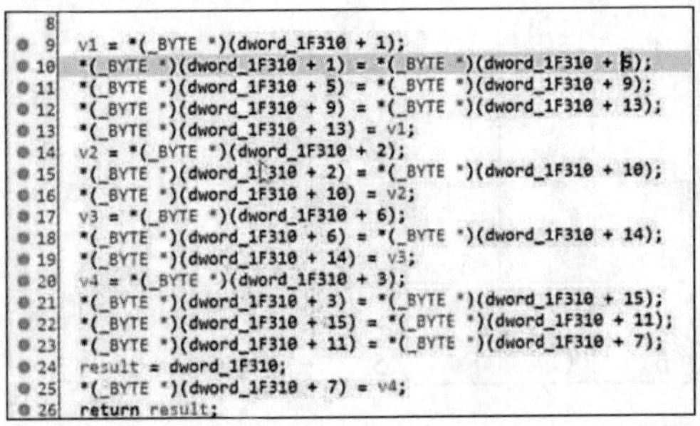

图 14-26　在 IDA 中，循环左移的实现方式 2 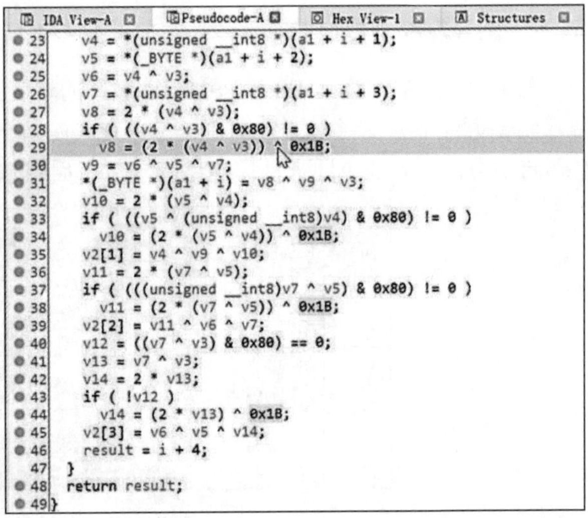　　　　图 14-27　列混淆的特征

4. 将 state 和密钥 K1~K9 进行异或并进行第 10 轮运算

第 10 轮运算相比之前 9 轮完全相同的运算少了一个列混淆，只包含以下 3 个步骤：

（1）S 盒替换。

（2）循环左移。

（3）将 state 和密钥 K10 进行异或。

成为一个优秀的逆向从业人员需要有自身的优势，至少要满足以下条件之一：

- C/C++写得特别棒，可以从 IDA 中复制出相应的代码并运行。
- 密码学学得很好（懂原理，懂实现）。
- IDA 动态调试特别有经验。
- 时间特别多。

通过上述讲解，我们知道，密码学在保证安全的前提下，要尽可能做到简单。

在密钥编排流程中，通过输入的种子密钥获得 10 个子密钥。这里常以字（Word）为单位来衡量密钥长度，1 字=4 字节，因此 AES 的密钥长度可以为 4 字、6 字或者 8 字。对于 AES-128（4 字密钥），需要迭代 10 轮，加上第 0 轮，共需要进行 11 次轮密钥加，而每一次轮密钥加都需要一个长度为 4 字的轮密钥，因此所需要的扩展密钥长度为 11 次×4 字=44 字，如图 14-28 所示。

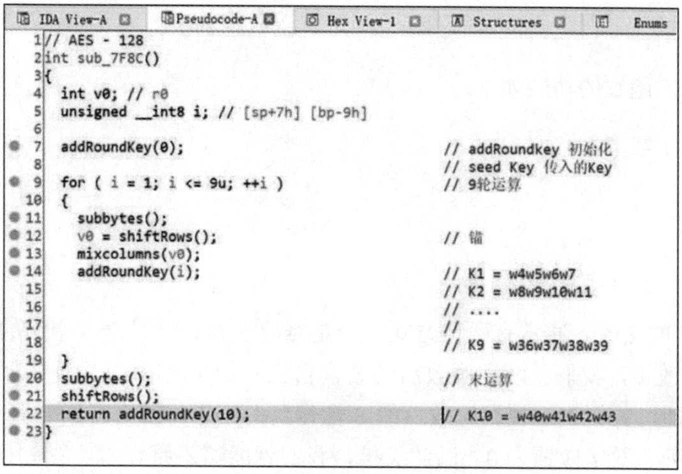

图 14-28 AES-128 代码实现

除了得到 AES 的各个步骤外，我们还可以根据循环条件来判断它是 AES-128（9+1=10 轮）、AES-192（11+1=12 轮）还是 AES-256（13+1=14 轮）。

Hook addRoundKey 可以得到种子密钥。下面介绍魔改 S 盒、魔改密钥的编排流程。

密钥的编排：

在种子密钥的表格中，每一竖列作为 1 字，这里用 wi 表示，如图 14-29 所示。

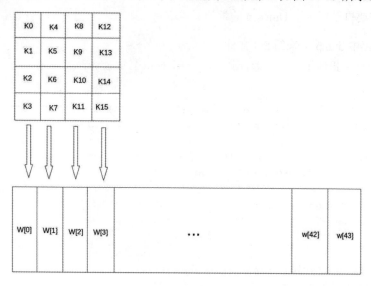

图 14-29 w0~w43

我们传入的密钥为 w0~w3，生成的密钥为 w4~w43。这是密钥扩展算法的执行过程。

其中：

w4 = T(w3)^w0

公式中使用的 T 函数的功能如下：

- 循环左移 1 字节。
- S 表替换。
- Rcon 异或。

```
wi = wi-1^wi-4 if(i%4)
```

针对 AES 算法的魔改，通常有两种方式：一是魔改 S 盒；二是魔改密钥的编排流程。

如果 S 盒被魔改了，我们就利用魔改后的 S 盒自己写一个 AES 的实现；如果想进行解密，则反向推导这个过程即可。例如，根据魔改后的 S 盒得到的结果，第 0 行第 0 列的数是 0x90，反向推理出第 9 行第 0 列的数字应该为 0。以此类推，我们就能够还原 S 盒。S 盒用来加密，逆 S 盒用来解密。

14.2.3　Unicorn 辅助分析

对于初始运算 addRoundKey，有些时候，我们无法直接从代码中找到密钥。这时可以看第一次异或，因为第一次异或的密钥就是我们输入的密钥。使用 Unicorn 验证 addRoundKey 是我们传入的 key。

基于 Unicorn 的模拟执行框架比较好的有两个：(ex)Androidnativemu 和 Unidbg。在以下 3 种情况中，我们需要自己写一个 Unicorn 的框架。

（1）从内存中 dump 下来的 SO 文件。

（2）SO 文件头损坏了，只想运行其中的一部分代码。

```java
private LinuxModule loadInternal(LibraryFile libraryFile) throws IOException {
    final ElfFile elfFile = ElfFile.fromBytes(libraryFile.mapBuffer());

    if (emulator.is32Bit() && elfFile.objectSize != ElfFile.CLASS_32) {
        throw new ElfException("Must be 32-bit");
    }
    if (emulator.is64Bit() && elfFile.objectSize != ElfFile.CLASS_64) {
        throw new ElfException("Must be 64-bit");
    }

    if (elfFile.encoding != ElfFile.DATA_LSB) {
        throw new ElfException("Must be LSB");
    }

    if (emulator.is32Bit() && elfFile.arch != ElfFile.ARCH_ARM) {
        throw new ElfException("Must be ARM arch.");
```

```
    }

    if (emulator.is64Bit() && elfFile.arch != ElfFile.ARCH_AARCH64) {
        throw new ElfException("Must be ARM64 arch.");
    }
```

在上述代码中，Unidbg 对 ELF 文件头进行了校验：大端序、32 位/64 位等。

（3）自定义算法。

当遇到一些自定义算法的时候，用自己的 Unicorn 可能比用(ex)Androidnativemu、Unidbg 更快、更顺手，参考网址 49（参见配书资源文件）。

```python
# 导入
import uTracer

# 初始化模拟器
emulator = uTracer.uTrace()
# 加载 SO 文件
emulator.load(r"C:\Users\pr0214\Desktop\DTA\SO 进阶课
\PycharmProjects\uTracer\example\lib\libcrack.so")

# 起始地址和终止地址
target = 0x7D60 + 1
target_end = 0x7D9E

# 参数准备以及填充，这里还缺封装
base = emulator.DATA_ADDR

# 参数1
# 明文
message = b"0123456789abcdef"
message_addr = base + 0x50
emulator[message_addr] = message

# 参数2
# 密钥
# key = b"0123456789abcdef"
key = b"23456789abcdef10"
key_addr = base + 0x100
emulator[key_addr] = key

# 参数3
out_buf = base + 0x200

# 传入参数
```

```
emulator["r0"] = message_addr
emulator["r1"] = key_addr
emulator["r2"] = out_buf

emulator.start(target, target_end)

# result = emulator[out_buf:out_buf+16]
# print(binascii.b2a_hex(result))

emulator.hexdump(out_buf, 0x10)
```

运行结果如图 14-30 所示。

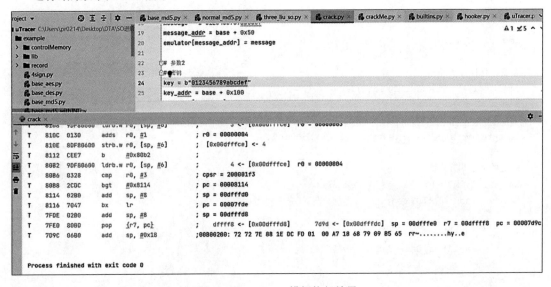

图 14-30　uTracer 模拟执行结果

从 IDA 中可知，第一个 addRoundKey 的异或出现在 .text 段的 0x80f4 处，从模拟执行的结果中查找 80f4，一共出现了 176 次。这 176 次是怎么来的呢？首先，addRoundKey 的一轮运算需要进行 16 次异或，密钥编排用到一轮 addRoundKey，而明文的运算用到了 10 轮，一共 11 轮 addRoundKey，所以总共进行了 176 次异或运算。

我们现在需要前 16 次异或运算。从图 14-30 可以看到，结果打印出了寄存器的值。第一次是 r0 xor r3，r0 = 00000000 指的是当前行代码运行之后的结果。我们需要看的是异或之前的寄存器 r0 和 r3。r0 = 00000030，r3 = 00000030，这是因为我们把明文和密钥的值设置得太接近了。将 key 设置为 23456789abcdef10，重新运行。

```
|r0|r3|
|----|----|
|32|30|
|33|31|
```

```
|34|32|
|...|...|
```

我们发现，r0 和 r3 的值分别对应密钥和明文。

然而，分析商用样本的时候，如果我们既不知道明文，也不知道密钥，是否可以通过 Hook addRoundKey 来获取明文和密钥呢？事实上是不可以的。我们可以获得种子密钥（初始密钥），但是却得不到明文。这是因为工作模式导致的。

14.2.4　AES 的工作模式

当 AES 的输入超过一个分组长度时，采用 ECB 模式，相同的分组输入会带来相同的输出，但不需要 IV。而采用 CBC 模式需要输入一个 IV，但相同的输入不会带来相同的结果，如图 14-31 所示。这就是 CBC 模式的优势。

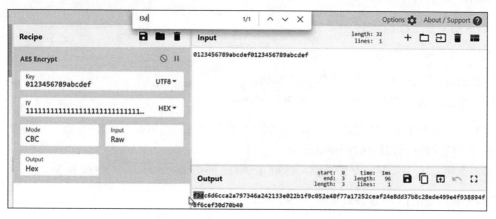

图 14-31　CBC 模式示例

CBC 模式的实现方式非常简单，即 CBC 模式下的每个分组数据在加密前都会和上一个分组计算出来的密文数据进行异或。那么问题来了，第一个明文分组哪来的上一个密文分组呢？因为没有上一个密文分组，所以我们要人为地传入一个分组，这个传入的分组就称为初始化向量（IV）。它的作用就是与第一个明文分组进行异或。CBC 加密的工作模式如图 14-32 所示。

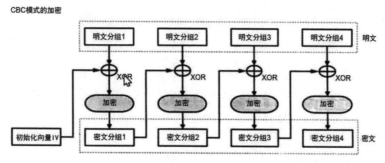

图 14-32　CBC 模式的加密

对于 CBC 模式的加密，下面我们主要讲解两点，即 CyberChef 验证和 PKCS#7 标准。

1. CyberChef 验证

明文：0123456789abcdef0123456789abcdef（str）。

密钥：0123456789abcdef（str）。

初始化向量：IV：1111111111111111（str）。

结果（结果取 64 个十六进制数）：e5b72c539aae45dea57686570b0b515667529ac954d9dee130a26b116dc601c4。

结果取 64 个十六进制数是因为我们的输入是两个分组长度，希望得到两个分组长度的输出，而不需要填充位。

手动计算过程：

（1）明文分组与 IV 异或：

`01 00 03 02 05 04 07 06 09 08 50 53 52 55 54 57`

（2）将结果进行 ECB 模式下的 AES 加密。

第一个密文块：e5b72c539aae45dea57686570b0b5156。

（3）将第一个密文块的结果与第二个明文块做异或。

第二个明文块：0123456789abcdef。

第二个明文块 xor 第一个密文块：d5 86 1e 60 ae 9b 73 e9 9d 4f e7 35 68 6f 34 30。

（4）将结果进行 ECB 模式下的 AES 加密：

`67529ac954d9dee130a26b116dc601c4`

结果完全正确。因此，采用 CBC 模式，在初始变换之前，有明文异或 IV 这样一步操作。

2. PKCS#7 标准

为什么我们用 AES 加密输入的正好是一个分组的长度，而输出的却是两个分组长度的密文呢？

假如输入的是：000102030405060708090a0b0c0d0e0f（HEX），输出是：

`a07999f0e2bfbe16f99593e984a449b7377222e061a924c591cd9c27ea163ed4`（hex）

按照 PKCS#7 标准，Padding 是必须存在的。PKCS#7 标准举例说明：如果输入的是 01，那么填充后的数据为 01 0f 0f 0f 0f 0f 0f 0f 0f 0f 0f 0f 0f 0f 0f；如果输入的是 01 02，那么填充后的数据为 01 02 0e 0e 0e 0e 0e 0e 0e 0e 0e 0e 0e 0e 0e；如果输入的是 10101010101010101010101010101010，那么填充后的数据为 101010101010101010101010101010101010101010101010101010101010101010。加密后的结果如图 14-33 所示。

图 14-33　验证 PKCS#7 标准

由于 ECB 模式下，相同的明文分组会被加密成相同的密文，我们可以得证上述填充方式确实存在。

一般的商用样本中的 AES 加密都是采用 CBC 模式。因此，在加密之前，明文分组和初始化向量做了一次异或。这样，我们就理解了样本 1 中使用的 AES 算法，如图 14-34 所示。

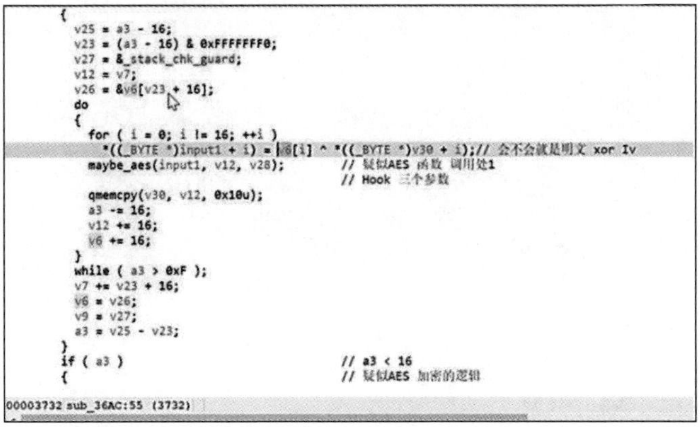

图 14-34　样本 1 中使用的 AES 算法

根据以上原理，我们推测：

```
for ( i = 0; i != 16; ++i )
    *((_BYTE *)v29 + i) = v6[i] ^ *((_BYTE *)v30 + i);
```

就是明文异或 IV。为了求证，我们需要 Hook 寄存器 R1、R2 的值。R1 是从 R8 中取得的，R0 是 index，按字节向后取。R2 以相同的方式从 R6 中获取，如图 14-35 所示。

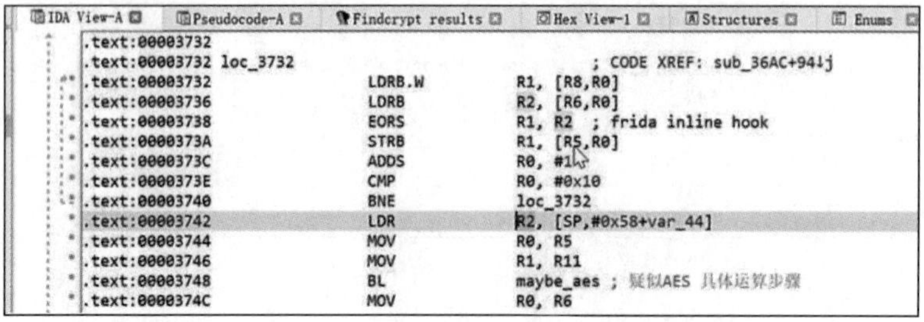

```
.text:00003732                          ; CODE XREF: sub_36AC+94↓j
.text:00003732 loc_3732
.text:00003732          LDRB.W   R1, [R8,R0]
.text:00003736          LDRB     R2, [R6,R0]
.text:00003738          EORS     R1, R2    ; frida inline hook
.text:0000373A          STRB     R1, [R5,R0]
.text:0000373C          ADDS     R0, #1
.text:0000373E          CMP      R0, #0x10
.text:00003740          BNE      loc_3732
.text:00003742          LDR      R2, [SP,#0x58+var_44]
.text:00003744          MOV      R0, R5
.text:00003746          MOV      R1, R11
.text:00003748          BL       maybe_aes  ; 疑似AES 具体运算步骤
.text:0000374C          MOV      R0, R6
```

图 14-35 Hook 寄存器 R1、R2 的值

我们可以在循环外看寄存器 R8 和 R6 的地址所指向的内容，代码如下：

```js
function hook_inline_3742(base_addr, offset_addr){
    var real_addr = base_addr.add(offset_addr);
    Interceptor.attach(real_addr, {
        onEnter: function (args) {
            console.log("Maybe IV xor plaintext");
            console.log(hexdump(this.context.r8));
            console.log(hexdump(this.context.r6));
        }
    });
}

function main() {
    // 确定 SO 的基地址
    var base_addr = Module.findBaseAddress("libmtguard.so");
    // hook_2F88(base_addr, 0x2F88+1); // thumb
    hook_inline_3742(base_addr, 0x3742+1);
}
```

寄存器 R8 的地址所指向的内容如图 14-36 所示。

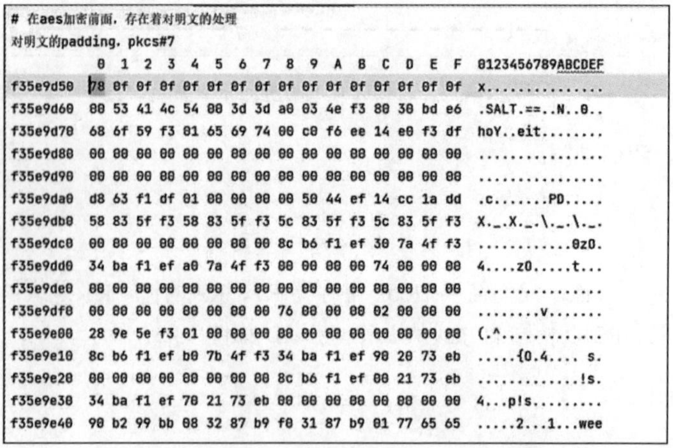

```
# 在aes加密前面，存在着对明文的处理
对明文的padding，pkcs#7
            0  1  2  3  4  5  6  7  8  9  A  B  C  D  E  F  0123456789ABCDEF
f35e9d50   78 0f 0f 0f 0f 0f 0f 0f 0f 0f 0f 0f 0f 0f 0f 0f  x...............
f35e9d60   80 53 41 4c 54 80 3d 3d a0 03 4e f3 80 30 bd e6  .SALT.==..N..0..
f35e9d70   68 6f 59 f3 81 65 69 74 80 c0 f6 ee 14 e0 f3 df  hoY..eit........
f35e9d80   00 00 00 00 00 00 00 00 00 00 00 00 00 00 00 00  ................
f35e9d90   00 00 00 00 00 00 00 00 00 00 00 00 00 00 00 00  ................
f35e9da0   d8 63 f1 df 01 00 00 00 00 50 44 ef 14 cc 1a dd  .c.......PD.....
f35e9db0   58 83 5f f3 58 83 5f f3 5c 83 5f f3 5c 83 5f f3  X._.X._.\._.\._.
f35e9dc0   00 00 00 00 00 00 00 8c b6 f1 ef 30 7a 4f f3  ...........0z0.
f35e9dd0   34 ba f1 ef a0 7a 4f f3 80 00 00 00 74 80 00 00  4....zO.....t...
f35e9de0   00 00 00 00 00 00 00 00 00 00 00 00 00 00 00 00  ................
f35e9df0   00 00 00 00 00 00 00 00 76 00 00 00 02 00 00 00  ........v.......
f35e9e00   28 9e 5e f3 01 00 00 00 00 00 00 00 00 00 00 00  (.^.............
f35e9e10   8c b6 f1 ef b0 7b 4f f3 34 ba f1 ef 90 20 73 eb  .....{0.4.... s.
f35e9e20   00 00 00 00 00 00 00 00 8c b6 f1 ef 00 21 73 eb  .............!s.
f35e9e30   34 ba f1 ef 70 21 73 eb 00 00 00 00 00 00 00 00  4...p!s.........
f35e9e40   90 b2 99 bb 08 32 87 b9 f0 31 87 b9 01 77 65 65  .....2...1...wee
```

图 14-36 寄存器 R8 的地址所指向的内容

我们发现，0f 0f…像是对明文的 PKCS#7 填充。传入的就是一字节，此处也显示填充了 15 字节，刚好可以对上。但是，我们传入的不是十六进制字符串，不是 78，所以猜测，在 AES 加密之前，存在对明文的处理。也就是说，到这一步，明文已经从 30 被处理成 78 了。再来看寄存器 R6 的地址所指向的内容，如图 14-37 所示。

```
# 0102030405060708
            0  1  2  3  4  5  6  7  8  9  A  B  C  D  E  F  0123456789ABCDEF
adf07918   30 31 30 32 30 33 30 34 30 35 30 36 30 37 30 38  0102030405060708
adf07928   a3 26 38 27 7c 31 1f f1 50 9d 5e f3 01 00 00 00  .&8'|1..P.^.....
adf07938   11 00 00 00 60 79 f0 ad 10 00 00 00 40 a1 5e f3  ....`y......@.^.
adf07948   78 7a f0 ad 23 aa 64 b6 60 ce 67 b6 01 00 00 00  xz..#.d.`.g.....
adf07958   27 b5 0e 50 e0 75 4d f3 74 69 65 6d 30 6e 61 75  '..P.uM.tiem0nau
adf07968   6b 6e 61 73 31 69 61 75 e9 ae 9c 83 d9 c0 fd f6  knas1iau........
adf07978   b2 ae 9c 85 83 c7 fd f0 65 42 5a d5 bc 82 a7 23  ........eBZ....#
adf07988   0e 2c 3b a6 8d eb c6 56 d4 1f b3 65 68 9d 14 46  .,;....V...eh..F
adf07998   66 b1 2f e0 eb 5a e9 b6 9a f6 0d 73 f2 6b 19 35  f./..Z.....s.k.5
adf079a8   94 da 36 d5 7f 80 df 63 61 24 c0 fd 93 4f d9 c8  ..6....ca$...O..
```

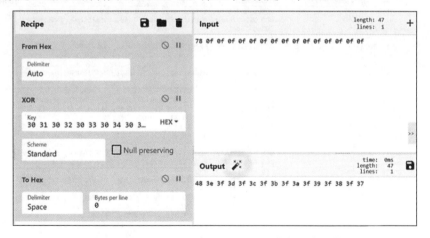

图 14-37　寄存器 R6 的地址所指向的内容

0102030405060708 像是一个 IV。需要注意的是，有两处调用了 maybe_aes，分别是大于一个分组和小于一个分组两种情况。我们虽然传入的是一字节的参数字符串 0，也就是 0x30，但进入的是大于或等于一个分组的分支。通过 CyberChef 异或验证，如图 14-38 所示，可以确定 maybe_aes 的第一个参数是经过处理的明文 78 0f 0f…0f，第二个参数是一个 buffer。

图 14-38　CyberChef 验证 maybe_aes 的第一个参数

目前，我们已经得知了 AES 算法传入的明文、IV，并且使用 CBC 模式，还差密钥未知。因此，我们合理猜测，第三个参数是密钥。

按照我们对 AES 的了解，循环之前应该是 addRoundKey。所以，while 之前的代码应该是 addRoundKey，如图 14-39 所示。

14

```
if ( key && plainText && buffer )
{
    v3 = *(_DWORD *)(key + 240);           // 不知道
    v4 = 1 - (v3 >> 1);                     // AES
                                            // 1.addRoundkey
    v5 = v3 >> 1;
    v36 = v4;
    v6 = *(_QWORD *)(key + 8);
    v35 = bswap32(plainText[3]) ^ HIDWORD(v6);
    v34 = bswap32(plainText[2]) ^ v6;
    v7 = *(_QWORD *)(key + 4) ^ bswap32(plainText[1]);
    v8 = bswap32(*plainText) ^ *(_DWORD *)key;
    v30 = (_DWORD *)(key + 32 * v5 - 32);
    v9 = key + 44;
    v29 = 8 * v5;
    while ( 1 )
    {
        v33 = *(_DWORD *)(v9 - 28);
        v11 = *(_DWORD *)(v9 - 24);
        v32 = (_DWORD *)v9;
        v12 = dword_36E9C[(unsigned __int8)v34] ^ RijnDael_AES_3629C[HIBYTE(v35)] ^ dword_3669C[BYTE2(v8)] ^ …
        v13 = HIDWORD(v11) ^ RijnDael_AES_3629C[HIBYTE(v34)] ^ dword_3669C[BYTE2(v35)] ^ dword_36A9C[BYTE1(v8)]
        v14 = v11 ^ RijnDael_AES_3629C[HIBYTE(v7)] ^ dword_3669C[BYTE2(v34)] ^ dword_36A9C[BYTE1(v35)] ^ dword…
        v15 = dword_36E9C[(unsigned __int8)v35] ^ dword_36A9C[BYTE1(v34)] ^ dword_3669C[BYTE2(v7)] ^ RijnDael_…
```

图 14-39　while 之前的代码是 addRoundKey

而与 plainText、plainText[1]、plainText[2]、plainText[3]分别进行异或的是种子密钥 w0、w1、w2、w3。接下来，我们进行算法还原。在 while 循环中，AES 应该有 4 个步骤：subbytes、shiftrows、mixcolumn、addRoundKey。在 while 循环之后，分别是 sbox、循环左移、addRoundKey。我们一步一步验证。首先是 subbytes，进入 dword_36e9c 查看其中是不是 S 盒，如图 14-40 所示。

```
.rodata:00036A9C         DCD 0xE631D7E6, 0x42C68442, 0x68B8D068, 0x41C38241, 0x99B02999
.rodata:00036A9C         DCD 0x2D775A2D, 0xF111E0F, 0xB0CB7BB0, 0x54FCA854, 0xBBD66DBB
.rodata:00036A9C         DCD 0x163A2C16
.rodata:00036E9C ; _DWORD dword_36E9C[256]
.rodata:00036E9C dword_36E9C  DCD 0x6363A5C6, 0x7C7C84F8, 0x777799EE, 0x7B7B8DF6, 0xF2F20DFF
.rodata:00036E9C                                   ; DATA XREF: maybe_aes+172↑o
.rodata:00036E9C                                   ; maybe_aes+178↑o ...
.rodata:00036E9C         DCD 0x6B6BBDD6, 0x6F6FB1DE, 0xC5C55491, 0x30305060, 0x1010302
.rodata:00036E9C         DCD 0x6767A9CE, 0x2B2B7D56, 0xFEFE19E7, 0xD7D762B5, 0xABABE64D
.rodata:00036E9C         DCD 0x76769AEC, 0xCACA458F, 0x82829D1F, 0xC9C94089, 0x7D7D87FA
.rodata:00036E9C         DCD 0xFAFA15EF, 0x5959EB82, 0x4747C98E, 0xF0F00BFB, 0xADADEC41
.rodata:00036E9C         DCD 0xD4D467B3, 0xA2A2FD5F, 0xAFAFEA45, 0x9C9CBF23, 0xA4A4A4F753
.rodata:00036E9C         DCD 0x727296E4, 0xC0C05B9B, 0xB7B7C275, 0xFDFD1CE1, 0x9393AE3D
.rodata:00036E9C         DCD 0x26266A4C, 0x36365A6C, 0x3F3F417E, 0xF7F702F5, 0xCCCC4F83
.rodata:00036E9C         DCD 0x34345C68, 0xA5A5F451, 0xE5E534D1, 0xF1F108F9, 0x717193E2
.rodata:00036E9C         DCD 0xD8D873AB, 0x31315362, 0x15153F2A, 0x4040C08, 0xC7C75295
.rodata:00036E9C         DCD 0x23236546, 0xC3C35E9D, 0x18182830, 0x9696A137, 0x5050F0A
.rodata:00036E9C         DCD 0x9A9AB52F, 0x707090E, 0x12123624, 0x80809B1B, 0xE2E23DDF
.rodata:00036E9C         DCD 0xEBEB26CD, 0x2727694E, 0xB2B2CD7F, 0x75759FEA, 0x9091B12
.rodata:00036E9C         DCD 0x83839E1D, 0x2C2C7458, 0x1A1A2E34, 0x1B1B2D36, 0x6E6EB2DC
.rodata:00036E9C         DCD 0x5A5AEEB4, 0xA0A0FB5B, 0x5252F6A4, 0x3B3B4D76, 0xD6D66187
.rodata:00036E9C         DCD 0xB3B3CE7D, 0x29297B52, 0xE3E33EDD, 0x2F2F715E, 0x84849713
.rodata:00036E9C         DCD 0x5353F5A6, 0xD1D168B9, 0, 0xEDED2CC1, 0x20206040
.rodata:00036E9C         DCD 0xFCFC1FE3, 0xB1B1C879, 0x5B5BEDB6, 0x6A6ABED4, 0xCBCB468D
.rodata:00036E9C         DCD 0xBEBED967, 0x39394B72, 0x4A4ADE94, 0x4C4CD498, 0x5858E8B0
.rodata:00036E9C         DCD 0xCFCF4A85, 0xD0D06BBB, 0xFEFEF2AC5, 0xAAAAE54F, 0xFBFB16ED
.rodata:00036E9C         DCD 0x4343C586, 0x4D4DD79A, 0x33335566, 0x85859411, 0x4545CF8A
00036E9C 00036E9C: .rodata:dword_36E9C (Synchronized with Hex View-1)
```

图 14-40　dword_36e9c

我们发现，这并不是 S 盒。从 GitHub 上搜索 0x6363a5c6，发现它是一个 T 表中的数据。为什么会在我们以为有 S 盒的地方出现的不是 S 盒，而是其他的表呢？我们再查看循环结束之后的 byte_34E74，如图 14-41 所示。

```
.rodata:00034E74 ; unsigned __int8 byte_34E74[1024]
.rodata:00034E74 byte_34E74    DCB 0x63, 0x63, 0x63, 0x63, 0x7C, 0x7C, 0x7C, 0x77
.rodata:00034E74                                     ; DATA XREF: sub_2B34+64↑o
.rodata:00034E74                                     ; sub_2B34+6A↑o ...
.rodata:00034E74              DCB 0x77, 0x77, 0x77, 0x7B, 0x7B, 0x7B, 0x7B, 0xF2, 0xF2
.rodata:00034E74              DCB 0xF2, 0xF2, 0x6B, 0x6B, 0x6B, 0x6B, 0x6F, 0x6F, 0x6F
.rodata:00034E74              DCB 0x6F, 0xC5, 0xC5, 0xC5, 0xC5, 0x30, 0x30, 0x30, 0x30
.rodata:00034E74              DCB 1, 1, 1, 1, 0x67, 0x67, 0x67, 0x67, 0x2B, 0x2B, 0x2B
.rodata:00034E74              DCB 0x2B, 0xFE, 0xFE, 0xFE, 0xFE, 0xD7, 0xD7, 0xD7, 0xD7
.rodata:00034E74              DCB 0xAB, 0xAB, 0xAB, 0xAB, 0x76, 0x76, 0x76, 0xCA
.rodata:00034E74              DCB 0xCA, 0xCA, 0xCA, 0x82, 0x82, 0x82, 0x82, 0xC9, 0xC9
.rodata:00034E74              DCB 0xC9, 0xC9, 0x7D, 0x7D, 0x7D, 0x7D, 0xFA, 0xFA, 0xFA
.rodata:00034E74              DCB 0xFA, 0x59, 0x59, 0x59, 0x59, 0x47, 0x47, 0x47, 0x47
.rodata:00034E74              DCB 0xF0, 0xF0, 0xF0, 0xF0, 0xAD, 0xAD, 0xAD, 0xAD, 0xD4
.rodata:00034E74              DCB 0xD4, 0xD4, 0xD4, 0xA2, 0xA2, 0xA2, 0xAF, 0xAF
.rodata:00034E74              DCB 0xAF, 0xAF, 0x9C, 0x9C, 0x9C, 0x9C, 0xA4, 0xA4, 0xA4
.rodata:00034E74              DCB 0xA4, 0x72, 0x72, 0x72, 0x72, 0xC0, 0xC0, 0xC0, 0xC0
.rodata:00034E74              DCB 0xB7, 0xB7, 0xB7, 0xB7, 0xFD, 0xFD, 0xFD, 0xFD, 0x93
.rodata:00034E74              DCB 0x93, 0x93, 0x93, 0x26, 0x26, 0x26, 0x26, 0x36, 0x36
.rodata:00034E74              DCB 0x36, 0x36, 0x3F, 0x3F, 0x3F, 0x3F, 0xF7, 0xF7, 0xF7
.rodata:00034E74              DCB 0xF7, 0xCC, 0xCC, 0xCC, 0xCC, 0x34, 0x34, 0x34, 0x34
.rodata:00034E74              DCB 0xA5, 0xA5, 0xA5, 0xA5, 0xE5, 0xE5, 0xE5, 0xE5, 0xF1
.rodata:00034E74              DCB 0xF1, 0xF1, 0xF1, 0x71, 0x71, 0x71, 0x71, 0xD8, 0xD8
.rodata:00034E74              DCB 0xD8, 0xD8, 0x31, 0x31, 0x31, 0x31, 0x15, 0x15, 0x15
```

图 14-41 byte_34E74

这非常像 S 盒，但一字节变成了 4 字节。我们发现整个逻辑变得不好理解了。这时就需要补充一个关于 AES 的知识点——算法实现方式。

我们通常看到的样本的实现方式都是按照算法本身的描述来实现 AES。这又涉及算法实现的两种方式：

（1）按照算法描述来实现。

（2）查表法。

作用：以空间换时间，加快 AES 的运算。查表法将 S 盒替换、循环左移、列混淆这三个步骤合三为一，整合成一个步骤，最后就有了 T 表。

T 表的原理需要一定的数学基础，这里就不展开讲了。笔者举一个比较形象的例子，例如我们去超市购物。选项如下：

牛奶类：{安慕希、纯甄、养乐多}。

水果类：{黄桃、蜜瓜、苹果、蓝莓}。

肉类：{牛肉、猪肉、羊肉、草鱼}。

查表法就是将所有的选项组合起来：

{安慕希、黄桃、牛肉}、{安慕希、蜜瓜、牛肉}、{安慕希、苹果、牛肉}…

一共有 3×4×4=48 种选项。于是，我们进入超市就看到了 48 个礼盒，每个礼盒中都有这样的一个组合，我们选择需要的拿走就行了。

类比这样的过程，我们也可以将 S 盒替换、循环左移、列混淆这三个步骤组合起来构成 T 表，从而加快 AES 的运算。这样做的坏处是什么呢？首先，存储 T 表占用了一定的空间。一般来说，我们会认为用几十字节的空间换取时间是值得的，所以在很多的 AES 实现中都会选择采用查表法，参考网址 50、网址 51（参见配书资源文件）。

为什么前 9 轮用 T 表，最后一轮用 S 盒呢？这其实也很好理解。因为在最后一轮中，只有 S 盒替换、循环左移，没有列混淆，而 T 表是这三个步骤的组合，所以不能使用 T 表。这时对于最后一轮又有两种处理方法：一种是将 S 盒替换、循环左移两步组合成一个小的 T 表；另一种是按正常的步骤执行。

再回到样本，我们就能够得知它在循环中用的是 T 表（T0~T3），出循环之后，似乎用的是 S 盒。但是笔者更倾向于它将 S 盒替换、循环左移两步组合到了一起，从而得到一个似是而非的 S' 盒。这个本身问题不大。

读者可能会有疑问，为什么不把 addRoundKey 也合并到一起呢？因为密钥是我们自己生成的，它可能会变化。但确实有一种情况是将密钥也放进去，这种情况可以称为 AES 白盒。我们现在通过 IDA 反编译 SO 得到的结果已经非常接近源码了。而 AES 白盒可以当作一种"源码"，它的目的是使逆向分析人员即使在有源码的情况下都无法简单地获得密钥。

样本 1 就是采用了这样的方法。它将 addRoundKey 也合并到了 T 表中，从而在反汇编的代码中找不到任何密钥出现的位置。所以整个 AES 算法基本上就变成了一个"查表"的过程。我们能够做的就是自上而下分析明文字节流的走向。通过这种操作，即使算法被魔改，也必定能找到魔改点以及处理手段。

现阶段，如果我们遇到白盒算法，有两种方法可以尝试：

（1）通过 Frida 或者 Xposed 主动调用。
（2）使用 Unicorn 或者 AndroidNativeEmu 模拟执行。

读者也可以将算法提取出来，研读并参考一些论文中的方法对密钥进行解析。

14.3 本章小结

本章主要讲解了以 DES 和 AES 为代表的对称加密算法及其逆向分析，详细介绍了相应的算法原理及其算法流程，最重要的是，分析了三个 AES 的样本并演示了逆向分析的过程。读者需要多分析样本，多从实战中获取经验，才能真正做到学以致用，对逆向分析产生实质性的帮助。

第 15 章 读懂 DEX 并了解 DexDump 解析过程

本章带领读者了解 DEX 文件结构，当我们了解 DEX 的文件结构后，对 Android 逆向工作中遇到的问题就会有更深入的理解。

在 Android 逆向的过程中，主要有两种技术会用到 DEX：

（1）函数抽取。

（2）VMP。

本章通过理论+实践的方式深入了解 DEX 的文件格式，从而进一步学习 DexDump 的解析过程。

15.1 环境及开发工具

既然都看到了 SO 的进阶部分，说明读者已经是 r0env 的老用户了，实验过程直接使用 r0env 即可。

但是还是要明确一下会用到哪些工具：

- 010 Editor。

注：若 010editor 过期，则可通过以下方法无限续期：

```
bash
# 先关闭软件，再执行以下命令
rm -rf ~/.config/SweetScape/
rm -rf ~/.local/share/SweetScape/
```

- VS Code：代码编辑器。

- OpenJDK 11。

安装命令如下:

```
apt install openjdk-11-jdk
```

- IDA Pro（Interactive Disassembler Professional）：交互式反汇编器专业版。
- Android Studio: Android IDEA 开发工具。

15.2 认识 DEX 文件结构

DEX 文件是 Android 平台上 Java 源码文件经编译和处理后得到的字节码文件。读者可能会有疑问，为什么 Android 平台不直接使用 Class 文件，而是另起炉灶重新实现了一种文件格式呢？此问题的答案有很多，笔者仅举一例。Android 系统主要针对移动设备，而移动设备的内存、存储空间相对计算机平台而言较小，并且主要使用 ARM 的 CPU。这种 CPU 有一个显著特点，就是通用寄存器比较多。在这种情况下，Class 格式的文件在移动设备上不能扬长避短。介绍 invokevirtual 指令的时候，我们看到 Class 文件中指令码执行的时候需要存取操作数栈（Operand Stack）。而在移动设备上，由于 ARM 的 CPU 有很多通用寄存器，DEX 中的指令码可以利用它们来存取参数。显然，寄存器内数据的存取速度比位于内存中的操作数栈的存取速度要快得多。

15.2.1 DEX 文件格式概貌

DEX 文件中存在一个个结构体指向的内容，下面我们一一详细讲解这些结构体及其内部变量。

- **header**:一般占 0x70 字节。
- **string_ids**：数组，元素类型为 string_id_item，它存储和字符串相关的信息。
- **type_ids**：数组，元素类型为 type_id_item。存储类型相关的信息（由 TypeDescriptor 描述）。
- **field_ids**：数组，元素类型为 field_id_item，存储成员变量信息，包括变量名、类型等。
- **method_ids**：数组，元素类型为 method_id_item，存储成员函数信息，包括函数名、参数和返回值类型等。
- **class_defs**：数组，元素类型为 class_def_item，存储类的信息。
- **data**: DEX 文件重要的数据内容都存在 data 区域中。一些数据结构会通过如 xx_off 这样的成员变量指向文件的某个位置，从该位置开始，存储了对应数据结构的内容，而 xx_off 的位置一般落在 data 区域中。
- **link_data**：理论上是预留区域，没有特别的作用。

15.2.2　DEX 文件格式项目搭建

步骤如下：

（1）在 Android Studio 中新建一个 myDex 项目，如图 15-1 所示。

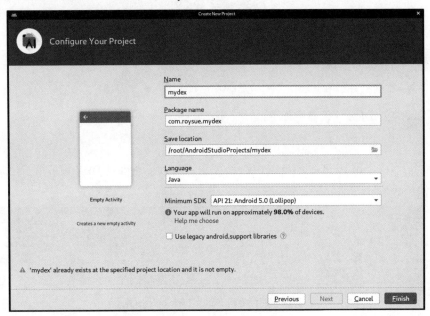

图 15-1　新建 myDex 项目

（2）新建项目后，等待项目同步，然后编译（Build）成一个 APK，如图 15-2 所示。

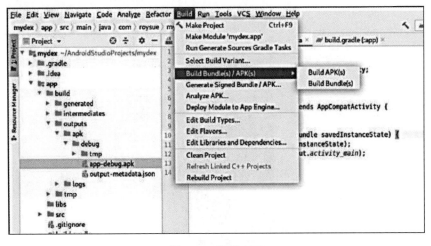

图 15-2　编译 APK

（3）编译完成的 APK 在如图 15-3 所示的目录下。

```
 ┌─(root    r0env)-[~/AndroidStudioProjects/mydex/app/build/outputs/apk/debug]
 └─# pwd
/root/AndroidStudioProjects/mydex/app/build/outputs/apk/debug
 ┌─(root    r0env)-[~/AndroidStudioProjects/mydex/app/build/outputs/apk/debug]
 └─# ls
app-debug.apk  output-metadata.json  tmp
```

图 15-3 编译 APK 的路径

（4）先用 unzip 解包 app-debug.apk，然后使用 010 Editor 打开 classes.dex 文件。

unzip 解包 app-debug.apk 命令如下：

```
unzip app-debug.apk
```

010 Editor 打开 DEX 文件命令如下：

```
010editor classes.dex
```

（5）使用 010 Editor 打开 classes.dex 文件，如图 15-4 所示为通过 DEX 模板解析出来的内容，和 DEX 文件格式契合。

Name	Value	Start	Size	Color		Comment
▶ struct header_item dex_header		0h	70h	Fg:	Bg:	Dex file header
▶ struct string_id_list dex_string_ids	32020 strings	70h	1F450h	Fg:	Bg:	String ID list
▶ struct type_id_list dex_type_ids	3145 types	1F4C0h	3124h	Fg:	Bg:	Type ID list
▶ struct proto_id_list dex_proto_ids	4924 prototypes	225E4h	E6D0h	Fg:	Bg:	Method prototype ID list
▶ struct field_id_list dex_field_ids	20464 fields	30CB4h	27F80h	Fg:	Bg:	Field ID list
▶ struct method_id_list dex_method_ids	25419 methods	58C34h	31A58h	Fg:	Bg:	Method ID list
▶ struct class_def_item_list dex_class_defs	2309 classes	8A68Ch	120A0h	Fg:	Bg:	Class definitions list
▶ struct map_list_type dex_map_list	18 items	37ABA4h	DCh	Fg:	Bg:	Map list

图 15-4 使用 010 Editor 打开 classes.dex

15.2.3 DEX 文件详细分析

DEX 文件格式如图 15-5 所示。

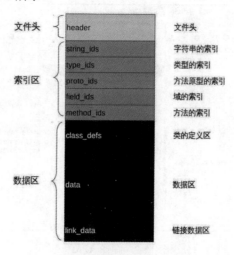

图 15-5 DEX 格式说明图

1. **header** 分析

在源码网站找到 DexFile 的头文件，如图 15-6 所示。

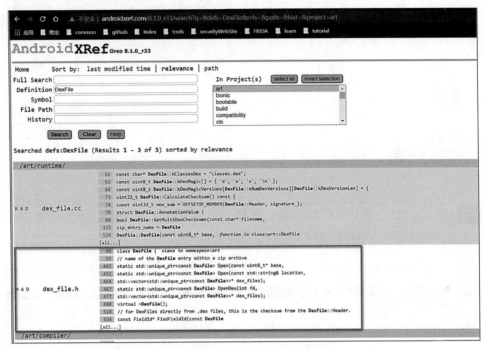

图 15-6　DexFile 头文件

在头文件中可以看到如图 15-7 所示的结构体。

```
// Raw header_item.
struct Header {
  uint8_t magic_[8];
  uint32_t checksum_;  // See also location_checksum_
  uint8_t signature_[kSha1DigestSize];
  uint32_t file_size_;  // size of entire file
  uint32_t header_size_;  // offset to start of next section
  uint32_t endian_tag_;
  uint32_t link_size_;  // unused
  uint32_t link_off_;  // unused
  uint32_t map_off_;  // unused
  uint32_t string_ids_size_;  // number of StringIds
  uint32_t string_ids_off_;  // file offset of StringIds array
  uint32_t type_ids_size_;  // number of TypeIds, we don't support more than 65535
  uint32_t type_ids_off_;  // file offset of TypeIds array
  uint32_t proto_ids_size_;  // number of ProtoIds, we don't support more than 65535
  uint32_t proto_ids_off_;  // file offset of ProtoIds array
  uint32_t field_ids_size_;  // number of FieldIds
  uint32_t field_ids_off_;  // file offset of FieldIds array
  uint32_t method_ids_size_;  // number of MethodIds
  uint32_t method_ids_off_;  // file offset of MethodIds array
  uint32_t class_defs_size_;  // number of ClassDefs
  uint32_t class_defs_off_;  // file offset of ClassDef array
  uint32_t data_size_;  // size of data section
  uint32_t data_off_;  // file offset of data section

  // Decode the dex magic version
  uint32_t GetVersion() const;

 private:
  DISALLOW_COPY_AND_ASSIGN(Header);
};
```

图 15-7　头文件中的字段结构体

对比 010 Editor 中解析出来的 header，如图 15-8 所示。

Name	Value	Start	Size	Color		Comment
▼ struct header_item dex_header		0h	70h	Fg:	Bg:	Dex file header
▶ struct dex_magic magic	dex 035	0h	8h	Fg:	Bg:	Magic value
uint checksum	5550AC2Ch	8h	4h	Fg:	Bg:	Alder32 checksum of rest of file
▶ SHA1 signature[20]	69BFAEA35153DD932248F4F8DA0011BC1CEC2BA2	Ch	14h	Fg:	Bg:	SHA-1 signature of rest of file
uint file_size	3648640	20h	4h	Fg:	Bg:	File size in bytes
uint header_size	112	24h	4h	Fg:	Bg:	Header size in bytes
uint endian_tag	12345678h	28h	4h	Fg:	Bg:	Endianness tag
uint link_size	0	2Ch	4h	Fg:	Bg:	Size of link section
uint link_off	0	30h	4h	Fg:	Bg:	File offset of link section
uint map_off	3648420	34h	4h	Fg:	Bg:	File offset of map list
uint string_ids_size	32020	38h	4h	Fg:	Bg:	Count of strings in the string ID...
uint string_ids_off	112	3Ch	4h	Fg:	Bg:	File offset of string ID list
uint type_ids_size	3145	40h	4h	Fg:	Bg:	Count of types in the type ID list
uint type_ids_off	128192	44h	4h	Fg:	Bg:	File offset of type ID list
uint proto_ids_size	4924	48h	4h	Fg:	Bg:	Count of items in the method p...
uint proto_ids_off	140772	4Ch	4h	Fg:	Bg:	File offset of method prototype I...
uint field_ids_size	20464	50h	4h	Fg:	Bg:	Count of items in the field ID list
uint field_ids_off	199860	54h	4h	Fg:	Bg:	File offset of field ID list
uint method_ids_size	25419	58h	4h	Fg:	Bg:	Count of items in the method I...
uint method_ids_off	363572	5Ch	4h	Fg:	Bg:	File offset of method ID list
uint class_defs_size	2309	60h	4h	Fg:	Bg:	Count of items in the class defi...
uint class_defs_off	566924	64h	4h	Fg:	Bg:	File offset of class definitions list
uint data_size	3007828	68h	4h	Fg:	Bg:	Size of data section in bytes
uint data_off	640812	6Ch	4h	Fg:	Bg:	File offset of data section

图 15-8　头文件中的字段结构

magic：标识 DEX 文件，取值必须是字符串"dex\n035\0"，或者 byte 数组 {0x64,0x65,0x78,0x0a,0x30,0x33,0x35,0x00}。

如果在 010 Editor 中删除 classes.dex 中的前 8 字节，在 jadx-gui 中就无法打开，如图 15-9 所示。

图 15-9　jadx-gui 打开 classes.dex（缺少 magic）

完整的 classes.dex 用 jadx-gui 打开如图 15-10 所示。

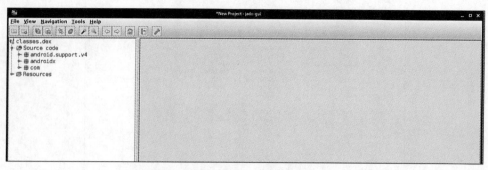

图 15-10　jadx-gui 打开 classes.dex（完整）

checksum：文件内容的校验和，不包括 magic 和 checksum 自己。该字段的内容用于检查文件是否损坏。JADX 可以正常打开，因为 JADX 不校验这个值，但是在正常的环境中是无法使用的。

修改 checksum 中的某个值，在手机上动态加载，检测是否可以正常加载。

修改 checksum 的值，如图 15-11 所示。

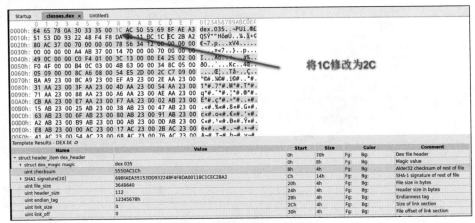

图 15-11 修改 checksum 的值

将 classes.dex 复制一份并传入手机的 sdcard 中，如图 15-12 所示。

```
┌──(root㉿r0env)-[~/AndroidStudioProjects/mydex/app/build/outputs/apk/debug/tmp]
└─# cp classes.dex classes2.dex
┌──(root㉿r0env)-[~/AndroidStudioProjects/mydex/app/build/outputs/apk/debug/tmp]
└─# adb push classes2.dex /sdcard/
classes2.dex: 1 file pushed, 0 skipped. 2.7 MB/s (3648640 bytes in 1.283s)
┌──(root㉿r0env)-[~/AndroidStudioProjects/mydex/app/build/outputs/apk/debug/tmp]
└─# adb shell cd /sdcard/
┌──(root㉿r0env)-[~/AndroidStudioProjects/mydex/app/build/outputs/apk/debug/tmp]
└─# adb shell
bullhead:/ $ cd sdcard/
bullhead:/sdcard $ ls -l
total 17288
drwxrwx--x  2 root sdcard_rw       4096 1970-03-31 06:54 Alarms
drwxrwx--x  4 root sdcard_rw       4096 1970-03-31 06:54 Android
drwxrwx--x  4 root sdcard_rw       4096 2018-01-01 20:01 DCIM
drwxrwx--x  2 root sdcard_rw       4096 2022-02-14 01:27 Download
drwxrwx--x  2 root sdcard_rw       4096 2022-02-12 22:36 MT2
-rw-rw----  1 root sdcard_rw    5942417 2020-09-06 15:27 Magisk-v20.4.zip
drwxrwx--x  2 root sdcard_rw       4096 1970-03-31 06:54 Movies
drwxrwx--x  2 root sdcard_rw       4096 1970-03-31 06:54 Music
drwxrwx--x  2 root sdcard_rw       4096 1970-03-31 06:54 Notifications
drwxrwx--x  3 root sdcard_rw       4096 2022-02-27 21:28 Pictures
drwxrwx--x  2 root sdcard_rw       4096 1970-03-31 06:54 Podcasts
drwxrwx--x  2 root sdcard_rw       4096 1970-03-31 06:54 Ringtones
drwxrwx--x  2 root sdcard_rw       4096 2022-02-07 19:56 TWRP
drwxrwx--x  2 root sdcard_rw       4096 2022-02-13 15:42 aaa
drwxrwx--x  3 root sdcard_rw       4096 2022-02-10 23:11 cache
-rw-rw----  1 root sdcard_rw    3648640 2022-03-09 01:05 classes2.dex
```

图 15-12 复制 classes.dex 并传入手机

在 AndroidManifest.xml 添加 sdcard 的读写权限，如图 15-13 所示。

```xml
<?xml version="1.0" encoding="utf-8"?>
<manifest xmlns:android="http://schemas.android.com/apk/res/android"
    package="com.roysue.mydex">

    <uses-permission android:name="android.permission.READ_EXTERNAL_STORAGE" />
    <uses-permission android:name="android.permission.WRITE_EXTERNAL_STORAGE" />

    <application
        android:allowBackup="true"
        android:icon="@mipmap/ic_launcher"
        android:label="mydex"
        android:roundIcon="@mipmap/ic_launcher_round"
        android:supportsRtl="true"
        android:theme="@style/Theme.Mydex">
        <activity android:name=".MainActivity">
            <intent-filter>
                <action android:name="android.intent.action.MAIN" />

                <category android:name="android.intent.category.LAUNCHER" />
            </intent-filter>
        </activity>
    </application>

</manifest>
```

图 15-13　添加 sdcard 的读写权限

安装 APK，在手机设置中给目标 App 打开权限，如图 15-14 所示。

图 15-14　给 sdcard 权限

在项目 MainActivity.java 中添加方框 2 中的代码，运行后显示方框 2 报错，并得知需要的 checksum 参数，如图 15-15 所示。之所以会报错，是因为 DexClassLoader 在加载 DEX 文件的时候会验证 checksum。

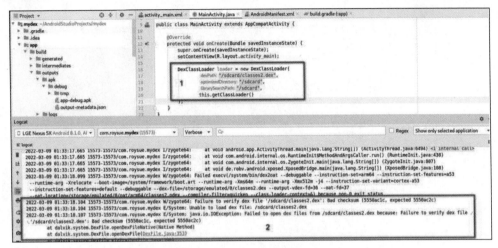

图 15-15　验证结果

接下来我们用一个 checksum.py 脚本来检验一下这个 checksum 字段，checksum.py 如代码清单 15-1 所示。

代码清单 15-1　chenksum.py

```python
#! /usr/bin/python
# -*- coding: utf8 -*-
import binascii   #删除缩进(Tab)
import sys

def CalculationVar(srcByte,vara,varb):#删除缩进(Tab)
    varA = vara
    varB = varb
    icount = 0
    listAB = []

    while icount < len(srcByte):
        varA = (varA + srcByte[icount]) % 65521
        varB = (varB + varA) % 65521
        icount += 1

    listAB.append(varA)
    listAB.append(varB)

    return listAB
```

```python
    def getCheckSum(varA,varB):    #删除缩进(Tab)
        Output = (varB << 16) + varA
        return Output

if __name__ == '__main__':
    filename = sys.argv[1]
    f = open(filename, 'rb', True)
    f.seek(0x0c)
    VarA = 1
    VarB = 0
    flag = 0
    CheckSum = 0
    while True:
        srcBytes = []
        for i in range(1024):           #一次只读1024字节，防止内存占用过大
            ch = f.read(1)
            if not ch:                  #如果读取到末尾，则设置标识符，然后退出读取循环
                flag = 1
                break
            else:
                ch = binascii.b2a_hex(ch) #将字节转为int类型，然后添加到数组中
                ch = str(ch)
                ch = int(ch,16)
                srcBytes.append(ch)
        varList = CalculationVar(srcBytes,VarA,VarB)
        VarA = varList[0]
        VarB = varList[1]
        if flag == 1:
            CheckSum = getCheckSum(VarA,VarB)
            break
    print('[*] DEX FILENAME: '+filename)
    print('[+] CheckSum = '+hex(CheckSum))
```

此脚本用 Python 2 运行，结果如图 15-16 所示。

```
─(root r0env)-[~/AndroidStudioProjects/mydex/app/build/outputs/apk/debug/tmp]
└─# python checksum.py classes2.dex
[*] DEX FILENAME: classes2.dex
[+] CheckSum = 0x5550ac1c
```

图 15-16   checksum.py 计算出来的正确结果

- **signature**：签名信息，不包括 magic、checksum 和 signature。该字段的内容用于检查文件是否被篡改。
- **file_size**：整个文件的长度，单位为字节，包括所有内容。
- **header_size**：默认是 0x70 字节。

- ****endian_tag****：表示文件内容应该按什么字节序来处理，默认取值为 0x12345678，Little Endian 格式。如果为 Big Endian，则该字段取值为 0x78563412。
- ****link_size、link_off****：默认为 0，表示无加固。若有值，则说明此 DEX 被加固。

header 中其实保存的就是一些信息的索引和偏移量。

2. **string_ids** 分析

string_id_list 结构如图 15-17 所示。

图 15-17　string_id_list 结构

在 id_list 中有多个 id，这些 id 其实就是一个指针，指向 item 结构体，一个 item 结构体中包含 string_data_off 和 string_item_string_data 两个字段。string_id_list 中存放着 DEX 文件的所有字符串内容。

3. **type_ids** 分析

在 type_id_list 中随意找一个值，如图 15-18 所示。

图 15-18　在 type_id_list 中随意找一个值

将 0x10A2 转为十进制的 4258，4258 是一个相对偏移量。在 string_id_list 中查找这个值，如图 15-19 所示。这样两者之间的关系就对应起来了。

struct string_id_item string_id[4254]	LZIZ	42E8h	4h	Fg:	Bg:	String ID
struct string_id_item string_id[4255]	LZZ	42ECh	4h	Fg:	Bg:	String ID
struct string_id_item string_id[4256]	LabelVisibility	42F0h	4h	Fg:	Bg:	String ID
struct string_id_item string_id[4257]	LabelVisibilityMode.java	42F4h	4h	Fg:	Bg:	String ID
struct string_id_item string_id[4258]	Landroid/accessibilityservice/AccessibilityServiceInfo;	42F8h	4h	Fg:	Bg:	String ID
struct string_id_item string_id[4259]	Landroid/accounts/AccountManager;	42FCh	4h	Fg:	Bg:	String ID
struct string_id_item string_id[4260]	Landroid/animation/Animator$AnimatorListener;	4300h	4h	Fg:	Bg:	String ID
struct string_id_item string_id[4261]	Landroid/animation/Animator$AnimatorPauseListener;	4304h	4h	Fg:	Bg:	String ID
struct string_id_item string_id[4262]	Landroid/animation/Animator;	4308h	4h	Fg:	Bg:	String ID
struct string_id_item string_id[4263]	Landroid/animation/AnimatorInflater;	430Ch	4h	Fg:	Bg:	String ID
struct string_id_item string_id[4264]	Landroid/animation/AnimatorListenerAdapter;	4310h	4h	Fg:	Bg:	String ID
struct string_id_item string_id[4265]	Landroid/animation/AnimatorSet$Builder;	4314h	4h	Fg:	Bg:	String ID
struct string_id_item string_id[4266]	Landroid/animation/AnimatorSet;	4318h	4h	Fg:	Bg:	String ID
struct string_id_item string_id[4267]	Landroid/animation/ArgbEvaluator;	431Ch	4h	Fg:	Bg:	String ID
struct string_id_item string_id[4268]	Landroid/animation/FloatEvaluator;	4320h	4h	Fg:	Bg:	String ID

图 15-19　查找相对偏移量

4. **proto_ids**分析

proto_id 结构如图 15-20 所示。

struct proto_id_item proto_id[14]	double (int, int)	2268Ch	Ch	Fg:	Bg:	Prototype ID
uint shorty_idx	(0xAA0) "DII"	2268Ch	4h	Fg:	Bg:	String ID of short-form descriptor
uint return_type_idx	(0x2) D	22690h	4h	Fg:	Bg:	Type ID of the return type
uint parameters_off	2309960	22694h	4h	Fg:	Bg:	File offset of parameter type list
struct type_item_list parameters	I, I	233F48h	8h	Fg:	Bg:	Prototype parameter data

图 15-20　proto_id 结构

- shorty_idx：指向 string_list 的一个偏移。查找规则同 type_ids 的分析。
- return_type_idx：指向 type_list。
- parameters_off：参数位置偏移量。
- type_item_list parameters：参数列表，如果不为 0，则说明有参数。

5. **field_ids**分析

field_ids 是类中的参数。图 15-21 是 field_id_item 的结构。class_idx 和 name_idx 组成了一个完整的类结构。type_idx 决定了其类型。

struct field_id_item field_id[9]	android.widget.RemoteViews android.app.Notification.contentView	30CFCh	8h	Fg:	Bg:	Field ID
ushort class_idx	(0x37) android.app.Notification	30CFCh	2h	Fg:	Bg:	Type ID of the class that defines...
ushort type_idx	(0x216) android.widget.RemoteViews	30CFEh	2h	Fg:	Bg:	Type ID for the type of this field
uint name_idx	(0x3823) "contentView"	30D00h	4h	Fg:	Bg:	String ID for the field's name

图 15-21　field_ids 结构

- class_idx：指向 type_id 的指针。
- name_idx：指向 string_id 的指针。
- type_idx：指向 type_id 的指针。

6. **method_ids**分析

method_ids 的结构如图 15-22 所示。

struct method_id_item method_id[2]	java.lang.String android.accessibilityservice.AccessibilityServiceInfo.getDesc...	58C44h	8h	Fg:	Bg:	Method ID
ushort class_idx	(0x6) android.accessibilityservice.AccessibilityServiceInfo	58C44h	2h	Fg:	Bg:	Type ID of the class that defines...
ushort proto_idx	(0x884) java.lang.String ()	58C46h	2h	Fg:	Bg:	Prototype ID for this method
uint name_idx	(0x441C) "getDescription"	58C48h	4h	Fg:	Bg:	String ID for the method's name

图 15-22　method_ids 结构

- class_idx：指向 type_id 的指针。
- proto_idx：指向 proto_id 的指针。
- name_idx：指向 string_id 的指针。

7. **class_defs** 分析（重点）

class_defs 结构如图 15-23 所示。

▼ struct class_def_item class_def[21]	public interface abstract annotation androidx.annotation.AnimRes	8A92Ch	20h	Fg:	Bg:	Class ID
uint class_idx	(0x240) androidx.annotation.AnimRes	8A92Ch	4h	Fg:	Bg:	Type ID for this class
enum ACCESS_FLAGS access_flags	(0x2601) ACC_PUBLIC ACC_INTERFACE ACC_ABSTRACT ACC_ANNOTATION	8A930h	4h	Fg:	Bg:	Access flags
uint superclass_idx	(0x864) java.lang.Object	8A934h	4h	Fg:	Bg:	Type ID for this class's superclass
uint interfaces_off	2307712	8A938h	4h	Fg:	Bg:	File offset to interface list
▼ struct type_item_list interfaces	java.lang.annotation.Annotation	233680h	6h	Fg:	Bg:	Interface data
uint size	1	233680h	2h	Fg:	Bg:	Number of entries in type list
▼ struct type_item_list[1]		233684h	2h	Fg:	Bg:	Type entry
▼ struct type_item_list[0]	Ljava/lang/annotation/Annotation;	233684h	2h	Fg:	Bg:	Type entry
ushort type_idx	2931	233684h	2h	Fg:	Bg:	Index into type_ids list
uint source_file_idx	(0x496) "AnimRes.java"	8A93Ch	4h	Fg:	Bg:	String ID for the name of the fil...
uint annotations_off	3568108	8A940h	4h	Fg:	Bg:	File offset to the annotation stru...
▼ struct annotations_directory_item annotations	3 class annotations, 0 field annotations, 0 method annotations, 0 parameter...	3671ECh	10h	Fg:	Bg:	Annotation data
uint class_annotations_off	3539480	3671ECh	4h	Fg:	Bg:	File offset to class annotations
▶ struct annotation_set_item class_annotations	3 annotation entries	360218h	10h	Fg:	Bg:	Class annotations
uint fields_size	0	3671F0h	4h	Fg:	Bg:	Number of fields annotated by t...
uint methods_size	0	3671F4h	4h	Fg:	Bg:	Number of methods annotated ...
uint parameters_size	0	3671F8h	4h	Fg:	Bg:	Number of method parameter li...
uint class_data_off	0	8A944h	4h	Fg:	Bg:	File offset to the class data for t...
uint static_values_off	0	8A948h	4h	Fg:	Bg:	File offset to static field data

图 15-23 class_defs 结构

- class_idx：指向 type_ids，代表本类的类型。
- access_flags：访问标志，比如 private、public 等。
- superclass_idx：指向 type_ids，代表基类类型，如果没有基类，则取值为 NO_INDEX（值为 -1）。00
- interfaces_off：如果本类实现了某些接口，则 interfaces_off 指向文件对应的位置，那里存储了一个 type_list。该 type_list 的 list 数组存储了每一个接口类的 type_idx 索引。

本例实现了一个接口，查看其在 DEX 文件中的引用，如图 15-24 所示。

在 myDex 项目中新建一个接口类，并在 MainActivity 中实现它，如图 15-25 所示。如果只是创建但并未引用，则 interfaces_off 仍为 0。

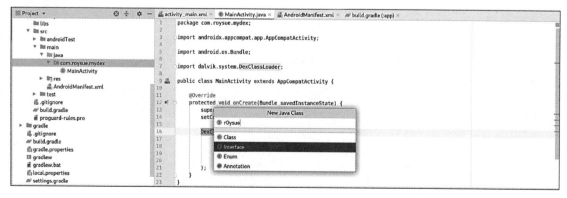

图 15-24 创建 r0ysue 接口

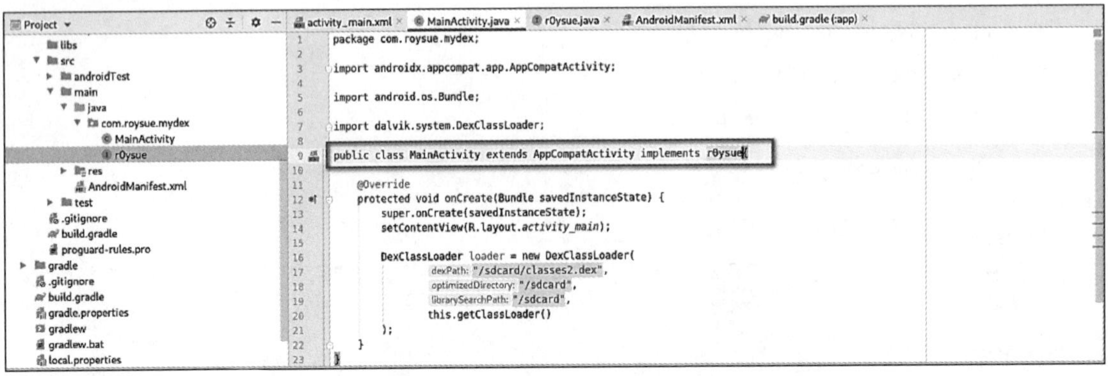

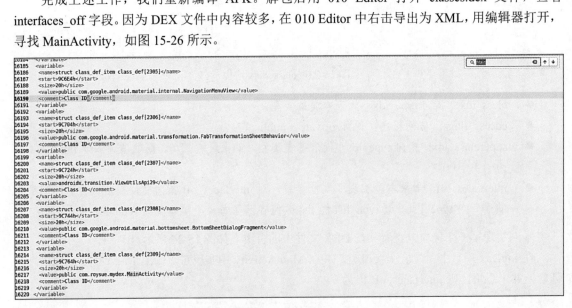

图 15-25　接口实现

完成上述工作，我们重新编译 APK。解包后用 010 Editor 打开 classes.dex 文件，查看 interfaces_off 字段。因为 DEX 文件中内容较多，在 010 Editor 中右击导出为 XML，用编辑器打开，寻找 MainActivity，如图 15-26 所示。

图 15-26　在 XML 文件中查找对应的偏移量

找到接口后，根据上面的偏移值在 class_defs 中查找，如图 15-27 所示。

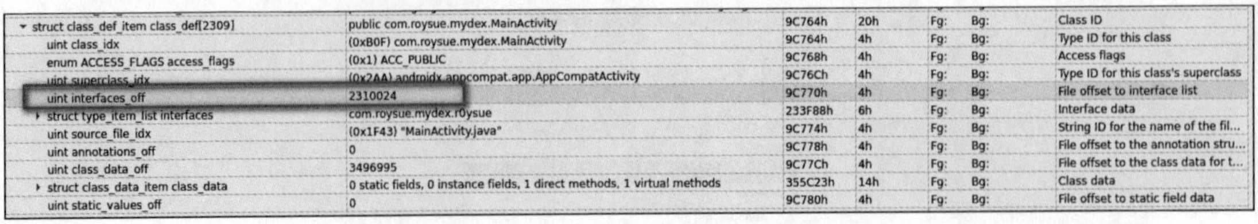

图 15-27　在 class_defs 中查找接口偏移量

- source_file_idx：指向 string_ids，该类对应的源文件名。

- annotations_off：存储和注解有关的信息。
- class_data_off：指向文件对应的位置，那里将存储更细节的信息，由 class_data_item 类型来描述。
- static_values_off：存储用来初始化类的静态变量的值，静态变量如果没有显示设置初值，则默认是 0 或者 null。如果有初值，初值信息就存储在文件 static_values_off 的地方，对应的数据结构名为 encoded_array_item（要加 final）。

8. **class_data_item**分析

class_data_item 结构如图 15-28 所示。

struct class_def_item class_def[2297]	public androidx.appcompat.app.AppCompatDialogFragment	9C5E4h	20h	Fg:	Bg:	Class ID
struct class_def_item class_def[2298]	androidx.transition.ViewUtilsApi23	9C604h	20h	Fg:	Bg:	Class ID
struct class_def_item class_def[2299]	androidx.viewpager2.widget.ViewPager2$RecyclerViewImpl	9C624h	20h	Fg:	Bg:	Class ID
▼ struct class_def_item class_def[2300]	public com.google.android.material.appbar.AppBarLayout$Behavior	9C644h	20h	Fg:	Bg:	Class ID
uint class_idx	(0x92E) com.google.android.material.appbar.AppBarLayout$Behavior	9C644h	4h	Fg:	Bg:	Type ID for this class
enum ACCESS_FLAGS access_flags	(0x1) ACC_PUBLIC	9C648h	4h	Fg:	Bg:	Access flags
uint superclass_idx	(0x92B) com.google.android.material.appbar.AppBarLayout$BaseBehavior	9C64Ch	4h	Fg:	Bg:	Type ID for this class's superclass
uint interfaces_off	0	9C650h	4h	Fg:	Bg:	File offset to interface list
uint source_file_idx	(0x4D0) "AppBarLayout.java"	9C654h	4h	Fg:	Bg:	String ID for the name of the fil...
uint annotations_off	3648072	9C658h	4h	Fg:	Bg:	File offset to the annotation stru...
▶ struct annotations_directory_item annotations	4 class annotations, 0 field annotations, 0 method annotations, 9 parameter...	37AA48h	58h	Fg:	Bg:	Annotation data
uint class_data_off	3496078	9C65Ch	4h	Fg:	Bg:	File offset to the class data for t...
▼ struct class_data_item class_data	0 static fields, 0 instance fields, 2 direct methods, 17 virtual methods	35588Eh	7Ch	Fg:	Bg:	Class data
struct uleb128 static_fields_size	0x0	35588Eh	1h	Fg:	Bg:	The number of static fields
struct uleb128 instance_fields_size	0x0	35588Fh	1h	Fg:	Bg:	The number of instance fields
struct uleb128 direct_methods_size	0x2	355890h	1h	Fg:	Bg:	The number of direct methods
struct uleb128 virtual_methods_size	0x11	355891h	1h	Fg:	Bg:	The number of virtual methods
▶ struct encoded_method_list direct_methods	2 methods	355892h	1h	Fg:	Bg:	Encoded sequence of direct me...
▶ struct encoded_method_list virtual_methods	17 methods	3558A2h	68h	Fg:	Bg:	Encoded sequence of virtual m...
uint static_values_off	0	9C660h	4h	Fg:	Bg:	File offset to static field data
▶ struct class_def_item class_def[2301]	public com.google.android.material.circularreveal.coordinatorlayout.Circular...	9C664h	20h	Fg:	Bg:	Class ID
▶ struct class_def_item class_def[2302]	public final com.google.android.material.datepicker.MaterialCalendar	9C684h	20h	Fg:	Bg:	Class ID
▶ struct class_def_item class_def[2303]	public final com.google.android.material.datepicker.MaterialDatePicker	9C6A4h	20h	Fg:	Bg:	Class ID
▶ struct class_def_item class_def[2304]	public final com.google.android.material.datepicker.MaterialTextInputPicker	9C6C4h	20h	Fg:	Bg:	Class ID

图 15-28 class_data_item 结构

- static_fields：类的静态成员信息，元素类型为 encoded_field。
- instance_fields：类的非静态成员信息，元素类型为 encoded_field。
- direct_methods：非虚函数信息，元素类型为 encoded_method。
- virtual_methods：虚函数信息，元素类型为 encoded_method。而 encoded_field 和 encoded_method 用于描述类的成员变量和成员函数的信息。
- encoded_field：field_idx_diff 指向 field_ids。注意这里是 field_idx_diff，它表示除数组里第一个元素的 field_idx_diff 取值为索引值外，该数组后续元素 field_idx_diff 的取值为和前一个索引值的差。
- access_flags：表示成员域的访问标志。
- encoded_method：method_idx_diff 指向 method_ids。diff 的含义与 field_idx_diff 一样。

9. **code_item**分析

code_item 结构如图 15-28 所示。

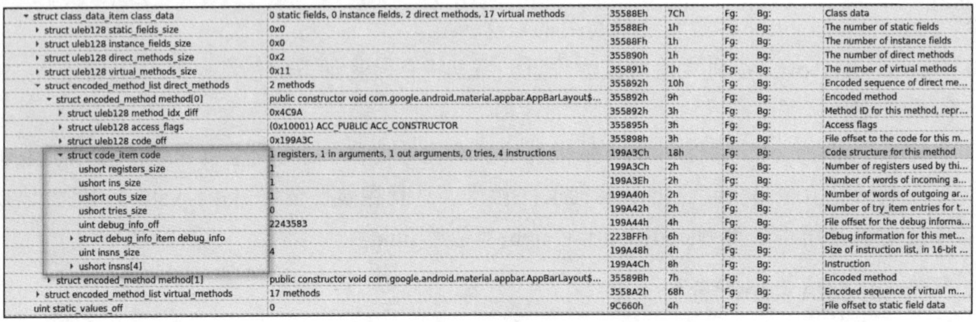

图 15-29 code_item 结构

- **registers_size**：此函数需要用到的寄存器个数。
- **ins_size**：输入参数所占的空间，以双字节为单位。
- **outs_size**：该函数表示内部调用其他函数时，所需参数占用的空间。同样以双字节为单位。
- **insns_size 和 insns 数组**：指令码数组的长度和指令码的内容。DEX 文件格式中的 JVM 指令码长度为 2 字节，而 Class 文件中 JVM 指令码长度为 1 字节。
- **tries_size 和 tries 数组**：如果该函数内部有 try 语句块，则 tries_size 和 tries 数组用于描述 try 语句块相关的信息。注意，tries 数组是可选项，如果 tries_size 为 0，则此 code_item 不包含 tries 数组。
- **padding**：用于将 tries 数组（如果有，并且 insns_size 是奇数长度的话）进行 4 字节对齐。
- **handlers**：catch 语句对应的内容，也是可选项。如果 tries_size 不为 0，才有 handlers 域。
- **debug_info_off**：给读者推荐网址 52（参见配书资源文件）的一篇文章，读者可以查阅此文章深入了解 debug_info_off 字段。此字段的好处是可以记录调用栈的行数。

接下来深入分析 debug_info_off。在 MainActivity.java 中编写 add 函数并运行，显示结果如图 15-30 所示。

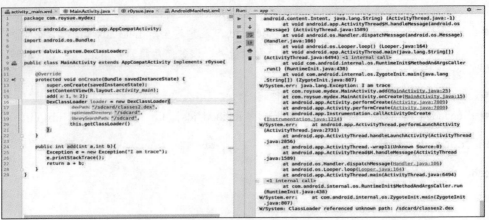

图 15-30 App 启动运行结果

第 15 章 读懂 DEX 并了解 DexDump 解析过程 329

我们可以看到运行结果对应的函数有行号显示。接下来在 010 Editor 中更改 debug_info_off 字段为 0，如图 15-31 所示。

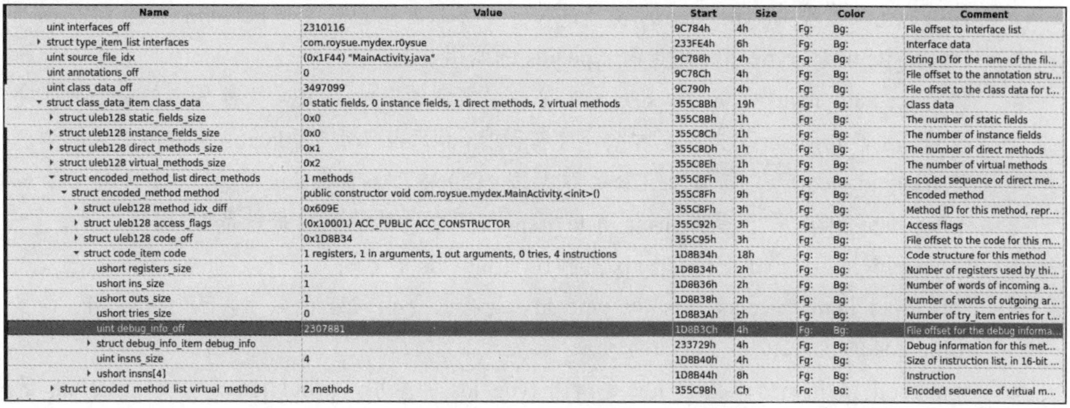

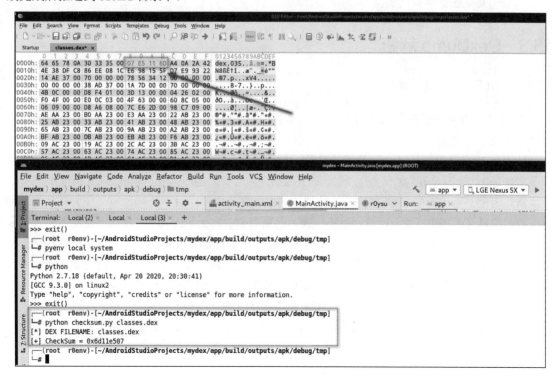

图 15-31 debug_info_off 置 0

置 0 后计算 checksum 的值，并在 010 Editor 中更改 checksum 字段，操作如图 15-32 所示。更改完成后推送到 sdcard 目录下。

图 15-32 计算 checksum 并更改相应位置

- **string_list**：绝对指针，指向 String_data_item。
- **type_list**：相对指针，指向 string_list 的结构体索引。size 表示 list 数组的个数，而 list 数组的元素类型为 type_item。函数的每一个参数都对应一个 type_item 元素。
- **type_item**：type_idx 是指向 type_ids 的索引。
- **string_data_item**：utf16_size 是字符串中字符的个数。DEX 文件的字符串采用了变种的 UTF8 格式，对于英文字符和数字字符而言，都只占一字节。data 是字符串对应的内容。
- **string_id_item**：类似于 Class 文件的 CONSTANT_String 类型，它只有一个成员 string_data_off，用于指明 string_data_item 位于文件的位置，也就是索引。
- **type_id_item**：descriptor_idx 是指向 string_ids 的索引。
- **field_id_item**：class_idx 和 type_idx 是指向 type_ids 的索引，而 name_idx 是指向 string_ids 的索引。
- **method_id_item**：class_idx 是指向 type_ids 的索引，proto_idx 是指向 proto_ids 的索引，name_idx 是指向 string_ids 的索引。
- **proto_id_item**：shorty_idx 是指向 string_ids 的索引，return_type_idx 是指向 type_ids 的索引，如果 parameters_off 不为 0，则文件对应的地方存储类型为 type_list 的结构，用于描述函数参数的类型。

15.3 DexDump 解析

15.3.1 ULEB128 格式讲解

在 15.2 节中讲解了 DEX 文件的格式，但是其中最基础的 ULEB128 格式，一直没有讲解。本小节先补充一下这部分知识。

ULEB128 的读取规则：先读取第一字节，并判断其是否大于 0x7f，如果大于，则代表该字节的最高位是 1，而不是 0。如果是 1，则代表还要读下一字节；如果是 0，则代表 ULEB128 编码的数值到此为止。而代表原来整型值的 32 位数据就嵌入在每字节中的 7 比特位中（应为最高位，被用来表示编码是否结束了）。所以，如果要表示的整型值非常大，就需要 5 字节表示（反而比原来多一字节）。而其实 ULEB128 可以表示的数值范围比 32 位整型值要大，有 3 位被浪费掉了。不过，一般程序中使用的整型值都不会太大，经常是小于 100 的，在这种情况下，只需要使用一字节就可以表示了，比普通 32 位整型表示法少用了 3 字节，节省的空间还是非常可观的。

接下来我们用一个实例来讲解 ULEB128 是如何计算的。

首先在 010 Editor 中选取一个数据格式为 ULEB128 的测试数据，如图 15-33 所示。

第 15 章 读懂 DEX 并了解 DexDump 解析过程

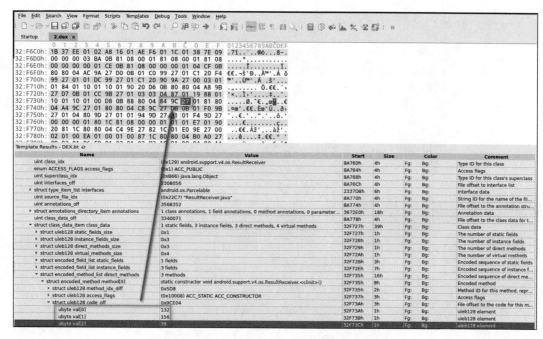

图 15-33 ULEB128 测试数据

然后提取相应的数据 27 9C 84 使用在线进制转换，如图 15-34 所示。

将转换后的结果进行提取，提取规则为：去除第 1 位，将后 7 位拼接到一起，进行进制转换，即可得到最后的结果，如图 15-35 所示。

图 15-34 进制转换　　　　　　　　图 15-35 提取转换结果

接下来用官方解析 ULEB128 的代码测试一下。官方源码参考网址 53（参见配书资源文件）。复制解析源码到 VS Code 中，具体代码如代码清单 15-2 所示。

代码清单 15-2　解析 ULEB128

```
#include<iostream>

using namespace std;
```

```c
typedef uint8_t u1;


int readUnsignedLeb128(u1** pStream) {
    u1* ptr = *pStream;
    int result = *(ptr++);

    if (result > 0x7f) {
        int cur = *(ptr++);
        result = (result & 0x7f) | ((cur & 0x7f) << 7);
        if (cur > 0x7f) {
            cur = *(ptr++);
            result |= (cur & 0x7f) << 14;
            if (cur > 0x7f) {
                cur = *(ptr++);
                result |= (cur & 0x7f) << 21;
                if (cur > 0x7f) {
                    /*
                     * Note: We don't check to see if cur is out of
                     * range here, meaning we tolerate garbage in the
                     * high four-order bits.
                     */
                    cur = *(ptr++);
                    result |= cur << 28;
                }
            }
        }
    }

    *pStream = ptr;
    return result;
}

int main()
{
    u1 * a = (u1*)malloc(100);
    a[0] = 132;
    a[1] = 156;
    a[2] = 39;
    printf("%x",readUnsignedLeb128(&a));
    return 0;
}
```

代码运行结果如图 15-36 所示，同前面的结果一致。

```
[Running] cd "/root/Desktop/myfile/code/c_code/test/" && g++ main.cpp -o main && "/root/Desktop/myfile/code/c_code/test/"main
9ce04
[Done] exited with code=0 in 0.204 seconds
```

图 15-36　ULEB128 解析结果

15.3.2　DexDump 解析过程

　　DEX 解析器在 Android SDK 中，在平常的 DEX 解析工作中主要用的是 Android Studio 的工具 DexDump，接下来我们就依托 DexDump 的源码来了解一下 DexDump 的解析过程。网址 54（参见配书资源文件）为 DexDump 源码。

　　首先定位到 DexDump.cpp 的入口 main 函数，函数体中有一个 while 循环，此处是对参数进行匹配，下面的 process 是处理的核心，代码如下：

```
int main(int argc, char* const argv[])
{
    bool wantUsage = false;
    int ic;
    memset(&gOptions, 0, sizeof(gOptions));
    gOptions.verbose = true;
    while (1) {
        // 参数的匹配
    }
    int result = 0;
    while (optind < argc) {
        result |= process(argv[optind++]);  // process 是处理的核心
    }
}
```

　　追踪 process，定义了 DexFile 指针，跟着这个指针走，发现它被传到 processDexFile 中处理，代码如下：

```
int process(const char* fileName)
{
    DexFile* pDexFile = NULL;   // 定义 DexFile 指针
    MemMapping map;
    bool mapped = false;
    int result = -1;

    if (dexOpenAndMap(fileName, gOptions.tempFileName, &map, false) != 0) {
        return result;
    }
    mapped = true;

    int flags = kDexParseVerifyChecksum;    // 是否校验 checksum
    if (gOptions.ignoreBadChecksum)
```

```
        flags |= kDexParseContinueOnError;

    pDexFile = dexFileParse((u1*)map.addr, map.length, flags);
    if (pDexFile == NULL) {
        fprintf(stderr, "ERROR: DEX parse failed\n");
        goto bail;
    }

    if (gOptions.checksumOnly) {
        printf("Checksum verified\n");
    } else {
        processDexFile(fileName, pDexFile);    // 处理 DEX 的位置
    }

    result = 0;

bail:
    if (mapped)
        sysReleaseShmem(&map);
    if (pDexFile != NULL)
        dexFileFree(pDexFile);
    return result;
}
```

追踪 processDexFile 函数，函数内的代码是基于 DexFile 指针进行解析工作的，有类的解析、Method 的解析以及不同选项下的解析规则，代码如下：

```
void processDexFile(const char* fileName, DexFile* pDexFile)
{
    char* package = NULL;
    int i;
    // 选项处理
    for (i = 0; i < (int) pDexFile->pHeader->classDefsSize; i++) {
        if (gOptions.showSectionHeaders)    // 节头处理
            dumpClassDef(pDexFile, i);
        dumpClass(pDexFile, i, &package);   // 拿到所有的类
    }
    dumpMethodHandles(pDexFile);        // 拿到获取 Method 的 Handle
    dumpCallSites(pDexFile);

    /* free the last one allocated */
    if (package != NULL) {
        printf("</package>\n");
        free(package);
    }
    if (gOptions.outputFormat == OUTPUT_XML)
```

```
        printf("</api>\n");
}
```

  dumpClass 函数中主要解析了 Class 的相关属性，包括编号、它继承了什么类以及实现的接口有哪些、Accessflag 是什么，解析得非常全面。accessFlag 就是修饰符，如 abstrace、static、final。

  下面介绍 dumpMethod，它是最核心的部分。dumpMethod 位于 dumpClass 中，它的作用是把所有类中的内部方法 dump 下来，实现的细节如下：

```
void dumpMethod(DexFile* pDexFile, const DexMethod* pDexMethod, int i)
{
    const DexMethodId* pMethodId;
    const char* backDescriptor;
    const char* name;
    char* typeDescriptor = NULL;
    char* accessStr = NULL;
    if (gOptions.exportsOnly &&
        (pDexMethod->accessFlags & (ACC_PUBLIC | ACC_PROTECTED)) == 0)
    {
        return;
    }
    pMethodId = dexGetMethodId(pDexFile, pDexMethod->methodIdx);
    name = dexStringById(pDexFile, pMethodId->nameIdx);
    typeDescriptor = dexCopyDescriptorFromMethodId(pDexFile, pMethodId);
    backDescriptor = dexStringByTypeIdx(pDexFile, pMethodId->classIdx);
    accessStr = createAccessFlagStr(pDexMethod->accessFlags,
            kAccessForMethod);
    if (gOptions.outputFormat == OUTPUT_PLAIN) {
        printf("    #%d              : (in %s)\n", i, backDescriptor);
        printf("      name          : '%s'\n", name);
        printf("      type          : '%s'\n", typeDescriptor);
        printf("      access        : 0x%04x (%s)\n",
            pDexMethod->accessFlags, accessStr);

        if (pDexMethod->codeOff == 0) {
            printf("      code          : (none)\n");
        } else {
            printf("      code          -\n");
            dumpCode(pDexFile, pDexMethod);
        }

        if (gOptions.disassemble)
            putchar('\n');
    } else if (gOptions.outputFormat == OUTPUT_XML) {
        bool constructor = (name[0] == '<');

        if (constructor) {
```

```c
        char* tmp;

        tmp = descriptorClassToDot(backDescriptor);
        printf("<constructor name=\"%s\"\n", tmp);
        free(tmp);

        tmp = descriptorToDot(backDescriptor);
        printf(" type=\"%s\"\n", tmp);
        free(tmp);
    } else {
        printf("<method name=\"%s\"\n", name);

        const char* returnType = strrchr(typeDescriptor, ')');
        if (returnType == NULL) {
            fprintf(stderr, "bad method type descriptor '%s'\n",
                typeDescriptor);
            goto bail;
        }

        char* tmp = descriptorToDot(returnType+1);
        printf(" return=\"%s\"\n", tmp);
        free(tmp);

        printf(" abstract=%s\n",
            quotedBool((pDexMethod->accessFlags & ACC_ABSTRACT) != 0));
        printf(" native=%s\n",
            quotedBool((pDexMethod->accessFlags & ACC_NATIVE) != 0));

        bool isSync =
            (pDexMethod->accessFlags & ACC_SYNCHRONIZED) != 0 ||
            (pDexMethod->accessFlags & ACC_DECLARED_SYNCHRONIZED) != 0;
        printf(" synchronized=%s\n", quotedBool(isSync));
    }

    printf(" static=%s\n",
        quotedBool((pDexMethod->accessFlags & ACC_STATIC) != 0));
    printf(" final=%s\n",
        quotedBool((pDexMethod->accessFlags & ACC_FINAL) != 0));
    // "deprecated=" not knowable w/o parsing annotations
    printf(" visibility=%s\n",
        quotedVisibility(pDexMethod->accessFlags));

    printf(">\n");

    /*
     * Parameters.
```

```
            */
            if (typeDescriptor[0] != '(') {
                fprintf(stderr, "ERROR: bad descriptor '%s'\n", typeDescriptor);
                goto bail;
            }
            char tmpBuf[strlen(typeDescriptor)+1];      /* more than big enough */
            int argNum = 0;
            const char* base = typeDescriptor+1;
            while (*base != ')') {
                char* cp = tmpBuf;
                while (*base == '[')
                    *cp++ = *base++;
                if (*base == 'L') {
                    /* copy through ';' */
                    do {
                        *cp = *base++;
                    } while (*cp++ != ';');
                } else {
                    /* primitive char, copy it */
                    if (strchr("ZBCSIFJD", *base) == NULL) {
                        fprintf(stderr, "ERROR: bad method signature '%s'\n", base);
                        goto bail;
                    }
                    *cp++ = *base++;
                }
                /* null terminate and display */
                *cp++ = '\0';
                char* tmp = descriptorToDot(tmpBuf);
                printf("<parameter name=\"arg%d\" type=\"%s\">\n</parameter>\n",
                    argNum++, tmp);
                free(tmp);
            }
            if (constructor)
                printf("</constructor>\n");
            else
                printf("</method>\n");
    }
bail:
    free(typeDescriptor);
    free(accessStr);
}
```

根据开头的定义，知道此函数中产出了什么内容，分别是 MethodID、类型描述、方法全称以及 accessFlag。在此函数中还包含一个 dumpCode 函数，这里面是解释器相关的操作，核心是对 pDexMethod 的处理。

```
void dumpCode(DexFile* pDexFile, const DexMethod* pDexMethod)
{
    const DexCode* pCode = dexGetCode(pDexFile, pDexMethod);

    if (gOptions.disassemble)
        dumpBytecodes(pDexFile, pDexMethod);

    dumpCatches(pDexFile, pCode);
    /* both of these are encoded in debug info */
    dumpPositions(pDexFile, pCode, pDexMethod);
    dumpLocals(pDexFile, pCode, pDexMethod);
}
```

上面的代码涉及 Dalvik 解析器的内容，本质上还是比较晦涩难懂的，我们将在后面的 DEX VMP 章节中展开讲解。

15.4 本章小结

本章通过 010 Editor 介绍了 DEX 的文件格式，并对每个部分进行了详细的讲解，尤其是在讲解 DEX 头部时，强调了其重要性。因为如果 DEX 头被抹掉，那么后面的解析工作也就无法展开。当然，对于 DEX 的解析，网上有很多脚本供我们使用，但是了解其规则，才能把 DEX 用到炉火纯青。最后，简要介绍了 Android 源码对 DEX 的处理过程。整个处理框架太大了，主要目的是引导读者找到解决问题的思路。本书的目标不是手把手教读者怎么做，而是总结出一些规律，让读者能自主学习和成长。

第 16 章

ELF 文件格式解读及其生成过程

在第 15 章中，我们学习了 DEX 文件的格式，并了解了它的解析器 DexDump。本章将学习另一种重要的文件，即 ELF（Executable and Linkable Format）文件。ELF 是一种对象文件的格式，用于定义不同类型的对象文件（Object files）中存放什么内容以及以何种格式存放这些内容。

Executable 和 Linkable 标识 ELF 文件的两大特性：可执行和可链接。ELF 文件由各种各样的结构体连接在一起。

1. **Executable（执行态）**

ELF 文件将参与文件的执行工作，包括二进制程序的运行以及动态 SO 的加载。它保持一个 Execution View 执行态。它的单位是段（Segment）。

2. **Linkable（链接态）**

链接态的单位是节（Section），链接视图在一个个节中，它的组成就是一个个节，我们介绍的就是这些节，了解这些节就可以清楚地认识链接视图。

接下来从 ELF 文件头来理解段和节。

16.1 ELF 文件头

在日常的二进制文件中，主要靠偏移来获取相应的值。所以文件头在运行态之前，我们可以手动指定一些偏移来去掉文件头。本节通过具体的案例来分析 ELF 文件头。

16.1.1 分析环境搭建

在 Android Studio 中新建以 Native C++为模板的项目，如图 16-1 所示。

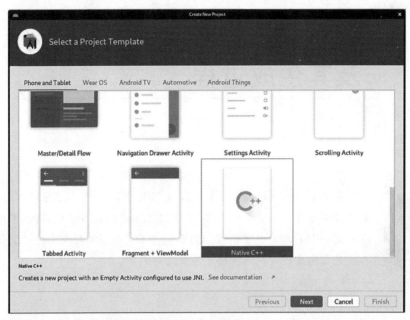

图 16-1 新建项目

等待 Gradle 完成，编译 APK 后，用 unzip 解压。详细操作步骤请回到第 15 章中查看。解压后有一个 libnative-lib.so 文件，通过 010 Editor 打开。

```
010editor libnative-lib.so
```

打开后，选择 ELF 模板来解析文件，头文件的格式如图 16-2 所示。

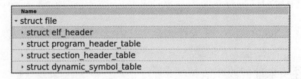

图 16-2 ELF 文件

- **elf_header**：ELF 文件头。
- **program_header_table**：程序头表。
- **section_header_table**：节区头表。
- **dynamic_symbol_table**：动态符号表。

下面详细分析上述文件。

16.1.2 elf_header

在 010 Editor 中打开 elf_header，如图 16-3 所示。

struct file	
▸ struct elf_header	
▸ struct e_ident_t e_ident	
enum e_type64_e e_type	ET_DYN (3)
enum e_machine64_e e_machine	183
enum e_version64_e e_version	EV_CURRENT (1)
Elf64_Addr e_entry_START_ADDRESS	0x000000000000EF30
Elf64_Off e_phoff_PROGRAM_HEADER_OFFSET_IN_FILE	64
Elf64_Off e_shoff_SECTION_HEADER_OFFSET_IN_FILE	213496
Elf32_Word e_flags	0
Elf64_Half e_ehsize_ELF_HEADER_SIZE	64
Elf64_Half e_phentsize_PROGRAM_HEADER_ENTRY_SIZE_IN_FILE	56
Elf64_Half e_phnum_NUMBER_OF_PROGRAM_HEADER_ENTRIES	8
Elf64_Half e_shentsize_SECTION_HEADER_ENTRY_SIZE	64
Elf64_Half e_shnum_NUMBER_OF_SECTION_HEADER_ENTRIES	25
Elf64_Half e_shtrndx_STRING_TABLE_INDEX	24

图 16-3 elf_header

同时寻找源码，看一下源码是如何声明 elf_header 字段的。网址 55（参见配书资源文件）为源码。

如图 16-4 所示，Elf32_Ehdr 为 32 位的头，Elf64_Ehdr 为 64 位的头。

```
struct Elf32_Ehdr {
    unsigned char e_ident[EI_NIDENT]; // ELF Identification bytes
    Elf32_Half    e_type;       // Type of file (see ET_* below)
    Elf32_Half    e_machine;    // Required architecture for this file (see EM_*)
    Elf32_Word    e_version;    // Must be equal to 1
    Elf32_Addr    e_entry;      // Address to jump to in order to start program
    Elf32_Off     e_phoff;      // Program header table's file offset, in bytes
    Elf32_Off     e_shoff;      // Section header table's file offset, in bytes
    Elf32_Word    e_flags;      // Processor-specific flags
    Elf32_Half    e_ehsize;     // Size of ELF header, in bytes
    Elf32_Half    e_phentsize;  // Size of an entry in the program header table
    Elf32_Half    e_phnum;      // Number of entries in the program header table
    Elf32_Half    e_shentsize;  // Size of an entry in the section header table
    Elf32_Half    e_shnum;      // Number of entries in the section header table
    Elf32_Half    e_shstrndx;   // Sect hdr table index of sect name string table
    bool checkMagic() const {
        return (memcmp(e_ident, ElfMagic, strlen(ElfMagic))) == 0);
    }
    unsigned char getFileClass() const { return e_ident[EI_CLASS]; }
    unsigned char getDataEncoding() const { return e_ident[EI_DATA]; }
};

// 64-bit ELF header. Fields are the same as for ELF32, but with different
// types (see above).
struct Elf64_Ehdr {
    unsigned char e_ident[EI_NIDENT];
    Elf64_Half    e_type;
    Elf64_Half    e_machine;
    Elf64_Word    e_version;
    Elf64_Addr    e_entry;
    Elf64_Off     e_phoff;
    Elf64_Off     e_shoff;
    Elf64_Word    e_flags;
    Elf64_Half    e_ehsize;
    Elf64_Half    e_phentsize;
    Elf64_Half    e_phnum;
    Elf64_Half    e_shentsize;
    Elf64_Half    e_shnum;
    Elf64_Half    e_shstrndx;
    bool checkMagic() const {
        return (memcmp(e_ident, ElfMagic, strlen(ElfMagic))) == 0);
    }
    unsigned char getFileClass() const { return e_ident[EI_CLASS]; }
    unsigned char getDataEncoding() const { return e_ident[EI_DATA]; }
};
```

图 16-4 查看源码字段

1. e_ident

e_ident 中给出的是 ELF 的一些标识信息，此字段下还有一些细分的结构。

1）file_identification

前 4 字节是 ELF 文件标识，也是我们常说的 magic 字段，ELF 文件头为（7F 45 4C 46）。

2）ei_class

- 01 表示 32 位。
- 02 表示 64 位。

3）ei_data

- 01 表示 Little Endian（小端排序）。
- 02 表示 Big Endian（大端排序）。

4）ei_version

表示 ELF 的版本，01 为默认版本。

其他字段均为 0，是填充位，无实际意义。

2. e_type

目标文件类型，它的取值如图 16-5 所示。

名称	取值	含义
ET_NONE	0	未知目标文件格式
ET_REL	1	可重定位文件
ET_EXEC	2	可执行文件
ET_DYN	3	共享目标文件
ET_CORE	4	Core 文件（转储格式）
ET_LOPROC	0xff00	特定处理器文件
ET_HIPROC	0xffff	特定处理器文件
ET_LOPROC 和 ET_HIPROC 之间的取值用来标识与处理器相关的文件格式		

图 16-5　e_type 取值

3. e_machine

CPU 架构，183 标识 ARM64，40 标识 ARM32。还有其他的架构，这里只讨论和 Android 相关的 ARM 架构。

4. e_version

目标文件的版本，取值同 e_ident[version]，如图 16-6 所示。

▼ struct file		0h	4160h
▼ struct elf_header		0h	40h
▼ struct e_ident_t e_ident		0h	10h
▶ char file_identification[4]	"ELF	0h	4h
enum ei_class_2_e ei_class_2	ELFCLASS64 (2)	4h	1h
enum ei_data_e ei_data	ELFDATA2LSB (1)	5h	1h
enum ei_version_e ei_version	E_CURRENT (1)	6h	1h
enum ei_osabi_e ei_osabi	ELFOSABI_NONE (0)	7h	1h
uchar ei_abiversion	0	8h	1h
▶ uchar ei_pad[6]		9h	6h
uchar ei_nident_SIZE	0	Fh	1h
enum e_type64_e e_type	ET_DYN (3)	10h	2h
enum e_machine64_e e_machine	183	12h	2h
enum e_version64_e e_version	EV_CURRENT (1)	14h	4h
Elf64_Addr e_entry_START_ADDRESS	0x000000000000EF30	18h	8h
Elf64_Off e_phoff_PROGRAM_HEADER_OFFSET_IN_FILE	64	20h	8h
Elf64_Off e_shoff_SECTION_HEADER_OFFSET_IN_FILE	213496	28h	8h
Elf32_Word e_flags	0	30h	4h
Elf64_Half e_ehsize_ELF_HEADER_SIZE	64	34h	2h
Elf64_Half e_phentsize_PROGRAM_HEADER_ENTRY_SIZE_IN_FILE	56	36h	2h
Elf64_Half e_phnum_NUMBER_OF_PROGRAM_HEADER_ENTRIES	8	38h	2h
Elf64_Half e_shentsize_SECTION_HEADER_ENTRY_SIZE	64	3Ah	2h
Elf64_Half e_shnum_NUMBER_OF_SECTION_HEADER_ENTRIES	25	3Ch	2h
Elf64_Half e_shtrndx_STRING_TABLE_INDEX	24	3Eh	2h

图 16-6　e_version 取值

5. e_entry

如果 ELF 是一个可执行程序，那么操作系统加载它后将跳转到 e_entry 的位置来执行该程序的代码。简单来说，对可执行程序而言，e_entry 是这个程序的入口地址。这里要特别指出的是，e_entry 是虚拟内存地址，不是实际的内存地址。

6. e_phoff

程序头表（Program Header Table）的偏移量（按字节计算）。如果文件没有程序头部表格，那么程序头表的偏移量一般是 64（0x40）。

7. e_shoff

节区头表（Section Header Table）的偏移量（按字节计算）。

8. e_flags

和处理器相关的文件标识。

9. e_ehsize

eh 是 elf header 的缩写，该字段表示 ELF 文件头的结构长度，64 位 ELF 文件头结构的长度为 64。

10. e_phentsize

每张程序表的大小。

11. e_phnum

程序表的数量。

12. e_shentsize

节区头表大小。

13. e_shnum

节区头表数量。

16.1.3 program_header_table

执行视图必须包含程序头，程序头中是各种各样的段。

16.1.4 section_header_table

IDA 打开 ELF 文件就是使用 section_header_table 来解析的。执行视图是没有节头的，也就是说，我们直接从内存中 dump SO 文件是没有节区表的。所以 IDA 不能正常解析。我们从内存中 dump 一个 SO 文件看一下效果。

安装 16.1.1 节中搭建的项目到手机中，并查看 SO 文件的内存分布，如图 16-7 所示。

```
(root@ r0env)-[~]
 # adb shell ps -e | grep r0ysue
(root@ r0env)-[~]
 # adb shell ps -e | grep roysue
u0_a115    30080  3572 4320160  61840 SyS_epoll_wait    0 S com.roysue.demoso
(root@ r0env)-[~]
 # adb shell
bullhead:/ $ su
bullhead:/ # cat /proc/30080/maps | grep libnative-lib.so
6f25647000-6f25677000 r-xp 00000000 fd:00 1534    /data/app/com.roysue.demoso-zEnFid19-fIK-93VpyiDlg==/lib/arm64/libnative-lib.so
6f25678000-6f2567c000 r--p 00030000 fd:00 1534    /data/app/com.roysue.demoso-zEnFid19-fIK-93VpyiDlg==/lib/arm64/libnative-lib.so
6f2567c000-6f2567d000 rw-p 00034000 fd:00 1534    /data/app/com.roysue.demoso-zEnFid19-fIK-93VpyiDlg==/lib/arm64/libnative-lib.so
```

图 16-7 SO 文件的内存分布

从 SO 文件的内存分布来看，6f25677000 和 6f25678000 缺少 0x1000 的偏移。

```bash
6f25647000-6f25677000 0x30000
6f25678000-6f2567c000 0x4000
6f2567c000-6f2567d000 0x1000
```

使用 Frida 从内存中 dump 一个执行视图下的 SO 文件，动静态对照看一下两者的效果。

首先给 App 添加 sdcard 的读写权限，具体操作请看第 15 章。

启动 frida-server，编写如下代码：

```
function main(){
    var file = new File("/sdcard/2.so","a+");
    file.write(ptr(0x6f25647000).readByteArray(0x30000));
    file.write(ptr(0x6f25678000).readByteArray(0x5000));
    file.flush();
    file.close();
}
```

在 sdcard 中找到 2.so 文件，如图 16-8 所示。

```
-rw-rw---- 1 root sdcard_rw  217088 2022-03-13 14:33 1.so
-rw-rw---- 1 root sdcard_rw  217088 2022-03-13 14:34 2.so
drwxrwx--x 2 root sdcard_rw    4096 1970-03-31 06:54 Alarms
drwxrwx--x 4 root sdcard_rw    4096 1970-03-31 06:54 Android
drwxrwx--x 4 root sdcard_rw    4096 2018-01-01 20:01 DCIM
```

图 16-8　生成 2.so 文件

通过 ADB 将 2.so 文件拉取（pull）到虚拟机上，并用 010 Editor 打开。可以从图 16-9 中看到，节区表全部都是空的。产生这种结果的原因是什么呢？这是因为 010 Editor 的 ELF 解析模板不正确，它是按偏移来解析的，所以是空的。

```
41F0h: 00 00 00 00 00 00 00 00 00 00 00 00 00 00 00 00  ................
4200h: 00 00 00 00 00 00 00 00 00 00 00 00 00 00 00 00  ................
4210h: 00 00 00 00 00 00 00 00 00 00 00 00 00 00 00 00  ................
4220h: 00 00 00 00 00 00 00 00 00 00 00 00 00 00 00 00  ................
4230h: 00 00 00 00 00 00 00 00 00 00 00 00 00 00 00 00  ................
4240h: 00 00 00 00 00 00 00 00 00 00 00 00 00 00 00 00  ................
4250h: 00 00 00 00 00 00 00 00 00 00 00 00 00 00 00 00  ................
4260h: 00 00 00 00 00 00 00 00 00 00 00 00 00 00 00 00  ................
4270h: 00 00 00 00 00 00 00 00 00 00 00 00 00 00 00 00  ................
4280h: 00 00 00 00 00 00 00 00 00 00 00 00 00 00 00 00  ................
4290h: 00 00 00 00 00 00 00 00 00 00 00 00 00 00 00 00  ................
42A0h: 00 00 00 00 00 00 00 00 00 00 00 00 00 00 00 00  ................
42B0h: 00 00 00 00 00 00 00 00 00 00 00 00 00 00 00 00
```

mplate Results - ELF.bt

Name	Value	Start	Size	Color
struct section_header_table		341F8h	640h	Fg: B
struct section_table_entry64_t section_table_element[0]	SHN_UNDEF	341F8h	40h	Fg: B
struct section_table_entry64_t section_table_element[1]	SHN_UNDEF	34238h	40h	Fg: B
struct section_table_entry64_t section_table_element[2]	SHN_UNDEF	34278h	40h	Fg: B
struct section_table_entry64_t section_table_element[3]	SHN_UNDEF	342B8h	40h	Fg: B
struct section_table_entry64_t section_table_element[4]	SHN_UNDEF	342F8h	40h	Fg: B
struct section_table_entry64_t section_table_element[5]	SHN_UNDEF	34338h	40h	Fg: B
struct section_table_entry64_t section_table_element[6]	SHN_UNDEF	34378h	40h	Fg: B
struct section_table_entry64_t section_table_element[7]	SHN_UNDEF	343B8h	40h	Fg: B
struct section_table_entry64_t section_table_element[8]	SHN_UNDEF	343F8h	40h	Fg: B
struct section_table_entry64_t section_table_element[9]	SHN_UNDEF	34438h	40h	Fg: B
struct section_table_entry64_t section_table_element[10]	SHN_UNDEF	34478h	40h	Fg: B
struct section_table_entry64_t section_table_element[11]	SHN_UNDEF	344B8h	40h	Fg: B
struct section_table_entry64_t section_table_element[12]	SHN_UNDEF	344F8h	40h	Fg: B
struct section_table_entry64_t section_table_element[13]	SHN_UNDEF	34538h	40h	Fg: B
struct section_table_entry64_t section_table_element[14]	SHN_UNDEF	34578h	40h	Fg: B
struct section_table_entry64_t section_table_element[15]	SHN_UNDEF	345B8h	40h	Fg: B
struct section_table_entry64_t section_table_element[16]	SHN_UNDEF	345F8h	40h	Fg: B
struct section_table_entry64_t section_table_element[17]	SHN_UNDEF	34638h	40h	Fg: B
struct section_table_entry64_t section_table_element[18]	SHN_UNDEF	34678h	40h	Fg: B
struct section_table_entry64_t section_table_element[19]	SHN_UNDEF	346B8h	40h	Fg: B

图 16-9　节区表全部是空

接下来我们看一下节区表中具体的 item 下的字段分布，010 Editor 解析如图 16-10 所示。

struct section_header_table	
struct section_table_entry64_t section_table_element[0]	SHN_UNDEF
struct s_name64_t s_name	SHN_UNDEF
enum s_type64_e s_type	SHT_NULL (0)
enum s_flags64_e s_flags	SF64_None (0)
Elf64_Addr s_addr	0x0000000000000000
Elf64_Off s_offset	0h
Elf64_Xword s_size	0
Elf64_Word s_link	0
Elf64_Word s_info	0
Elf64_Xword s_addralign	0
Elf64_Xword s_entsize	0

图 16-10　节区表 item

节区表中包含目标文件中的所有信息，目标文件中的每个节区都有对应的节区头部来描述它，反过来，有节区头部不意味着有节区。如图 16-11 所示是节区表的源码结构体。

```c
// Section header.
struct Elf32_Shdr {
    Elf32_Word sh_name;         // Section name (index into string table)
    Elf32_Word sh_type;         // Section type (SHT_*)
    Elf32_Word sh_flags;        // Section flags (SHF_*)
    Elf32_Addr sh_addr;         // Address where section is to be loaded
    Elf32_Off  sh_offset;       // File offset of section data, in bytes
    Elf32_Word sh_size;         // Size of section, in bytes
    Elf32_Word sh_link;         // Section type-specific header table index link
    Elf32_Word sh_info;         // Section type-specific extra information
    Elf32_Word sh_addralign;    // Section address alignment
    Elf32_Word sh_entsize;      // Size of records contained within the section
};

// Section header for ELF64 - same fields as ELF32, different types.
struct Elf64_Shdr {
    Elf64_Word  sh_name;
    Elf64_Word  sh_type;
    Elf64_Xword sh_flags;
    Elf64_Addr  sh_addr;
    Elf64_Off   sh_offset;
    Elf64_Xword sh_size;
    Elf64_Word  sh_link;
    Elf64_Word  sh_info;
    Elf64_Xword sh_addralign;
    Elf64_Xword sh_entsize;
};
```

图 16-11　节区表的源码结构体

各个字段的含义解释如表 16-1 所示。

表16-1　节区表各个字段的含义

成员	说明
sh_name	给出节区名称，是节区头部字符串节区（Section Header String Table Section）的索引。名字是一个 NULL 结尾的字符串
sh_type	为节区的内容和语义进行分类
sh_flags	节区支持 1 位形式的标志，这些标志描述了多种属性
sh_addr	如果节区将出现在进程的内存映像中，则此成员给出节区的第一字节应处的位置；否则，此字段为 0
sh_offset	此成员的取值给出节区的第一字节与文件头之间的偏移。不过，SHT_NOBITS 类型的节区不占用文件的空间，因此其 sh_offset 成员给出的是其概念性的偏移
sh_size	节区的长度（字节数）。除非节区的类型是 SHT_NOBITS，否则节区占用文件中的 sh_size 字节。类型为 SHT_NOBITS 的节区长度可能非 0，不过却不占用文件中的空间
sh_link	此成员给出节区头部表索引链接。其具体的解释依赖于节区类型
sh_info	此成员给出附加信息，其解释依赖于节区类型
sh_addralign	某些节区带有地址对齐约束。例如，如果一个节区保存一个 doubleword，那么系统必须保证整个节区能够按双字对齐。sh_addr 对 sh_addralign 取模，结果必须为 0。目前仅允许取值为 0 和 2 的幂次数。数值 0 和 1 表示节区没有对齐约束
sh_entsize	某些节区中包含固定大小的项目，如符号表。对于这类节区，此成员给出每个表项的长度字节数。如果节区中并不包含固定长度表项的表格，则此成员取值为 0

从图 16-10 中我们看到索引为 0 的首个节区头表中的数据都为 0。但是它的确存在，这个是固定节区，含义如图 16-12 所示。

字段名称	取值	说明
sh_name	0	无名称
sh_type	SHT_NULL	非活动
sh_flags	0	无标志
sh_addr	0	无地址
sh_offset	0	无文件偏移
sh_size	0	无尺寸大小
sh_link	SHN_UNDEF	无链接信息
sh_info	0	无辅助信息
sh_addralign	0	无对齐要求
sh_entsize	0	无表项

图 16-12　首个节区头表

下面认识一下各个节区的含义。
通过 readelf 查看节区表：

```
# readelf -S libnative-lib.so
There are 25 section headers, starting at offset 0x341f8:

Section Headers:
  [Nr] Name              Type             Address           Offset
       Size              EntSize          Flags  Link  Info  Align
  [ 0]                   NULL             0000000000000000  00000000
       0000000000000000  0000000000000000           0     0     0
  [ 1] .note.gnu.build-i NOTE             0000000000000200  00000200
       0000000000000024  0000000000000000   A       0     0     4
  [ 2] .hash             HASH             0000000000000228  00000228
       00000000000000d4  0000000000000004   A       4     0     8
  [ 3] .gnu.hash         GNU_HASH         0000000000000d00  00000d00
       0000000000000c40  0000000000000000   A       4     0     8
  [ 4] .dynsym           DYNSYM           0000000000001940  00001940
       0000000000002820  0000000000000018   A       5     3     8
  [ 5] .dynstr           STRTAB           0000000000004160  00004160
       00000000000020ba  0000000000000000   A       0     0     1
  [ 6] .gnu.version      VERSYM           000000000000621a  0000621a
       0000000000000358  0000000000000002   A       4     0     2
  [ 7] .gnu.version_r    VERNEED          0000000000006578  00006578
       0000000000000040  0000000000000000   A       5     2     8
  [ 8] .rela.dyn         RELA             00000000000065b8  000065b8
       0000000000007d70  0000000000000018   A       4     0     8
  [ 9] .rela.plt         RELA             000000000000e328  0000e328
       0000000000000720  0000000000000018   AI      4    20     8
```

```
  [10] .plt              PROGBITS        000000000000ea50  0000ea50
       00000000000004e0  0000000000000010  AX       0     0     16
  [11] .text             PROGBITS        000000000000ef30  0000ef30
       0000000000019310  0000000000000000  AX       0     0     4
  [12] .rodata           PROGBITS        0000000000028240  00028240
       0000000000002591  0000000000000000  A        0     0     8
  [13] .eh_frame_hdr     PROGBITS        000000000002a7d4  0002a7d4
       0000000000000f74  0000000000000000  A        0     0     4
  [14] .eh_frame         PROGBITS        000000000002b748  0002b748
       0000000000004298  0000000000000000  A        0     0     8
  [15] .gcc_except_table PROGBITS        000000000002f9e0  0002f9e0
       0000000000000404  0000000000000000  A        0     0     4
  [16] .note.android.ide NOTE            000000000002fde4  0002fde4
       0000000000000098  0000000000000000  A        0     0     4
  [17] .fini_array       FINI_ARRAY      0000000000031d58  00030d58
       0000000000000010  0000000000000008  WA       0     0     8
  [18] .data.rel.ro      PROGBITS        0000000000031d68  00030d68
       0000000000002d48  0000000000000000  WA       0     0     8
  [19] .dynamic          DYNAMIC         0000000000034ab0  00033ab0
       00000000000001f0  0000000000000010  WA       5     0     8
  [20] .got              PROGBITS        0000000000034ca0  00033ca0
       0000000000000360  0000000000000008  WA       0     0     8
  [21] .data             PROGBITS        0000000000035000  00034000
       0000000000000028  0000000000000000  WA       0     0     8
  [22] .bss              NOBITS          0000000000035030  00034028
       0000000000000480  0000000000000000  WA       0     0     16
  [23] .comment          PROGBITS        0000000000000000  00034028
       00000000000000dc  0000000000000001  MS       0     0     1
  [24] .shstrtab         STRTAB          0000000000000000  00034104
       00000000000000ef  0000000000000000           0     0     1
Key to Flags:
  W (write), A (alloc), X (execute), M (merge), S (strings), I (info),
  L (link order), O (extra OS processing required), G (group), T (TLS),
  C (compressed), x (unknown), o (OS specific), E (exclude),
  p (processor specific)
```

1. .shstrtab

所有节的一个名称,非常重要,有一些判断是根据名称来进行的。如图16-13所示,同C语言找字符串的逻辑类似,开头是指定的,碰到'\0'停止。此节区的作用就是指定节的名称。

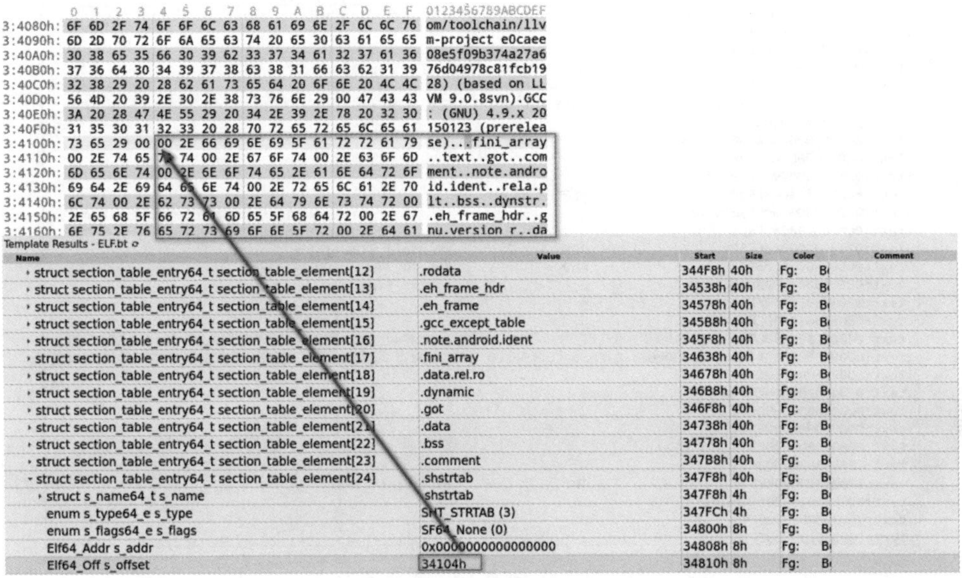

图 16-13 .shstrtab

2. .text

用于存储程序的指令。简单来说，程序的机器指令就放在这个节中。根据规范，.text 节的 sh_type 为 SHT_PROGBITS（取值为 1）；意为 Program Bits，即完全由应用程序自己决定（程序的机器指令当然是由程序自己决定的），sh_flags 为 SHT_ALLOC（当 ELF 文件加载到内存时，表示该节会分配内存）和 SHT_EXECINSTR（表示该节包含可执行的机器指令）。

在 010 Editor 中查看相应的段，如图 16-14 所示。

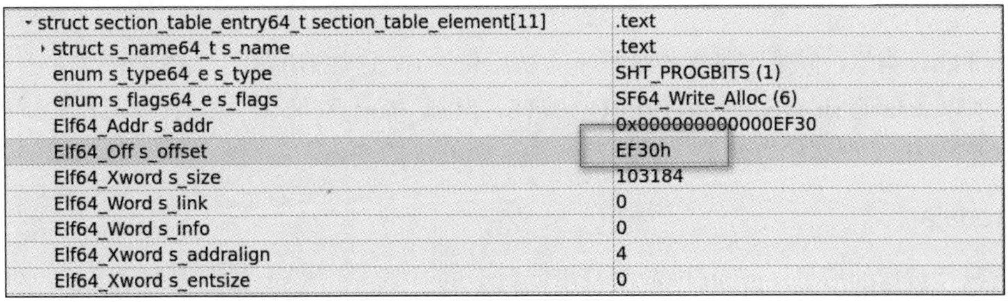

struct section_table_entry64_t section_table_element[11]	.text
struct s_name64_t s_name	.text
enum s_type64_e s_type	SHT_PROGBITS (1)
enum s_flags64_e s_flags	SF64_Write_Alloc (6)
Elf64_Addr s_addr	0x000000000000EF30
Elf64_Off s_offset	EF30h
Elf64_Xword s_size	103184
Elf64_Word s_link	0
Elf64_Word s_info	0
Elf64_Xword s_addralign	4
Elf64_Xword s_entsize	0

图 16-14 .text 起始偏移

在 IDA 中跳到指定的地址，如图 16-15 所示。这正是.text 的起始位置。

```
.text:000000000000EF30 ; ===============================================================
.text:000000000000EF30
.text:000000000000EF30 ; Segment type: Pure code
.text:000000000000EF30                 AREA .text, CODE
.text:000000000000EF30                 ; ORG 0xEF30
.text:000000000000EF30                 CODE64
.text:000000000000EF30
.text:000000000000EF30 ; =============== S U B R O U T I N E =======================================
.text:000000000000EF30
.text:000000000000EF30
.text:000000000000EF30                 EXPORT start
.text:000000000000EF30 start                                           ; DATA XREF: LOAD:0000000000000018↑o
.text:000000000000EF30                                                 ; LOAD:0000000000001958↑o ...
.text:000000000000EF30                 ADRP            X0, #off_31D68@PAGE
.text:000000000000EF34                 ADD             X0, X0, #off_31D68@PAGEOFF
.text:000000000000EF38                 B               .__cxa_finalize
.text:000000000000EF38 ; End of function start
.text:000000000000EF38
.text:000000000000EF3C ; [00000004 BYTES: COLLAPSED FUNCTION nullsub_2. PRESS CTRL-NUMPAD+ TO EXPAND]
.text:000000000000EF40 ; [00000004 BYTES: COLLAPSED FUNCTION j_nullsub_2. PRESS CTRL-NUMPAD+ TO EXPAND]
.text:000000000000EF44
.text:000000000000EF44 ; =============== S U B R O U T I N E =======================================
.text:000000000000EF44
.text:000000000000EF44
.text:000000000000EF44 sub_EF44                                        ; DATA XREF: .text:000000000000EF50↓o
.text:000000000000EF44                                                 ; .text:000000000000EF54↓o
.text:000000000000EF44                 CBZ             X0, locret_EF4C
.text:000000000000EF48                 BR              X0
.text:000000000000EF4C ; ---------------------------------------------------------------------------
.text:000000000000EF4C
```

图 16-15　查看 IDA 指定位置

3. .bss

bss 是 block storage segment 的缩写（bss 一词很有些历史，感兴趣的读者可自行了解）。在 ELF 规范中，.bss 节包含一块内存区域，这块区域在 ELF 文件被加载到进程空间时会由系统创建并设置这块内存的内容为 0。注意，.bss 节在 ELF 文件中不占据任何文件的空间，所以其 sh_type 为 SHT_NOBITS（取值为 8），它只是在 ELF 加载到内存的时候会分配一块由 sh_size 指定大小的内存。.bss 的 sh_flags 取值必须为 SHT_ALLOC 和 SHT_WRITE（表示该区域的内存是可写的）。

4. .data

.data 和.bss 类似，但是它包含的数据不会初始化为 0。这种情况下，就需要在文件中包含对应的信息。所以.data 的 sh_type 为 SHT_PROGBITS，但 sh_flags 和.bss 一样。读者可以尝试在 main.c 中定义一个"char c='f'"这样的变量，就能看到.data section 的变化了。

5. .rodata

包含只读数据的信息，printf 中的字符串就属于这一类。它的 sh_flags 只能为 SHF_ALLOC。

6. .systab

往往包含全部的符号表信息，但不是其中所有符号信息都会参与动态链接，所以 ELF 还专门定义了一个.dynsym 节（类型为 SHT_DYNSYM），这个节存储的仅是动态链接需要的符号信息。

7. .dynstr

专属的字符串表。

8. .got

全局偏移表。

9. .dynstr

此节区包含用于动态链接的字符串，大多数情况下这些字符串代表与符号表项相关的名称。

10. .hash

此节区包含一个符号 hash 表。

11. .interp

此节区包含程序解释器的路径名。如果程序包含一个可加载的段，段中包含此节区，那么节区的属性将包含 SHF_ALLOC 位，否则该位为 0。

12. .line

此节区包含符号调试的行号信息，其中描述了源程序与机器指令之间的对应关系。其内容是未定义的。

13. .plt

此节区包含过程链接表。

14. .dynsym

此节区包含动态链接符号表。查看此节区表：

```
bash
# readelf -s libnative-lib.so
Symbol table '.dynsym' contains 428 entries:
   Num:    Value          Size Type    Bind   Vis      Ndx Name
     0: 0000000000000000     0 NOTYPE  LOCAL  DEFAULT  UND
     1: 000000000000ef30     0 SECTION LOCAL  DEFAULT   11
     2: 0000000000031d68     0 SECTION LOCAL  DEFAULT   18
     3: 0000000000000000     0 FUNC    WEAK   DEFAULT  UND pthread_create@LIBC (2)
     4: 0000000000000000     0 FUNC    GLOBAL DEFAULT  UND realloc@LIBC (2)
     5: 0000000000000000     0 FUNC    GLOBAL DEFAULT  UND pthread_key_create@LIBC (2)
     6: 0000000000000000     0 FUNC    GLOBAL DEFAULT  UND __vsnprintf_chk@LIBC (2)
     7: 0000000000000000     0 FUNC    GLOBAL DEFAULT  UND pthread_once@LIBC (2)
     8: 0000000000000000     0 FUNC    GLOBAL DEFAULT  UND __cxa_finalize@LIBC (2)
     9: 0000000000000000     0 FUNC    GLOBAL DEFAULT  UND calloc@LIBC (2)
    10: 0000000000000000     0 FUNC    GLOBAL DEFAULT  UND syslog@LIBC (2)
    11: 0000000000000000     0 FUNC    GLOBAL DEFAULT  UND posix_memalign@LIBC (2)
    12: 0000000000000000     0 FUNC    GLOBAL DEFAULT  UND __strlen_chk@LIBC (2)
```

```
    13: 0000000000000000     0 FUNC    GLOBAL DEFAULT  UND abort@LIBC (2)
    14: 0000000000000000     0 FUNC    GLOBAL DEFAULT  UND __memcpy_chk@LIBC (2)
    15: 0000000000000000     0 FUNC    GLOBAL DEFAULT  UND dl_iterate_phdr@LIBC (3)
    16: 0000000000000000     0 FUNC    GLOBAL DEFAULT  UND
android_set_abort_message@LIBC (2)
    17: 0000000000000000     0 FUNC    GLOBAL DEFAULT  UND islower@LIBC (2)
    18: 0000000000000000     0 FUNC    GLOBAL DEFAULT  UND __memmove_chk@LIBC (2)
    19: 0000000000000000     0 OBJECT  GLOBAL DEFAULT  UND __sF@LIBC (2)
    20: 0000000000000000     0 FUNC    GLOBAL DEFAULT  UND fputc@LIBC (2)
    21: 0000000000000000     0 FUNC    GLOBAL DEFAULT  UND __stack_chk_fail@LIBC (2)
    22: 0000000000000000     0 FUNC    GLOBAL DEFAULT  UND pthread_setspecific@LIBC (2)
    23: 0000000000000000     0 FUNC    GLOBAL DEFAULT  UND openlog@LIBC (2)
    24: 0000000000000000     0 FUNC    GLOBAL DEFAULT  UND strcmp@LIBC (2)
    25: 0000000000000000     0 FUNC    GLOBAL DEFAULT  UND pthread_mutex_lock@LIBC (2)
    26: 0000000000000000     0 FUNC    GLOBAL DEFAULT  UND closelog@LIBC (2)
    27: 0000000000000000     0 FUNC    GLOBAL DEFAULT  UND pthread_getspecific@LIBC (2)
    28: 0000000000000000     0 FUNC    GLOBAL DEFAULT  UND memmove@LIBC (2)
    29: 0000000000000000     0 FUNC    GLOBAL DEFAULT  UND isxdigit@LIBC (2)
    30: 0000000000000000     0 FUNC    GLOBAL DEFAULT  UND strlen@LIBC (2)
    31: 0000000000000000     0 FUNC    GLOBAL DEFAULT  UND vasprintf@LIBC (2)
    32: 0000000000000000     0 FUNC    GLOBAL DEFAULT  UND malloc@LIBC (2)
    33: 0000000000000000     0 FUNC    GLOBAL DEFAULT  UND memcpy@LIBC (2)
    34: 0000000000000000     0 FUNC    GLOBAL DEFAULT  UND memset@LIBC (2)
    35: 0000000000000000     0 FUNC    GLOBAL DEFAULT  UND free@LIBC (2)
    36: 0000000000000000     0 FUNC    GLOBAL DEFAULT  UND vfprintf@LIBC (2)
    37: 0000000000000000     0 FUNC    GLOBAL DEFAULT  UND pthread_mutex_unlock@LIBC (2)
    38: 0000000000000000     0 FUNC    GLOBAL DEFAULT  UND __cxa_atexit@LIBC (2)
```

16.2 ELF 可执行文件的生成过程与执行视图

16.1 节讲解了 ELF 文件的基本格式,但是程序头表还没有讲解,本节讲述 ELF 文件的生成过程以及 ELF 文件的程序头表。

16.2.1 ARM 可执行文件的生成过程

r0env 系统中的 Clang 不支持 ARM 编译,采用 Android 工具包的 NDK 中的 Clang 编译。NDK 中的 Clang 路径为:

```
/root/Android/Sdk/ndk/22.1.7171670/toolchains/llvm/prebuilt/linux-x86_64/bin
```

为了方便使用,将其加入系统环境变量中。Kali Linux 最新版本的默认 Shell 是 Zsh, Zsh 虽然方便,但其并不兼容,且不能运行诸多编译系统,我们应该回到 Bash。按照官网的方式,执行 chsh

-s /bin/bash 后重启，回退到 Bash。打开虚拟机，按快捷键 Ctrl+Alt+T，默认打开的就是 Bash。环境变量添加：

```
# vim .zshrc
# 在文件后面追加如下代码
export PATH="/root/Android/Sdk/ndk/22.1.7171670/toolchains/llvm/prebuilt/linux-x86_64/bin:$PATH"
# source .bashrc
```

ARM 编译的过程分为 4 步：预编译、编译、汇编、链接。新建一个 demo.c 文件，编写如下代码：

```c
#include <stdio.h>
void add()
{
    printf("add ");
}

void sub()
{
    printf("sub ");
}

int main()
{
    add();
    sub();
    printf("Hello, World!\n");
    return 0;
}
```

1. 预编译

处理以 "#" 开头的关键字，删除注释，添加行号和文件标识，命令如下：

```
clang -target armv8a-linux-androideabi21 -E demo.c -o demo.i
```

生成的 demo.i 文件如图 16-16 所示。

```
1  # 1 "demo.c"
2  # 1 "<built-in>" 1
3  # 1 "<built-in>" 3
4  # 376 "<built-in>" 3
5  # 1 "<command line>" 1
6  # 1 "<built-in>" 2
7  # 1 "demo.c" 2
8  # 1 "/root/Android/Sdk/ndk/22.1.7171670/toolchains/llvm/prebuilt/linux-x86_64/bin/../sysroot/usr/include/stdio.h" 1 3 4
9  # 41 "/root/Android/Sdk/ndk/22.1.7171670/toolchains/llvm/prebuilt/linux-x86_64/bin/../sysroot/usr/include/stdio.h" 3 4
10 # 1 "/root/Android/Sdk/ndk/22.1.7171670/toolchains/llvm/prebuilt/linux-x86_64/bin/../sysroot/usr/include/sys/cdefs.h" 
11 # 361 "/root/Android/Sdk/ndk/22.1.7171670/toolchains/llvm/prebuilt/linux-x86_64/bin/../sysroot/usr/include/sys/cdefs.h"
12 # 1 "/root/Android/Sdk/ndk/22.1.7171670/toolchains/llvm/prebuilt/linux-x86_64/bin/../sysroot/usr/include/android/versio
13 # 362 "/root/Android/Sdk/ndk/22.1.7171670/toolchains/llvm/prebuilt/linux-x86_64/bin/../sysroot/usr/include/sys/cdefs.h"
14 # 1 "/root/Android/Sdk/ndk/22.1.7171670/toolchains/llvm/prebuilt/linux-x86_64/bin/../sysroot/usr/include/android/api-le
15 # 176 "/root/Android/Sdk/ndk/22.1.7171670/toolchains/llvm/prebuilt/linux-x86_64/bin/../sysroot/usr/include/android/api-
16 # 1 "/root/Android/Sdk/ndk/22.1.7171670/toolchains/llvm/prebuilt/linux-x86_64/bin/../sysroot/usr/include/bits/get_devic
17 # 38 "/root/Android/Sdk/ndk/22.1.7171670/toolchains/llvm/prebuilt/linux-x86_64/bin/../sysroot/usr/include/bits/get_devi
18 int __system_property_get(const char* __name, char* __value);
19 int atoi(const char* __s) __attribute__((__pure__));
20
21 static __inline int android_get_device_api_level() {
22   char value[92] = { 0 };
23   if (__system_property_get("ro.build.version.sdk", value) < 1) return -1;
24   int api_level = atoi(value);
25   return (api_level > 0) ? api_level : -1;
26 }
27 # 177 "/root/Android/Sdk/ndk/22.1.7171670/toolchains/llvm/prebuilt/linux-x86_64/bin/../sysroot/usr/include/android/api-
28 # 363 "/root/Android/Sdk/ndk/22.1.7171670/toolchains/llvm/prebuilt/linux-x86_64/bin/../sysroot/usr/include/sys/cdefs.h"
29
30 # 1 "/root/Android/Sdk/ndk/22.1.7171670/toolchains/llvm/prebuilt/linux-x86_64/bin/../sysroot/usr/include/android/ndk-ve
31 # 365 "/root/Android/Sdk/ndk/22.1.7171670/toolchains/llvm/prebuilt/linux-x86_64/bin/../sysroot/usr/include/sys/cdefs.h"
```

图 16-16 demo.i 文件

2. 编译

".i" 文件生成 ".s" 文件，编译是对于预处理完的文件进行一系列的词法分析、语法分析、语义分析，以及优化后产生相应的汇编代码文件。内联函数的替换就发生在这一阶段。

编译命令：

```
clang -target armv8a-linux-androideabi21 -S demo.i -o demo.s
```

生成如图 16-17 所示的 demo.s 文件。

```
1   .text
2   .syntax unified
3   .eabi_attribute 67, "2.09"     @ Tag_conformance
4   .eabi_attribute 6, 14          @ Tag_CPU_arch
5   .eabi_attribute 7, 65          @ Tag_CPU_arch_profile
6   .eabi_attribute 8, 1           @ Tag_ARM_ISA_use
7   .eabi_attribute 9, 2           @ Tag_THUMB_ISA_use
8   .fpu    neon
9   .eabi_attribute 12, 3          @ Tag_Advanced_SIMD_arch
10  .eabi_attribute 42, 1          @ Tag_MPextension_use
11  .eabi_attribute 34, 1          @ Tag_CPU_unaligned_access
12  .eabi_attribute 68, 3          @ Tag_Virtualization_use
13  .eabi_attribute 15, 1          @ Tag_ABI_PCS_RW_data
14  .eabi_attribute 16, 1          @ Tag_ABI_PCS_RO_data
15  .eabi_attribute 17, 2          @ Tag_ABI_PCS_GOT_use
16  .eabi_attribute 20, 1          @ Tag_ABI_FP_denormal
17  .eabi_attribute 21, 0          @ Tag_ABI_FP_exceptions
18  .eabi_attribute 23, 3          @ Tag_ABI_FP_number_model
19  .eabi_attribute 24, 1          @ Tag_ABI_align_needed
20  .eabi_attribute 25, 1          @ Tag_ABI_align_preserved
21  .eabi_attribute 38, 1          @ Tag_ABI_FP_16bit_format
22  .eabi_attribute 18, 4          @ Tag_ABI_PCS_wchar_t
23  .eabi_attribute 26, 2          @ Tag_ABI_enum_size
24  .eabi_attribute 14, 0          @ Tag_ABI_PCS_R9_use
25  .file    "demo.c"
26  .globl   main                  @ -- Begin function main
27  .p2align  2
28  .type    main,%function
29  .code    32                    @ @main
```

图 16-17 demo.s 文件

3. 汇编

汇编器是将汇编代码转换成机器可以执行的命令，每一条汇编语句都对应一条机器指令，并生成可重定位目标程序的 .o 文件，该文件为二进制文件。汇编相对于编译的过程比较简单，根据汇编指令表和机器指令表一一进行翻译即可。所以汇编器的汇编过程相对于编译器是比较简单的。

汇编命令：

```
lang -target armv8a-linux-androideabi21 -c demo.s -o demo.o
```

查看文件类型，如图 16-18 所示。

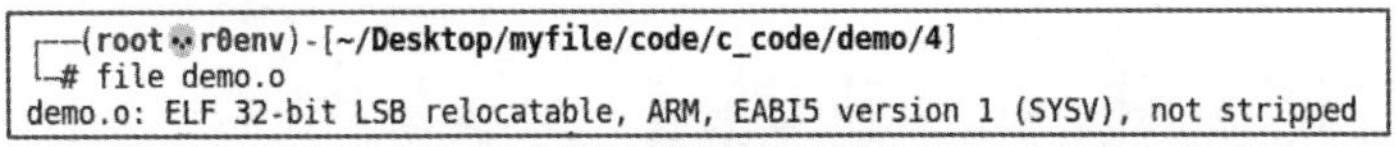

图 16-18　demo.o 文件类型

我们将 demo.o 文件推送（push）到手机中，赋予权限并运行，结果如图 16-19 所示，发现它并不能运行。

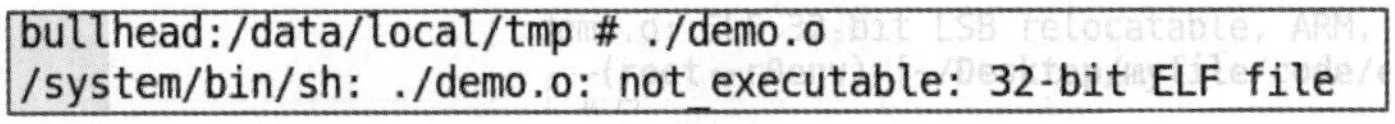

图 16-19　在手机中执行 demo.o 文件

用 010 Editor 打开 demo.o 文件，如图 16-20 所示，此文件没有程序头表。这就是因为少了最重要的链接步骤。

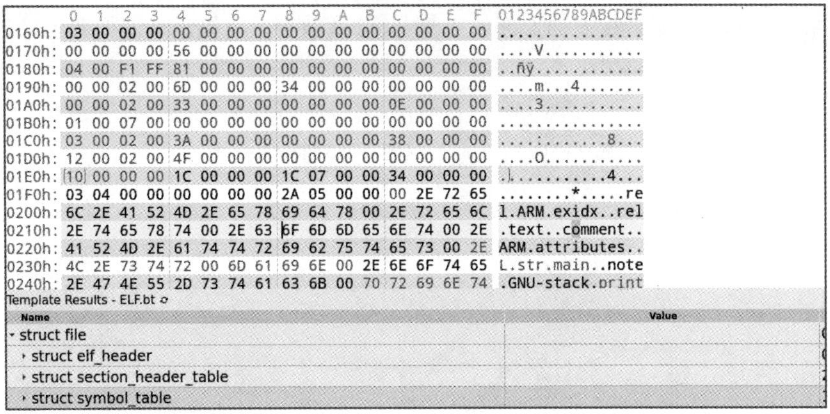

图 16-20　用 010 Editor 打开 demo.o 文件

4. 链接

编译器将生成的多个 .o 文件链接到一起并生成一个可执行文件，在这个过程中，编译器做的一个重要的事情是将每个文件中 call 指令后面的地址补充上，方式是从当前文件的函数地址符表中开始找，如果没有，继续从别的文件的函数地址符表中找，找到后填补在 call 指令后面，如果找不到，则链接失败。

用 objdump 打开文件，用系统中的 objdump 是打不开的，因为其不能解析 ARM 文件。我们使用 NDK 套件中的 objdump。路径如下：

```
/root/Android/Sdk/ndk/22.1.7171670/toolchains/llvm/prebuilt/linux-x86_64/arm-linux
-androideabi/bin/
```

读者可以自行添加到环境变量并重命名，与系统中的 objdump 进行区分。执行 objdump 命令，查看 demo.o 文件的二进制信息，如图 16-21 所示。可以看到 add 并无链接地址。

```
00000038 <main>:
      38:   e92d4800    push    {fp, lr}
      3c:   e1a0b00d    mov     fp, sp
      40:   e24dd008    sub     sp, sp, #8
      44:   e3000000    movw    r0, #0
      48:   e58d0004    str     r0, [sp, #4]
      4c:   ebffffff    bl      0 <add>
      50:   ebffffff    bl      1c <sub>
      54:   e59f0018    ldr     r0, [pc, #24]   ; 74 <main+0x3c>
      58:   e08f0000    add     r0, pc, r0
      5c:   ebffffff    bl      0 <printf>
      60:   e3001000    movw    r1, #0
      64:   e58d0000    str     r0, [sp]
      68:   e1a00001    mov     r0, r1
      6c:   e1a0d00b    mov     sp, fp
      70:   e8bd8800    pop     {fp, pc}
      74:   00000014    .word   0x00000014
```

图 16-21　查看 demo.o 文件的二进制信息

使用下面的命令进行链接操作：

```
clang -target armv8a-linux-androideabi21 demo.c -o demo
```

再次使用 objdump 查看链接好的 Demo。此时 add 函数有了链接地址，如图 16-22 所示。

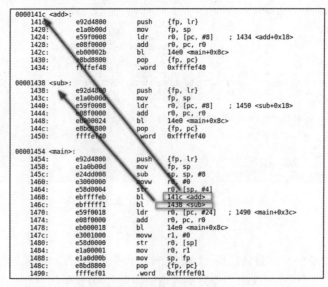

图 16-22　查看 Demo 的二进制信息

16.2.2 执行视图

在 16.1.4 节中，我们从运行的内存中 dump 出一个 2.so 文件，它就是一个执行视图（Execution View），而我们最初编译 App 生成的 libnative-lib.so 是一个完整的文件，称其为链接视图（Linking View）。图 16-23 是链接视图和执行视图的大致结构。

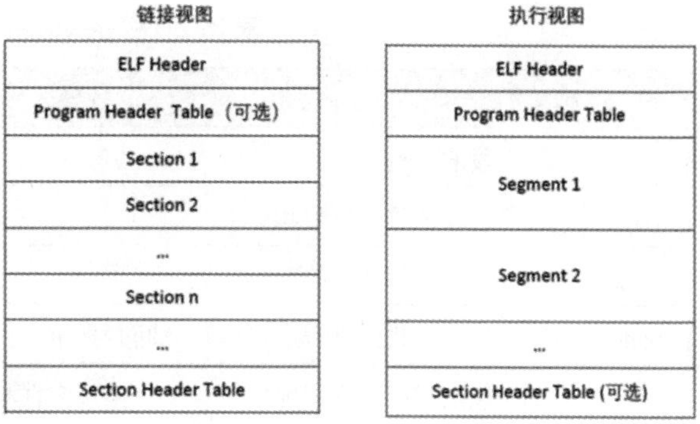

1. 链接视图

静态链接器（编译后参与生成最终 ELF 过程的链接器，如 ld）会以链接视图解析 ELF 文件。编译时生成的.o 文件（目标文件）以及链接后的.so 文件（共享库）均可通过链接视图解析，链接视图可以没有段表（如目标文件不会有段表）。

图 16-23　链接视图和执行视图

2. 执行视图

动态链接器（加载器，如 x86 架构 Linux 下的/lib/ld-linux.so.2 和 Android 系统下的/system/linker 均为动态链接器）会以执行视图解析 ELF 文件并动态链接，执行视图可以没有节表。

图 16-24 展示了程序头表的结构体。

```
// Program header for ELF32.
struct Elf32_Phdr {
    Elf32_Word  p_type;   // Type of segment
    Elf32_Off   p_offset; // File offset where segment is located, in bytes
    Elf32_Addr  p_vaddr;  // Virtual address of beginning of segment
    Elf32_Addr  p_paddr;  // Physical address of beginning of segment (OS-specific)
    Elf32_Word  p_filesz; // Num. of bytes in file image of segment (may be zero)
    Elf32_Word  p_memsz;  // Num. of bytes in mem image of segment (may be zero)
    Elf32_Word  p_flags;  // Segment flags
    Elf32_Word  p_align;  // Segment alignment constraint
};

// Program header for ELF64.
struct Elf64_Phdr {
    Elf64_Word  p_type;   // Type of segment
    Elf64_Word  p_flags;  // Segment flags
    Elf64_Off   p_offset; // File offset where segment is located, in bytes
    Elf64_Addr  p_vaddr;  // Virtual address of beginning of segment
    Elf64_Addr  p_paddr;  // Physical addr of beginning of segment (OS-specific)
    Elf64_Xword p_filesz; // Num. of bytes in file image of segment (may be zero)
    Elf64_Xword p_memsz;  // Num. of bytes in mem image of segment (may be zero)
    Elf64_Xword p_align;  // Segment alignment constraint
};
```

图 16-24　程序头表的结构体

程序头表结构体字段说明如表 16-2 所示。

表16-2 程序头表结构体字段说明

字段	说明
p_type	段的类型
p_flags	段标记符（权限相关）
p_offset	该段位于文件的起始位置
p_vaddr	该段加载到进程虚拟内存空间时指定的内存地址，是该段进入文件的一个相对偏移
p_paddr	该段对应的物理地址。对于可执行文件和动态库文件而言，这个值并没有什么意义，因为系统用的是虚拟地址，不会使用物理地址
p_filesz	该段在文件中占据的大小，其值可以为 0。因为段是由节组成的，有些节在文件中不占据空间，比如前文提到的.bss 节
p_memsz	该段在内存中占据的空间，其值可以为 0
p_align	段加载到内存后，其首地址需要按 p_align 的要求进行对齐

通过 readelf 来看一下实际的值：

```
readelf -l libnative-lib.so

Elf file type is DYN (Shared object file)
Entry point 0xef30
There are 8 program headers, starting at offset 64

Program Headers:
  Type           Offset              VirtAddr            PhysAddr
                 FileSiz             MemSiz                        Flags   Align
  LOAD           0x0000000000000000  0x0000000000000000  0x0000000000000000
                 0x000000000002fe7c  0x000000000002fe7c            R E     1000
  LOAD           0x0000000000030d58  0x0000000000031d58  0x0000000000031d58
                 0x00000000000032d0  0x0000000000003758            RW      1000
  DYNAMIC        0x0000000000033ab0  0x0000000000034ab0  0x0000000000034ab0
                 0x00000000000001f0  0x00000000000001f0            RW      8
  NOTE           0x0000000000000200  0x0000000000000200  0x0000000000000200
                 0x0000000000000024  0x0000000000000024            R       4
  NOTE           0x000000000002fde4  0x000000000002fde4  0x000000000002fde4
                 0x0000000000000098  0x0000000000000098            R       4
  GNU_EH_FRAME   0x000000000002a7d4  0x000000000002a7d4  0x000000000002a7d4
                 0x0000000000000f74  0x0000000000000f74            R       4
  GNU_STACK      0x0000000000000000  0x0000000000000000  0x0000000000000000
                 0x0000000000000000  0x0000000000000000            RW      10
  GNU_RELRO      0x0000000000030d58  0x0000000000031d58  0x0000000000031d58
                 0x00000000000032a8  0x00000000000032a8            R       1
```

程序头表中的重要字段说明如表 16-3 所示。

表16-3　程序头表中的重要字段说明

标记符名	取值	说明
PT_LOAD	1	可加载到内存的段，加载方式由该 ELF 文件指定位置（由 p_offset 决定），指定文件大小（由 p_filesz 决定）映射到内存的指定位置（由 p_vaddr 决定），指定内存大小（由 p_memsz 决定）。注意，p_filesz 可以比 p_memsz 小，多余的部分填 0。一个 ELF 文件可以包含多个 PT_LOAD 的段，在程序头表中、PT_LOAD 段按 p_vaddr 排序
PT_DYNAMIC	2	和动态链接有关
PT_INTERP	3	可执行程序使用动态库的时候，动态库的加载工作是由程序解释器来完成的，程序解释器其实就是链接器（ld）该段用于说明此程序使用的链接器在系统中的绝对路径
PT_NOTE	4	和 Note 节有关，这里不详细介绍
PT_PHDR	6	代表程序头表本身的信息，当文件加载到内存的时候，系统根据该项找到程序头表的位置
PT_LOPROC	0x70000000	Processor Specific 的定义，例如 GNU_EH_FRAME 取值为 0x6474e550
PT_HIPROC	0x7FFFFFFF	

16.2.3　GOT 和 PLT

GOT 表的说明如图 16-25 所示。

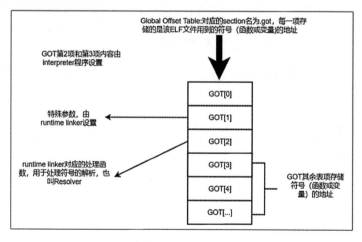

图 16-25　GOT 表

1. GOT

一般而言，符号地址计算有两个时机：

（1）ld 将控制权交给可执行程序之前。此时，ld 已经加载了依赖的动态库，而且也知道这些动态库加载到内存的虚拟地址，这样就可以计算出所有需要的符号的地址。如果要使用这种方法，就需要设置环境变量 LD_BIND_NOW（exportLD_BIND_NOW=1）。对于运行中调用 dlopen 来加载 SO 文件的程序而言，就需设置 dlopen 的 flag 参数为 RTLOAD_NOW。这种做法的主要缺点在于它使得那些大量依赖动态库的程序的加载时间变长。

（2）用的时候再计算。相比第一种方式，这种方式可以让 ld 尽快把控制权交给可执行程序本身，从而提升程序启动速度。如果要使用这种方式，就需要设置环境变量 LD_BIND_NOT（ld 默认采用这种策略）。而对于 dlopen 来说，设置 flag 为 RTLOAD_LAZY 即可。程序本身不知道 ld 到底会使用哪种方法，所以编译器生成的二进制文件必须同时支持这两种方法。这是如何做到的呢？接下来介绍第二个辅助手段——PLT。

2. PLT

PLT 表的逻辑图如图 16-26 所示。

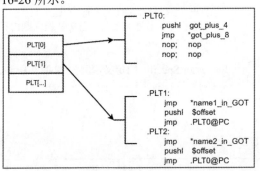

图 16-26　PLT 表

PLT[0]存储的是跳转到 GOT 表 Resolver 的指令。pushl got_plus_4 是 GOT[1]的元素压栈，jmp *got_plus_8 则是跳转到 GOT[2]所存储的地址上执行，也就是前面讲的 Resolver。

再看 PLT[1]，*name1_in_GOT 表示符号 name1 在 GOT 对应项的内容，我们以 GOT[name1]表示。如果该符号的地址还没有计算，GOT[name1]存储的就是其后一条指令（此处是 pushl $offset）的地址。pushl $offset 的目的是将计算这个符号地址所需的参数压栈（offset 的含义后面再介绍）。然后 Trampoline Code 执行到 jmp .PLT0@PC，即跳转到 PLT[0]的代码处执行（其结果就是跳转到 Resolver）。

接下来通过 GDB 动态调试来看这两张表之间的应用关系。普通 GDB 调试单一，我们安装 gdb-multiarch。

```
apt install gdb-multiarch
```

再安装一个可视化的调试工具 GEF，参考网址 56（参见配书资源文件）。

读者可以使用手动方式安装。然后启动，启动后界面如图 16-27 所示。

第 16 章　ELF 文件格式解读及其生成过程

```
┌──(root㉿r0env)-[~]
└─# gdb-multiarch
GNU gdb (Debian 10.1-2) 10.1.90.20210103-git
Copyright (C) 2021 Free Software Foundation, Inc.
License GPLv3+: GNU GPL version 3 or later <http://gnu.org/licenses/gpl.html>
This is free software: you are free to change and redistribute it.
There is NO WARRANTY, to the extent permitted by law.
Type "show copying" and "show warranty" for details.
This GDB was configured as "x86_64-linux-gnu".
Type "show configuration" for configuration details.
For bug reporting instructions, please see:
<https://www.gnu.org/software/gdb/bugs/>.
Find the GDB manual and other documentation resources online at:
    <http://www.gnu.org/software/gdb/documentation/>.

For help, type "help".
Type "apropos word" to search for commands related to "word".
GEF for linux ready, type `gef` to start, `gef config` to configure
93 commands loaded for GDB 10.1.90.20210103-git using Python engine 3.9
[*] 3 commands could not be loaded, run `gef missing` to know why.
gef➤
```

图 16-27　启动 gdb-multiarch

在 r0env 中找到 gdbserver，并推送到手机中。gdbserver 路径如下：

/root/Android/Sdk/ndk/22.1.7171670/prebuilt/android-arm/gdbserver

启动 gdbserver，并开启端口监听，操作如图 16-28 所示。

```
bullhead:/data/local/tmp # chmod 777 gdbserver_arm
bullhead:/data/local/tmp # ./gdb
gdbserver           gdbserver_arm
bullhead:/data/local/tmp # ./gdbserver_arm 0.0.0.0:11678 demo
warning: Found custom handler for signal 39 (Unknown signal 39) preinstalled.
Some signal dispositions inherited from the environment (SIG_DFL/SIG_IGN)
won't be propagated to spawned programs.
gdbserver: Unable to determine the number of hardware watchpoints available.
gdbserver: Unable to determine the number of hardware breakpoints available.
Process /data/local/tmp/demo created; pid = 15584
Listening on port 11678
```

图 16-28　开启 gdbserver 端口监听

在 GEF 开启远程连接，如图 16-29 所示。

```
gef➤  gef-remote 192.168.18.140:11678
Reading /data/local/tmp/demo from remote target...
warning: File transfers from remote targets can be slow. Use "set sysroot" to access files locally instead.
Reading /data/local/tmp/demo from remote target...
Reading symbols from target:/data/local/tmp/demo...
(No debugging symbols found in target:/data/local/tmp/demo)
Reading /system/bin/linker from remote target...
Reading /system/bin/linker from remote target...
0xf772af18 in __dl_start () from target:/system/bin/linker
[+] Connected to '192.168.18.140:11678'
[+] Remote information loaded to temporary path '/tmp/gef/15584'
gef➤
```

图 16-29　GEF 远程连接

反编译 Demo 文件，如图 16-30 所示。

设置断点有两种方式。

（1）b main。

（2）b *（函数地址）。

对 main 函数设置断点并执行到断点位置，如图 16-31 所示。输入 c 运行到断点位置，效果如图 16-32 所示。

```
gef➤  disassemble main
Dump of assembler code for function main:
   0xab0f7454 <+0>:     push    {r11, lr}
   0xab0f7458 <+4>:     mov     r11, sp
   0xab0f745c <+8>:     sub     sp, sp, #8
   0xab0f7460 <+12>:    movw    r0, #0
   0xab0f7464 <+16>:    str     r0, [sp, #4]
   0xab0f7468 <+20>:    bl      0xab0f741c <add>
   0xab0f746c <+24>:    bl      0xab0f7438 <sub>
   0xab0f7470 <+28>:    ldr     r0, [pc, #24]   ; 0xab0f7490 <main+60>
   0xab0f7474 <+32>:    add     r0, pc, r0
   0xab0f7478 <+36>:    bl      0xab0f74e0
   0xab0f747c <+40>:    movw    r1, #0
   0xab0f7480 <+44>:    str     r0, [sp]
   0xab0f7484 <+48>:    mov     r0, r1
   0xab0f7488 <+52>:    mov     sp, r11
   0xab0f748c <+56>:    pop     {r11, pc}
   0xab0f7490 <+60>:                    ; <UNDEFINED> instruction: 0xffffef01
End of assembler dump.
```

图 16-30　反编译

```
gef➤  disassemble main
Dump of assembler code for function main:
   0xaac86454 <+0>:     push    {r11, lr}
   0xaac86458 <+4>:     mov     r11, sp
   0xaac8645c <+8>:     sub     sp, sp, #8
   0xaac86460 <+12>:    movw    r0, #0
   0xaac86464 <+16>:    str     r0, [sp, #4]
   0xaac86468 <+20>:    bl      0xaac8641c <add>
   0xaac8646c <+24>:    bl      0xaac86438 <sub>
   0xaac86470 <+28>:    ldr     r0, [pc, #24]   ; 0xaac86490 <main+60>
   0xaac86474 <+32>:    add     r0, pc, r0
   0xaac86478 <+36>:    bl      0xaac864e0
   0xaac8647c <+40>:    movw    r1, #0
   0xaac86480 <+44>:    str     r0, [sp]
   0xaac86484 <+48>:    mov     r0, r1
   0xaac86488 <+52>:    mov     sp, r11
   0xaac8648c <+56>:    pop     {r11, pc}
   0xaac86490 <+60>:                    ; <UNDEFINED> instruction: 0xffffef01
End of assembler dump.
gef➤  b *0xaac86454
Breakpoint 1 at 0xaac86454
```

图 16-31　断点

```
$r1  : 0xfffefa34  →  0xfffefb48  →  "/data/local/tmp/demo"
$r2  : 0xfffefa3c  →  0xfffefb5d  →  "_=/data/local/tmp/demo"
$r3  : 0x0
$r4  : 0xaac86454  →  <main+0> push {r11,  lr}
$r5  : 0xfffefa34  →  0xfffefb48  →  "/data/local/tmp/demo"
$r6  : 0x1
$r7  : 0xfffefa3c  →  0xfffefb5d  →  "_=/data/local/tmp/demo"
$r8  : 0x0
$r9  : 0x0
$r10 : 0x0
$r11 : 0xfffefa28  →  0x00000000
$r12 : 0xf7661630  →  0xf7618e01  →  <pthread_mutex_unlock+1> stmdb sp!,  {r4,  r5,  r6,  r7,  r8,  lr}
$sp  : 0xfffefa00  →  0x00000000
$lr  : 0xf764a35f  →  <__libc_init+51> bl 0xf764d7e4
$pc  : 0xaac86454  →  <main+0> push {r11,  lr}
$cpsr: [NEGATIVE zero carry overflow interrupt fast thumb]
──────────────────────────────────────────────────────────────── stack ────
0xfffefa00|+0x0000: 0x00000000   ← $sp
0xfffefa04|+0x0004: 0x00000000
0xfffefa08|+0x0008: 0x00000000
0xfffefa0c|+0x000c: 0x00000000
0xfffefa10|+0x0010: 0x00000000
0xfffefa14|+0x0014: 0xaac863e0  →  <_start_main+68> andeq r1,  r0,  r8,  lsr r2
0xfffefa18|+0x0018: 0x00000000
0xfffefa1c|+0x001c: 0xaac874f0  →  0xffffffff
─────────────────────────────────────────────────────────── code:arm:ARM ────
   0xaac86448 <sub+16>     bl     0xaac864e0
   0xaac8644c <sub+20>     pop    {r11,  pc}
   0xaac86450 <sub+24>     ;      <UNDEFINED> instruction: 0xffffef40
●→ 0xaac86454 <main+0>     push   {r11,  lr}
   0xaac86458 <main+4>     mov    r11,  sp
   0xaac8645c <main+8>     sub    sp,  sp,  #8
   0xaac86460 <main+12>    movw   r0,  #0
   0xaac86464 <main+16>    str    r0,  [sp,  #4]
   0xaac86468 <main+20>    bl     0xaac8641c <add>
──────────────────────────────────────────────────────────────── threads ────
[#0] Id 1, Name: "demo", stopped 0xaac86454 in main (), reason: BREAKPOINT
────────────────────────────────────────────────────────────────── trace ────
[#0] 0xaac86454 → main()
```

图 16-32　运行断点

熟悉了 GDB 的基本使用方法，我们编写一个函数来观察这两张表之间的对应关系。

编写如图 16-33 所示的代码。

```c
// hello.c
#include "stdio.h"
#include "a.h"

int main(int argc,char const *argv[]){

    int a = add(1,1);
    int s = sub(3,1);

    printf("i am from main %d %d",a,s);
}
```

```c
// a.h
int add(int,int);
int sub(int,int);
```

```c
// a.c
int add(int a,int b){
    return a+b;
}
int sub(int a,int b){
    return a-b;
}
```

图 16-33　编写代码

编译代码，结果如图 16-34 所示，显示找不到 linker。

```
┌──(root㉿r0env)-[~/Desktop/myfile/code/c_code/demo/4]
└─# clang -target armv7a-linux-androideabi21 hello.c -o hello
ld: error: undefined symbol: add
>>> referenced by hello.c
>>>               /tmp/hello-d4ea79.o:(main)

ld: error: undefined symbol: sub
>>> referenced by hello.c
>>>               /tmp/hello-d4ea79.o:(main)
clang: error: linker command failed with exit code 1 (use -v to see invocation)
```

图 16-34　编译结果

回溯到汇编状态查看结果，如图 16-35 所示。bl 段为 0，无法回填这个地址。所以我们要编译一个链接的 SO 文件去链接这个地址，编译的 SO 文件如图 16-36 所示。编译完成后推送到手机中。

```
┌──(root㉿r0env)-[~/Desktop/myfile/code/c_code/demo/4]
└─# objdump_arm -d hello.s

hello.s:     file format elf32-littlearm


Disassembly of section .text:

00000000 <main>:
   0:   e92d4800        push    {fp, lr}
   4:   e1a0b00d        mov     fp, sp
   8:   e24dd020        sub     sp, sp, #32
   c:   e50b0004        str     r0, [fp, #-4]
  10:   e50b1008        str     r1, [fp, #-8]
  14:   e3000001        movw    r0, #1
  18:   e58d000c        str     r0, [sp, #12]
  1c:   e59d100c        ldr     r1, [sp, #12]
  20:   ebfffffe        bl      0 <add>
  24:   e50b000c        str     r0, [fp, #-12]
  28:   e3000003        movw    r0, #3
  2c:   e3001001        movw    r1, #1
  30:   ebfffffe        bl      0 <sub>
  34:   e59f1030        ldr     r1, [pc, #48]   ; 6c <main+0x6c>
  38:   e08f1001        add     r1, pc, r1
  3c:   e58d0010        str     r0, [sp, #16]
  40:   e51b000c        ldr     r0, [fp, #-12]
  44:   e59d2010        ldr     r2, [sp, #16]
  48:   e58d0008        str     r0, [sp, #8]
  4c:   e1a00001        mov     r0, r1
  50:   e59d1008        ldr     r1, [sp, #8]
  54:   ebfffffe        bl      0 <printf>
  58:   e3001000        movw    r1, #0
  5c:   e58d0004        str     r0, [sp, #4]
  60:   e1a00001        mov     r0, r1
  64:   e1a0d00b        mov     sp, fp
  68:   e8bd8800        pop     {fp, pc}
  6c:   0000002c        .word   0x0000002c
```

图 16-35　查看结果

```
┌──(root㉿r0env)-[~/Desktop/myfile/code/c_code/demo/4]
└─# clang -target armv7a-linux-androideabi21 a.c -fPIC -shared -o liba.so
┌──(root㉿r0env)-[~/Desktop/myfile/code/c_code/demo/4]
└─# clang -target armv7a-linux-androideabi21 hello.c -L. -la -Wl,-rpath,./ -o hello
┌──(root㉿r0env)-[~/Desktop/myfile/code/c_code/demo/4]
└─# adb push hello /data/local/tmp
hello: 1 file pushed, 0 skipped. 17.5 MB/s (4296 bytes in 0.000s)
┌──(root㉿r0env)-[~/Desktop/myfile/code/c_code/demo/4]
└─# adb push liba.so
adb: push requires an argument
┌──(root㉿r0env)-[~/Desktop/myfile/code/c_code/demo/4]
└─# adb push liba.so /data/local/tmp
liba.so: 1 file pushed, 0 skipped. 17.9 MB/s (3480 bytes in 0.000s)
```

图 16-36　编译的 SO 文件

使用 objdump 反编译查看结果，如图 16-37 所示。这时就有了 PLT 表，add 和 sub 就有了回填地址。add 和 sub 分别用相对于 main 函数的偏移地址表示。

```
00001470 <main>:
    1470:   e92d4800    push    {fp, lr}
    1474:   e1a0b00d    mov     fp, sp
    1478:   e24dd020    sub     sp, sp, #32
    147c:   e50b0004    str     r0, [fp, #-4]
    1480:   e50b1008    str     r1, [fp, #-8]
    1484:   e3000001    movw    r0, #1
    1488:   e58d000c    str     r0, [sp, #12]
    148c:   e59d100c    ldr     r1, [sp, #12]
    1490:   eb000022    bl      1520 <main+0xb0>
    1494:   e50b000c    str     r0, [fp, #-12]
    1498:   e3000003    movw    r0, #3
    149c:   e3001001    movw    r1, #1
    14a0:   eb000022    bl      1530 <main+0xc0>
    14a4:   e59f1030    ldr     r1, [pc, #48]   ; 14dc <main+0x6c>
    14a8:   e08f1001    add     r1, pc, r1
    14ac:   e58d0010    str     r0, [sp, #16]
    14b0:   e51b000c    ldr     r0, [fp, #-12]
    14b4:   e59d2010    ldr     r2, [sp, #16]
    14b8:   e58d0008    str     r0, [sp, #8]
    14bc:   e1a00001    mov     r0, r1
    14c0:   e59d1008    ldr     r1, [sp, #8]
    14c4:   eb00001d    bl      1540 <main+0xd0>
    14c8:   e3001000    movw    r1, #0
    14cc:   e58d0004    str     r0, [sp, #4]
    14d0:   e1a00001    mov     r0, r1
    14d4:   e1a0d00b    mov     sp, fp
    14d8:   e8bd8800    pop     {fp, pc}
    14dc:   ffffef20    .word   0xffffef20

Disassembly of section .plt:

000014e0 <.plt>:
    14e0:   e52de004    push    {lr}            ; (str lr, [sp, #-4]!)
    14e4:   e28fe600    add     lr, pc, #0, 12
    14e8:   e28eea01    add     lr, lr, #4096   ; 0x1000
    14ec:   e5bef18c    ldr     pc, [lr, #396]! ; 0x18c
    14f0:   d4d4d4d4    .word   0xd4d4d4d4
```

图 16-37　反编译结果

再次使用 GDB 调试工具对 hello 调试，可以看到第一个加载的就是我们自己编译的 liba.so 文件，如图 16-38 所示。

```
gef> b *0xab201470
Breakpoint 1 at 0xab201470
gef> c
Continuing.
Reading /data/local/tmp/liba.so from remote target...
Reading /system/lib/libdl.so from remote target...
Reading /system/lib/libc.so from remote target...
Reading /system/lib/libnetd_client.so from remote target...
Reading /system/lib/libcutils.so from remote target...
Reading /system/lib/libc++.so from remote target...
Reading /system/lib/libm.so from remote target...
Reading /system/lib/liblog.so from remote target...
```

图 16-38　加载结果

接着在 main 中的 add 函数调用处设置断点，如图 16-39 所示。

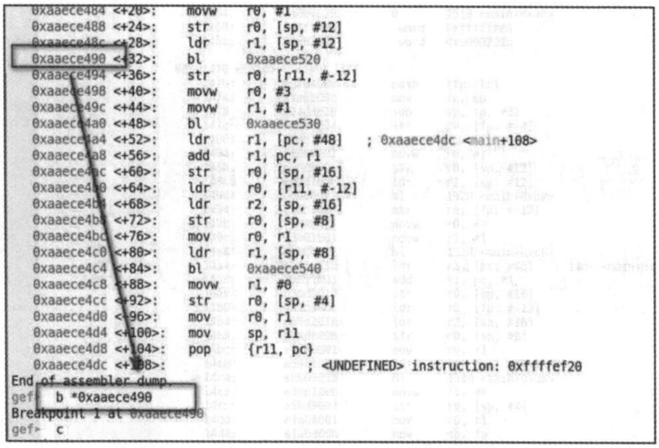

图 16-39 在 add 函数调用处设置断点

进入 add 函数内部，操作如图 16-40 所示。

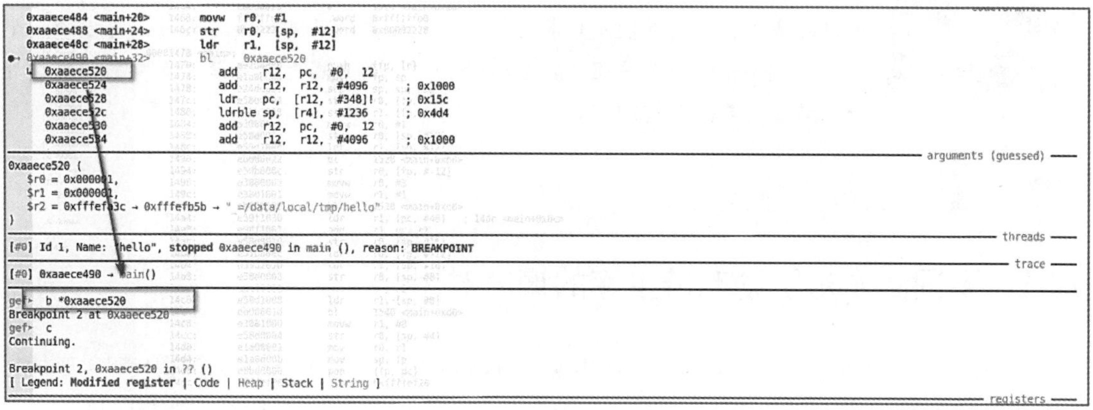

图 16-40 add 函数内部

16.3 本章小结

在本章中，我们对 ELF 文件的格式有了初步的了解，认识了 ELF 文件头、节区表、程序表以及它们的细分字段的含义和作用，并通过项目实战一步一步地观察了文件内部的格式引用关系。本章内容涉及的知识点较多，实战部分也比较复杂，需要读者耐心阅读并进行实践。

第 17 章 高版本 Android 函数地址索引彻底解决方案

第 16 章介绍了 ELF 文件格式以及 ARM 可执行文件的生成过程。本章将着手解决 dlopen 高版本无法加载白名单外 SO 文件的问题。dlopen 是一个动态加载的库函数，其功能是以指定模式打开指定的动态链接库文件，并返回一个句柄给 dlsym 的调用进程。使用 dlclose 可以卸载已打开的库。本章的内容读者无须实操，看懂原理即可。

17.1 不同版本对于动态链接库的调用对比

首先，新建一个项目，在 native-lib.cpp 中编写如图 17-1 所示的测试代码，测试代码主要是对高版本的库函数的引用和低版本的库函数的引用进行对比。此代码分别打开了低版本的 libc.so 库和高版本的 libart.so 库。

```
void* libcdlopen = dlopen( filename: "libc.so", flag: RTLD_NOW);
void* puts = dlsym(libcdlopen, symbol: "fputs");
// 日志打印
__android_log_print( prio: 6, tag: "r0ysue", fmt: "%s",strerror(errno));
void* libartdlopen = dlopen( filename: "libart.so", flag: RTLD_NOW);
void* artFindNativeMethod=dlsym(libartdlopen, symbol: "artFindNativeMethod");
// 日志打印
__android_log_print( prio: 6, tag: "r0ysue", fmt: "%s",strerror(errno));
```

图 17-1 测试代码

运行代码后，结果如图 17-2 所示，低版本的 libc.so 库可以正常打开，而高版本的 libart.so 库显示 No such file or directory。出现这样的结果是因为在较高的 Android 版本上，dlopen 不能直接调用打开库。在这种情况下，对于调用和 Hook 都产生了很大的影响。所以我们需要通过一些手段来

第 17 章 高版本 Android 函数地址索引彻底解决方案

绕过这个问题，直接从 SO 文件中获取其符号地址。

```
2022-03-16 20:54:41.023 581-581/com.roysue.fakedlsym E/r0ysue: Socket operation on non-socket
2022-03-16 20:54:41.024 581-581/com.roysue.fakedlsym E/r0ysue: No such file or directory
```

图 17-2 运行结果

17.2 高版本加载 SO 文件

17.2.1 自定义库查看库函数的偏移

对于无法加载库的情况，网上有很多脚本可用。我们直接把脚本封装成了 .cpp 文件，通过这种方式来加载高版本的库。首先在项目中导入封装好的 .cpp 文件，如图 17-3 所示，并在 CMakeList.txt 中导入。

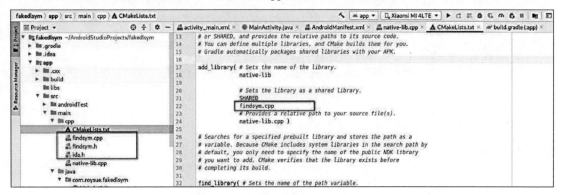

图 17-3 导入自定义库

我们先定义一个函数，如图 17-4 所示。

```
int a = findsym("/system/lib64/libart.so","artFindNativeMethod");
__android_log_print( prio: 6, tag: "r0ysue", fmt: "%p",a);
```

图 17-4 定义函数

查看库函数的偏移，如图 17-5 所示。

```
/com.r0ysue.fakedlsym E/r0ysue: 0x517770
```

图 17-5 查看库函数的偏移

这里的 libart.so 并不是谷歌官方镜像中的原版，而是笔者为了自己刷源码编译的 libart.so，因为 libart.so 中的函数都不允许外部执行，所以自己编译了一个，它包括一些自定义的函数，以方便讲解。

17.2.2 自定义库实现的背景

编译的函数名为 mygetmethodname，首先演示如何编译源码刷自定义函数。源码目录：

aosp819r1/art/runtime/art_method.cc

自定义函数的逻辑比较简单，参数类型为 ArtMethod，返回值是一个 inline 函数。后面的章节会带领读者刷源码，本章主要带领读者体验一下。

```
extern "C" const char* mygetmethodname(ArtMethod* m)REQUIRES_SHARED(Locks::mutator_lock_){
    return m->GetName();
}
```

我们要找到这个函数，先要找到地址，可直接在内存中查找，如图 17-6 所示。

```
sailfish:/ # cat /proc/4383/maps |grep libart.so
7ddc615000-7ddcd86000 r--p 00000000 103:12 1824      /system/lib64/libart.so
7df39d0000-7df3fda000 r-xp 00000000 103:12 1824      /system/lib64/libart.so
7df3fdb000-7df3feb000 r--p 0060a000 103:12 1824      /system/lib64/libart.so
7df3feb000-7df3fee000 rw-p 0061a000 103:12 1824      /system/lib64/libart.so
```

图 17-6 查找函数偏移

当然，也可以使用代码遍历名称来匹配首地址，首地址加上函数的偏移就是函数地址，代码如图 17-7 所示。

```
char line[1024];
int *start;
int *end;
int n=1;
FILE *fp=fopen("/proc/self/maps","r");
while (fgets(line, sizeof(line), fp) {
if (strstr(line, "linker64") ) {
    __android_log_print(6,"r0ysue","%s", line);
    if(n==1){
    start = reinterpret_cast<int *>(strtoul(strtok(line, "-"), NULL, 16));
    end = reinterpret_cast<int *>(strtoul(strtok(NULL, " "), NULL, 16));
    }
    else{
    strtok(line, "-");
    end = reinterpret_cast<int *>(strtoul(strtok(NULL, " "), NULL, 16));
    }
    n++;
}
}
```

图 17-7 代码遍历获取地址

接下来尝试获取这个函数地址，并输出函数名。在第 16 章的 ELF 中讲到了首地址就是段的加载地址，我们遍历获取地址后，减去 0x25000 就是函数的真实地址。如图 17-8 所示，libart.so 的首地址 0x25000 就是段的加载地址。

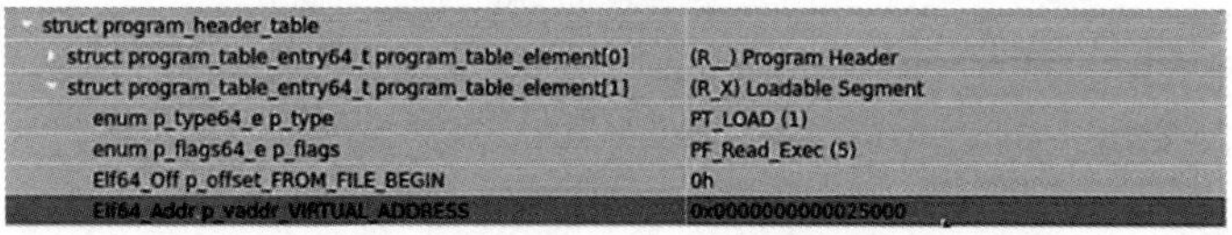

第 17 章　高版本 Android 函数地址索引彻底解决方案　369

图 17-8　ibart.so 的首地址 0x25000 就是段的加载地址

获取 mygetmethodname 的真实地址：

```
long* mygetmethodname = reinterpret_cast<long *>((char *) start + a -0x25000); // 获
取函数的真实地址
```

下一步是寻找 ArtMethod，这一步比较难。我们可以看一下源码是如何解析的，源码如图 17-9 所示。

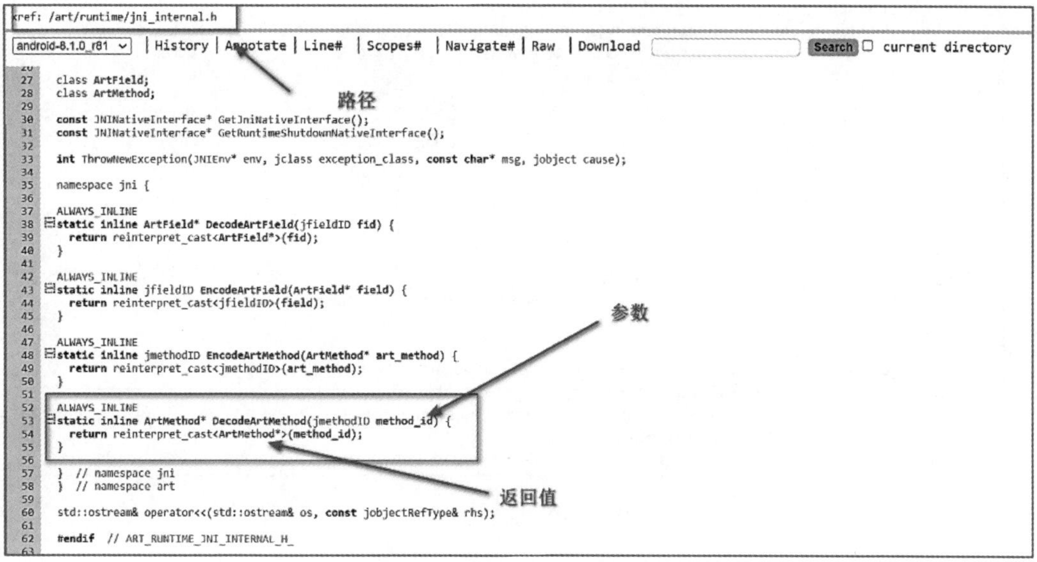

图 17-9　ArtMethod 解析

可以清楚地看到 EncodeArtMethod 是一个 inline 函数，它对 jmethodID 类型的参数直接通过 art_method 类型返回，这就说明 jmethodID 和 art_method 是一样的，我们只要获取 jmethodID 就可以获取 art_method。

使用反射的方式获取 jmethodID：

```
// MainActivity.java
public int testMehtod(){ return 1;}
// mative-lib.cpp
// 获取自定义的 testMehtod
jclass myclass=env->FindClass("com/roysue/fakedlsym/MainActivity");
jmethodID testMehtod = env->GetMethodID(myclass,"testMehtod","()I");
```

完整的代码如图 17-10 所示。

```cpp
typedef const char* (*getartmethodname)(void*);

extern "C" JNIEXPORT jstring JNICALL
Java_com_roysue_fakedlsym_MainActivity_stringFromJNI(
        JNIEnv* env,
        jobject /* this */) {
    std::string hello = "Hello from C++";

    char line[1024];
    int *start;  // 首地址
    int *end;    // 末地址
    int n=1;
    FILE *fp=fopen( path: "/proc/self/maps", mode: "r");
    while (fgets(line, sizeof(line), fp)) {
        if (strstr(line, n: "libart.so") ) {
            __android_log_print( prio: 6, tag: "r0ysue", fmt: "%s", line);
            if(n==1){
                start = reinterpret_cast<int *>(strtoul( s: strtok(line, delimiter: "-"), end_ptr: NULL, base: 16));
                end = reinterpret_cast<int *>(strtoul( s: strtok( s: NULL, delimiter: " "), end_ptr: NULL, base: 16));
            }
            else{
                strtok(line, delimiter: "-");
                end = reinterpret_cast<int *>(strtoul( s: strtok( s: NULL, delimiter: " "), end_ptr: NULL, base: 16));
            }
            n++;
        }
    }

    int a = findsym("/system/lib64/libart.so","mygetmethodname");
    long* mygetmethodname = reinterpret_cast<long *>((char *) start + a -0x25000); // 获取函数的真实地址

    // 获取自定义的 testMehtod
    jclass myclass=env->FindClass( name: "com/roysue/fakedlsym/MainActivity");
    jmethodID testMehtod = env->GetMethodID(myclass, name: "testMehtod", sig: "()I");
    getartmethodname aaa = reinterpret_cast<getartmethodname>(mygetmethodname);
    const char* methodname1 = aaa(testMehtod);
    __android_log_print( prio: 6, tag: "r0ysue", fmt: "%s",methodname1);
```

图 17-10 完整的代码

从图 17-11 中可以看到结果输出了我们自定义的 testmethod。

```
E/r0ysue: 7df39d0000-7df3fda000 r-xp 00000000 103:12 1824
E/r0ysue: 7df3fdb000-7df3feb000 r--p 0060a000 103:12 1824
E/r0ysue: 7df3feb000-7df3fee000 rw-p 0061a000 103:12 1824
E/r0ysue: testmethod
```

图 17-11 结果输出

至此，我们可以总结一下 ART 的调用原则：先定义一个类型，把地址赋值给一个定义好的变量，再强制转换。

17.2.3 自定义库 findsym 的实现

接下来实现一个函数，找到我们自定义的 mygetmethodname 函数，本质就是模仿 dlopen 函数和 dlsym 函数，dlopen 函数 libart.so 已经帮我们加载到内存中了，其实只需要模仿 dlsym 函数即可。dlopen 函数和 dlsym 函数的最终实现都在 linker 中，如图 17-12 所示。

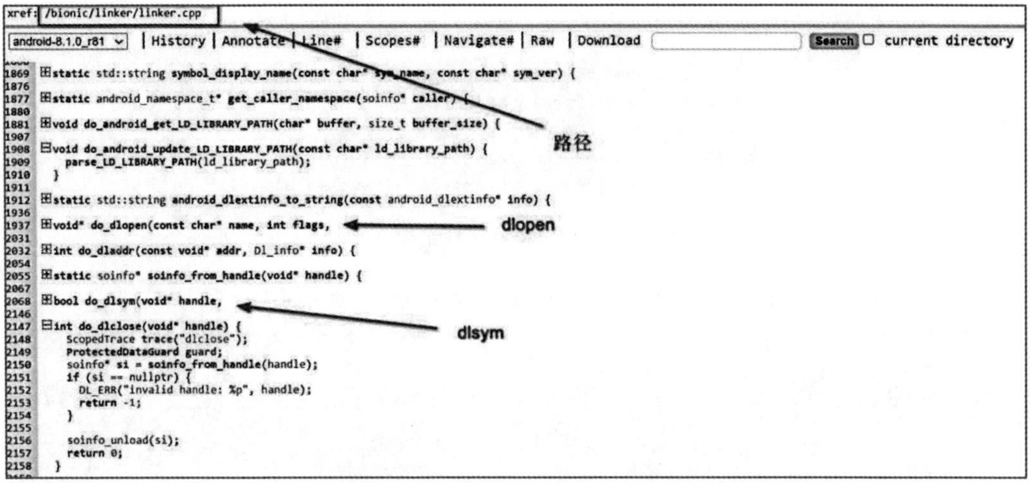

第 17 章　高版本 Android 函数地址索引彻底解决方案

```
xref: /bionic/linker/linker.cpp
android-8.1.0_r81    History  Annotate  Line#  Scopes#  Navigate#  Raw  Download          Search   current directory
1869  static std::string symbol_display_name(const char* sym_name, const char* sym_ver) {
1876
1877  static android_namespace_t* get_caller_namespace(soinfo* caller) {
1881  void do_android_get_LD_LIBRARY_PATH(char* buffer, size_t buffer_size) {                    路径
1908  void do_android_update_LD_LIBRARY_PATH(const char* ld_library_path) {
1909     parse_LD_LIBRARY_PATH(ld_library_path);
1910  }
1911
1912  static std::string android_dlextinfo_to_string(const android_dlextinfo* info) {
1936
1937  void* do_dlopen(const char* name, int flags,                          dlopen
2031
2032  int do_dladdr(const void* addr, Dl_info* info) {
2054
2055  static soinfo* soinfo_from_handle(void* handle) {
2067
2068  bool do_dlsym(void* handle,
2146
2147  int do_dlclose(void* handle) {                                         dlsym
2148     ScopedTrace trace("dlclose");
2149     ProtectedDataGuard guard;
2150     soinfo* si = soinfo_from_handle(handle);
2151     if (si == nullptr) {
2152       DL_ERR("invalid handle: %p", handle);
2153       return -1;
2154     }
2155
2156     soinfo_unload(si);
2157     return 0;
2158  }
```

图 17-12　dlopen 函数和 dlsym 函数的最终实现

函数调用最终还是要回归 ELF 的文件格式，ELF 文件有 ELF 文件头、程序头表、节区头表，最重要的是程序头表，节区头表在后面的加壳中会再详细介绍。dlsym 的最终实现和程序头表有很大的关系。其中 number 和 offset 关系到 dlsym 的解析。

寻找字符串表和符号表，我们可以编写程序来读取这些偏移的值，代码如图 17-13 所示。

```
int findSymoff(){
    char line[1024];
    int *start;  // 首地址
    int *end;    // 末地址
    int n=1;
    FILE *fp=fopen( path: "/proc/self/maps", mode: "r");
    while (fgets(line, sizeof(line), fp)) {
        if (strstr(line, n: "libart.so") ) {
            __android_log_print( prio: 6, tag: "r0ysue", fmt: "%s", line);
            if(n==1){
                start = reinterpret_cast<int *>(strtoul( s: strtok(line, delimiter: "-"), end_ptr: NULL, base: 16));
                end = reinterpret_cast<int *>(strtoul( s: strtok( s: NULL, delimiter: " "), end_ptr: NULL, base: 16));
            }
            else{
                strtok(line, delimiter: "-");
                end = reinterpret_cast<int *>(strtoul( s: strtok( s: NULL, delimiter: " "), end_ptr: NULL, base: 16));
            }
            n++;
        }
    }
    // 导入elf.h
    Elf64_Ehdr header;
    memcpy(&header,start,sizeof(Elf64_Ehdr));
    // 找到程序头表的偏移
    int phoff = header.e_phoff;
    int phsize = header.e_phentsize;
    int phnum = header.e_phnum;

    Elf64_Phdr myphdr;
    memcpy(&myphdr, src: (char*)start+phoff,sizeof(Elf64_Phdr));
    for(int n=0;n<phnum;n++){
        memcpy(&myphdr, src: (char*)(start)+header.e_phoff+sizeof(Elf64_Phdr)*n,sizeof(Elf64_Phdr));
        __android_log_print( prio: 6, tag: "r0ysue", fmt: "%d",myphdr.p_type);
    }
}
```

图 17-13　读取偏移值的代码

这段代码读取到了程序头表中每个段的 type 取值，读取结果如图 17-14 所示。

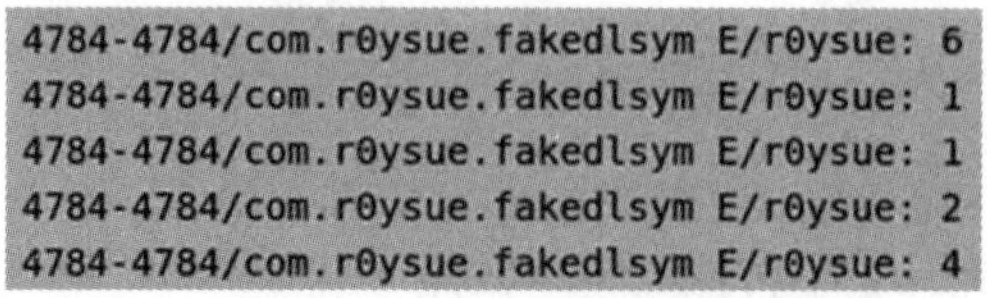

```
4784-4784/com.r0ysue.fakedlsym E/r0ysue: 6
4784-4784/com.r0ysue.fakedlsym E/r0ysue: 1
4784-4784/com.r0ysue.fakedlsym E/r0ysue: 1
4784-4784/com.r0ysue.fakedlsym E/r0ysue: 2
4784-4784/com.r0ysue.fakedlsym E/r0ysue: 4
```

图 17-14　偏移读取结果

在 010 Editor 中查看 libart.so 的程序头表 type 字段的取值，如图 17-15 所示。

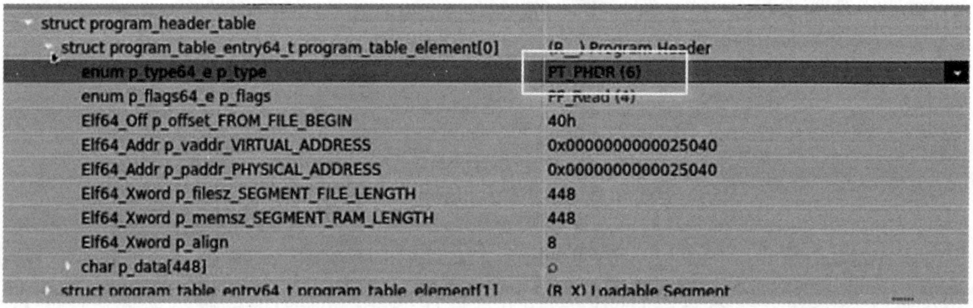

图 17-15　010 Editors 查看 type 取值

17.3　SO 符号地址寻找

在 17.2 节中展示了用程序头表的方式获得符号地址，本节用另外两种方式展示 SO 符号地址的获取。

17.3.1　通过节头获取符号地址

通过 ELF 头结构也可以找到节头的地址，也就是 e_shoff，节头表比程序头表友好许多，它的项非常多，唯一不好的一点是它不会加载到内存中，所以执行视图中没有 e_shoff，只能通过绝对路径找到它，手动解析文件并加载，以此来得到我们想要的值。节头解析主要是搬抄源码来魔改，创新就是模仿加修改。

加载文件并映射到内存中，代码如下：

```
int fd;
void *start;
struct stat sb;
fd = open(lib, O_RDONLY);
fstat(fd, &sb);
start = mmap(NULL, sb.st_size, PROT_READ, MAP_PRIVATE, fd, 0);
```

在这种解析方式中，在 ELF 头中需要的值是 e_shoff、e_shentsize、e_shnum、e_shstrndx，即

第 17 章　高版本 Android 函数地址索引彻底解决方案

节头表偏移、节大小、节个数、节头表字符串，这些值都在 elf header 中有体现，不过我们的最终目标是拿到符号表和字符串表，也就是如图 17-16 所示的 symtab 和 strtab 中的 sh_offset。

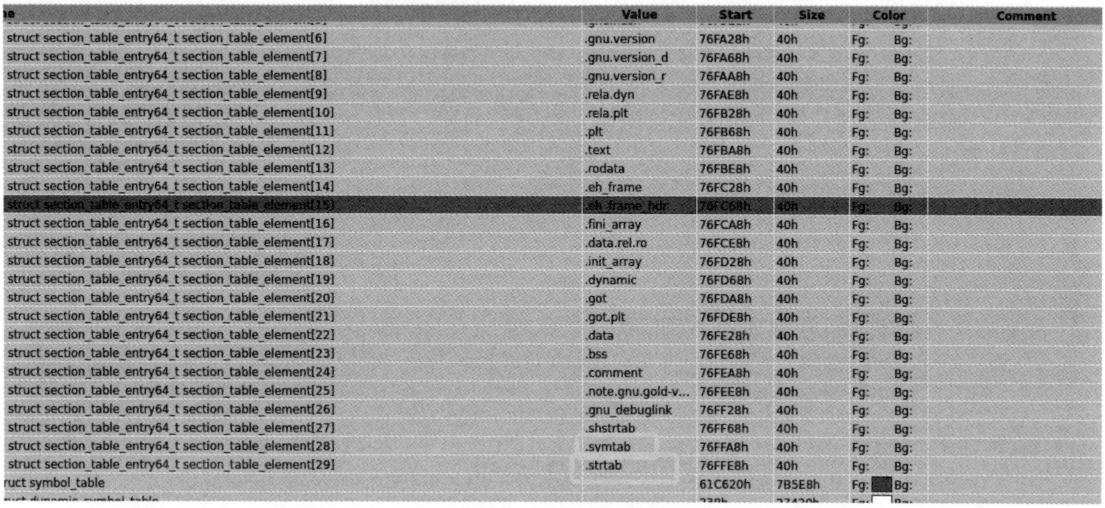

图 17-16　symtab 和 strtab

通过自定义代码来遍历获取要查找的项目，代码如下：

```
Elf64_Ehdr header;
memcpy(&header, start, sizeof(Elf64_Ehdr));
int secoff = header.e_shoff;
int secsize = header.e_shentsize;
int secnum = header.e_shnum;
int secstr = header.e_shstrndx;
Elf64_Shdr strtab;
memcpy(&strtab, (char *) start + secoff + secstr * secsize, sizeof(Elf64_Shdr));
int strtaboff = strtab.sh_offset;
char strtabchar[strtab.sh_size];

memcpy(&strtabchar, (char *) start + strtaboff, strtab.sh_size);
Elf64_Shdr enumsec;
int symoff = 0;
int symsize = 0;
int strtabsize = 0;
int stroff = 0;
for (int n = 0; n < secnum; n++) {

    memcpy(&enumsec, (char *) start + secoff + n * secsize, sizeof(Elf64_Shdr));

    if (strcmp(&strtabchar[enumsec.sh_name], ".symtab") == 0) {
```

```
            symoff = enumsec.sh_offset;
            symsize = enumsec.sh_size;

        }
        if (strcmp(&strtabchar[enumsec.sh_name], ".strtab") == 0) {
            stroff = enumsec.sh_offset;
            strtabsize = enumsec.sh_size;

        }


}
```

最后遍历符号表即可获取项的物理偏移,代码如下:

```
int realoff=0;
char relstr[strtabsize];
Elf64_Sym tmp;
memcpy(&relstr, (char *) start + stroff, strtabsize);

for (int n = 0; n < symsize; n = n + sizeof(Elf64_Sym)) {
    memcpy(&tmp, (char *)start + symoff+n, sizeof(Elf64_Sym));
    if(tmp.st_name!=0&&strstr(relstr+tmp.st_name,sym)){
        realoff=tmp.st_value;
        break;
    }
}
return realoff;
```

17.3.2　模仿 Android 通过哈希寻找符号

这种方式就是 dlsym 的官方写法,由于 libart.so 会自动加载到内存中,因此不需要 dlopen,我们只需要在 map 中找到它的首地址即可,代码和 17.2 节的代码相同,读者可自行翻阅。这里我们主要来看官方是如何实现的,一路追踪 do_dlopen,最终找到了 soinfo::gnu_lookup 函数,这里面是 do_dlopen 的主要实现逻辑,我们只需要实现它即可。这里多了 4 项之前没有提到,就是它的导出表的 4 项,所以这种方法只能找到导出表中的函数或者变量。

在代码中可以看到 dd.d_tag==5 和 dd.d_tag==6,5 代表的是字符串表,而 6 代表的是符号表,这里的任务就是把这两张表筛选出来。

```
size_t gnu_nbucket_ = 0;
    // skip symndx
    uint32_t gnu_maskwords_ = 0;
    uint32_t gnu_shift2_ = 0;
    ElfW(Addr) *gnu_bloom_filter_ = nullptr;
    uint32_t *gnu_bucket_ = nullptr;
```

```cpp
            uint32_t *gnu_chain_ = nullptr;
            int phof = 0;
            Elf64_Ehdr header;
            memcpy(&header, startr, sizeof(Elf64_Ehdr));
            uint64 rel = 0;
            size_t size = 0;
            long *plt = nullptr;
            char *strtab_ = nullptr;
            Elf64_Sym *symtab_ = nullptr;
            Elf64_Phdr cc;
            memcpy(&cc, ((char *) (startr) + header.e_phoff), sizeof(Elf64_Phdr));
            for (int y = 0; y < header.e_phnum; y++) {
                memcpy(&cc, (char *) (startr) + header.e_phoff + sizeof(Elf64_Phdr) * y,
                    sizeof(Elf64_Phdr));
                if (cc.p_type == 6) {
                    phof = cc.p_paddr - cc.p_offset;//改用程序头的偏移获得首段偏移, 使用之前的方法也
行
                }
            }
            for (int y = 0; y < header.e_phnum; y++) {
                memcpy(&cc, (char *) (startr) + header.e_phoff + sizeof(Elf64_Phdr) * y,
                    sizeof(Elf64_Phdr));
                if (cc.p_type == 2) {
                    Elf64_Dyn dd;
                    for (y = 0; y == 0 || dd.d_tag != 0; y++) {
                        memcpy(&dd, (char *) (startr) + cc.p_offset + y * sizeof(Elf64_Dyn) +
0x1000,
                            sizeof(Elf64_Dyn));

                        if (dd.d_tag == 0x6ffffef5) {//0x6ffffef5 为导出表项

                            gnu_nbucket_ = reinterpret_cast<uint32_t *>((char *) startr +
dd.d_un.d_ptr -
                                                                        phof)[0];
                            // skip symndx
                            gnu_maskwords_ = reinterpret_cast<uint32_t *>((char *) startr +
dd.d_un.d_ptr -
                                                                        phof)[2];
                            gnu_shift2_ = reinterpret_cast<uint32_t *>((char *) startr +
dd.d_un.d_ptr -
                                                                        phof)[3];

                            gnu_bloom_filter_ = reinterpret_cast<ElfW(Addr) *>((char *) startr +
                                                                        dd.d_un.d_ptr + 16 -
phof);
                            gnu_bucket_ = reinterpret_cast<uint32_t *>(gnu_bloom_filter_ +
```

```cpp
gnu_maskwords_);
                // amend chain for symndx = header[1]
                gnu_chain_ = reinterpret_cast<uint32_t *>( gnu_bucket_ +
                                                gnu_nbucket_ -
                                                reinterpret_cast<uint32_t *>(
                                                    (char *) startr +
                                                    dd.d_un.d_ptr -
phof)[1]);

            }
            if (dd.d_tag == 5) {
                strtab_ = reinterpret_cast< char *>((char *) startr + dd.d_un.d_ptr
- phof);
            }
            if (dd.d_tag == 6) {
                symtab_ = reinterpret_cast<Elf64_Sym *>((
                    (char *) startr + dd.d_un.d_ptr - phof));
            }

        }
      }
    }
```

之后模仿 gnu_lookup 函数即可，hashmap 的查询方法如下：

```cpp
char* name_=symname;//直接抄的Android源码
  uint32_t h = 5381;
  const uint8_t* name = reinterpret_cast<const uint8_t*>(name_);
  while (*name != 0) {
      h += (h << 5) + *name++; // h*33 + c = h + h * 32 + c = h + h << 5 + c
  }
  int index=0;
  uint32_t h2 = h >> gnu_shift2_;
  uint32_t bloom_mask_bits = sizeof(ElfW(Addr))*8;
  uint32_t word_num = (h / bloom_mask_bits) & gnu_maskwords_;
  ElfW(Addr) bloom_word = gnu_bloom_filter_[word_num];
  n = gnu_bucket_[h % gnu_nbucket_];
  do {
     Elf64_Sym * s = symtab_ + n;
     char * sb=strtab_+ s->st_name;
     if (strcmp(sb ,reinterpret_cast<const char *>(name_)) == 0 ) {
        break;
     }
  } while ((gnu_chain_[n++] & 1) == 0);
  Elf64_Sym * mysymf=symtab_+n;
  long* finaladdr= reinterpret_cast<long*>(sb->st_value + (char *) start-phof);
  return finaladdr;
```

17.4　本章小结

在本章中，我们了解了高版本的库函数不会主动加载函数索引地址的问题，在第 17.2 节中，提供了 3 种不同的解决方案，虽然它们都比较简单，但是这些解决方案的思想可以帮我们理解和解决逆向过程的问题。实际上，这些函数也可以使用自定义加载方式来加载，从而绕过其自身的加载逻辑，这就涉及后面的 GOT Hook 和 PLT Hook 技术。

第 18 章

从 findExportByName
源码分析到 anti-frida 新思路

本章从 Frida 源码的角度出发，结合前面对 ELF 知识的讲解，为读者提供了一个新的 anti-frida 思路。在第 16 章和第 17 章，读者应该对 ELF 的各种文件头和文件的内部结构有了基本的认知，并通过 Android 源码封装了一个 findsym 工具包。在本章中，我们将使用该工具包，让它的功能大放异彩。

Frida 存在两种操作模式：一种是 attach，另一种是 spawn。attach 模式用于在应用启动后进行代码的注入，而 spawn 模式则将启动的控制权交给 Frida。采用 spawn 模式时，即使应用已经启动，Frida 也会重新启动应用。Frida 通过加上 -f 参数指定包名，以 spawn 模式操作 App。Frida 的 Hook 操作都是基于 linker 进行的，因此加载行为也是通过 linker 来实现的。

18.1 两种模式下 anti-frida 的演示

本节为读者演示如何通过两种模式来 anti-frida。

18.1.1 Frida attach 模式下的 anti-frida

新建工程 anti-frida，在 native-lib.cpp 中编写代码（参照代码清单 18-1 的代码），并在主函数中调用。在 ELF 加载时，Frida 对 ELF 的 magic 校验不是很强，在这里我们对 magic 异或，抹掉头部校验。直接修改内存中的保护属性是禁止的，这时就可以使用 mprotect 来解除保护，修改内存中某一指定区域的属性。

代码清单 18-1　anti_1

```c
void anti_1(){
```

第 18 章 从 findExportByName 源码分析到 anti-frida 新思路

```cpp
    char line[1024];
    int *start;
    int *end;
    int n=1;
    FILE *fp=fopen("/proc/self/maps","r");
    while (fgets(line, sizeof(line), fp)) {
        if (strstr(line, "linker64") ) {
            __android_log_print(6,"r0ysue","%s", line);
            if(n==1){
                start = reinterpret_cast<int *>(strtoul(strtok(line, "-"), NULL, 16));
                end = reinterpret_cast<int *>(strtoul(strtok(NULL, " "), NULL, 16));
            }
            else{
                strtok(line, "-");
                end = reinterpret_cast<int *>(strtoul(strtok(NULL, " "), NULL, 16));
            }
            n++;
        }
    }

    mprotect(start,PAGE_SIZE,PROT_WRITE|PROT_READ|PROT_EXEC);
    __android_log_print(6,"r0ysue","%p", *(long*)start);
    *(long*)start=*(long*)start^0x7f;
    __android_log_print(6,"r0ysue","%p", *(long*)start);
}
```

代码编写完成后,在手机中运行 App。然后在手机中启动 frida-server,如图 18-1 所示。

```
┌──(root㉿r0env)-[~]
└─# adb shell
bullhead:/ $ su
bullhead:/ # cd /data/local/tmp
bullhead:/data/local/tmp # ./fr
frida-server-12.8.0-android-arm64        frida-server-14.2.17-android-arm64
bullhead:/data/local/tmp # ./frida-server-12.8.0-android-arm64
```

图 18-1　启动 frida-server

在 attach 模式下,使用 Frida 对此 App 进行 Hook,操作如图 18-2 所示。

```
┌──(root㉿r0env)-[~]
└─# frida -UF
     ____
    / _  |   Frida 12.8.0 - A world-class dynamic instrumentation toolkit
   | (_| |
    > _  |   Commands:
   /_/ |_|       help      -> Displays the help system
   . . . .       object?   -> Display information about 'object'
   . . . .       exit/quit -> Exit
   . . . .
   . . . .   More info at https://www.frida.re/docs/home/

Failed to attach: timeout was reached
```

图 18-2　在 attach 模式下启动 Frida

Android Studio 报错，如图 18-3 所示。在手机端，App 直接崩溃，如图 18-4 所示。

```
A/Frida: Unable to locate the Android linker; please file a bug

--------- beginning of crash
A/libc: Fatal signal 6 (SIGABRT), code -6 in tid 12581 (sue.antifrida_7), pid 11518 (sue.antifrida_7)
```

图 18-3　Android 报错截图

图 18-4　App 崩溃

通过对头部的处理，做到了 anti-frida 的效果，后面的源码分析会看到这样做的原理。Frida 对其他部分的校验比较健全，但是对头部不太看重，这就成为我们操作的突破点。

18.1.2　Frida spawn 模式下的 anti-frida

编写代码（参照清单 18-2 的代码），实现在 spawn 模式下的 anti-frida。在阅读源码的过程中，发现 somain 未校验是否为空。通过 IDA 打开/system/bin/linker64，查找 somain，发现确实有 get_somain 函数可以获取 somain 的值，如图 18-5 所示。

```
E/r0ysue: 72f7664000-72f7758000 r-xp 00000000 103:09 554
E/r0ysue: 72f7759000-72f775d000 r--p 000f4000 103:09 554
E/r0ysue: 72f775d000-72f775e000 rw-p 000f8000 103:09 554
E/r0ysue: 0x72f7655010
```

图 18-5　get_somain

代码清单 18-2　anti_2

```cpp
void anti_2(){
    // 方法2
    // 查看 Frida 源码发现 somainv 未校验是否为空
    // linker 中有一个函数 get_somainv
    char line[1024];
    int *start;
    int *end;
    int n=1;
    FILE *fp=fopen("/proc/self/maps","r");
    while (fgets(line, sizeof(line), fp)) {
        if (strstr(line, "linker64") ) {
            __android_log_print(6,"r0ysue","%s", line);
            if(n==1){
                start = reinterpret_cast<int *>(strtoul(strtok(line, "-"), NULL, 16));
                end = reinterpret_cast<int *>(strtoul(strtok(NULL, " "), NULL, 16));
            }
            else{
                strtok(line, "-");
                end = reinterpret_cast<int *>(strtoul(strtok(NULL, " "), NULL, 16));
            }
            n++;
        }
    }

    // 查看 get_somain 函数的值
    // int getsomainoff = findsym("/system/bin/linker64","__dl__Z17solist_get_somainv");
    // void* somain = ((getsomain)((char*)start+getsomainoff))();
    // __android_log_print(6,"r0ysue","%p", somain);

    int getsomainoff = findsym("/system/bin/linker64","__dl__ZL6somain");
    *(long*)((char*)start+getsomainoff)=0;

    // frida spawn 直接 crash
    // 在结构体为空的情况下调用了里面的值就会崩溃

}
```

在主函数中调用 anti_2()，查看一下 get_somain 返回的值，如图 18-6 所示。

图 18-6 get_somain 返回的值

既然有这个值，那么把这个值置空会实现 anti-frida 的效果吗？即代码清单 18-2 未注释的部分。再次启动，并开启 Frida 的 spawn 模式。操作如图 18-7 所示。

```
└─# frida -U -f com.roysue.antifrida_7 --no-pause
     ____
    / _  |   Frida 12.8.0 - A world-class dynamic instrumentation toolkit
   | (_| |
    > _  |   Commands:
   /_/ |_|       help      -> Displays the help system
   . . . .       object?   -> Display information about 'object'
   . . . .       exit/quit -> Exit
   . . . .
   . . . .   More info at https://www.frida.re/docs/home/
Spawned `com.roysue.antifrida_7`. Resuming main thread!
[LGE Nexus 5X::com.roysue.antifrida_7]-> Module.getBaseAddress("libart.so")
Process crashed: Bad access due to invalid address

***
*** *** *** *** *** *** *** *** *** *** *** *** *** *** *** ***
Build fingerprint: 'google/bullhead/bullhead:8.1.0/OPM7.181205.001/5080180:user/release-keys'
Revision: 'rev_1.0'
ABI: 'arm64'
pid: 16769, tid: 16787, name: gum-js-loop  >>> com.roysue.antifrida_7 <<<
signal 11 (SIGSEGV), code 1 (SEGV_MAPERR), fault addr 0x1a0
Cause: null pointer dereference
    x0   0000000000000000  x1   0000000000000000  x2   000000725d2c9f38  x3   000000725b1cd758
    x4   0000000000000008  x5   8080808080808000  x6   fefefefefefeff6e  x7   7f7f7f7f7f7f7f7f
    x8   00000072f7683e5c  x9   eb2fb91fdad717fa  x10  0000000000000000  x11  0000000000004193
    x12  00000072f3dae098  x13  00000072f30a71e0  x14  0000000000000000  x15  0000000000000006
    x16  000000725d2a4ae8  x17  00000072f5f3371c  x18  0000000000000000  x19  000000725d2c9f38
    x20  000000725b1cd738  x21  0000000000000000  x22  000000725b1cd758  x23  0000000000000000
    x24  0000000000000001  x25  0000000000000006  x26  0000000000000000  x27  0000000000000280
    x28  0000000000000000  x29  000000725b1cd720  x30  000000725c25abd0
    sp   000000725b1cd6f0  pc   00000072f7683e5c  pstate 0000000060000000

backtrace:
    #00 pc 000000000001fe5c  /system/bin/linker64 (__dl__ZNK6soinfo12get_realpathEv)
    #01 pc 0000000000235bcc  /data/local/tmp/re.frida.server/frida-agent-64.so
***
[LGE Nexus 5X::com.roysue.antifrida_7]->

Thank you for using Frida!
```

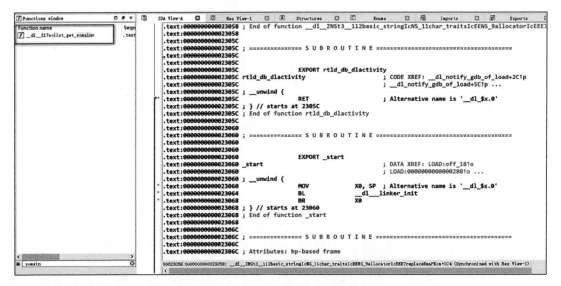

图 18-7 在 spawn 模式下启动 App

在图 18-7 中可以看到，抛出了 signal 11 的异常。signal 11 代表空指针异常。这就是因为我们把 somain 置 0 了,而 Frida 要引用 somain 中的结构体,找不到就抛出了空指针异常。下一节将通过源码来剖析两种 anti-frida 的新思路。

18.2 源码分析

下面以 findExportByName 为入口点进入源码的分析。首先查看哪些源码中定义了此文件,查看方式如图 18-8 所示。

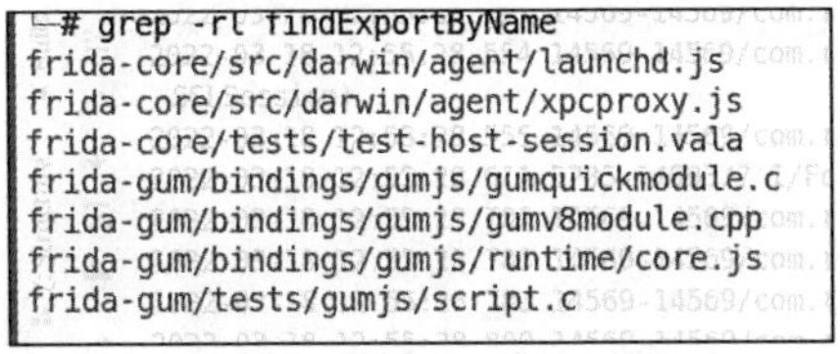

```
~# grep -rl findExportByName
frida-core/src/darwin/agent/launchd.js
frida-core/src/darwin/agent/xpcproxy.js
frida-core/tests/test-host-session.vala
frida-gum/bindings/gumjs/gumquickmodule.c
frida-gum/bindings/gumjs/gumv8module.cpp
frida-gum/bindings/gumjs/runtime/core.js
frida-gum/tests/gumjs/script.c
```

图 18-8 查找定义的文件

直接查看 C 文件,有 gumquickmodule.c 和 gumv8module.cpp 两处定义了。具体使用的是哪个定义呢?我们直接在源码中写日志,编译源码,看一下使用的是哪个文件。首先打开两个文件,并查看定义,如图 18-9 所示。

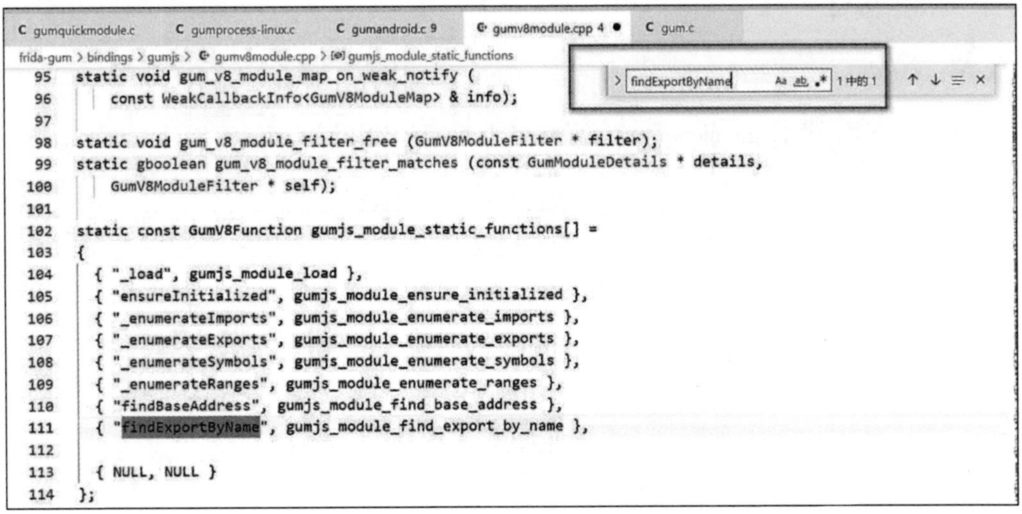

```
 95  static void gum_v8_module_map_on_weak_notify (
 96      const WeakCallbackInfo<GumV8ModuleMap> & info);
 97
 98  static void gum_v8_module_filter_free (GumV8ModuleFilter * filter);
 99  static gboolean gum_v8_module_filter_matches (const GumModuleDetails * details,
100      GumV8ModuleFilter * self);
101
102  static const GumV8Function gumjs_module_static_functions[] =
103  {
104      { "_load", gumjs_module_load },
105      { "ensureInitialized", gumjs_module_ensure_initialized },
106      { "_enumerateImports", gumjs_module_enumerate_imports },
107      { "_enumerateExports", gumjs_module_enumerate_exports },
108      { "_enumerateSymbols", gumjs_module_enumerate_symbols },
109      { "_enumerateRanges", gumjs_module_enumerate_ranges },
110      { "findBaseAddress", gumjs_module_find_base_address },
111      { "findExportByName", gumjs_module_find_export_by_name },
112
113      { NULL, NULL }
114  };
```

图 18-9 查看定义

找到声明函数的地方,并加入日志打印,如图 18-10 所示。对另一个文件进行同样的操作,日志打印显示 quick 即可。

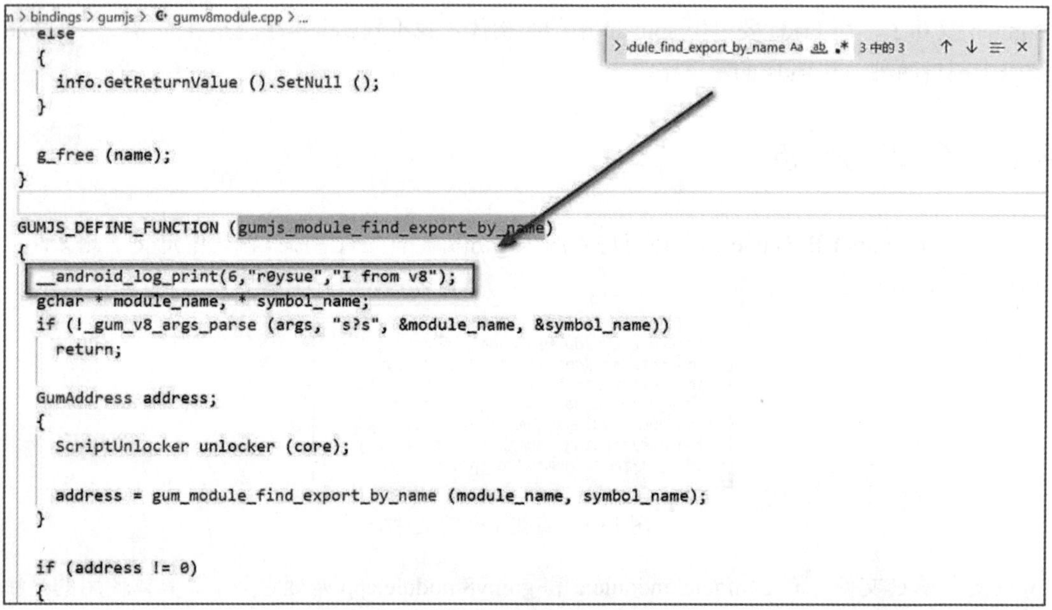

图 18-10　加入日志

18.2.1　Frida 编译

编译过程如下：

（1）源码下载。直接 git clone 不会下载引用文件，我们采用递归下载，具体方法如图 18-11 所示。

图 18-11　源码下载

（2）NDK 下载，最新的源码要求的 NDK 版本是 r24，请读者自行下载安装。

（3）加入环境变量，如图 18-12 所示。

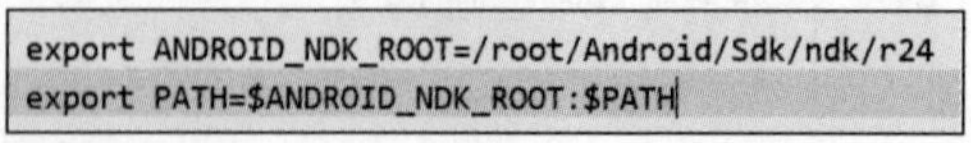

图 18-12　NDK 加入环境变量

（4）输入 proxychains make core-android-arm64 开始编译。

（5）在编译过程中如果遇到 File 重复的问题，如图 18-13 所示，那么在报错路径下更改即可，具体更改如图 18-14 所示。

```
FAILED: bindings/gumjs/gumquickscript-runtime.h bindings/gumjs/gumquickscript-objc.h bindings/gumjs/gumquickscript-java.h bindings/gumjs/gumv8script-runtime.h bindings/gumjs/gumv8script-obj
c.h bindings/gumjs/gumv8script-java.h bindings/gumjs/gumcmodule-runtime.h
/root/Desktop/myfile/sourceCode/frida/build/tmp-android-arm/frida-gum/bindings/gumjs/generate-runtime.py arm /root/Desktop/myfile/sourceCode/frida /root/Desktop/myfile/sourceCode/f
rida-gum /root/Desktop/myfile/sourceCode/frida/build/sdk-android-arm/include/capstone /root/Desktop/myfile/sourceCode/frida/build/sdk-android-arm/lib/tcc/include /root/Desktop/myfile/so
urceCode/frida/build/tmp-android-arm/frida-gum/bindings/gumjs/quickcompile /root/Desktop/myfile/sourceCode/frida/build/tmp-android-arm/frida-gum/bindings/gumjs
[proxychains] DLL init: proxychains-ng 4.15
[proxychains] DLL init: proxychains-ng 4.15
[proxychains] DLL init: proxychains-ng 4.15
[proxychains] DLL init: proxychains-ng 4.15
TypeScript error: /root/Desktop/myfile/sourceCode/frida/build/tmp-android-arm/frida-gum/bindings/gumjs/node_modules/typescript/lib/lib.dom.d.ts(5058,11): Error TS2300: Duplicate identifier
'File'.
fileName: /root/Desktop/myfile/sourceCode/frida/build/tmp-android-arm/frida-gum/bindings/gumjs/node_modules/typescript/lib/lib.dom.d.ts:
 line: 5058,
 column: 11,
 inputs: Set(7) {
   '/root/Desktop/myfile/sourceCode/frida/build/tmp-android-arm/frida-gum/bindings/gumjs/runtime/entrypoint-quickjs.js',
   '/root/node_modules/@types/frida-gum/index.d.ts',
   '/root/Desktop/myfile/sourceCode/frida/build/tmp-android-arm/frida-gum/bindings/gumjs/node_modules/typescript/lib/lib.d.ts',
   '/root/Desktop/myfile/sourceCode/frida/build/tmp-android-arm/frida-gum/bindings/gumjs/node_modules/typescript/lib/lib.es5.d.ts',
   '/root/Desktop/myfile/sourceCode/frida/build/tmp-android-arm/frida-gum/bindings/gumjs/node_modules/typescript/lib/lib.dom.d.ts',
   '/root/Desktop/myfile/sourceCode/frida/build/tmp-android-arm/frida-gum/bindings/gumjs/node_modules/typescript/lib/lib.webworker.importscripts.d.ts',
   '/root/Desktop/myfile/sourceCode/frida/build/tmp-android-arm/frida-gum/bindings/gumjs/node_modules/typescript/lib/lib.scripthost.d.ts'
 }
}
Command '[PosixPath('/root/Desktop/myfile/sourceCode/frida/build/tmp-android-arm/frida-gum/bindings/gumjs/node_modules/.bin/frida-compile'), PosixPath('runtime/entrypoint-quickjs.js'), '-o'
, PosixPath('runtime-build-quick/frida.js'), '-c']' returned non-zero exit status 1.
ninja: build stopped: subcommand failed.
Could not rebuild /root/Desktop/myfile/sourceCode/frida/build/tmp-android-arm/frida-gum
make[1]: *** [Makefile.linux.mk:111: build/frida-android-arm/lib/pkgconfig/frida-gum-1.0.pc] Error 1
make[1]: Leaving directory '/root/Desktop/myfile/sourceCode/frida'
make: *** [Makefile:4: core-android-arm64] Error 2
```

图 18-13 编译 File 错误

```
/** Provides information about files and allows JavaScript in a web page to access their content. */
interface File extends Blob {
    readonly lastModified: number;
    readonly name: string;
    readonly webkitRelativePath: string;
}

declare var File1: {          File --> File1
    prototype: File;
    new(fileBits: BlobPart[], fileName: string, options?: FilePropertyBag): File;
};
```

图 18-14 报错更改

（6）完成编译后，编译产物在/root/Desktop/myfile/sourceCode/frida/build/frida-android-arm64/bin 中。

（7）通过 Frida 启动，我们可以得出最终使用的源码是 gumquickmodule.c。

18.2.2 源码追踪分析

我们以 findExportByName 为切入点来理清如何来 anti-frida。源码路径：frida/frida-gum/bindings/gumjs/gumquickmodule.c。

观察如图 18-15 所示的报错信息，这是上面 anti-frida 的报错信息。根据报错信息来追踪上下文的源码。

```
const GumModuleDetails *
gum_android_get_linker_module_details (void)
{
    static GOnce once = G_ONCE_INIT;

    g_once (&once, (GThreadFunc) gum_try_init_linker_details, NULL);

    if (once.retval == NULL)
        gum_panic ("Unable to locate the Android linker; please file a bug");

    return once.retval;
}
```

图 18-15 报错信息

继续往下追踪，在 gum_try_init_linker_details 函数中有一个[vdso]的值，它代表什么呢？我们可以看如图 18-16 所示的注释。

```c
static const GumModuleDetails *
gum_try_init_linker_details (void)
{
  const GumModuleDetails * result = NULL;
  gchar * linker_path;
  GRegex * linker_path_pattern;
  gchar * maps, ** lines;
  gint num_lines, vdso_index, i;

  linker_path = gum_find_linker_path ();
  linker_path_pattern = gum_find_linker_path_pattern ();

  /*
   * Using /proc/self/maps means there might be false positives, as the
   * application – or even Frida itself – may have mmap()ed the module.
   *
   * Knowing that the linker is mapped right around the vdso, with no
   * empty space between, we just have to find the vdso, and we can
   * count on the next or previous linker mapping being the actual
   * linker.
   */
  g_file_get_contents ("/proc/self/maps", &maps, NULL, NULL);
  lines = g_strsplit (maps, "\n", 0);
  num_lines = g_strv_length (lines);

  vdso_index = -1;
  for (i = 0; i != num_lines; i++)
  {
    const gchar * line = lines[i];

    if (g_str_has_suffix (line, " [vdso]"))
    {
      vdso_index = i;
      break;
    }
  }
```

图 18-16　[vdso]注释

提示　Using /proc/self/maps means there might be false positives, as the application – or even Frida itself – may have mmap()ed the module.
使用/proc/self/maps 意味着可能会有误报，因为应用程序甚至是 Frida 本身可能已经对模块进行了 mmap()。
Knowing that the linker is mapped right around the vdso, with no empty space between, we just have to find the vdso, and we can count on the next or previous linker mapping being the actual linker.
知道链接器映射在 vdso 周围，中间没有空格，我们只需要找到 vdso，可以指望下一个或上一个链接器映射是实际的链接器。

我们不用把 mmaps 的映射全部遍历一遍，而是通过[vdso]。可以这样理解[vdso]，它相当于一个分叉点，在这个分叉点上有两个加载方向，只要我们找到这个[vdso]，向两边遍历即可找到我们需要的 linker。这是一个比较固定的写法，也是 Frida 比较巧妙的地方，我们可以形象地称之为"劈叉"。

在系统的内存中查看这个值，如图 18-17 所示。

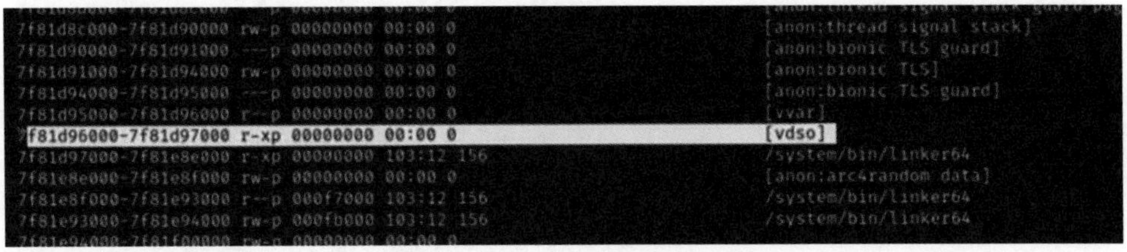

第 18 章 从 findExportByName 源码分析到 anti-frida 新思路   387

```
7f81d8c000-7f81d90000 rw-p 00000000 00:00 0                [anon:thread signal stack]
7f81d90000-7f81d91000 ---p 00000000 00:00 0                [anon:bionic TLS guard]
7f81d91000-7f81d94000 rw-p 00000000 00:00 0                [anon:bionic TLS]
7f81d94000-7f81d95000 ---p 00000000 00:00 0                [anon:bionic TLS guard]
7f81d95000-7f81d96000 r--p 00000000 00:00 0                [vvar]
7f81d96000-7f81d97000 r-xp 00000000 00:00 0                [vdso]
7f81d97000-7f81e8e000 r-xp 00000000 103:12 156             /system/bin/linker64
7f81e8e000-7f81e8f000 rw-p 00000000 00:00 0                [anon:arc4random data]
7f81e8f000-7f81e93000 r--p 000f7000 103:12 156             /system/bin/linker64
7f81e93000-7f81e94000 rw-p 000fb000 103:12 156             /system/bin/linker64
```

图 18-17 查看系统[vdso]

再往下就可以追踪到 attach 模式下 anti-frida 的关键函数 gum_try_parse_linker_proc_maps_line，如图 18-18 所示。

```
static gboolean
gum_try_parse_linker_proc_maps_line (const gchar * line,
                                     const gchar * linker_path,
                                     const GRegex * linker_path_pattern,
                                     GumModuleDetails * module,
                                     GumMemoryRange * range)
{
  GumAddress start, end;
  gchar perms[5] = { 0, };
  gchar path[PATH_MAX];
  gint n;
  const guint8 elf_magic[] = { 0x7f, 'E', 'L', 'F' };

  n = sscanf (line,
      "%" G_GINT64_MODIFIER "x-%" G_GINT64_MODIFIER "x "
      "%4c "
      "%*x %*s %*d "
      "%s",
      &start, &end,
      perms,
      path);
  if (n != 4)
    return FALSE;

  if (!g_regex_match (linker_path_pattern, path, 0, NULL))
    return FALSE;

  if (perms[0] != 'r')
    return FALSE;

  if (memcmp (GSIZE_TO_POINTER (start), elf_magic, sizeof (elf_magic)) != 0)
    return FALSE;

  module->name = strrchr (linker_path, '/') + 1;  // 使用 linker_path 做一个判断
  module->range = range;
  module->path = linker_path;
```

图 18-18 gum_try_parse_linker_proc_maps_line 函数

在代码中可以清楚地看到对 ELF 魔术字段的校验、魔术字段长度的校验、linker_path 的校验，如果这些校验不通过，那么可以直接返回 FALSE，我们直接返回 FALSE 是否就可以实现 anti-frida 呢？

继续追踪，下面看到 soinfo 函数的一个枚举，如图 18-19 所示。此函数中有一个 API，这个 API 获取的是 linker 加载需要用到的 dlopen、dlsym、header。此外，还有一个 get_somain 函数，该函数就是 spawn 模式下 anti-frida 的关键切入点。

```
gum_enumerate_soinfo (GumFoundSoinfoFunc func,
                      gpointer user_data)
{
  GumLinkerApi * api;
  GumSoinfo * somain, * sovdso, * solinker, * si, * next;
  GHashTable * ranges;
  GumSoinfoDetails details;
  gboolean carry_on;

  api = gum_linker_api_get ();

  pthread_mutex_lock (api->dl_mutex);

  somain = api->solist_get_somain ();
  sovdso = NULL;
  solinker = NULL;

  ranges = NULL;
```

图 18-19　soinfo 函数的枚举

在如图 18-20 所示的 gum_soinfo_get_body 函数中，它没有校验 somain 的值是否为空就直接获取 body，这势必会引起空指针异常。这就是 spawn 模式下 anti-frida 的新思路，我们可以称之为地址错误法。从这里可以看到，Frida 对这块的处理不是很严谨。

```
1622  gum_soinfo_get_body (GumSoinfo * self)
1623  {
1624    guint api_level = gum_android_get_api_level ();
1625    if (api_level >= 26)
1626      return &self->modern.body;
1627    else if (api_level >= 23)
1628      return &self->legacy23.body;
1629    else
1630      return &self->legacy.legacy23.body;
1631  }
```

图 18-20　gum_soinfo_get_body

18.3　本章小结

在本章中，我们向读者演示了 Frida 的 attach 和 spawn 模式下的 anti-frida 效果，并通过追踪源码分析了 anti-frida 的原理。在 attach 模式下，Frida 只是简单地校验了头部信息，因此我们只需直接抹掉头部即可实现 anti-frida。而在 spawn 模式下，Frida 会引用 somain 结构体中的字段，我们将其置空。当 Frida 引用这些字段时，由于找不到对应的字段，就会触发空指针异常，最终实现 anti-frida 的效果。

第 19 章

PLT 和 GOT 的 Hook

本章讲解 PLT 和 GOT 的 Hook。我们首先回顾一下 GOT 和 PLT 的概念：GOT 是 Linux ELF 文件中的全局偏移表，用于定位全局变量和函数。PLT 是 Linux ELF 文件中的过程链接表，用于延迟绑定函数，即只有在函数第一次被调用时才进行绑定（包括符号查找和重定位等操作），如果函数从未被使用过，则不进行绑定。基于延迟绑定的机制可以显著提高程序的启动速度，特别有利于一些引用了大量函数的程序。这也是进行 GOT 和 PLT 的 Hook 的核心原理。

19.1 GOT 的 Hook

GOT 的 Hook 的原理：将 GOT 中的表项通过 Hook 进行替换，以实现我们自定义的函数。
实现 GOT 的 Hook 有两种方式。

1. 根据节头实现 GOT 的 Hook

节是一个静态的表现形式，因为内存中是没有节的。我们可以通过把 SO 文件传到目录中，然后加载这个 SO 文件，并解析这个节，而不是动态解析它加载到内存中的那一块地址。

2. 根据程序头实现 GOT 的 Hook

程序头 Hook 比节头 Hook 好很多，因为程序头是一个运行态，我们可以动态地获取和更改里面的内容。
首先，新建一个以 C++为模板的项目 myGotHook。在 native-lib.cpp 中编写如图 19-1 所示的代码。

```
void myr1() {

    __android_log_print( prio: 6,  tag: "r0ysue",  fmt: "i am from myr1");
}

void myr0() {

    __android_log_print( prio: 6,  tag: "r0ysue",  fmt: "i am from myr0");

}

extern "C" JNIEXPORT jstring JNICALL
Java_com_r0ysue_mygothook_MainActivity_stringFromJNI(
        JNIEnv *env,
        jobject /* this */) {
    std::string hello = "Hello from C++";
    return env->NewStringUTF(hello.c_str());
}
```

图 19-1　初始代码

然后将解压生成的 APK 文件在 010 Editor 中打开，找到.got 表的起始偏移位置，如图 19-2 所示。

Name	Value	Start	Size	Color	
› struct section_table_entry64_t section_table_element[17]	.fini_array	1554...	40h	Fg:	B
› struct section_table_entry64_t section_table_element[18]	.data.rel.ro	1554...	40h	Fg:	B
› struct section_table_entry64_t section_table_element[19]	.dynamic	1554...	40h	Fg:	B
˅ struct section_table_entry64_t section_table_element[20]	.got	1555...	40h	Fg:	B
› struct s_name64_t s_name	.got	1555...	4h	Fg:	B
enum s_type64_e s_type	SHT_PROGBITS (1)	1555...	4h	Fg:	B
enum s_flags64_e s_flags	SF64_Alloc_Exec (3)	1555...	8h	Fg:	B
Elf64_Addr s_addr	0x0000000000035C90	1555...	8h	Fg:	B
Elf64_Off s_offset	34C90h	1555...	8h	Fg:	B
Elf64_Xword s_size	880	1555...	8h	Fg:	B
Elf64_Word s_link	0	1555...	4h	Fg:	B
Elf64_Word s_info	0	1555...	4h	Fg:	B

图 19-2　.got 表

通过.got 表在 IDA 中找到其偏移，并下滑找到 myr0 函数的地址（并不是.got 表的起始位置偏移），如图 19-3 所示。

```
.got:0000000000035D58                                   ; DATA XREF: .__stack_chk
.got:0000000000035D58                                   ; .__stack_chk_fail+4↑r .
.got:0000000000035D60 fgets_ptr       DCQ fgets         ; DATA XREF: .fgets↑o
.got:0000000000035D60                                   ; .fgets+4↑r ...
.got:0000000000035D68 _Z4myr0v_ptr    DCQ _Z4myr0v      ; DATA XREF: myr0(void)↑o
.got:0000000000035D68                                   ; myr0(void)+4↑r ...
.got:0000000000035D68                                   ; myr0(void)
.got:0000000000035D70 _ZNSt6__ndk112basic_stringIcNS_11char_traitsIcEENS_9allocatorIcEE
.got:0000000000035D70                                   ; DATA XREF: std::__ndk1:
.got:0000000000035D70                                   ; std::__ndk1::basic_stri
.got:0000000000035D70                                   ; std::__ndk1::basic_stri
```

图 19-3　myr0 的偏移

可以看到 myr0 函数的偏移，尝试改掉这个偏移（将 myr0 函数的地址改变为 myr1 的地址），查看替换的效果，代码如图 19-4 所示。这段代码的作用是找到 SO 文件在内存中的分布，运行程序并查看程序的改变会不会对.got 的偏移产生影响。

第 19 章　PLT 和 GOT 的 Hook

```cpp
void hook(){
    hook();
    char line[1024];
    int *start;
    int *end;
    int n=1;
    FILE *fp=fopen( path: "/proc/self/maps", mode: "r");
    while (fgets(line, sizeof(line), fp)) {
        if (strstr(line, n: "libnative-lib.so") ) {
            __android_log_print( prio: 6, tag: "r0ysue", fmt: "%s", line);
            if(n==1){
                start = reinterpret_cast<int *>(strtoul( s: strtok(line, delimiter: "-"), end_ptr: NULL, base: 16));
                end = reinterpret_cast<int *>(strtoul( s: strtok( s: NULL, delimiter: " "), end_ptr: NULL, base: 16));
            }
            else{
                strtok(line, delimiter: "-");
                end = reinterpret_cast<int *>(strtoul( s: strtok( s: NULL, delimiter: " "), end_ptr: NULL, base: 16));
            }
            n++;
        }
    }
    mprotect(start, size: end-start, prot: PROT_EXEC|PROT_READ|PROT_EXEC);
//     *(long*)((char*)start+0x34f08) = reinterpret_cast<long>(mytest);
}
```

图 19-4　Hook 代码

重新对生成的 APK 解包，并使用 010 Editor 查看，在图 19-5 中，我们可以看到 .got 表的地址偏移发生了改变。

图 19-5　.got 表的地址偏移

同时，查看新生成的 SO 文件中的 myr0 的地址，如图 19-6 所示。

图 19-6　myr0 的地址

对 Hook 代码更新，使用硬偏移的方式来替换。将图 19-4 中的注释打开替换后运行 App，结果如图 19-7 所示。

图 19-7　硬偏移替换结果

可以看到这个替换失败了。这是因为当我们完成代码的编写后，编译之后它的偏移就变了，所以不能使用地址偏移的方式替换来完成Hook。这种情况下，需要通过节头和程序头来Hook GOT。

19.1.1　根据节头实现Hook

通过节头Hook需要静态的SO文件在内存中加载。首先看一下代码的更改会不会对节头的首地址产生影响，如果影响太大，那么通过节头来Hook就没有意义了。

两次运行App，第一次将myr0函数置空调用，第二次将myr0加入几条日志打印，将得到的两个SO文件（1.so和2.so）分别拖入010 Editor中查看，如图19-8所示。可以看到，.got的首地址的偏移并未改变。

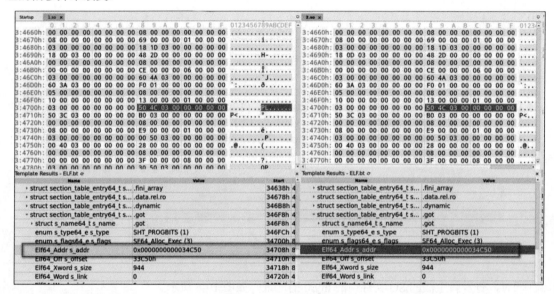

图19-8　.got表的偏移

获取SO文件的基本信息

首先要做的是拿到静态SO文件的首地址。将SO文件推送到/data/local/tmp目录下，并在Hook函数中编写如图19-9所示的代码。

图19-9　SO文件首地址获取代码

可以看到顺利拿到了 SO 文件的首地址，如图 19-10 所示。

```
2022-03-26 20:32:57.908 28381-28381/? D/JustTrustMe: OKHTTP 3.x not found in com.roysue.mygothook -- not hooking OkHostnameVerifier.verify(String, X509)(
2022-03-26 20:32:58.383 28381-28381/com.roysue.mygothook E/r0ysue: 7021008000-702103a000 r-xp 00000000 fd:00 4828                    /data/app/com.roysue.mygotho
2022-03-26 20:32:58.384 28381-28381/com.roysue.mygothook E/r0ysue: 702103a000-702103e000 r--p 00031000 fd:00 4828                    /data/app/com.roysue.mygotho
2022-03-26 20:32:58.384 28381-28381/com.roysue.mygothook E/r0ysue: 702103e000-702103f000 rw-p 00035000 fd:00 4828                    /data/app/com.roysue.mygotho
2022-03-26 20:32:58.393 28381-28381/com.roysue.mygothook E/r0ysue: 0x70b843b000
2022-03-26 20:32:58.443 28381-28409/com.roysue.mygothook D/OpenGLRenderer: HWUI GL Pipeline
```

图 19-10 静态 SO 文件的首地址

拿到 SO 文件的首地址之后，还要通过偏移拿到 GOT 的位置，GOT 的位置计算方式如图 19-11 所示。

Name	Value	Start	Size	Color	Comment
struct section_table_entry64_t section...	.data.rel.ro	34678h	40h	Fg: B	
struct section_table_entry64_t section...	.dynamic	346B8h	40h	Fg: B	
struct section_table_entry64_t section...	.got	346F8h	40h	Fg: B	
struct s_name64_t s_name	.got	346F8h	4h	Fg: B	
enum s_name32_e s_name_off	13h	346F8h	4h	Fg: B	
string s_name_str[5]	.got	34117h	5h	Fg: B	
enum s_type64_e s_type	SHT_PRO...	346FCh	4h	Fg: B	
enum s_flags64_e s_flags	SF64_Allo...	34700h	8h	Fg: B	
Elf64_Addr s_addr	0x000000...	34708h	8h	Fg: B	
Elf64_Off s_offset	33C50h	34710h	8h	Fg: B	
Elf64_Xword s_size	944	34718h	8h	Fg: B	
Elf64_Word s_link	0	34720h	4h	Fg: B	
Elf64_Word s_info	0	34724h	4h	Fg: B	
Elf64_Xword s_addralign	8	34728h	8h	Fg: B	
Elf64_Xword s_entsize	8	34730h	8h	Fg: B	
char data[944]		33C50h	3B0h	Fg: B	
struct section_table_entry64_t section...	.data	34738h	40h	Fg: B	
struct section_table_entry64_t section...	.bss	34778h	40h	Fg: B	
struct section_table_entry64_t section...	.comment	347B8h	40h	Fg: B	
struct section_table_entry64_t section...	.shstrtab	347F8h	40h	Fg: B	
struct s_name64_t s_name	.shstrtab	347F8h	4h	Fg: B	
enum s_type64_e s_type	SHT_STRT...	347FCh	4h	Fg: B	
enum s_flags64_e s_flags	SF64_Non...	34800h	8h	Fg: B	
Elf64_Addr s_addr	0x000000...	34808h	8h	Fg: B	
Elf64_Off s_offset	34104h	34810h	8h	Fg: B	
Elf64_Xword s_size	239	34818h	8h	Fg: B	
Elf64_Word s_link	0	34820h	4h	Fg: B	
Elf64_Word s_info	0	34824h	4h	Fg: B	
Elf64_Xword s_addralign	1	34828h	8h	Fg: B	
Elf64_Xword s_entsize	0	34830h	8h	Fg: B	
char data[239]		34104h	EFh	Fg: B	

a
got表的位置 = a+b
b

图 19-11 GOT 的位置

接下来遍历 strtab 表获取所有节的名称，并过滤出.got 表（通过上述计算方式），编写如图 19-12 所示的代码，获取结果如图 19-13 所示。

19

```c
int fd;
void *start;
struct stat sb;
fd = open( pathname: "/data/local/tmp/1.so", O_RDONLY);
fstat(fd, &sb);
start = mmap( addr: NULL, sb.st_size, PROT_READ, MAP_PRIVATE, fd, offset: 0);
__android_log_print( prio: 6, tag: "r0ysue", fmt: "%p", start);
Elf64_Ehdr header;
memcpy(&header, start, sizeof(Elf64_Ehdr));
int secoff = header.e_shoff;
int secsize = header.e_shentsize;
int secnum = header.e_shnum;
int secstr = header.e_shstrndx;
Elf64_Shdr strtab;
memcpy(&strtab, src: (char *) start + secoff + secstr * secsize, sizeof(Elf64_Shdr));
int strtaboff = strtab.sh_offset;
char strtabchar[strtab.sh_size];

memcpy(&strtabchar, src: (char *) start + strtaboff, strtab.sh_size);
Elf64_Shdr enumsec;
int gotoff = 0;
int gotsize = 0;
for (int n = 0; n < secnum; n++) {
    memcpy(&enumsec, src: (char *) start + secoff + n * secsize, sizeof(Elf64_Shdr));
    __android_log_print(6,"r0ysue","%s",&strtabchar[enumsec.sh_name]);
    // 过滤得到.got表
    if (strcmp(&strtabchar[enumsec.sh_name], ".got") == 0) {
        gotoff = enumsec.sh_addr;
        gotsize = enumsec.sh_size;
    }
}
```

图 19-12　获取.got 表的代码

```
com.r0ysue.mygothook E/r0ysue: 0x70b843b000
com.r0ysue.mygothook E/r0ysue: .note.gnu.build-id
com.r0ysue.mygothook E/r0ysue: .hash
com.r0ysue.mygothook E/r0ysue: .gnu.hash
com.r0ysue.mygothook E/r0ysue: .dynsym
com.r0ysue.mygothook E/r0ysue: .dynstr
com.r0ysue.mygothook E/r0ysue: .gnu.version
com.r0ysue.mygothook E/r0ysue: .gnu.version_r
com.r0ysue.mygothook E/r0ysue: .rela.dyn
com.r0ysue.mygothook E/r0ysue: .rela.plt
com.r0ysue.mygothook E/r0ysue: .plt
com.r0ysue.mygothook E/r0ysue: .text
com.r0ysue.mygothook E/r0ysue: .rodata
com.r0ysue.mygothook E/r0ysue: .eh_frame_hdr
com.r0ysue.mygothook E/r0ysue: .eh_frame
com.r0ysue.mygothook E/r0ysue: .gcc_except_table
com.r0ysue.mygothook E/r0ysue: .note.android.ident
com.r0ysue.mygothook E/r0ysue: .fini_array
com.r0ysue.mygothook E/r0ysue: .data.rel.ro
com.r0ysue.mygothook E/r0ysue: .dynamic
com.r0ysue.mygothook E/r0ysue: .got
com.r0ysue.mygothook E/r0ysue: .data
com.r0ysue.mygothook E/r0ysue: .bss
com.r0ysue.mygothook E/r0ysue: .comment
com.r0ysue.mygothook E/r0ysue: .shstrtab
```

图 19-13　所有节的名称

拿到 GOT 的首地址后，我们知道 GOT 中存储着全局函数表，然后继续遍历找到 myr1 函数，进行替换，如图 19-14 所示。

```
for (int n = 0; n < gotsize; n = n + 8) {
    long *tmp = reinterpret_cast<long *>((char *) startr + gotoff + n);

    if (*tmp == (long)myr0) {
        mprotect( addr: reinterpret_cast<void *>((long)tmp & 0xfffffff000), PAGE_SIZE, // 0xfffffff000 页对齐
                  prot: PROT_WRITE | PROT_READ | PROT_EXEC);
        *tmp = reinterpret_cast<long>(myr1);
    }
}
```

图 19-14　函数替换

函数的调用关系如图 19-15 所示。

```
extern "C" JNIEXPORT jstring JNICALL
Java_com_r0ysue_mygothook_MainActivity_stringFromJNI(
        JNIEnv *env,
        jobject /* this */) {
    std::string hello = "Hello from C++";


    hook();
    myr0();


    return env->NewStringUTF(hello.c_str());
}
```

图 19-15　调用关系

最后得到的结果如图 19-16 所示。在这里可以看到 myr0 函数被成功地替换为 myr1。

```
2022-03-26 22:35:58.451 9072-9072/com.r0ysue.mygothook I/zygote64:      at void de.robv.android.xposed.XposedBridge.main(java
2022-03-26 22:35:58.672 9072-9072/com.r0ysue.mygothook E/r0ysue: 70213c8000-70213fa000 r-xp 00000000 fd:00 1035
2022-03-26 22:35:58.672 9072-9072/com.r0ysue.mygothook E/r0ysue: 70213fa000-70213fe000 r--p 00031000 fd:00 1035
2022-03-26 22:35:58.673 9072-9072/com.r0ysue.mygothook E/r0ysue: 70213fe000-70213ff000 rw-p 00035000 fd:00 1035
2022-03-26 22:35:58.682 9072-9072/com.r0ysue.mygothook E/r0ysue: 0x7021270000
2022-03-26 22:35:58.682 9072-9072/com.r0ysue.mygothook E/r0ysue: i am from myr1
2022-03-26 22:35:58.750 9072-9098/com.r0ysue.mygothook D/OpenGLRenderer: HWUI GL Pipeline
```

图 19-16　替换结果

19.1.2　根据程序头来实现 Hook

程序头的 Hook 其实和节头的 Hook 是相同的，都是通过找到 GOT 实现的，但是它们的搜索方式是不同的。使用程序头表的方式不用打开文件，而是直接在内存中找，所以这样的方式会更快、更高效。

首先拿到程序头表的一个首地址，并获取程序头表的数量、大小和偏移，如图 19-17 所示。

```
mprotect(startr, size: (long) end - (long) startr, prot: PROT_WRITE | PROT_READ | PROT_EXEC);
Elf64_Ehdr header;
memcpy(&header, startr, sizeof(Elf64_Ehdr));
int phnum = header.e_phnum;
int phsize = header.e_phentsize;
int phoff = header.e_phoff;
```

图 19-17　获取程序头基本信息

GOT 在程序头中的体现是动态段，动态段的类型是 2，如图 19-18 所示。所以我们需要找到动态段，再根据动态段寻找函数。

Elf64_Half e_shtrndx_STRING_TABLE_INDEX	24	3Eh	2h
struct program_header_table		40h	1C
struct program_table_entry64_t program_table_element[0]	(R_X) Loadable Segment	40h	38
struct program_table_entry64_t program_table_element[1]	(RW_) Loadable Segment	78h	38
struct program_table_entry64_t program_table_element[2]	(RW_) Dynamic Segment	B0h	38
enum p_type64_e p_type	PT_DYNAMIC (2)	B0h	4h
enum p_flags64_e p_flags	PF_Read_Write (6)	B4h	4h
Elf64_Off p_offset_FROM_FILE_BEGIN	33A60h	B8h	8h
Elf64_Addr p_vaddr_VIRTUAL_ADDRESS	0x0000000000034A60	C0h	8h
Elf64_Addr p_paddr_PHYSICAL_ADDRESS	0x0000000000034A60	C8h	8h
Elf64_Xword p_filesz_SEGMENT_FILE_LENGTH	496	D0h	8h
Elf64_Xword p_memsz_SEGMENT_RAM_LENGTH	496	D8h	8h
Elf64_Xword p_align	8	E0h	8h

图 19-18　动态段

GOT 的查找代码如图 19-19 所示。这里我们使用 elf.h 封装好的 d_tag 来找到 GOT。

```
for (int n = 0; n < dynamicsize; n = n + sizeof(Elf64_Dyn)) {
    memcpy(&tmp, src: (char *) startr + dynamicoff + n, sizeof(Elf64_Dyn));
    if (tmp.d_tag == DT_PLTGOT) {    //DT_PLTGOT==3
        int off = (int) (tmp.d_un.d_val);
        off = (off - off % 0x1000) + 0x1000;    // 找到页结尾
        __android_log_print( prio: 6, tag: "r0ysue", fmt: "%p", off);
        gotoff = tmp.d_un.d_val;  // 找到got的偏移
        gotsize = off - gotoff;   // got表的大小
        __android_log_print( prio: 6, tag: "r0ysue", fmt: "%x", gotoff);
        __android_log_print( prio: 6, tag: "r0ysue", fmt: "%x", gotsize);

    }

}
```

图 19-19　GOT 查找代码

在图 19-19 的注释中可以看到找到的页的结尾，为什么要找到页的结尾呢？如图 19-20 所示，GOT 的结尾就是一个页的结束，得到页结束的位置，就可以获取 GOT 的大小。

```
.got:0000000000035FF8                                         ; secgothook(void)+560↑r
.got:0000000000035FF8 ; .got          ends                    ; mytest(void)
.got:0000000000035FF8
.data:0000000000036000 ; ===========================================================================
.data:0000000000036000
.data:0000000000036000 ; Segment type: Pure data
.data:0000000000036000                AREA .data, DATA, ALIGN=3
```

图 19-20　页的结尾

下一步的操作是替换，将 myr0 的地址替换为 myr1 的地址，就达到了 Hook 的效果，代码如

图 19-21 所示。

```
for (n = 0; n < gotsize; n = n + 8) {
    long *tmp = reinterpret_cast<long *>((char *) startr + gotoff + n);
    if (*tmp == (long) myr0) {
        *tmp = reinterpret_cast<long>(myr1);
    }
}
```

图 19-21　地址替换

替换成功的结果如图 19-22 所示。

```
/com.r0ysue.mygothook E/r0ysue: 70213c1000-70213f3000 r-xp 00000000 fd:00 1046
/com.r0ysue.mygothook E/r0ysue: 70213f3000-70213f7000 r--p 00031000 fd:00 1046
/com.r0ysue.mygothook E/r0ysue: 70213f7000-70213f8000 rw-p 00035000 fd:00 1046
/com.r0ysue.mygothook E/r0ysue: 0x36000
/com.r0ysue.mygothook E/r0ysue: 35c28
/com.r0ysue.mygothook E/r0ysue: 3d8
/com.r0ysue.mygothook E/r0ysue: i am from myr1
```

图 19-22　替换结果

根据节头和程序头的替换都可以实现最终的 Hook 效果，它们各有特点，核心原理都是延迟绑定，SO 文件加载到内存中后不会立即全部绑定函数，而是在使用的时候才会进行初次绑定，这就给我们 Hook 留下了很好的操作空间。根据节头替换需要加载静态的 SO 文件，从节区表中找到 GOT，然后在这张表中对目标地址进行替换。程序头表的方式更加简单直接，直接从内存中获取，并得到其首地址和大小。最后遍历寻找目标函数进行替换。

节头的 GOT 的 Hook 可能不全面，不是所有的函数都在 GOT 中，如 libart.so 中的 LoadMethod 函数就不在 GOT 中，如图 19-23 所示。

```
art::ClassLinker::LoadMethod(art::DexFile const&,art::Clas...  0000000000133E50
art::DexFileVerifier::CheckLoadMethodId(uint,char const*)    00000000001A7034
```

图 19-23　LoadMethod 函数

19.2　PLT 的 Hook

处理函数时会通过一小段 PLT 代码到达 GOT 中，PLT 代码段决定要将地址跳到哪里，我们通过更改 PLT 代码即可达到 Hook 的效果。

PLT 的 Hook 也分为两个方面，与 GOT 表的方式一样，也是通过节头和程序头来跳转的。

19.2.1　根据节头来实现 Hook

首先要拿到 App 在内存中加载的位置，因为在 CPP 中编写代码比较冗长，所以先在 Java 层编写代码获取路径，如图 19-24 所示。

```java
@Override
protected void onCreate(Bundle savedInstanceState) {
    super.onCreate(savedInstanceState);
    setContentView(R.layout.activity_main);
    filesPath=getApplication().getPackageCodePath();
    filesPath =filesPath.substring(0,filesPath.length()-8);
    filesPath=filesPath+"lib/arm64/libnative-lib.so";
    // Example of a call to a native method
    Log.e( tag: "r0ysue",filesPath);
    TextView tv = findViewById(R.id.sample_text);
    tv.setText(stringFromJNI());
}
```

图 19-24 获取路径

然后在 Native 通过反射拿到路径，如图 19-25 所示。这样无论 SO 文件是否变化，都可以获取 SO 文件的位置。

```cpp
int fd;
void *start;
struct stat sb;
jclass myclass = myenv->FindClass( name: "com/r0ysue/myplthook/MainActivity");
jfieldID myfile = myenv->GetStaticFieldID(myclass,  name: "filesPath",  sig: "Ljava/lang/String;");
jstring myfilepath = static_cast<jstring>(myenv->GetStaticObjectField(myclass, myfile));
const char *mylibpath = myenv->GetStringUTFChars(myfilepath,  isCopy: 0);
fd = open(mylibpath, O_RDONLY);
__android_log_print( prio: 6,  tag: "r0ysue",  fmt: "mylibpath: %s", mylibpath);
fstat(fd, &sb);
start = mmap( addr: NULL, sb.st_size, PROT_READ, MAP_PRIVATE, fd,  offset: 0);
```

图 19-25 反射获取路径

接下来获取.plt 表和.rela.plt 表（重定向表）的基本信息，代码如图 19-26 所示。重定向表是函数的重定向的机制，在程序启动的时候需要完成重定向。

```cpp
memcpy(&strtabchar,  src: (char *) start + strtaboff, strtab.sh_size);
Elf64_Shdr enumsec;
int pltoff = 0;
int pltsize = 0;
int relaoff = 0;
int relasize = 0;
int thoff = 0;
for (int n = 0; n < secnum; n++) {

    memcpy(&enumsec,  src: (char *) start + secoff + n * secsize, sizeof(Elf64_Shdr));

    if (strcmp(&strtabchar[enumsec.sh_name], ".plt") == 0) {
        pltoff = enumsec.sh_addr;
        pltsize = enumsec.sh_size;
        __android_log_print( prio: 6,  tag: "r0ysue",  fmt: "%p", pltoff);
    }

    if (strcmp(&strtabchar[enumsec.sh_name], ".rela.plt") == 0) {
        relaoff = enumsec.sh_addr;
        relasize = enumsec.sh_size;
        __android_log_print( prio: 6,  tag: "r0ysue",  fmt: "%p", relaoff);
    }
}
```

图 19-26 获取基本信息

有了这张重定向表后,就可以遍历找到待跳转的函数位置,也就是 myr0 的位置,代码如图 19-27 所示。

```cpp
Elf64_Rela tmp;
for (int n = 0; n < relasize; n = n + sizeof(Elf64_Rela)) {
    memcpy(&tmp, src: (char *) startr + relaoff + n, sizeof(Elf64_Rela));
    long *ptr = reinterpret_cast<long *>((char *) startr + tmp.r_offset);
    if (*ptr == (long) myr0) {

        thoff = n / sizeof(Elf64_Rela);
        __android_log_print( prio: 6, tag: "r0ysue", fmt: "%d ", thoff);
    }
}
```

图 19-27　遍历重定向表

其中 thoff 是一个位置顺序,也就是加载顺序,在 IDA 中打开查看,如图 19-28 所示。

```
* LOAD:000000000000E698            Elf64_Rela <0x34CD0, 0xC00000402, 0>  ; R_AARCH64_JUMP_SLOT __strlen_chk
* LOAD:000000000000E6B0            Elf64_Rela <0x34CD8, 0xD00000402, 0>  ; R_AARCH64_JUMP_SLOT strtoul
* LOAD:000000000000E6C8            Elf64_Rela <0x34CE0, 0xE00000402, 0>  ; R_AARCH64_JUMP_SLOT fstat
* LOAD:000000000000E6E0            Elf64_Rela <0x34CE8, 0x4600000402, 0> ; R_AARCH64_JUMP_SLOT _ZdaPvSt11align_val_t
* LOAD:000000000000E6F8            Elf64_Rela <0x34CF0, 0x6500000402, 0> ; R_AARCH64_JUMP_SLOT _ZN7_JNIEnv12NewStringUTFEPKc
* LOAD:000000000000E710            Elf64_Rela <0x34CF8, 0x1300000402, 0> ; R_AARCH64_JUMP_SLOT _ZNSt6__ndk117__compressed_pair
* LOAD:000000000000E728            Elf64_Rela <0x34D00, 0xF000000402, 0> ; R_AARCH64_JUMP_SLOT abort
* LOAD:000000000000E740            Elf64_Rela <0x34D08, 0x3200000402, 0> ; R_AARCH64_JUMP_SLOT __cxa_call_unexpected
* LOAD:000000000000E758            Elf64_Rela <0x34D10, 0x7D00000402, 0> ; R_AARCH64_JUMP_SLOT _ZNSt6__ndk112basic_stringIcNS_1
* LOAD:000000000000E770            Elf64_Rela <0x34D18, 0x1200000402, 0> ; R_AARCH64_JUMP_SLOT __memcpy_chk
* LOAD:000000000000E788            Elf64_Rela <0x34D20, 0x1100000402, 0> ; R_AARCH64_JUMP_SLOT dl_iterate_phdr
* LOAD:000000000000E7A0            Elf64_Rela <0x34D28, 0x1200000402, 0> ; R_AARCH64_JUMP_SLOT android_set_abort_message
* LOAD:000000000000E7B8            Elf64_Rela <0x34D30, 0x1300000402, 0> ; R_AARCH64_JUMP_SLOT islower
* LOAD:000000000000E7D0            Elf64_Rela <0x34D38, 0xBC00000402, 0> ; R_AARCH64_JUMP_SLOT _ZnwmSt11align_val_t
* LOAD:000000000000E7E8            Elf64_Rela <0x34D40, 0x1400000402, 0> ; R_AARCH64_JUMP_SLOT __memmove_chk
* LOAD:000000000000E800            Elf64_Rela <0x34D48, 0xE500000402, 0> ; R_AARCH64_JUMP_SLOT _ZNSt6__ndk117_DeallocateCallerS
* LOAD:000000000000E818            Elf64_Rela <0x34D50, 0x4100000402, 0> ; R_AARCH64_JUMP_SLOT _ZNSt13bad_exceptionD1Ev
* LOAD:000000000000E830            Elf64_Rela <0x34D58, 0x10800000402, 0>; R_AARCH64_JUMP_SLOT _ZNSt9type_infoD2Ev
* LOAD:000000000000E848            Elf64_Rela <0x34D60, 0x1500000402, 0> ; R_AARCH64_JUMP_SLOT _ZSt15get_new_handlerv
* LOAD:000000000000E860            Elf64_Rela <0x34D68, 0x1600000402, 0> ; R_AARCH64_JUMP_SLOT fputc
* LOAD:000000000000E878            Elf64_Rela <0x34D70, 0x1700000402, 0> ; R_AARCH64_JUMP_SLOT __stack_chk_fail
* LOAD:000000000000E890            Elf64_Rela <0x34D78, 0x1800000402, 0> ; R_AARCH64_JUMP_SLOT fgets
* LOAD:000000000000E8A8            Elf64_Rela <0x34D80, 0x1510000402, 0> ; R_AARCH64_JUMP_SLOT _Z4myr0v
* LOAD:000000000000E8C0            Elf64_Rela <0x34D88, 0x1A300000402, 0>; R_AARCH64_JUMP_SLOT _ZNSt6__ndk112basic_stringIcNS_
* LOAD:000000000000E8D8            Elf64_Rela <0x34D90, 0x1900000402, 0> ; R_AARCH64_JUMP_SLOT pthread_setspecific

0000E830 000000000000E830: LOAD:000000000000E830 (Synchronized with Hex View-1)
```

图 19-28　加载顺序

通过日志打印可以看到,myr0 的位次为 38,如图 19-29 所示。

```
04.223 3948-3948/com.r0ysue.myplthook E/r0ysue: 0xe6c0
04.223 3948-3948/com.r0ysue.myplthook E/r0ysue: 0xef60
04.223 3948-3948/com.r0ysue.myplthook E/r0ysue: 38
04.223 3948-3948/com.r0ysue.myplthook E/r0ysue: i am from myr0
```

图 19-29　myr0 的位次

PLT 的 Hook 的核心代码如图 19-30 所示,因为我们不知道 myr0 和 myr1 哪个先被加载,所以需要通过地址大小进行判断。

```
long *myr0plt = reinterpret_cast<long *>((char *) startr + pltoff + 0x20 + thoff * 0x10);
mprotect( addr: (void *) PAGE_START( x: (long) myr0plt), PAGE_SIZE, prot: PROT_WRITE | PROT_READ | PROT_EXEC);
__android_log_print( prio: 6, tag: "r0ysue", fmt: "%s", strerror(errno));
int faniloff = 0;
int flag;
if ((unsigned long) myr0 > (unsigned long) myr1) {
    faniloff = (unsigned long) myr0plt - (unsigned long) myr1;
    flag = 0;
} else {
    faniloff = (unsigned long) myr1 - (unsigned long) myr0plt;
    flag = 1;
}

unsigned long code;
if (flag == 1) {
    code = 0x14000000 + faniloff / 4;

} else {
    code = 0x18000000 - faniloff / 4;
}
__android_log_print( prio: 6, tag: "r0ysue", fmt: "%p", code);
*(myr0plt) = code;
```

图 19-30 Hook 的核心代码

替换后的结果如图 19-31 所示。

```
/com.r0ysue.myplthook E/r0ysue: 0xe748
/com.r0ysue.myplthook E/r0ysue: 0xf030
/com.r0ysue.myplthook E/r0ysue: 41
/com.r0ysue.myplthook E/r0ysue: Try again
/com.r0ysue.myplthook E/r0ysue: 0x140000f1
/com.r0ysue.myplthook E/r0ysue: i am from myr1
```

图 19-31 替换后的结果

通过节头实现 PLT 的 Hook 如代码清单 19-1 所示。

代码清单 19-1　节头 Hook PLT 完成代码

```C++
void secplthook() {
    char line[1024];
    int *startr;
    int *end;
    int n = 1;
    FILE *fp = fopen("/proc/self/maps", "r");
    while (fgets(line, sizeof(line), fp)) {
        if (strstr(line, "libnative-lib.so")) {
            __android_log_print(6, "r0ysue", "%s", line);
            if (n == 1) {
                startr = reinterpret_cast<int *>(strtoul(strtok(line, "-"), NULL, 16));
                end = reinterpret_cast<int *>(strtoul(strtok(NULL, " "), NULL, 16));
            } else {
                strtok(line, "-");
                end = reinterpret_cast<int *>(strtoul(strtok(NULL, " "), NULL, 16));
```

```
            }
            n++;
        }
    }

    int fd;
    void *start;
    struct stat sb;
    jclass myclass = myenv->FindClass("com/r0ysue/myplthook/MainActivity");
    jfieldID myfile = myenv->GetStaticFieldID(myclass, "filesPath",
"Ljava/lang/String;");
    jstring myfilepath = static_cast<jstring>(myenv->GetStaticObjectField(myclass,
myfile));
    const char *mylibpath = myenv->GetStringUTFChars(myfilepath, 0);
    fd = open(mylibpath, O_RDONLY);
    __android_log_print(6, "r0ysue", "mylibpath: %s", mylibpath);
    fstat(fd, &sb);
    start = mmap(NULL, sb.st_size, PROT_READ, MAP_PRIVATE, fd, 0);
    __android_log_print(6, "r0ysue", "%p", start);
    Elf64_Ehdr header;
    memcpy(&header, start, sizeof(Elf64_Ehdr));
    int secoff = header.e_shoff;
    int secsize = header.e_shentsize;
    int secnum = header.e_shnum;
    int secstr = header.e_shstrndx;
    Elf64_Shdr strtab;
    memcpy(&strtab, (char *) start + secoff + secstr * secsize, sizeof(Elf64_Shdr));
    int strtaboff = strtab.sh_offset;
    char strtabchar[strtab.sh_size];

    memcpy(&strtabchar, (char *) start + strtaboff, strtab.sh_size);
    Elf64_Shdr enumsec;
    int pltoff = 0;
    int pltsize = 0;
    int relaoff = 0;
    int relasize = 0;
    int thoff = 0;
    for (int n = 0; n < secnum; n++) {

        memcpy(&enumsec, (char *) start + secoff + n * secsize, sizeof(Elf64_Shdr));

        if (strcmp(&strtabchar[enumsec.sh_name], ".plt") == 0) {
            pltoff = enumsec.sh_addr;
            pltsize = enumsec.sh_size;
            __android_log_print(6, "r0ysue", "%p", pltoff);

        }
```

```cpp
        if (strcmp(&strtabchar[enumsec.sh_name], ".rela.plt") == 0) {
            relaoff = enumsec.sh_addr;
            relasize = enumsec.sh_size;
            __android_log_print(6, "r0ysue", "%p", relaoff);
        }
    }


    Elf64_Rela tmp;
    for (int n = 0; n < relasize; n = n + sizeof(Elf64_Rela)) {
        memcpy(&tmp, (char *) startr + relaoff + n, sizeof(Elf64_Rela));
        long *ptr = reinterpret_cast<long *>((char *) startr + tmp.r_offset);
        if (*ptr == (long) myr0) {

            thoff = n / sizeof(Elf64_Rela);
            __android_log_print(6, "r0ysue", "%d ", thoff);
        }
    }


    long *myr0plt = reinterpret_cast<long *>((char *) startr + pltoff + 0x20 + thoff * 0x10);
    mprotect((void *) PAGE_START((long) myr0plt), PAGE_SIZE, PROT_WRITE | PROT_READ | PROT_EXEC);
    __android_log_print(6, "r0ysue", "%s", strerror(errno));
    int faniloff = 0;
    int flag;
    if ((unsigned long) myr0 > (unsigned long) myr1) {
        faniloff = (unsigned long) myr0plt - (unsigned long) myr1;
        flag = 0;

    } else {
        faniloff = (unsigned long) myr1 - (unsigned long) myr0plt;
        flag = 1;
    }

    unsigned long code;
    if (flag == 1) {
        code = 0x14000000 + faniloff / 4;

    } else {
        code = 0x18000000 - faniloff / 4;
    }
    __android_log_print(6, "r0ysue", "%p", code);
    *(myr0plt) = code;



}
```

19.2.2 根据程序头来实现 Hook

本小节介绍根据程序头来实现 PLT 的 Hook，基础信息的代码与 19.2.1 节的代码相同，后面附完整的代码。

首先定位到 PLT，代码如图 19-32 所示。PLT 的位置信息在 elf.h 头文件中都有定义，可以直接获取。

```
for (int n = 0; n < phnum; n++) {
    memcpy(&tmp, src: (char *) start + phoff + n * phsize, sizeof(Elf64_Phdr));
    if (tmp.p_type == 2) {
        Elf64_Dyn tmp1;
        __android_log_print(6,"r0ysue","%x",tmp.p_memsz);
        for (int n = 0; n < tmp.p_filesz; n = n + sizeof(Elf64_Dyn)) {
            memcpy(&tmp1, src: (char *) start + tmp.p_paddr + n, sizeof(Elf64_Dyn));
            if (tmp1.d_tag == DT_JMPREL) {//DT_JMPREL==0X17
                reloff = tmp1.d_un.d_val;
            }
            if (tmp1.d_tag == DT_PLTRELSZ) {//DT_PLTRELSZ==2
                relsize = tmp1.d_un.d_val;
            }
        }
    }
}
```

图 19-32 PLT 定位

后面的思路也是基于重定向表的搜查替换，代码如图 19-33 所示。

```
pltoff = (reloff + relsize);
if (pltoff % 0x10 != 0) {
    pltoff = pltoff + 8;
}
int thoff=0;
Elf64_Rela tmp2;
for (int n = 0; n < relsize; n = n + sizeof(Elf64_Rela)) {
    memcpy(&tmp2, src: (char *) start + reloff + n, sizeof(Elf64_Rela));
    long *ptr = reinterpret_cast<long *>((char *) start + tmp2.r_offset);
    if (*ptr == (long) myr0) {
        thoff = n / sizeof(Elf64_Rela);
    }
}

long *myr0plt = reinterpret_cast<long *>((char *) start + pltoff + 0x20 + thoff * 0x10);
mprotect( addr: (void *) PAGE_START( x: (long) myr0plt), PAGE_SIZE, prot: PROT_WRITE | PROT_READ | PROT_EXEC);
__android_log_print( prio: 6, tag: "r0ysue", fmt: "%s", strerror(errno));
int faniloff = 0;
int flag;
if ((unsigned long) myr0 > (unsigned long) myr1) {
    faniloff = (unsigned long) myr0plt - (unsigned long) myr1;
    flag = 0;
} else {
    faniloff = (unsigned long) myr1 - (unsigned long) myr0plt;
    flag = 1;
}
```

图 19-33 基于重定向表的搜查替换

程序头表的完整 PLT 的 Hook 代码如代码清单 19-2 所示。

代码清单 19-2　程序头 Hook PLT

```C++
void proplthook() {
    char line[1024];
    int *start;
    int *end;
    int n = 1;
    FILE *fp = fopen("/proc/self/maps", "r");
    while (fgets(line, sizeof(line), fp)) {
        if (strstr(line, "libnative-lib.so")) {
            __android_log_print(6, "r0ysue", "%s", line);
            if (n == 1) {
                start = reinterpret_cast<int *>(strtoul(strtok(line, "-"), NULL, 16));
                end = reinterpret_cast<int *>(strtoul(strtok(NULL, " "), NULL, 16));
            } else {
                strtok(line, "-");
                end = reinterpret_cast<int *>(strtoul(strtok(NULL, " "), NULL, 16));
            }
            n++;
        }
    }
    Elf64_Ehdr header;
    memcpy(&header, start, sizeof(Elf64_Ehdr));
    int phoff = header.e_phoff;
    int phsize = header.e_phentsize;
    int phnum = header.e_phnum;
    Elf64_Phdr tmp;
    int reloff = 0;
    int relsize = 0;
    int pltoff = 0;

    for (int n = 0; n < phnum; n++) {
        memcpy(&tmp, (char *) start + phoff + n * phsize, sizeof(Elf64_Phdr));
        if (tmp.p_type == 2) {
            Elf64_Dyn tmp1;
//          __android_log_print(6,"r0ysue","%x",tmp.p_memsz);
            for (int n = 0; n < tmp.p_filesz; n = n + sizeof(Elf64_Dyn)) {
                memcpy(&tmp1, (char *) start + tmp.p_paddr + n, sizeof(Elf64_Dyn));
                if (tmp1.d_tag == DT_JMPREL) {//DT_JMPREL==0X17
                    reloff = tmp1.d_un.d_val;
                }
                if (tmp1.d_tag == DT_PLTRELSZ) {//DT_PLTRELSZ==2
                    relsize = tmp1.d_un.d_val;
                }
            }
```

```cpp
        }
    }

    pltoff = (reloff + relsize);

    if (pltoff % 0x10 != 0) {
        pltoff = pltoff + 8;
    }
    int thoff=0;
    Elf64_Rela tmp2;
    for (int n = 0; n < relsize; n = n + sizeof(Elf64_Rela)) {
        memcpy(&tmp2, (char *) start + reloff + n, sizeof(Elf64_Rela));
        long *ptr = reinterpret_cast<long *>((char *) start + tmp2.r_offset);
        if (*ptr == (long) myr0) {
            thoff = n / sizeof(Elf64_Rela);
        }
    }


    long *myr0plt = reinterpret_cast<long *>((char *) start + pltoff + 0x20 + thoff * 0x10);
    mprotect((void *) PAGE_START((long) myr0plt), PAGE_SIZE, PROT_WRITE | PROT_READ | PROT_EXEC);
        __android_log_print(6, "r0ysue", "%s", strerror(errno));
    int faniloff = 0;
    int flag;
    if ((unsigned long) myr0 > (unsigned long) myr1) {
        faniloff = (unsigned long) myr0plt - (unsigned long) myr1;
        flag = 0;

    } else {
        faniloff = (unsigned long) myr1 - (unsigned long) myr0plt;
        flag = 1;
    }

//    010000D4
    unsigned long code;
    if (flag == 1) {
        code = 0x14000000 + faniloff / 4;

    } else {
        code = 0x18000000 - faniloff / 4;
    }
    __android_log_print(6, "r0ysue", "%p", code);
    *(myr0plt) = code;
}
```

19.3 从 GOT 和 PLT 的 Hook 到 xHook 原理剖析

在 19.1 节和 19.2 节中，为读者演示了对 GOT 和 PLT 的 Hook 的简易版本，其中包含两张表 Hook 的核心原理。当然，两张表的 Hook 已经有完整的框架来完成，那就是 xHook。

xHook 是一个针对 Android 平台 ELF（可执行文件和动态库）的 PLT Hook 库。

官网有 xHook 详细的原理概述及案例代码，参考网址 57（参见配书资源文件）。

19.3.1 xHook 的优点

xHook 的优点如下：

- 支持 Android 4.0~10（API level 14~29）。
- 支持 armeabi、armeabi-v7a、arm64-v8a、x86 和 x86_64。
- 支持 ELF HASH 和 GNU HASH 索引的符号。
- 支持 SLEB128 编码的重定位信息。
- 支持通过正则表达式批量设置 Hook 信息。
- 不需要 root 权限或任何系统权限。
- 不依赖于任何的第三方动态库。

19.3.2 源码赏析

xHook 框架的核心代码位于 xh_core.c 中，xh_core.h 头文件中核心函数的定义如图 19-34 所示，其中最重要的是 xhook_refresh 函数。

```
#ifdef __cplusplus
extern "C" {
#endif

int xh_core_register(const char *pathname_regex_str, const char *symbol,
                     void *new_func, void **old_func);

int xh_core_ignore(const char *pathname_regex_str, const char *symbol);

int xh_core_refresh(int async);

void xh_core_clear();

void xh_core_enable_debug(int flag);

void xh_core_enable_sigsegv_protection(int flag);

#ifdef __cplusplus
}
#endif
```

图 19-34　核心源码

在 Hook 之前，需要对 symbol 表中的每一个函数进行注册操作。注册之后，才可以在

xhook_refresh 中操作。从图 19-34 中可以看到，在 xhook_refresh 中传入的参数是 int 类型的，在这里只传入两个参数，分别是 0 和 1。给 async 参数传 1 表示执行异步的 Hook 操作，传 0 表示执行同步的 Hook 操作。如果注入成功则返回 0，失败则返回非 0。

进入 xhook_refresh 函数内部，如图 19-35 所示。

```
int xh_core_refresh(int async)
{
    //init
    xh_core_init_once();
    if(!xh_core_init_ok) return XH_ERRNO_UNKNOWN;

    if(async)
    {
        //init for async
        xh_core_init_async_once();
        if(!xh_core_async_init_ok) return XH_ERRNO_UNKNOWN;

        //refresh async
        pthread_mutex_lock(&xh_core_mutex);
        xh_core_refresh_thread_do = 1;
        pthread_cond_signal(&xh_core_cond);
        pthread_mutex_unlock(&xh_core_mutex);
    }
    else
    {
        //refresh sync
        pthread_mutex_lock(&xh_core_refresh_mutex);
        xh_core_refresh_impl();
        pthread_mutex_unlock(&xh_core_refresh_mutex);
    }

    return 0;
}
```

图 19-35　xhook_refresh 函数

在 xh_core_init_once() 中进行初始化的操作，开启一个线程，回填 GOT。

19.4　本 章 小 结

本章讲解了 GOT 和 PLT 的 Hook 原理以及实践操作。通过利用延迟绑定机制，我们可以在内存中完美替换自定义函数。函数的绑定并不是一开始就完成的，而是在使用时进行第一次绑定，这为 Hook 操作提供了便利。我们介绍了自己编写的 GOT 和 PLT 的 Hook 代码，并介绍了 PLT 的 Hook 框架 xHook，包括对 xHook 的特点和源码的初步追踪。

第 20 章

番外篇——另类方法寻找 SO 文件首地址

在前面的章节中,我们介绍了通过 dlopen 和 dlsym 的方式来查找 SO 文件的首地址,并讨论了在高版本 Android 中不允许查找白名单之外的 SO 文件首地址的问题。这些 SO 文件的查找都是基于 maps 的方式进行的。而在本章中,我们将介绍一种新奇的思路来查找 SO 文件的首地址。

20.1 项目搭建

首先新建一个以 C++ 为模板的 findso 项目。导入第 17 章项目中的 ida.h、findsym.h、findsym.cpp 文件,如图 20-1 所示。本节会使用之前编写的函数。

在 CMakeLists.txt 中加入相应的配置,如图 20-2 所示。

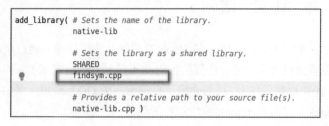

图 20-1 导入文件　　　　　　图 20-2 CMakeLists.txt 文件配置

我们需要知道 linker 的首地址,如果 linker 都获取不到的话,就不能获取链表头,也无法进行后续的解析工作。

可以尝试通过绝对地址和相对地址来获取 linker 的地址,观察是否可以正常获取 linker 的地址,代码如图 20-3 所示。

第 20 章 番外篇——另类方法寻找 SO 文件首地址

```cpp
void* findsobase(const char *soname) {
    // 测试 so 是否在白名单中
    // 首先需要拿到linker的首地址
    void *linkeraddr_0 = dlopen( filename: "linker", flag: RTLD_NOW);
    __android_log_print( prio: 6, tag: "r0ysue", fmt: "%p %s", linkeraddr_0, strerror(errno));
    void *linkeraddr_1 = dlopen( filename: "/system/bin/linker", flag: RTLD_NOW);
    __android_log_print( prio: 6, tag: "r0ysue", fmt: "%p %s", linkeraddr_1, strerror(errno));


    return reinterpret_cast<void *>(1);


}
```

图 20-3 获取 linker 地址

运行代码后，在日志中会打印出如图 20-4 所示的报错。

```
2022-03-30 21:19:29.203 29847-29847/com.r0ysue.findso E/r0ysue: 0x0 No such file or directory
2022-03-30 21:19:29.204 29847-29847/com.r0ysue.findso E/linker: library "/system/bin/linker" ("/system/bin/linker") needed or dlopened by "/data/app/com.r0ysue
.findso-m9tKzURdcuTBFnvuMkdkIA==/lib/arm64/libnative-lib.so" is not accessible for the namespace: [name="classloader-namespace", ld_library_paths="",
default_library_paths="/data/app/com.r0ysue.findso-m9tKzURdcuTBFnvuMkdkIA==/lib/arm64:/data/app/com.r0ysue.findso-m9tKzURdcuTBFnvuMkdkIA==/base.apk!/lib/arm64-v8a",
permitted_paths="/data:/mnt/expand:/data/data/com.r0ysue.findso"]
2022-03-30 21:19:29.205 29847-29847/com.r0ysue.findso E/r0ysue: 0x0 No such file or directory
```

图 20-4 linker 获取失败

可以看到 linker 不可以直接打开，所以只剩下一种方式了，就是通过 maps 扫描。当然，Frida 的开发者也是通过方式二来打开 SO 文件的，可见 maps 扫描寻找是比较可靠的方式。

方式一：从 maps 中获取 linker 的地址

在这里直接展示之前用到的 linker 查找方式，如图 20-5 所示。

方式二：遍历 soinfo 链表

通过阅读 IDA 解析的 linker，可以发现如图 20-6 所示的规律。a1 表示 soinfo 指针，指针加上固定的偏移就是某些特征字段。

```cpp
char line[1024];
int *start;
int *end;
long* base;
int n = 1;
FILE *fp = fopen( path: "/proc/self/maps", mode: "r");
while (fgets(line, sizeof(line), fp)) {
    if (strstr(line, n: "linker64")) {
        __android_log_print( prio: 6, tag: "r0ysue", fmt: "%s", line);
        if (n == 1) {
            start = reinterpret_cast<int *>(strtoul( s: strtok(line, delimiter: "-"), end_ptr: NULL, base: 16));
            end = reinterpret_cast<int *>(strtoul( s: strtok( s: NULL, delimiter: " "), end_ptr: NULL, base: 16));
        } else {
            strtok(line, delimiter: "-");
            end = reinterpret_cast<int *>(strtoul( s: strtok( s: NULL, delimiter: " "), end_ptr: NULL, base: 16));
        }
        n++;
    }
}
```

soinfo 映射表
a1+56->dynstrtab_
a1=phdr
a1+8=phdr_num
a1+16=base
a1+24=size
a1+256=load_bias
a1+64=symtab
a1+72=nbucket_
a1+80=nchain_
a1+88=bucket_
a1+96=chain_

图 20-5 linker 获取方式   图 20-6 linker

我们可以解析 linker，观察一下它的规律，如图 20-7 所示。

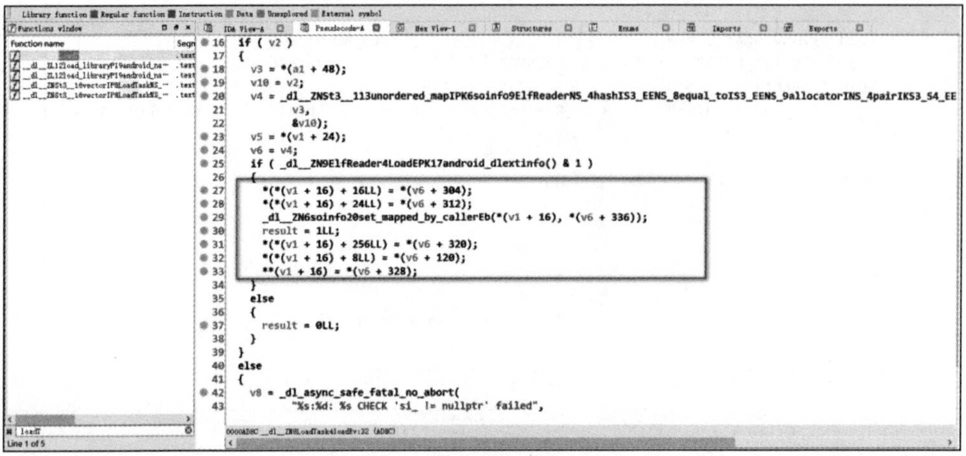

图 20-7　IDA 解析 linker

通过 soinfo 获取 linker，首先需要获取 so_list。获取 header 后，继续参考 Android 源码进行遍历，如图 20-8 所示。

```
soinfo* find_containing_library(const void* p) {
  ElfW(Addr) address = reinterpret_cast<ElfW(Addr)>(p);
  for (soinfo* si = solist_get_head(); si != nullptr; si = si->next) {
    if (address >= si->base && address - si->base < si->size) {
      return si;
    }
  }
  return nullptr;
}
```

图 20-8　遍历 soinfoAndroid 源码

如图 20-9 所示，在 IDA 中查看 find_containing_library 的实现，并移植到我们的 Android 程序中。

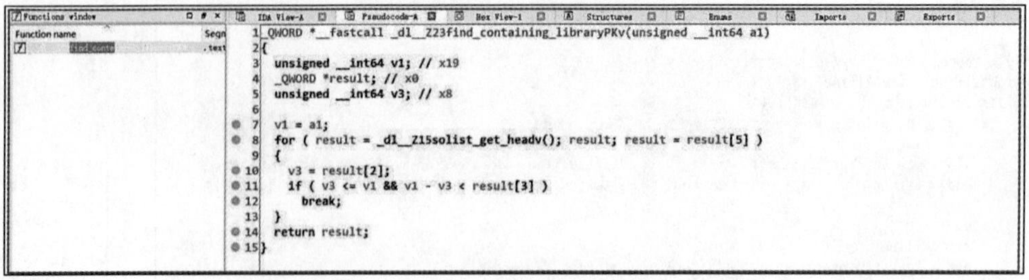

图 20-9　find_containing_library 的实现

移植好的遍历 SO 文件的代码如图 20-10 所示。在图 20-11 中可以看到所有加载的 SO 文件。

```cpp
int soheaderoff = findsym("/system/bin/linker64", "__dl__ZL6solist");
long *soheader = reinterpret_cast<long *>((char *) start + soheaderoff);

n=0;
for (_QWORD * result = reinterpret_cast<uint64 *>(*(_QWORD *) soheader); result; result = (_QWORD *)result[5] ){
    const char* name = reinterpret_cast<const char *>(*(_QWORD *) ((__int64) result + 408));
    __android_log_print( prio: 6, tag: "r0ysue", fmt: " soname:%s ", name);
}
```

图 20-10　遍历 SO 文件

```
com.r0ysue.findso E/r0ysue:    soname:ld-android.so
com.r0ysue.findso E/r0ysue:    soname:(null)
com.r0ysue.findso E/r0ysue:    soname:libcutils.so
com.r0ysue.findso E/r0ysue:    soname:libutils.so
com.r0ysue.findso E/r0ysue:    soname:liblog.so
com.r0ysue.findso E/r0ysue:    soname:libbinder.so
com.r0ysue.findso E/r0ysue:    soname:libandroid_runtime.so
com.r0ysue.findso E/r0ysue:    soname:libselinux.so
com.r0ysue.findso E/r0ysue:    soname:libhwbinder.so
com.r0ysue.findso E/r0ysue:    soname:libwilhelm.so
com.r0ysue.findso E/r0ysue:    soname:libc++.so
com.r0ysue.findso E/r0ysue:    soname:libc.so
com.r0ysue.findso E/r0ysue:    soname:libm.so
com.r0ysue.findso E/r0ysue:    soname:libdl.so
com.r0ysue.findso E/r0ysue:    soname:libbacktrace.so
com.r0ysue.findso E/r0ysue:    soname:libvndksupport.so
com.r0ysue.findso E/r0ysue:    soname:libbase.so
com.r0ysue.findso E/r0ysue:    soname:libmemtrack.so
com.r0ysue.findso E/r0ysue:    soname:libandroidfw.so
com.r0ysue.findso E/r0ysue:    soname:libappfuse.so
com.r0ysue.findso E/r0ysue:    soname:libcrypto.so
com.r0ysue.findso E/r0ysue:    soname:libnativehelper.so
```

图 20-11　遍历结果

下一步获取过滤出的目标 SO 文件，以便开展后续的工作。可以在日志栏看到我们过滤出来的目标 SO 文件，如图 20-12 所示。

```cpp
int soheaderoff = findsym("/system/bin/linker64", "__dl__ZL6solist");
long *soheader = reinterpret_cast<long *>((char *) start + soheaderoff);

n=0;
for (_QWORD * result = reinterpret_cast<uint64 *>(*(_QWORD *) soheader); result; result = (_QWORD *)result[5] ){
    if(*(_QWORD *) ((__int64) result + 408)!= 0&&strcmp(reinterpret_cast<const char *>(*(_QWORD *) ((__int64) result + 408)), soname)==0) {
        const char* name= reinterpret_cast<const char *>(*(_QWORD *) ((__int64) result + 408));
        __android_log_print( prio: 6, tag: "r0ysue", fmt: " soname:%s ", name);
    }
}
```

```
32.854 19592-19592/? I/zygote64:     at void com.android.internal.os.RuntimeInit$MethodAndArgsCaller.run() (RuntimeInit.java:438)
32.854 19592-19592/? I/zygote64:     at void com.android.internal.os.ZygoteInit.main(java.lang.String[]) (ZygoteInit.java:807)
32.854 19592-19592/? I/zygote64:     at void de.robv.android.xposed.XposedBridge.main(java.lang.String[]) (XposedBridge.java:108)
33.026 19592-19592/? E/r0ysue: 7491eb000-7491fac000 r-xp 00000000 103:09 554      /system/bin/linker64
33.026 19592-19592/? E/r0ysue: 7491fad000-7491fb1000 r--p 000f4000 103:09 554      /system/bin/linker64
33.026 19592-19592/? E/r0ysue: 7491fb1000-7491fb2000 rw-p 000f8000 103:09 554      /system/bin/linker64
33.029 19592-19592/? E/r0ysue:     soname:libart.so
33.078 19592-19619/? D/OpenGLRenderer: HWUI GL Pipeline
```

图 20-12　过滤结果

在 linker 中有一个 load 函数，可以获取一些信息字段，如图 20-13 所示。里面的字段与图 20-7

中的字段都是一一对应的。需要注意的是，这些值不是通过指针相加的，而是通过'char*'的方式。

```
bool load() {
    ElfReader& elf_reader = get_elf_reader();
    if (!elf_reader.Load(extinfo_)) {
        return false;
    }
    si_->base = elf_reader.load_start();
    si_->size = elf_reader.load_size();
    si_->set_mapped_by_caller(elf_reader.is_mapped_by_caller());
    si_->load_bias = elf_reader.load_bias();
    si_->phnum = elf_reader.phdr_count();
    si_->phdr = elf_reader.loaded_phdr();
```

图 20-13　linker 中的 load 函数

编写代码，输出关键字段，如图 20-14 所示。

```
n=0;
for (_QWORD * result = reinterpret_cast<uint64 *>(*(_QWORD *) soheader); result; result = (_QWORD *)result[5] ){
    if(*(_QWORD *) ((__int64) result + 408)!= 0&&strcmp(reinterpret_cast<const char *>(*(_QWORD *) ((__int64) result + 408)), soname)==0) {
        base= reinterpret_cast<long *>(*(_QWORD *) ((char *) result + 16));
        long* size= reinterpret_cast<long *>(*(_QWORD *) ((char *) result + 24));
        long* load_bias= reinterpret_cast<long *>(*(_QWORD *) ((char *) result + 256));
        const char* name= reinterpret_cast<const char *>(*(_QWORD *) ((__int64) result + 408));
        __android_log_print( prio: 6, tag: "r0ysue", fmt: " soname:%s ", name);
        __android_log_print( prio: 6, tag: "r0ysue", fmt: " base:%p ", *base);
        __android_log_print( prio: 6, tag: "r0ysue", fmt: " size:%x ",size);
        __android_log_print( prio: 6, tag: "r0ysue", fmt: " loadbias:%p ", load_bias);
    }
}
```

图 20-14　输出关键字段

运行 App，查看结果，如图 20-15 所示。

将 base 和 load_bias 通过十六进制相减可以看到，结果为 25000，如图 20-16 所示，就是 libart.so 的首段偏移。

图 20-15　运行结果

图 20-16　base 和 load_bias 的差值

当然，我们也可以到内存中的 maps 查看加载的 libart.so，比对一下我们找出来的地址，看看是否是一样的。在图 20-17 的比对中，可以发现通过程序查找出来的地址与内存中的首地址是相同的。

图 20-17 结果比对

使用第二种方式有一个好处，就是所有通过 linker 的 SO 文件都能被找到。不好的地方在于：如果 SO 文件加壳，那么这个 SO 文件不经过 linker，就不会被记录到 soinfo 的链表中。

20.2 封装成库

通过 20.1 节中 linker 的两种查找方式的结合就可以完美找到 SO 文件的首地址，我们可以把这个查找过程的代码封装成一个函数，以便后续调用。

将 linker 查找的第一种方式封装为 initlinker()，第二种方式封装为 findsobase()，并补齐相关的头文件，如图 20-18 所示。代码如代码清单 20-1 所示。

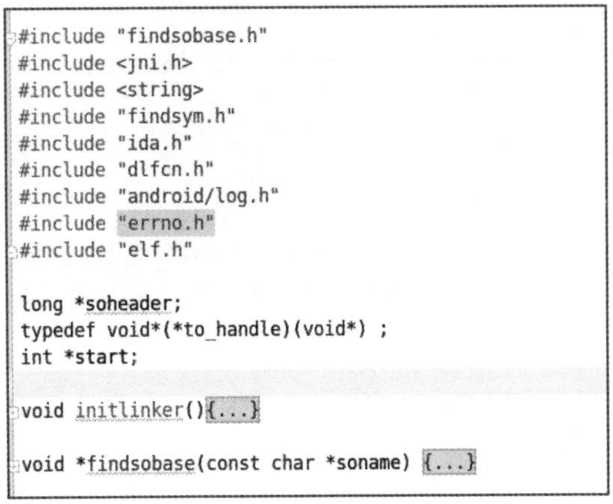

图 20-18 函数封装

代码清单 20-1 linker 寻找 SO 文件的首地址

```c
#include "findsobase.h"
#include <jni.h>
#include <string>
#include "findsym.h"
#include "ida.h"
#include "dlfcn.h"
#include "android/log.h"
```

```cpp
#include "errno.h"
#include "elf.h"

long *soheader;
typedef void*(*to_handle)(void*) ;
int *start;

void initlinker(){
    char line[1024];
    int *end;
    long* base;
    int n = 1;
    FILE *fp = fopen("/proc/self/maps", "r");
    while (fgets(line, sizeof(line), fp)) {
        if (strstr(line, "linker64")) {
            __android_log_print(6, "r0ysue", "%s", line);
            if (n == 1) {
                start = reinterpret_cast<int *>(strtoul(strtok(line, "-"), NULL, 16));
                end = reinterpret_cast<int *>(strtoul(strtok(NULL, " "), NULL, 16));

            } else {
                strtok(line, "-");
                end = reinterpret_cast<int *>(strtoul(strtok(NULL, " "), NULL, 16));
            }
            n++;
        }
    }
    int soheaderoff = findsym("/system/bin/linker64", "__dl__ZL6solist");
    soheader = reinterpret_cast<long *>((char *) start + soheaderoff);
}

void *findsobase(const char *soname) {

    long* base;
    long* soinfo;
    int n=0;
    for (_QWORD * result = reinterpret_cast<uint64 *>(*(_QWORD *) soheader); result; result = (_QWORD *)result[5] ){
        if(*(_QWORD *) ((__int64) result + 408)!= 0&&strcmp(reinterpret_cast<const char *>(*(_QWORD *) ((__int64) result + 408)), soname)==0) {
            base= reinterpret_cast<long *>(*(_QWORD *) ((char *) result + 16));
            long* size= reinterpret_cast<long *>(*(_QWORD *) ((char *) result + 24));
            long* load_bias= reinterpret_cast<long *>(*(_QWORD *) ((char *) result + 256));
            const char* name= reinterpret_cast<const char *>(*(_QWORD *) ((__int64) result + 408));
            __android_log_print(6, "r0ysue", " soname:%s ", name);
```

```
            __android_log_print(6, "r0ysue", " base:%p ", *base);
            __android_log_print(6, "r0ysue", " size:%x ",size);
            __android_log_print(6, "r0ysue", " loadbias:%p ", load_bias);
            soinfo= reinterpret_cast<long *>(result);
            break;
        }


    }

    return soinfo;
}
...
```

通过 linker 查找的第二种方式，可以直接通过如图 20-6 所示的方式来构造，而不用去找其结构。

20.3　通过 soinfo 的映射表遍历符号

在 linker 的第二种查找方式中，可以获取 strtab 和 symtab 表，通过这种方式也可以遍历符号表。

要遍历 symtab，就需要知道其大小，通过 IDA 分析观察发现，symtab 后面就是 strtab，所以获取其大小可以通过两者相减得到，如图 20-19 所示。

图 20-19　symtab 和 strtab 的位置

这里就可以用到前面的 linker 原理，通过 soinfo 来实现。首先获取 strtab、symtab 和 strsize，代码如图 20-20 所示。

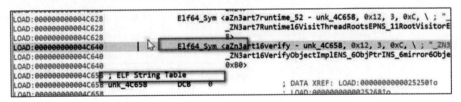

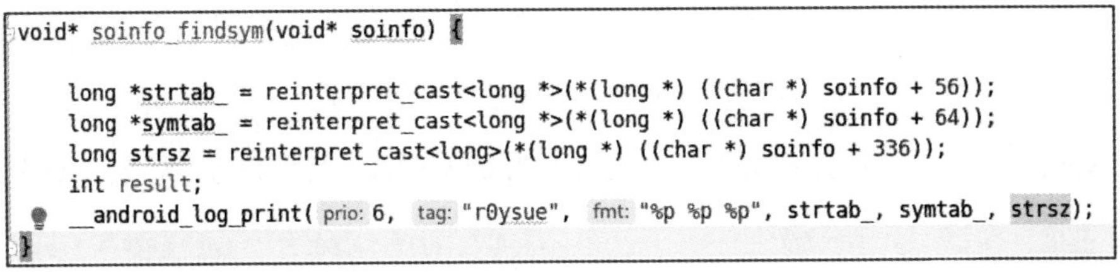

图 20-20　获取 symtab、strtab 和 strsize

打印结果如图 20-21 所示。

$$\boxed{\texttt{0x6f8c404658 0x6f8c3dd238 0x70415}}$$

图 20-21　打印结果

让 symtab 和 load_bias 相减的结果如图 20-22 所示。

```
输入十六进制数（A）  6f8c3dd238
输入十六进制数（B）  6f8c3b8000
              计算
结果：
A-B（十六进制）  25238
A-B（十进制）    152120
```

图 20-22　symtab 和 load_bias 相减结果

在 IDA 中查看这个地址，即可发现这就是 ELF Symbol Table 的起始位置，如图 20-23 所示。

```
LOAD:0000000000025228        DCB 0x93, 0x62, 0xCD, 0x71, 0x83, 0x3F, 0x1C
LOAD:0000000000025238        ; ELF Symbol Table
LOAD:0000000000025238        Elf64_Sym <0>
LOAD:0000000000025250        Elf64_Sym <aPthreadMutexLo - byte_4C658, 0x22, 0,
LOAD:0000000000025268        Elf64_Sym <aPthreadMutexUn - byte_4C658, 0x22, 0,
LOAD:0000000000025280        Elf64_Sym <aCxaFinalize - byte_4C658, 0x12, 0, 0,
LOAD:0000000000025298        Elf64_Sym <aRegisterAtfork - byte_4C658, 0x12, 0,
LOAD:00000000000252B0        Elf64_Sym <aCxaAtexit - byte_4C658, 0x12, 0, 0,
```

图 20-23　ELF Symbol Table 的起始位置

让 strtab 和 load_bias 相减，结果如图 20-24 所示，即可得到 strtab 的开头。

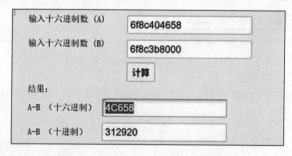

图 20-24　strtab 和 load_bias 相减结果

获取 symtab 的首地址和 symsize 的大小便可以开始遍历查找目标符号，完整的代码如图 20-25 所示。

```cpp
void* soinfo_findsym(void* soinfo) {
    long *strtab_ = reinterpret_cast<long *>(*(long *) ((char *) soinfo + 56));
    long *symtab_ = reinterpret_cast<long *>(*(long *) ((char *) soinfo + 64));
    long strsz = reinterpret_cast<long>(*(long *) ((char *) soinfo + 336));
    int result;
    __android_log_print( prio: 6, tag: "r0ysue", fmt: "%p %p %p", strtab_, symtab_, strsz);

    char strtab[strsz];
    memcpy(&strtab, strtab_, strsz);
    Elf64_Sym mytmpsym;
    for (int n = 0; n < (long) strtab_ - (long) symtab_; n = n + sizeof(Elf64_Sym)) {
        memcpy(&mytmpsym,(char*)symtab_+n,sizeof(Elf64_Sym));
//            __android_log_print(6,"r0ysue","%p %s",mytmpsym.st_name,strtab+mytmpsym.st_name);
        if(strstr( h: reinterpret_cast<const char *>(strtab + mytmpsym.st_name), n: "artFindNativeMethod"))
            result=mytmpsym.st_value;
    }

    return reinterpret_cast<void *>(result);
}
```

图 20-25 遍历查找目标符号

20.4 dlopen 和 dlsym 获取符号地址

通过 dlopen 打开 SO 文件，获取 handle 传递给 dlsym 打开符号地址，代码如图 20-26 所示。

```cpp
void* libc=dlopen( filename: "libc.so", flag: RTLD_NOW);
__android_log_print( prio: 6, tag: "r0ysue", fmt: " dlopen %p",libc);

long* mystrstr= static_cast<long *>(dlsym(libc, symbol: "strstr"));
__android_log_print( prio: 6, tag: "r0ysue", fmt: "dlsym %p",mystrstr);
```

图 20-26 通过 dlopen 和 dlsym 打开符号地址

在源码中也可以看到 dlopen 和 dlsym 的传递过程，核心源码将 soinfo 变成一个 handle，如图 20-27 所示。

```cpp
ProtectedDataGuard guard;
soinfo* si = find_library(ns, translated_name, flags, extinfo, caller);
loading_trace.End();

if (si != nullptr) {
    void* handle = si->to_handle();
    LD_LOG(kLogDlopen,
           "... dlopen calling constructors: realpath=\"%s\", soname=\"%s\", handle=%p",
           si->get_realpath(), si->get_soname(), handle);
    si->call_constructors();
    failure_guard.Disable();
```

图 20-27 handle 的构造

handle 类似于哈希的查询与反查询的过程，我们现在通过 soinfo 尝试还原出 handle。直接将 soinfo 传入，调用 to_handle 函数即可，代码如图 20-28 所示。

```
void* getmyhandle(void* soinfo){
    int handleoff=findsym("/system/bin/linker64","__dl__ZN6soinfo9to_handleEv");
    to_handle func= reinterpret_cast<to_handle>((char *) start + handleoff);
    void* myhandle=func(soinfo);

    return myhandle;
}
```

图 20-28　handle 还原

20.5　本章小结

在本章中引入了 SO 文件首地址查找方式的新思路，该新思路基于对 IDA 中 linker 加载 SO 文件源码的分析，找到了一个单向链表 soinfo，通过这个链表可以获取一个 soinfo 映射表。通过这个映射表，我们可以获取一些信息字段，如 strtab（字符串表）、symtab（符号表）等。

第 21 章

Java Hook 的原理

前面的章节介绍了 SO 文件的格式以及追踪的原理和方法。在本章中，我们将重心转移到 Java 层，带领读者了解 Java 函数的执行流程，并从中总结出 Java Hook 的原理。

21.1 Java 函数源码追踪

21.1.1 什么是 Java Hook

新建一个以 C++为模板的 Android 工程，在 Java 层编写 add 函数。

Java Hook 就是通过 Hook 操作替换掉 add 函数，或者传入我们自己的参数，返回想要的返回值。

在项目中定义了 add 函数，我们可以替换 add 函数，实现 Java Hook，代码如图 21-1 所示。

```
@Override
protected void onCreate(Bundle savedInstanceState) {
    super.onCreate(savedInstanceState);
    setContentView(R.layout.activity_main);

    // Example of a call to a native method
    TextView tv = findViewById(R.id.sample_text);
    tv.setText(stringFromJNI());
    add();
}

/**
 * A native method that is implemented by the 'native-lib' native library,
 * which is packaged with this application.
 */
public native String stringFromJNI();

void add(){
    Log.e( tag: "roysue", msg: "I'm from add");
}
```

图 21-1　add 函数

当然，也可以使用 Frida 来实现 Hook，在第 21.1.2 节中将介绍如何使用 Frida 进行 Hook 操作。

我们可以在 Native 层实现 Java Hook，这样一方面可以用于加固我们的 App，在网址 58（参见配书资源文件）的看雪网站有一道关于这方面的题目，读者可以研究一下。另一方面，通过 Java Hook 可以实现一个简易的主动调用框架，实现类似于 Frida 的效果。可以说，如果没有 Java Hook，Frida 项目本身就是不完整的，因为只有 C 层的 Frida 用着也不舒服，所以我们有必要学习 Java Hook 的原理，并将其运用到自定义的主动调用框架中。

21.1.2　源码追踪

首先，我们需要知道 Java 中类加载的流程，我们将从 Android 8.1.0 的源码进行追踪。

类加载的核心函数在 ClassLoader.java 中，即 loadClass 函数。要加载一个类，我们可以使用 loadClass 函数来进行加载，源码如图 21-2 所示。

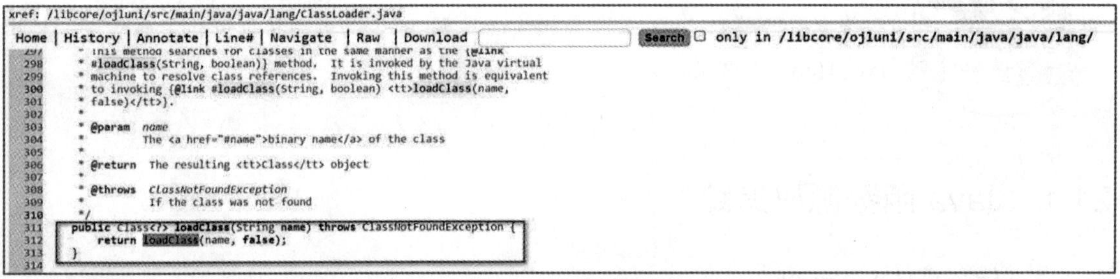

```
xref: /libcore/ojluni/src/main/java/java/lang/ClassLoader.java
Home | History | Annotate | Line# | Navigate | Raw | Download    Search  only in /libcore/ojluni/src/main/java/java/lang/
     * This method searches for classes in the same manner as the {@link
     * #loadClass(String, boolean)} method.  It is invoked by the Java virtual
     * machine to resolve class references.  Invoking this method is equivalent
     * to invoking {@link #loadClass(String, boolean) <tt>loadClass(name,
     * false)</tt>}.
     *
     * @param  name
     *         The <a href="#name">binary name</a> of the class
     *
     * @return  The resulting <tt>Class</tt> object
     *
     * @throws  ClassNotFoundException
     *          If the class was not found
     */
    public Class<?> loadClass(String name) throws ClassNotFoundException {
        return loadClass(name, false);
    }
```

图 21-2　loadClass 函数

loadClass 核心源码如图 21-3 所示，它使用的是双亲委派的机制，先判断是否为加载过的 class，如果加载过，就不再加载，如果没加载过，就使用它的 parent 加载器进行加载。

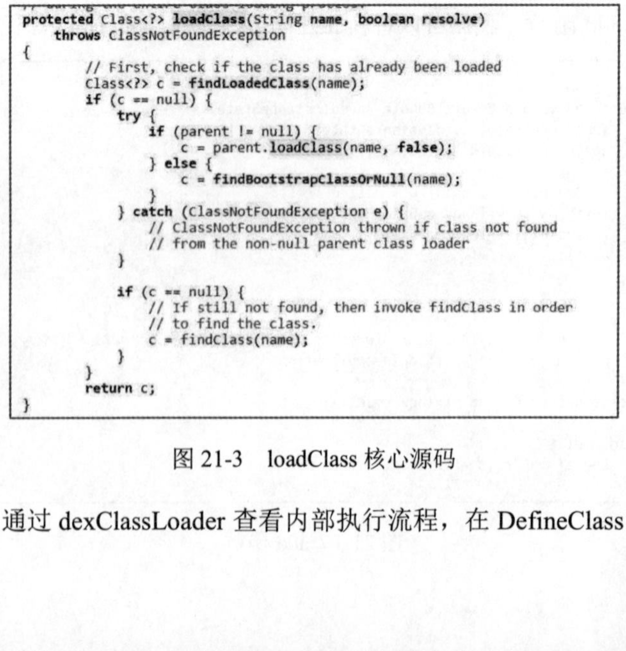

```
protected Class<?> loadClass(String name, boolean resolve)
    throws ClassNotFoundException
{
        // First, check if the class has already been loaded
        Class<?> c = findLoadedClass(name);
        if (c == null) {
            try {
                if (parent != null) {
                    c = parent.loadClass(name, false);
                } else {
                    c = findBootstrapClassOrNull(name);
                }
            } catch (ClassNotFoundException e) {
                // ClassNotFoundException thrown if class not found
                // from the non-null parent class loader
            }

            if (c == null) {
                // If still not found, then invoke findClass in order
                // to find the class.
                c = findClass(name);
            }
        }
        return c;
}
```

图 21-3　loadClass 核心源码

继续往下追踪，通过 dexClassLoader 查看内部执行流程，在 DefineClass 函数中做参数校验，如图 21-4 所示。

```
mirror::Class* ClassLinker::DefineClass(Thread* self,
                                  const char* descriptor,
                                  size_t hash,
                                  Handle<mirror::ClassLoader> class_loader,
                                  const DexFile& dex_file,
                                  const DexFile::ClassDef& dex_class_def) {
    StackHandleScope<3> hs(self);
    auto klass = hs.NewHandle<mirror::Class>(nullptr);

    // Load the class from the dex file.
    if (UNLIKELY(!init_done_)) {
        // finish up init of hand crafted class_roots_
        if (strcmp(descriptor, "Ljava/lang/Object;") == 0) {
            klass.Assign(GetClassRoot(kJavaLangObject));
        } else if (strcmp(descriptor, "Ljava/lang/Class;") == 0) {
            klass.Assign(GetClassRoot(kJavaLangClass));
        } else if (strcmp(descriptor, "Ljava/lang/String;") == 0) {
            klass.Assign(GetClassRoot(kJavaLangString));
        } else if (strcmp(descriptor, "Ljava/lang/ref/Reference;") == 0) {
            klass.Assign(GetClassRoot(kJavaLangRefReference));
        } else if (strcmp(descriptor, "Ljava/lang/DexCache;") == 0) {
            klass.Assign(GetClassRoot(kJavaLangDexCache));
        } else if (strcmp(descriptor, "Ldalvik/system/ClassExt;") == 0) {
            klass.Assign(GetClassRoot(kDalvikSystemClassExt));
        }
    }
```

图 21-4 DefineClass 参数校验

继续追踪，我们会发现 ClassLinker 中有一个 LoadMethod 函数，它负责加载 Method（方法），为它赋一些初始值。通过这个函数，ArtMethod 将完成初始化，源码如图 21-5 所示。在 LoadMethod 函数的参数中，ArtMethod 是 Method 在 ART 虚拟机中的一个表现形式，ART 虚拟机执行的就是这个 Method，同时这也是一个关键的脱壳点。

```
void ClassLinker::LoadMethod(const DexFile& dex_file,
                             const ClassDataItemIterator& it,
                             Handle<mirror::Class> klass,
                             ArtMethod* dst) {
    uint32_t dex_method_idx = it.GetMemberIndex();
    const DexFile::MethodId& method_id = dex_file.GetMethodId(dex_method_idx);
    const char* method_name = dex_file.StringDataByIdx(method_id.name_idx_);

    ScopedAssertNoThreadSuspension ants("LoadMethod");
    dst->SetDexMethodIndex(dex_method_idx);
    dst->SetDeclaringClass(klass.Get());
    dst->SetCodeItemOffset(it.GetMethodCodeItemOffset());

    dst->SetDexCacheResolvedMethods(klass->GetDexCache()->GetResolvedMethods(), image_pointer_size_);

    uint32_t access_flags = it.GetMethodAccessFlags();
```

图 21-5 LoadMethod 初始化

接续追踪，就到了本章的核心 LinkCode 的位置，LinkCode 源码如代码清单 21-1 所示。

代码清单 21-1 LinkCode 源码

```
static void LinkCode(ClassLinker* class_linker,
                     ArtMethod* method,
                     const OatFile::OatClass* oat_class,
                     uint32_t class_def_method_index)
```

```cpp
REQUIRES_SHARED(Locks::mutator_lock_) {
    Runtime* const runtime = Runtime::Current();
    if (runtime->IsAotCompiler()) {
      // The following code only applies to a non-compiler runtime.
      return;
    }
    // Method shouldn't have already been linked.
    DCHECK(method->GetEntryPointFromQuickCompiledCode() == nullptr);
    if (oat_class != nullptr) {
      // Every kind of method should at least get an invoke stub from the oat_method.
      // non-abstract methods also get their code pointers.
      const OatFile::OatMethod oat_method =
 oat_class->GetOatMethod(class_def_method_index);
      oat_method.LinkMethod(method);
    }

    // Install entry point from interpreter.
    const void* quick_code = method->GetEntryPointFromQuickCompiledCode();
    bool enter_interpreter = class_linker->ShouldUseInterpreterEntrypoint(method,
 quick_code);

    if (!method->IsInvokable()) {
      EnsureThrowsInvocationError(class_linker, method);
      return;
    }

    if (method->IsStatic() && !method->IsConstructor()) {
      // For static methods excluding the class initializer, install the trampoline.
      // It will be replaced by the proper entry point by
 ClassLinker::FixupStaticTrampolines
      // after initializing class (see ClassLinker::InitializeClass method).
      method->SetEntryPointFromQuickCompiledCode(GetQuickResolutionStub());
    } else if (quick_code == nullptr && method->IsNative()) {
      method->SetEntryPointFromQuickCompiledCode(GetQuickGenericJniStub());
    } else if (enter_interpreter) {
      // Set entry point from compiled code if there's no code or in interpreter only mode.
      method->SetEntryPointFromQuickCompiledCode(GetQuickToInterpreterBridge());
    }

    if (method->IsNative()) {
      // Unregistering restores the dlsym lookup stub.
      method->UnregisterNative();

      if (enter_interpreter || quick_code == nullptr) {
        // We have a native method here without code. Then it should have either the generic
 JNI
        // trampoline as entrypoint (non-static), or the resolution trampoline (static).
        // TODO: this doesn't handle all the cases where trampolines may be installed.
        const void* entry_point = method->GetEntryPointFromQuickCompiledCode();
        DCHECK(class_linker->IsQuickGenericJniStub(entry_point) ||
```

```
                class_linker->IsQuickResolutionStub(entry_point));
        }
    }
}
```

LinkCode 中区分了普通函数和 Native 函数,以及解释模式和 JIT 模式(Quick 模式)。我们的实验都是在解释模式下执行的,例如前面编写的 add 函数就是在解释模式下执行的。后续讲解 Frida 时再为读者详细阐述这两种模式。

在解释模式下,会先经过 GetEntryPointFromQuickCompiledCode(),它是加载 SO 文件的起始点。然后进入 GetQuickToInterpreterBridge(),源码如图 21-6 所示。

```
} else if (enter_interpreter) {
    // Set entry point from compiled code if there's no code or in interpreter only mode.
    method->SetEntryPointFromQuickCompiledCode(GetQuickToInterpreterBridge());
}
```

图 21-6　解释模式

继续往下追踪,GetQuickToInterpreterBridge()是一个 inline 函数,它返回的是一个地址,这个地址是使用汇编语言编写的函数,文件如图 21-7 所示。

```
/art/runtime/arch/arm64/
  A D   quick_entrypoints_arm64.S       2301 ENTRY art_quick_to_interpreter_bridge
                                        2318 END art_quick_to_interpreter_bridge
/art/runtime/arch/mips64/
```

图 21-7　文件位置

这个地址的功能就是跳转到解释器,汇编源码如图 21-8 所示。

```
ENTRY art_quick_to_interpreter_bridge
    SETUP_SAVE_REFS_AND_ARGS_FRAME         // Set up frame and save arguments.

    //  x0 will contain mirror::ArtMethod* method.
    mov x1, xSELF                          // How to get Thread::Current() ???
    mov x2, sp

    // uint64_t artQuickToInterpreterBridge(mirror::ArtMethod* method, Thread* self,
    //                                      mirror::ArtMethod** sp)
    bl     artQuickToInterpreterBridge

    RESTORE_SAVE_REFS_AND_ARGS_FRAME       // TODO: no need to restore arguments in this case.
    REFRESH_MARKING_REGISTER

    fmov d0, x0

    RETURN_OR_DELIVER_PENDING_EXCEPTION
END art_quick_to_interpreter_bridge
```

图 21-8　汇编源码

回到 LinkCode 中,如果是 Native 函数,就先调用 UnregisterNative()把所有的注册都解除,源码如图 21-9 所示。追踪进去有一个 GetJniDlsymLookupStub(),它是寻找 JNI 绑定地址的函数,源码如图 21-10 所示。

```
if (method->IsNative()) {
    // Unregistering restores the dlsym lookup stub.
    method->UnregisterNative();
}

if (enter_interpreter || quick_code == nullptr) {
    // We have a native method here without code. Then it should have either the generic JNI
    // trampoline as entrypoint (non-static), or the resolution trampoline (static).
    // TODO: this doesn't handle all the cases where trampolines may be installed.
    const void* entry_point = method->GetEntryPointFromQuickCompiledCode();
    DCHECK(class_linker->IsQuickGenericJniStub(entry_point) ||
           class_linker->IsQuickResolutionStub(entry_point));
}
```

图 21-9 UnregisterNative()

```
void ArtMethod::UnregisterNative() {
    CHECK(IsNative() && !IsFastNative()) << PrettyMethod();
    // restore stub to lookup native pointer via dlsym
    SetEntryPointFromJni(GetJniDlsymLookupStub());
}
```

图 21-10 寻找 JNI 绑定地址

解释模式和 JNI 模式的区别就是 EntryPoint 不同，如果想进行 Java Hook，就要把普通函数改为 Native 函数，在 C 中编写一段执行代码来模拟执行 Java 层的逻辑。

综上所述 LinkCode 函数中有一个重要的过程，就是 GetEntryPointFromQuickCompiledCode 函数，它的返回值是 quick_code，如果为空，则代表该方法没有进行编译，在 oat 文件中也没有对应的汇编代码，需要进入解释模式而不是 JIT 模式，这里有一个分支，如果 ArtMethod 是 JNI 方法，则会将类似于 ELF 文件的 ENTRYPOINT 设置为 art_quick_generic_jni_trampoline，普通方法会设置为 art_quick_to_interpreter_bridge，如图 21-11 所示。这里是关键，代表我们可以通过修改这里来完成 Java 或 JNI 函数的跳转，也就是每次 ART 虚拟机解释执行某个 Java 函数的时候，都会跳转到该 Java 函数的 EntryPoint，只要把这里替换成我们自己的 C 函数的地址，就可以实现拦截 Java 函数了，当然也要保留原函数的执行过程。

方法类型	机器码入口	JNI 机器码入口
JNI 方法	art_quick_generic_jni_trampoline 或对应的机器码	art_jni_dlsym_lookup_stub
非 JNI 方法	art_quick_to_interpreter_bridge 或对应的机器码	其他用途

图 21-11 方法说明

21.2 Java Hook 实践

在 21.1 节中，我们了解了 Java 函数的大致执行流程，那么对于 Java 函数的入口点，我们怎么修改呢？

是否可以通过 Hook 关键 art 函数的方式来获取 ArtMethod 呢？

当然可以尝试，但是需要注意的是我们希望 Hook 的是自身的 DEX 文件，而它的加载时间早于我们的 SO 文件加载到内存的时间。因此，像 LinkCode 或者 LoadMethod 这样的函数在 DEX 加载流程上是不行的。因此需要找到一个系统函数，它包含我们要 Hook 的函数的 ArtMethod，并且在 DEX 加载完成后才执行。这个就很难办了，首先通过函数名过滤 ArtMethod 就很困难，其次在 DEX 加载完成之后也很难找到一个函数能够覆盖所有 ArtMethod，所以这种方法不可行。

首先获取 ArtMethod，之前讲解过，jmethod ID 其实就是 ArtMethod，它们之间有一个转换函数，源码如图 21-12 所示。

```
static inline ArtMethod* DecodeArtMethod(jmethodID method_id) {
    return reinterpret_cast<artmethod*>(method_id);
}
```

图 21-12　jmethodID 和 ArtMethod 的转换

然后将它的 EntryPoint 替换为我们的自定义函数，此时产生了两个问题：

（1）怎么找到 EntryPoint？

在 Android 的源码中提供了相关函数，也就是前面提到的 GetEntryPointFromQuickCompiledCode，只要通过 IDA 查看它的实现就好了，遗憾的是这是一个 inline 函数，无法使用 IDA 查看，只能通过自己修改源码来查看偏移，然后直接脱出 libart.so，查看导出函数 getmynative。源码修改如图 21-13 所示。

```
//ArtMethod.cc
extern "C" const void * getmynative(ArtMethod*
m)REQUIRES_SHARED(Locks::mutator_lock_){

m->SetEntryPointFromQuickCompiledCode(GetQuickGenericJniStub());
m->UnregisterNative();
const void* entry_point = m->GetEntryPointFromQuickCompiledCode();
if(m->IsFastNative())
return GetQuickGenericJniStub();
return entry_point;

}
```

图 21-13　getmynative 源码

通过 IDA 解析查看，如图 21-14 所示，可以看到 EnterPoint 的偏移为 40，我们直接通过偏移更改即可。

```
1  __int64 (__fastcall *__fastcall getmynative(__int64 a1))()
2  {
3    __int64 (__fastcall *result)();  // x0
4
5    *(_QWORD *)(a1 + 40) = art_quick_generic_jni_trampoline;
6    art::ArtMethod::UnregisterNative((art::ArtMethod *)a1);
7    if ( (~*(_DWORD *)(a1 + 4) & 0x80100) != 0 )
8      result = *(__int64 (__fastcall **)())(a1 + 40);
9    else
10     result = art_quick_generic_jni_trampoline;
11   return result;
12 }
```

图 21-14　EnterPoint 偏移

（2）直接进行替换会减少很多系统函数的调用，能否成功呢？

我们不妨试一下，看一下是否会成功。理论上是不会成功的，因为还有一些参数、栈的保存都没有处理，代码如图 21-15 所示。

输出结果如图 21-16 所示，可以看到结果成功了，虽然成功了，但是要思考一个问题，就是如何执行原函数，如果这里通过反射直接执行的话，我们就获取不到参数，所以这种方案不可靠，还需要继续改进。

```
void* func(void* a,void* b,int c,int d){

    __android_log_print(6,"r0ysue","i am success");
    return reinterpret_cast<void *="">(5);

}

jclass a=env->FindClass(classname);
jmethodID b=env->GetMethodID(a,methodname, shoty);
*((long *)a1 + 5)= reinterpret_cast<long>(func);
```

```
/com.r0ysue.myjavahook E/zygote64: 1
/com.r0ysue.myjavahook E/zygote64: 0x74a1e941b4
/com.r0ysue.myjavahook E/r0ysue: i am success
/com.r0ysue.myjavahook E/r0ysue: 5
/com.r0ysue.myjavahook E/r0ysue: xxxxxxx
/com.r0ysue.myjavahook E/r0ysue: 100
```

图 21-15　直接替换　　　　　　　　　　图 21-16　替换结果

既然上面的方案有瑕疵，就是有很多函数没有执行，导致我们获取不到参数，那么可以采取另一种方案，就是模仿 Native 函数的方式，将 Java 函数改为 Native 函数，再使用 Native 函数注册，根据 21.1 节的内容，只要将 EntryPoint 改成 art_quick_generic_jni_trampoline 机器码入口即可。再次使用 IDA 打开 libart.so，发现 art_quick_generic_jni_trampoline 是一个函数，但它不是一个导出函数，使用导出表没有办法找到它，这时就需要用到第 19 章提到的使用节头表的索引方式来解决，最后用 RegisterNatives 注册函数地址，代码如图 21-17 所示。

```
int so=findsym("/system/lib64/libart.so","art_quick_generic_jni_trampoline");
*((long *)a1 + 5)= reinterpret_cast<long>((char *) startr +so - 0x25000);//需要-0x25000是因为libart.so的程序头在偏移为0x25000的位置
env->RegisterNatives(a,getMethods,1);
```

图 21-17　模仿 Native 函数

这样就完成了 Java 函数的 Native 化，效果如图 21-18 所示。

```
-74b93f1000 rw-p 00619000 103:12 1824                    /system/lib64/libart.so
p register non-native method com.r0ysue.myjavahook.MainActivity.add2(II)I as native
ext.cc:534] JNI DETECTED ERROR IN APPLICATION: JNI FindClass called with pending exception java.lang.NoSuchMethodError: no native method "Lcom/r0ysue/myjavahook/MainActivit
ext.cc:534]   at java.lang.String com.r0ysue.myjavahook.MainActivity.stringFromJNI() (MainActivity.java:-2)
ext.cc:534]   at void com.r0ysue.myjavahook.MainActivity.onCreate(android.os.Bundle) (MainActivity.java:26)
ext.cc:534]   at void android.app.Activity.performCreate(android.os.Bundle, android.os.PersistableBundle) (Activity.java:6999)
```

图 21-18　Java 函数的 Native 化

可以看到效果不是很好，报错提示不是 native 函数，无法完成注册，我们看一下它是如何判断是否为 Native 函数的：之前编译 ROM 的时候就加入了判断方法，公式为(~*(_DWORD *)(a1 + 4) & 0x80100)，只要结果不为 0，我们就认为函数是 Native 函数。

至此，我们就可以像动态注册的 JNI 函数一样来实现 C 层的 add 函数了，执行也就正常了。接下来解决最后一个问题——如何执行原函数。

C 调用 Java 只有一种方式，那就是反射，所以我们只能使用反射，使用反射就需要收集从参数。例如图 21-19 所示的函数就需要一个实例、一个 env 和两个参数。

幸运的是，我们是动态注册的，动态注册的实例函数会自带 env 和 this 实例，所以直接调用即可，代码如图 21-20 所示。

```
public int add2(int a,int b){
    return 7;
}
```

图 21-19　案例

```
void* add(void* a,void* b,int c,int d){
    JNIEnv *st=(JNIEnv *)a;
    jclass a2=st->FindClass("com/r0ysue/myjavahook/MainActivity");
    jmethodID b2=st->GetMethodID(a2,"add2", "(II)I");
    int yy = st->CallIntMethod(static_cast<jobject>(b), b2, c, d);
    __android_log_print(6,"r0ysue","%x",yy);
    return reinterpret_cast<void *="">(5);
}
```

图 21-20　动态注册

运行后查看结果，如图 20-21 所示，直接死循环爆栈，还是思路没找好，因为没有将函数恢复就反射，肯定会再次调用，所以要想一个办法将 ArtMethod 恢复为当初 Java 函数的样子，这里采用全局变量的方式。

```
    at com.r0ysue.myjavahook.MainActivity.add2(Native Method)
    at com.r0ysue.myjavahook.MainActivity.add2(Native Method)
    at com.r0ysue.myjavahook.MainActivity.add2(Native Method)
    at com.r0ysue.myjavahook.MainActivity.add2(Native Method)
    at com.r0ysue.myjavahook.MainActivity.add2(Native Method)
    at com.r0ysue.myjavahook.MainActivity.add2(Native Method)
2021-10-14 17:28:19.465 15953-15953/com.r0ysue.myjavahook E/AndroidRuntime:       at com.r0ysue.myjavahook.MainActivity.add2(Native Method)
    at com.r0ysue.myjavahook.MainActivity.add2(Native Method)
    at com.r0ysue.myjavahook.MainActivity.add2(Native Method)
    at com.r0ysue.myjavahook.MainActivity.add2(Native Method)
```

图 21-21　运行结果

新的方案是将 Hook 函数保存的 jump 和 native 直接复原，baocun 函数需要一个参数，那就是

ArtMethod 指针，代码如图 21-22 所示。这样操作就很完美了，无论怎么 Hook 也没有问题，也不会崩溃。

```
void hook(){
....
    nativ=*(_QWORD *)(a1 + 4);//isNative()?
    jump=*((long *)a1 + 5);//EntryPoint
....
}

void huifu(__int64 a,int n){

    *((long *)a + 5)=old1;
    *(_QWORD *)(a + 4)=old2;

}

void baocun(__int64 a){
 old1=*((long *)a + 5);
 old2=*(_QWORD *)(a + 4);
    *((long *)a + 5)= jump;
    *(_QWORD *)(a + 4)= nativ|0x80100;
}

void* add(void* a,void* b,int c,int d){
    JNIEnv *st=(JNIEnv *)a;
    jclass a2=st->FindClass("com/r0ysue/myjavahook/MainActivity");
    jmethodID b2=st->GetMethodID(a2,"add2", "(II)I");
    baocun((__int64)b2);
    int yy= st->CallIntMethod(static_cast<jobject>(b), b2, c, d);
    __android_log_print(6,"r0ysue","%x",yy);
    huifu((__int64)b2)
    return reinterpret_cast<void *="">(5);
}
```

图 21-22 复原代码

但是当 Hook 多个函数的时候，上述操作就会出现问题，上面使用的是全局变量，第二次 Hook 的时候会覆盖第一次的 baocun()和 huifu()，我们更新保存方式，采用数组来保存，代码如图 21-23 所示。

```
void baocun(__int64 a,int n){

 old1[n]=*((long *)a + 5);
 old2[n]=*(_QWORD *)(a + 4);

    *((long *)a + 5)= jump[n];
//    *(_QWORD *)(a + 4)=0x1d8ee408000101;
    *(_QWORD *)(a + 4)= nativ[n]|0x80100;
}

void huifu(__int64 a,int n){

    *((long *)a + 5)=old1[n];
    *(_QWORD *)(a + 4)=old2[n];

}
void hook(){
...
    nativ[ns]=*(_QWORD *)(a1 + 4);
    jump[ns]=*((long *)a1 + 5);
...
}
```

图 21-23 采用数组保存

这样一个 Java 层的 Hook 框架就完成了。开始编写时感觉比 inline Hook 难，写完后发现代码量很少，原理比较简单，主要思路是获得 ArtMethod 而不需要再 Hook。

21.3 Frida 中 Java Hook 的实现

前面阐述了 Java Hook 的核心原理，就是对 EntryPoint 入口点进行替换，核心是替换如图 21-11 所示的机器码入口。接下来寻找 Class，可以通过反射方式获取，如图 21-24 所示。当然，也可以把所需要的 Class 作为参数进行获取操作。

然后获取 ArtMethod。这里提供两种方式：一种是通过反射获取 jmethodID，通过之前章节的学习，我们知道，jmethodID 其实就可以看作是 ArtMethod；另一种是通过 Hook JNI 函数获取 ArtMethod。有两个 JNI 函数中含有 ArtMethod 参数，我们可以通过 Hook 操作获取。这两个函数分别是 UnregisterNative 和 LoadMethod，如图 21-25 和图 21-26 所示。

```
jclass a = env->FindClass(classname);
```

图 21-24　通过反射方式获取 Class

```
void __fastcall art::ArtMethod::UnregisterNative(art::ArtMethod *this)
{
  int v2; // w0
  int v3; // w0
  int v4; // w20
  _BYTE *v5; // x1
  char v6; // [xsp+8h] [xbp-48h]
  _BYTE v7[7]; // [xsp+9h] [xbp-47h] BYREF
  void *v8; // [xsp+18h] [xbp-38h]
  char v9[8]; // [xsp+20h] [xbp-30h] BYREF
  __int64 v10; // [xsp+28h] [xbp-28h]
```

图 21-25　UnregisterNative 函数

```
void ClassLinker::LoadMethod(const DexFile& dex_file,
                             const ClassDataItemIterator& it,
                             Handle<mirror::Class> klass,
                             ArtMethod* dst) {
  uint32_t dex_method_idx = it.GetMemberIndex();
  const DexFile::MethodId& method_id = dex_file.GetMethodId(dex_method_idx);
  const char* method_name = dex_file.StringDataByIdx(method_id.name_idx_);
```

图 21-26　LoadMethod 函数

最后将 EntryPoint 替换为 generic_jni_trampoline，具体的实现前面已经阐述过了，这里不再赘述。

本节主要是通过 Frida 的源码追踪来理解 Java Hook 的原理。

21.3.1 Frida perform 源码追踪

Frida 最常用的写法如图 21-27 所示。我们所编写的代码都位于 perform 函数内部，接下来追踪到源码中，看看 perform 到底做了什么。

```
function main(){
    Java.perform(function(args){
        var clazz = Java.use("com.roysue.demoso.MainActivity");
        clazz.add.implementation = function(args){
            return this.add(args);
        }
    })
}
```

图 21-27　Frida 的写法

perform 源码位于 frida-java-bridge/lib/index.js 中，网址 59（参见配书资源文件）为源码地址。源码如图 21-28 所示。

```
perform (fn) {
  this._checkAvailable();

  if (!this._isAppProcess() || this.classFactory.loader !== null) {
    try {
      this.vm.perform(fn);
    } catch (e) {
      Script.nextTick(() => { throw e; });
    }
  } else {
    this._pendingVmOps.push(fn);
    if (this._pendingVmOps.length === 1) {
      this._performPendingVmOpsWhenReady();
    }
  }
}
```

图 21-28　perform 源码

可以看到 fn 传入了 perform 中，主要工作都在 perform 中完成。我们继续追踪进去观察，这里首次 Hook 的操作进入的是 else，继续追踪 _performPendingVmOpsWhenReady 函数。进入函数内部，如图 21-29 所示。

```
_performPendingVmOpsWhenReady () {
  this.vm.perform(() => {
    const { classFactory: factory } = this;

    const ActivityThread = factory.use('android.app.ActivityThread');
    const app = ActivityThread.currentApplication();
    if (app !== null) {
      initFactoryFromApplication(factory, app);
      this._performPendingVmOps();
      return;
    }
```

图 21-29　_performPendingVmOpsWhenReady 函数

1. use 函数追踪

在图 21-29 中，通过 use 函数加载了 android.app.ActivityThread 类，我们继续追踪进去看做了什么，use 函数如图 21-30 所示。

```javascript
use (className, options = {}) {
  const allowCached = options.cache !== 'skip';

  let C = allowCached ? this._getUsedClass(className) : undefined;
  if (C === undefined) {
    try {
      const env = vm.getEnv();

      const { _loader: loader } = this;
      const getClassHandle = (loader !== null)
        ? makeLoaderClassHandleGetter(className, loader, env)
        : makeBasicClassHandleGetter(className);

      C = this._make(className, getClassHandle, env);
    } finally {
      if (allowCached) {
        this._setUsedClass(className, C);
      }
    }
  }

  return C;
}
```

图 21-30　use 函数

核心是类的加载，如果有了初始化的 loader，直接通过初始化的 loader 进行加载，进入 makeLoaderClassHandleGetter 内部，如图 21-31 所示，这里使用已有的 loader 进行初始化，以获取当前的线程 ID。

```javascript
function makeLoaderClassHandleGetter (className, usedLoader, callerEnv) {
  if (cachedLoaderMethod === null) {
    cachedLoaderInvoke = callerEnv.vaMethod('pointer', ['pointer']);
    cachedLoaderMethod = usedLoader.loadClass.overload('java.lang.String').handle;
  }

  callerEnv = null;

  return function (env) {
    const classNameValue = env.newStringUtf(className);

    const tid = getCurrentThreadId();
    ignore(tid);
    try {
      const result = cachedLoaderInvoke(env.handle, usedLoader.$h, cachedLoaderMethod,
        classNameValue);
      env.throwIfExceptionPending();
      return result;
    } finally {
      unignore(tid);
      env.deleteLocalRef(classNameValue);
    }
  };
}
```

图 21-31　makeLoaderClassHandleGetter 函数

如果初始化没有 loader，就需要进入 makeBasicClassHandleGetter，通过 env.FindClass 反射获取加载传入的类，这个 FindClass 反射是 env 中封装的反射，源码如图 21-32 所示。

```
function makeBasicClassHandleGetter (className) {
  const canonicalClassName = className.replace(/\./g, '/');

  return function (env) {
    const tid = getCurrentThreadId();
    ignore(tid);
    try {
      return env.findClass(canonicalClassName);
    } finally {
      unignore(tid);
    }
  };
}
```

图 21-32　makeBasicClassHandleGetter 函数

2. ActivityThread.currentApplication 源码追踪

我们获取的 ActivityThread 到底是什么呢？可以进入源码中深究一下，在 AOSP 源码网站搜索，如图 21-33 所示。

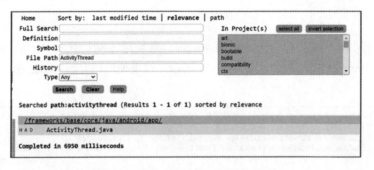

图 21-33　ActivityThread 源码追踪

在文件中搜索 currentApplication，可以看到如图 21-34 所示的 currentApplication 源码定义。继续追踪 currentActivityThread 函数，如图 21-35 所示。

图 21-34　currentApplication 源码　　　　图 21-35　currentActivityThread 函数

最后就到了 sCurrentActivityThread 的定义处，它是内存中当前的 ActivityThread 的一个静态变量，通过这样的方式获取当前的线程。

3. initFactoryFromApplication 函数

这里直接演示正常的处理情况，获取当前的线程之后，进入 initFactoryFromApplication 函数，源码如图 21-36 所示，这个函数的功能是初始化 App 的加载目录。

```
function initFactoryFromApplication (factory, app) {
  const Process = factory.use('android.os.Process');

  factory.loader = app.getClassLoader();

  if (Process.myUid() === Process.SYSTEM_UID.value) {
    factory.cacheDir = '/data/system';
    factory.codeCacheDir = '/data/dalvik-cache';
  } else {
    if ('getCodeCacheDir' in app) {
      factory.cacheDir = app.getCacheDir().getCanonicalPath();
      factory.codeCacheDir = app.getCodeCacheDir().getCanonicalPath();
    } else {
      factory.cacheDir = app.getFilesDir().getCanonicalPath();
      factory.codeCacheDir = app.getCacheDir().getCanonicalPath();
    }
  }
}
```

图 21-36　initFactoryFromApplication 函数

当然，这个过程也可以使用 Java 实现，可以给读者演示一下，代码如图 21-37 所示。

```
try{
    Class ActivityThread = getClassLoader().loadClass( name: "android.app.ActivityThread");
    Method currentApplication = ActivityThread.getDeclaredMethod( name: "currentApplication");
    android.app.Application mcurrentApplication = (Application) currentApplication.invoke( obj: null);
    Log.e( tag: "roysue", msg: "mcurrentApplication => "+mcurrentApplication.toString());
    ClassLoader mloader = mcurrentApplication.getClassLoader();
    Log.e( tag: "roysue", msg: "mloader => " + mloader);
    // hava cache
    Log.e( tag: "roysue", msg: "getCacheDir => " + mcurrentApplication.getCacheDir().getCanonicalPath());
    Log.e( tag: "roysue", msg: "getCodeCacheDir => " + mcurrentApplication.getCodeCacheDir().getCanonicalPath());
    // no cache
    Log.e( tag: "roysue", msg: "getFilesDir => " + mcurrentApplication.getFilesDir().getCanonicalPath());
    Log.e( tag: "roysue", msg: "getCacheDir => " + mcurrentApplication.getCacheDir().getCanonicalPath());
} catch (Exception e) {
    e.printStackTrace();
}
```

图 21-37　线程目录初始化（Java 实现）

运行结果如图 21-38 所示。

```
2022-04-07 21:11:27.844 20957-20957/? E/roysue: mcurrentApplication => android.app.Application@e5fed7b
2022-04-07 21:11:27.845 20957-20957/? E/roysue: mloader => dalvik.system.PathClassLoader[DexPathList[[zip file "/data/app/com.roysue.demoso-2/base.apk"],nativeLibraryDirectories=[/data/app/com.roysue.demoso-2/lib/arm, /data/app/com.roysue.demoso-2/lib/armeabi-v7a, /system/lib, /vendor/lib]]]
2022-04-07 21:11:27.845 20957-20957/? E/roysue: getCacheDir => /data/data/com.roysue.demoso/cache
2022-04-07 21:11:27.845 20957-20957/? E/roysue: getCodeCacheDir => /data/data/com.roysue.demoso/code_cache
2022-04-07 21:11:27.845 20957-20957/? E/roysue: getFilesDir => /data/data/com.roysue.demoso/files
2022-04-07 21:11:27.845 20957-20957/? E/roysue: getCacheDir => /data/data/com.roysue.demoso/cache
```

图 21-38　运行结果

4. _performPendingVmOps 函数

进行_performPendingVmOps 函数的追踪，这个函数的作用是将当前线程注册到虚拟机中，如果不注册当前线程，在后面就不可以使用 env。我们可以在 Android 程序中做个实验，编写函数 myr0，在其中通过 env 调用 FindClass，代码如图 21-39 所示，但是不进行线程注册，看是否可以正常调用。

```cpp
#include <jni.h>
#include <string>
#include <pthread.h>
#include "android/log.h"

JNIEnv *myenv;

void myr0(JNIEnv *env){
    __android_log_print( prio: 6, tag: "roysue", fmt: "enter this thread");
    jclass myclass = env->FindClass( name: "com/roysue/demoso/MainActivity");
    __android_log_print( prio: 6, tag: "roysue", fmt: "%p",myclass);
    __android_log_print( prio: 6, tag: "roysue", fmt: "finish this thread");
}

extern "C" JNIEXPORT jstring JNICALL
Java_com_roysue_demoso_MainActivity_stringFromJNI(
        JNIEnv* env,
        jobject /* this */) {
    std::string hello = "Hello from C++";

    myenv = env;
    pthread_t thread;
    __android_log_print( prio: 6, tag: "roysue", fmt: "before this thread");
    pthread_create(&thread, attr: nullptr, reinterpret_cast<void *(*)(void *)>(myr0), env);

    return env->NewStringUTF(hello.c_str());
}
```

图 21-39　不注册线程，调用 FindClass

运行结果如图 21-40 所示，程序直接崩溃。

```
com.roysue.demoso E/roysue: before this thread
com.roysue.demoso E/roysue: enter this thread
com.roysue.demoso A/m.roysue.demos: java_vm_ext.cc:570] JNI DETECTED ERROR IN APPLICATION: a thread (tid 15779 is making JNI calls without
com.roysue.demoso A/m.roysue.demos: java_vm_ext.cc:570]         in call to FindClass
com.roysue.demoso E/ActivityThread: start sleep......
? A/m.roysue.demos: runtime.cc:630] Runtime aborting...
? A/m.roysue.demos: runtime.cc:630] Dumping all threads without mutator lock held
? A/m.roysue.demos: runtime.cc:630] All threads:
? A/m.roysue.demos: runtime.cc:630] DALVIK THREADS (17):
? A/m.roysue.demos: runtime.cc:630] "Thread-2" prio=10 tid=17 Runnable
? A/m.roysue.demos: runtime.cc:630]   | group="" sCount=0 dsCount=0 flags=0 obj=0x12d82238 self=0x7a98b4a400
? A/m.roysue.demos: runtime.cc:630]   | sysTid=15779 nice=-10 cgrp=default sched=0/0 handle=0x7a41db5d50
? A/m.roysue.demos: runtime.cc:630]   | state=R schedstat=( 56991251 2561978 32 ) utm=1 stm=4 core=3 HZ=100
? A/m.roysue.demos: runtime.cc:630]   | stack=0x7a41cbf000-0x7a41cc1000 stackSize=991KB
? A/m.roysue.demos: runtime.cc:630]   | held mutexes= "abort lock" "mutator lock"(shared held)
? A/m.roysue.demos: runtime.cc:630]   native: #00 pc 00000000004148e4  /apex/com.android.runtime/lib64/libart.so (art::DumpNativeStack
:_1::char_traits<char>&, int, BacktraceMap*, char const*, art::ArtMethod*, void*, bool)+140)
? A/m.roysue.demos: runtime.cc:630]   native: #01 pc 00000000004fbc08  /apex/com.android.runtime/lib64/libart.so (art::Thread::DumpStack
:_1::char_traits<char>&, bool, BacktraceMap*, bool) const+504)
? A/m.roysue.demos: runtime.cc:630]   native: #02 pc 00000000005163a4  /apex/com.android.runtime/lib64/libart.so (art::DumpCheckpoint::Run
```

图 21-40　运行结果，程序直接崩溃

我们可以将 env->FindClass 放到 Native 层的主线程中，代码如图 21-41 所示，这样就可以正常运行了。

```cpp
extern "C" JNIEXPORT jstring JNICALL
Java_com_roysue_demoso_MainActivity_stringFromJNI(
        JNIEnv* env,
        jobject /* this */) {
    std::string hello = "Hello from C++";

    jclass myclass = env->FindClass( name: "com/roysue/demoso/MainActivity");
    __android_log_print( prio: 6, tag: "roysue", fmt: "%p",myclass);

//    myenv = env;
//    pthread_t thread;
//    __android_log_print(6,"roysue","before this thread");
//    pthread_create(&thread, nullptr, reinterpret_cast<void *(*)(void *)>(myr0), env);

    return env->NewStringUTF(hello.c_str());
}
```

图 21-41　主线程函数运行

运行结果如图 21-42 所示，可以正常获取偏移地址了。

```
roysue main: 0x79
```

图 21-42　运行结果

上述案例足以说明线程注册对整个程序正常执行的重要性。

综上所述，通过 Frida 的 perform 源码的追踪，可以发现 perform 函数做了两件事：

（1）得到线程目录。
（2）将线程注册到虚拟机中。

21.3.2　Frida implementation 源码追踪

在 perform 的源码中，我们知道了它主要进行线程的注册，为的是后面的函数调用。在图 21-27 的 Frida 主动调用的代码中，不仅有 perform 还有 implementation，implementation 的作用是什么呢？

implementation 源码位于 frida-java-bridge/lib/class-factory.js 中，如图 21-43 所示，源码比较长，这里只截函数的头部，后面的代码在讲解的过程中再为读者呈现。

```
implementation: {
  enumerable: true,
  get () {
    const replacement = this._r;
    return (replacement !== undefined) ? replacement : null;
  },
  set (fn) {
    const params = this._p;
    const holder = params[1];
    const type = params[2];

    if (type === CONSTRUCTOR_METHOD) {
      throw new Error('Reimplementing $new is not possible; replace implementatio
    }

    const existingReplacement = this._r;
    if (existingReplacement !== undefined) {
      holder.$f._patchedMethods.delete(this);

      const mangler = existingReplacement._m;
      mangler.revert(vm);

      this._r = undefined;
```

图 21-43　implementation 源码（部分）

如图 21-44 所示是 implementation 函数的核心，首先获取传入的参数。

```
if (fn !== null) {
  const [methodName, classWrapper, type, methodId, retType, argTypes] = params;

  const replacement = implement(methodName, classWrapper, type, retType, argTypes,
                                fn, this);
  const mangler = makeMethodMangler(methodId);
  replacement._m = mangler;
  this._r = replacement;

  mangler.replace(replacement, type === INSTANCE_METHOD, argTypes, vm, api);

  holder.$f._patchedMethods.add(this);
}
```

图 21-44 implementation 函数的核心

在 implement 中初始化所有的参数，核心就是 replace 函数，这步是 Java Hook 最核心的地方，也就是 EntryPoint 的机器码入口替换过程，源码如图 21-45 所示，可以看到这里也做了同样的事情，都要获取一个 ArtMethod。

```
replace (impl, isInstanceMethod, argTypes, vm, api) {
  this.originalMethod = fetchArtMethod(this.methodId, vm);

  const originalFlags = this.originalMethod.accessFlags;

  if ((originalFlags & kAccXposedHookedMethod) !== 0 && xposedIsSupported()) {
    const hookInfo = this.originalMethod.jniCode;
    this.hookedMethodId = hookInfo.add(2 * pointerSize).readPointer();
    this.originalMethod = fetchArtMethod(this.hookedMethodId, vm);
  }
}
```

图 21-45 replace 源码

继续往下追踪，在 patchArtMethod 中对 4 种模式进行修正，源码如图 21-46 所示。

```
patchArtMethod(replacementMethodId, {
  jniCode: impl,
  accessFlags: ((originalFlags & ~(kAccCriticalNative | kAccFastNative |
                  kAccNterpEntryPointFastPathFlag)) | kAccNative) >>> 0,
  quickCode: api.artClassLinker.quickGenericJniTrampoline,
  interpreterCode: api.artInterpreterToCompiledCodeBridge
}, vm);
```

图 21-46 patchArtMethod 源码

最后进行函数替换，通过 set 函数把原始函数替换为 JavaScript 编译好的函数，源码如图 21-47 所示。

```
artController.replacedMethods.set(hookedMethodId, replacementMethodId);
```

图 21-47 函数替换

21.4 本章小结

在本章中，我们介绍了什么是 Java Hook 的概念，并带领读者追踪了 Android 中 Java 函数的执行流程，并从中发现了函数的位置，通过替换 EntryPoint 的机器码入口来实现函数的替换，从而实现了 Hook 的操作。同时，我们还追踪了 Frida 的 perform 和 implementation 源码，通过对 Android 和 Frida 源码的追踪，我们可以彻底明白 Java Hook 的原理。

第 22 章

inline Hook 中用到的汇编指令

在前面的章节中，我们向读者介绍了 PLT Hook 和 GOT Hook，这两种方式都是通过使用 inline Hook 技术来替换函数地址的偏移。然而，常见的 Hook 不仅有函数的替换方式，还有函数的附加方式。

在本章中，我们将会总结这两种方式的特点，并说明在 inlink Hook 中使用的汇编指令和需要注意的事项。

22.1 两种 Hook 方式的介绍

1. 函数替换的方式

函数替换的方式是指直接跳转到目标函数，不需要再跳转回来。与 PLT Hook 一样，PLT 本身不需执行函数的原始逻辑，如果执行函数的原始逻辑，那么可以使用绝对地址指令进行替换，并不需要 PLT 中的函数。

2. 函数附加的方式

函数附加的方式是指需要执行函数的原始逻辑，跳转到自定义函数地址后，还需要跳转回来，这就需要寄存器来保存原始的参数，以便跳转回原始函数的时候，有参数可以继续执行原始函数。

比如我们要 Hook LoadMethod 函数，但是在 LoadClassMembers 函数中调用了 LoadMethod 函数，那么就不能打断 LoadMethod 函数的执行流程，必须让它的整个流程执行完成，不然 loadMethod 函数没有执行完成就跳转回 LoadClassMembers 函数，整个程序会崩溃，LoadClassMembers 函数 Android 源码如代码清单 22-1 所示。

代码清单 22-1 LoadClassMembers 函数源码

```
void ClassLinker::LoadClassMembers(Thread* self,
                                   const DexFile& dex_file,
                                   const uint8_t* class_data,
                                   Handle<mirror::Class> klass) {
  {
    // Note: We cannot have thread suspension until the field and method arrays are setup or
    // Class::VisitFieldRoots may miss some fields or methods
    ScopedAssertNoThreadSuspension nts(__FUNCTION__);
    // Load static fields
    // We allow duplicate definitions of the same field in a class_data_item
    // but ignore the repeated indexes here, b/21868015
    LinearAlloc* const allocator = GetAllocatorForClassLoader(klass->GetClassLoader());
    ClassDataItemIterator it(dex_file, class_data);
    LengthPrefixedArray<ArtField>* sfields = AllocArtFieldArray(self,
                                                                allocator,
                                                                it.NumStaticFields());
    size_t num_sfields = 0;
    uint32_t last_field_idx = 0u;
    for (; it.HasNextStaticField(); it.Next()) {
      uint32_t field_idx = it.GetMemberIndex();
      DCHECK_GE(field_idx, last_field_idx);  // Ordering enforced by DexFileVerifier
      if (num_sfields == 0 || LIKELY(field_idx > last_field_idx)) {
        DCHECK_LT(num_sfields, it.NumStaticFields());
        LoadField(it, klass, &sfields->At(num_sfields));
        ++num_sfields;
        last_field_idx = field_idx;
      }
    }

    // Load instance fields
    LengthPrefixedArray<ArtField>* ifields = AllocArtFieldArray(self,
                                                                allocator,
                                                                it.NumInstanceFields());
    size_t num_ifields = 0u;
    last_field_idx = 0u;
    for (; it.HasNextInstanceField(); it.Next()) {
      uint32_t field_idx = it.GetMemberIndex();
      DCHECK_GE(field_idx, last_field_idx);  // Ordering enforced by DexFileVerifier
      if (num_ifields == 0 || LIKELY(field_idx > last_field_idx)) {
        DCHECK_LT(num_ifields, it.NumInstanceFields());
        LoadField(it, klass, &ifields->At(num_ifields));
        ++num_ifields;
```

```cpp
            last_field_idx = field_idx;
          }
        }

        if (UNLIKELY(num_sfields != it.NumStaticFields()) ||
            UNLIKELY(num_ifields != it.NumInstanceFields())) {
          LOG(WARNING) << "Duplicate fields in class " << klass->PrettyDescriptor()
              << " (unique static fields: " << num_sfields << "/" << it.NumStaticFields()
              << ", unique instance fields: " << num_ifields << "/" << it.NumInstanceFields()
<<
          // NOTE: Not shrinking the over-allocated sfields/ifields, just setting size
          if (sfields != nullptr) {
            sfields->SetSize(num_sfields);
          }
          if (ifields != nullptr) {
            ifields->SetSize(num_ifields);
          }
        }
        // Set the field arrays
        klass->SetSFieldsPtr(sfields);
        DCHECK_EQ(klass->NumStaticFields(), num_sfields);
        klass->SetIFieldsPtr(ifields);
        DCHECK_EQ(klass->NumInstanceFields(), num_ifields);
        // Load methods
        bool has_oat_class = false;
        const OatFile::OatClass oat_class =
            (Runtime::Current()->IsStarted() && !Runtime::Current()->IsAotCompiler())
              ? OatFile::FindOatClass(dex_file, klass->GetDexClassDefIndex(),
&has_oat_class)
              : OatFile::OatClass::Invalid();
        const OatFile::OatClass* oat_class_ptr = has_oat_class ? &oat_class : nullptr;
        klass->SetMethodsPtr(
            AllocArtMethodArray(self, allocator, it.NumDirectMethods() +
it.NumVirtualMethods()),
            it.NumDirectMethods(),
            it.NumVirtualMethods());
        size_t class_def_method_index = 0;
        uint32_t last_dex_method_index = DexFile::kDexNoIndex;
        size_t last_class_def_method_index = 0;
        // TODO These should really use the iterators
        for (size_t i = 0; it.HasNextDirectMethod(); i++, it.Next()) {
          ArtMethod* method = klass->GetDirectMethodUnchecked(i, image_pointer_size_);
          LoadMethod(dex_file, it, klass, method);
          LinkCode(this, method, oat_class_ptr, class_def_method_index);
          uint32_t it_method_index = it.GetMemberIndex();
          if (last_dex_method_index == it_method_index) {
```

```
      // duplicate case
      method->SetMethodIndex(last_class_def_method_index);
    } else {
      method->SetMethodIndex(class_def_method_index);
      last_dex_method_index = it_method_index;
      last_class_def_method_index = class_def_method_index;
    }
    class_def_method_index++;
  }
  for (size_t i = 0; it.HasNextVirtualMethod(); i++, it.Next()) {
    ArtMethod* method = klass->GetVirtualMethodUnchecked(i, image_pointer_size_);
    LoadMethod(dex_file, it, klass, method);
    DCHECK_EQ(class_def_method_index, it.NumDirectMethods() + i);
    LinkCode(this, method, oat_class_ptr, class_def_method_index);
    class_def_method_index++;
  }
  DCHECK(!it.HasNext());
}
// Ensure that the card is marked so that remembered sets pick up native roots.
Runtime::Current()->GetHeap()->WriteBarrierEveryFieldOf(klass.Get());
self->AllowThreadSuspension();
}
```

两种不同的替换方式对应两种不同的解决方案，在函数的替换中，使用的是定向跳转的方案，而在函数的附加中，要使用寄存器来保存参数，以保证程序的正常执行。

22.2 定向跳转

在定向跳转中，常用的是 b 系列指令，如 b 指令、bl 指令、br 指令、blr 指令，它们的功能分别如下。

- b 指令：跳转到相对地址。
- bl 指令：跳转到相对地并处理 x30 寄存器。
- br 指令：跳转到绝对地址，是参数位寄存器。
- blr 指令：跳转到绝对地址并处理 x30 寄存器。

下面介绍 b 系列指令的优缺点。

1. b 指令

b 指令有位的大小限制，不能跳转太远，所以它只适合在单个 SO 文件之内跳转，如果需要做到跨 SO 文件跳转，b 指令就有它的局限性。请看下面的例子。

在前面的 PLT Hook 中，给读者演示了使用硬偏移的方式实现替换，代码如图 22-1 所示。

```
int code;
if(flag == 1){
    code = 0x14000000 + offsit/4;
}else{
    code = 0x18000000 - offsit/4;
}
```

图 22-1　硬偏移

以 0x14000000 为例，其中前两位表示指令位，后六位表示偏移位。然而，内存中的地址是 10 位的。我们之前编写的查看 libart.so 在内存中的分布的代码如图 22-2 所示。显然，b 指令无法表示 10 位的地址，因此无法实现对跨 SO 文件的函数替换。

当然，我们也可以尝试做一次硬偏移替换，观察替换的整个流程。

（1）新建以 C++为模板的项目 myinlinehook_demo。以 libart.so 为例，Android 程序的运行肯定会经过这个 SO 文件，首先要查找 libart.so 在内存中的分布，需要用到之前的 findsym.cpp、findsym.h、ida.h，找到之前的项目导入进来即可，然后在 CMakeLists.txt 中写入，文件如图 22-3 所示。

```
7b1c11b000-7b1c714000 r-xp 00000000 103:09 1973
7b1c714000-7b1c724000 r--p 005f8000 103:09 1973
7b1c724000-7b1c727000 rw-p 00608000 103:09 1973
```

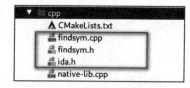

图 22-2　libart.so 在内存中的分布　　　　图 22-3　项目文件

（2）获取 libart.so 的内存分布后，以此来获取 libart.so 在内存中的首地址偏移，代码如图 22-4 所示。

```
// 动态获取 libart.so 首地址及内存中的分布
char line[1024];
int *start;
int *end;
int n=1;
FILE *fp=fopen( path: "/proc/self/maps", mode: "r");
while (fgets(line, sizeof(line), fp)) {
    if(strstr(line, n: "libart.so")) {
        __android_log_print( prio: 6, tag: "r0ysue", fmt: "%s", line);
        if(n==1){
            start = reinterpret_cast<int *>(strtoul( s: strtok(line, delimiter: "-"), end_ptr: NULL, base: 16));
            end = reinterpret_cast<int *>(strtoul( s: strtok( s: NULL, delimiter: " "), end_ptr: NULL, base: 16));
        }
        else{
            strtok(line, delimiter: "-");
            end = reinterpret_cast<int *>(strtoul( s: strtok( s: NULL, delimiter: " "), end_ptr: NULL, base: 16));
        }
        n++;
    }
}
```

图 22-4　libart.so 的内存分布

第 22 章　inline Hook 中用到的汇编指令

（3）寻找 loadMethod 方法在 libart.so 中的相对偏移。首先需要找到 libart.so 的位置，可以通过进程来查询，如图 22-5 所示。随便找一个进程，因为 libart.so 相当于 Java 的执行环境，所以用 Android 编写的程序肯定含有它。

```
1|bullhead:/ # ps -e | grep com.r
u0_a110       15094 30121 4303600   48808 ptrace_stop 7b9f80cdfc t com.roysue.myinlinehook_demo
bullhead:/ # cat proc/15094/maps | grep libart.so
7b05aa2000-7b06200000 r--p 00000000 103:09 1973                          /system/lib64/libart.so
7b1c11b000-7b1c228000 r-xp 00000000 103:09 1973                          /system/lib64/libart.so
7b1c228000-7b1c229000 rwxp 0010d000 103:09 1973                          /system/lib64/libart.so
7b1c229000-7b1c714000 r-xp 0010e000 103:09 1973                          /system/lib64/libart.so
7b1c714000-7b1c724000 r--p 005f8000 103:09 1973                          /system/lib64/libart.so
7b1c724000-7b1c727000 rw-p 00608000 103:09 1973                          /system/lib64/libart.so
```

图 22-5　libart.so 位置查询

然后使用 IDA 打开，查看 loadMethod 的导出符号，如图 22-6 所示。

Name	Address	Ordinal
art::ClassLinker::LoadMethod(art::DexFile const&,art::ClassDataItemIterator...	000000000013268C	
art::DexFileVerifier::CheckLoadMethodId(uint,char const*)	00000000001A42B0	

图 22-6　导出符号

通过 libart.so start 获取 loadMethod 在内存中的符号地址，代码如图 22-7 所示。

```
// 寻找loadMethod方法在libart.so中的相对偏移
int loadMethod_off = findsym("/system/lib64/libart.so",
       "_ZN3art11ClassLinker10LoadMethodERKNS_7DexFileERKNS_21ClassDataItemIteratorENS_6HandleI
// 通过 libart.so start 拿到 loadMethod 在内存中的地址
long *loadMethod = reinterpret_cast<long *>((char *)start + loadMethod_off - 0x25000);
__android_log_print( prio: 6, tag: "r0ysue", fmt: "loadMethod => %p",loadMethod);
__android_log_print( prio: 6, tag: "r0ysue", fmt: "loadMethod_off => %p",loadMethod_off);
```

图 22-7　获取符号地址

（4）编写替换函数 myreplace，代码如图 22-8 所示。

```
void myreplace(){
    __android_log_print( prio: 6, tag: "r0ysue", fmt: "I'm from myreplace");
}
```

图 22-8　替换函数 myreplace

（5）计算偏移大小，判断程序是向前跳还是向后跳，代码如图 22-9 所示。

```
// 判断大小 用大偏移- 小偏移
int flag = 0;        // 往前跳
long offsit = 0;
if((long)loadMethod > (long)myreplace){
    offsit = (long)loadMethod - (long)myreplace;
}else{
    flag = 1;        // 往后跳
    offsit = (long)myreplace - (long)loadMethod;
}
```

图 22-9　偏移计算

计算得到 offsit 的结果为 0x9adc82b0，我们在 armconverter 中直接进行 ARM 架构的机器码和汇编指令的互相转换，发现 0x9adc82b0 删除一位也不会有变化，说明内存越界了，如图 22-10 所示。指令代码一共 8 位，然而它使用 8 位来表示偏移，显然不可以这样做，这就是 b 指令的局限性。

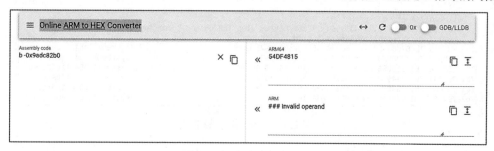

图 22-10　转换

（6）最后通过硬编码强制替换，代码如图 22-11 所示，最终的结果当然是程序崩溃。

```
/*
 * 使用强制跳转，看下是否会成功
 * 结果：崩溃
 */
int code;
if(flag == 1){
    code = 0x14000000 + offsit/4;
}else{
    code = 0x18000000 - offsit/4;
}
__android_log_print( prio: 6, tag: "r0ysue", fmt: "loadMethod => %p", loadMethod);
__android_log_print( prio: 6, tag: "r0ysue", fmt: "offsit => %p",offsit);
__android_log_print( prio: 6, tag: "r0ysue", fmt: "code => %p",code);
// loadMethod 为起始位置，给1页可读、可写、可执行的权限
mprotect( addr: (void *)PAGE_START, x: (long)loadMethod),PAGE_SIZE, prot: PROT_WRITE|PROT_READ|PROT_EXEC);
// 直接将 code 赋值给 loadmethod
*(int *)((char *)loadMethod) = code;
sleep( seconds: 30);
```

图 22-11　硬编码强制替换

2. bl 指令

在本节的开头介绍了 b 系列指令的跳转方式以及它们各自的功能，在硬编码强制替换的案例中，通过地址的比对可以发现 b 指令的局限性，所以如果我们想 Hook art 函数，一定不可以使用 b 指令来 Hook，Hook art 需要的是无条件跳转，当然无条件跳转指令只有 b、bl、br 和 blr 这 4 条，我们可以使用 bl 指令吗？再次使用汇编转机器码的方式观察一下，如图 22-12 所示。

图 22-12　bl 指令

当 bl 指令删除一位后，机器码并未发生变化，说明它也越界了，bl 指令只是处理了 x30 寄存器，而且我们也不需要处理 x30 寄存器，所以它也不能进行 Hook art 函数的操作，同 b 指令一样，它的表达位数太短了。那么 br 指令可以吗？

3. br 指令

br 指令可以跳转到寄存器的位置，如图 22-13 所示，x0 寄存器有 16 位，而地址有 10 位，所以 br 指令可以做到 Hook art 函数。

图 22-13　x0 寄存器位数

4. blr 指令

blr 指令可以 Hook 函数吗？其实它也不行，因为无论是函数替换还是函数附加,都要使用 return 指令。当然，函数的附加不一定使用，但是函数的替换一定要使用 return 指令，那么 x30 寄存器就一定要保存下来，因为 x30 寄存器中保存着函数的返回值。使用 blr 指令就要处理 x30 寄存器，所以它也不能 Hook art 函数。

22.3　寄存器保存

22.3.1　寄存器选择

这里有一个新的问题,我们使用哪个寄存器去做跳板呢？图 22-14 展示了各个寄存器的作用以及注意事项。

```
x0~x7: 传递子程序的参数和返回值，使用时不需要保存，多余的参数用堆栈传递，6
4位的返回结果保存在x0中。
x8: 用于保存子程序的返回地址，使用时不需要保存。
x9~x15: 临时寄存器，也叫可变寄存器，子程序使用时不需要保存。
x16~x17: 子程序内部调用寄存器（IPx），使用时不需要保存，尽量不要使用。
x18: 平台寄存器，它的使用与平台相关，尽量不要使用。
x19~x28: 临时寄存器，子程序使用时必须保存。
x29: 帧指针寄存器（FP），用于连接栈帧，使用时必须保存。
x30: 链接寄存器（LR），用于保存子程序的返回地址。
x31: 堆栈指针寄存器（SP），用于指向每个函数的栈顶。
```

图 22-14　寄存器说明

x0~x7 是不可以动的，这些寄存器中保存的是原始函数的参数和返回值，如果改变了这些寄存器，那么我们做的 Hook 就没有意义了。x8 寄存器保存着返回地址，同样不可以动。x9~x15 是临时寄存器，从它们的功能来看是可以使用的，如果我们进行 inline Hook，可以不在函数的开头 Hook，在中间的某一段 Hook 即可，如图 22-15 所示，但是如果我们改变了图中 x10 寄存器的值，那么 x10 寄存器就不准确了，所以这个方案也不可行。x19~x28 也是一样的道理。

```
libart.so:0000007B668E770C AA 06 40 F9 LDR    X10, [X21,#8]
libart.so:0000007B668E7710 28 79 68 B8 LDR    W8, [X9,X8,LSL#2]
libart.so:0000007B668E7714 48 01 08 8B ADD    X8, X10, X8
libart.so:0000007B668E7718 F6 03 08 AA MOV    X22, X8
```

图 22-15　在中间段进行 inline Hook

x29 是栈帧，不可以使用。前面讲过 x30 是返回值，也不可以使用。那么我们能使用的就是 x16~x18 寄存器。x16、x17 在 PLT Hook 中会用到，如图 22-16 所示。在 LLVM 中，x16、x17 寄存器可以做一些传值的处理。

```
.plt:0000000000012B50    ADRP    X16, #strlen_ptr@PAGE
.plt:0000000000012B54    LDR     X17, [X16,#strlen_ptr@PAGEOFF]
.plt:0000000000012B58    ADD     X16, X16, #strlen_ptr@PAGEOFF
.plt:0000000000012B5C    BR      X17
```

图 22-16　x16、x17 寄存器在 PLT Hook 中的应用

唯一没有被使用的是 x18 寄存器，它是一个平台寄存器，如果不是 LLVM，那么就使用 x16、x17 寄存器，否则使用 x18 寄存器。

本章使用 x16 寄存器作为跳板来操作。先来看它的汇编转机器码的过程，如图 22-17 所示。

图 22-17　x16 寄存器转码

当然，也可以使用 x17 寄存器，其转码如图 22-18 所示。x16 寄存器和 x17 寄存器之间就差了 2。

图 22-18　x17 寄存器转码

22.3.2　3 种寄存器赋值的方案

找到了可以使用的寄存器之后，就要给它进行赋值，这里有 3 种方案，如图 22-19 所示。

第 22 章　inline Hook 中用到的汇编指令

```
mov x16,
adrp
ldr x16,4
```

图 22-19　3 种寄存器赋值的方案

1. MOV

小位赋值，如图 22-20 所示。它可以给寄存器赋值，但是一旦赋值位数变长，它就会越界，如图 22-21 所示。

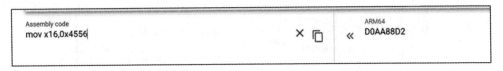

图 22-20　小位赋值

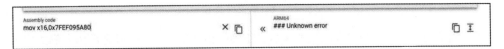

图 22-21　多位赋值

2. ADRP

使用 ADRP 的方式需要对页进行对齐，所以也不可行。

3. LDR

LDR 赋值方式为 LDR x16,4，4 代表偏移，汇编转机器码如图 22-22 所示。

图 22-22　转码结果

在 IDA 中对这个寄存器进行赋值操作，先把 SUB 指令改为 LDR 指令，如图 22-23 所示。

```
.text:0000000000019618 ; __unwind { // __gxx_personality_v0
.text:0000000000019618                 SUB     SP, SP, #0x80
.text:000000000001961C                 STR     X25, [SP,#0x70+var_40]
.text:0000000000019620                 STP     X24, X23, [SP,#0x70+var_30]
.text:0000000000019624                 STP     X22, X21, [SP,#0x70+var_20]
```

图 22-23　指令修改

转为 HEX 模式，编辑十六进制，如图 22-24 所示。

```
5E0  F6 57 41 A9 0E 49 00 39  4E 17 80 52 C0 00 07 2A   .....I.9N..R...*
5F0  6B 01 0E 4A 44 01 80 3D  04 45 00 39 41 89 00 AD   k..JD..=.E.9A...
600  00 01 80 3D 0B 51 00 39  49 09 01 39 40 0D 01 39   ...=.Q.9I..9@..9
610  F7 07 43 F8 C0 03 5F D6  FF 03 02 D1 F9 1B 00 F9   ..C..._.........
620  F8 5F 04 A9 F6 57 05 A9  F4 4F 06 A9 FD 7B 07 A9   ._...W...O...{..
630  FD C3 01 91 55 D0 3B D5  AA 13 40 F9 4B 01 00 B9   ....U.;...@.K...
640  6B 59 47 F9 4C 01 00 90  8C BD 46 F9 EA 17 00 F9   kYG.L.....F.....
650  6B 01 40 B9 CD 2A 81 52  6D 2C B7 72 1F 01 40 B9   k.@..*.Rm<.r..@.
660  6E 01 0D 0B CE 05 00 0D  CD 0D 00 F9 0B 01 00 39   n........K.....9
670  EB 03 2B 2A 6B 79 1F 32  7F 05 00 31 ED 07 9F 1A   ..+*ky.2...1....
680  EE 17 9F 1A 9F 25 00 71  EF D7 9F 1A 9F 29 00 71   .....%.q.....).q
690  FD 01 0D 2A FC A7 9F 1A  FE 03 00 39 8E 01 0F 4A   ...*.......9...J
```

图 22-24 十六进制修改

编辑完之后，就变为 LDR 指令了，如图 22-25 所示，这里把 F9 18 00 F9 赋值给 x16 寄存器。

```
019618                   ; __unwind { // __gxx_personality_v0
019618                   ; DATA XREF: LOA
019618 30 00 00 58       LDR      X16, loc_1961C
01961C
01961C          loc_1961C
01961C F9 1B 00 F9       STR      X25, [SP,#0x30]
019620 F8 5F 04 A9       STP      X24, X23, [SP,#0x40]
019624 F6 57 05 A9       STP      X22, X21, [SP,#0x50]
```

图 22-25 赋值给寄存器

选择正确的指令、寄存器以及赋值方式后，正确的指令构造方式就呼之欲出，形式如图 22-25 所示。指令构造完成之后就可以编码进行跳转了。

首先，我们需要编写代码来构造跳转的指令，并将指令进行汇编转换为机器码，然后获取 myreplace 函数的地址，再依次赋值给 loadmethod 的前 3 条指令，以完成函数的跳转操作，代码如图 22-26 所示。

```
int code1 = 0x50000058;    // ldr x16,8 => 0x50000058
int code2 = 0xdf160200;    // br x16 => 0x00021FD6
long code3 = reinterpret_cast<long>(myreplace);   // 待跳转的地址

/*
 * 将code1、code2、code3分别赋值给loadmethod的前3条指令
 */
*(int*)((char*)loadMethod) = code1;
*(int*)((char*)loadMethod+4) = code2;
*(long*)((char*)loadMethod+8) = code3;
```

图 22-26 函数跳转代码

对于打印出来的 loadmethod 地址，在 IDA 中远程连接手机查看结果，如图 22-27 所示，可以看到 loadmethod 函数被替换了。

```
libart.so:0000007B668E768C ;---------------------------------------------------
libart.so:0000007B668E768C LDR         X16, =_Z9myreplacev    ; myreplace(void)
libart.so:0000007B668E7690 BR          X16                    ; myreplace(void)
libart.so:0000007B668E7690 ;---------------------------------------------------
libart.so:0000007B668E7694 off_7B668E7694 DCQ _Z9myreplacev   ; DATA XREF: libart.so:0000007B668E768C↑r
libart.so:0000007B668E7694                                   ; myreplace(void)
```

图 22-27 函数替换结果

22.4 本章小结

本章介绍了两种方式的 inline Hook：函数替换方式和函数附加方式。针对这两种 Hook 方式，本章讲述了定向跳转中 b 系列指令中的选择问题，如果只在单个 SO 文件中进行跳转，使用 b 指令就足够了，如果要跨 SO 文件进行跳转，那么 b 指令所能表达的位数不足以支持它进行文件间的跳转操作，这时需要使用 br 指令。选择指令后，我们需要保存待跳转函数的地址，这时需要在各个寄存器之间进行选择。通过比较 x0~x31 寄存器的功能，我们最终选择 x16~x18 作为跳板寄存器。最后，我们演示了对寄存器进行赋值的操作。

第 23 章 基于 Capstone 处理特殊指令

本章将介绍一个新的工具 Capstone，它的作用是把十六进制指令转化为汇编指令。我们通过 Capstone 来处理特殊指令，并演示它的基本操作。尽管这个工具相对来说有些陈旧，但在关键时候它可以大放异彩。网址 60（参见配书资源文件）为 Capstone 的代码地址。

23.1 编译 Capstone 并配置测试环境

1. clone 源码

读者需要通过编译 Capstone 源码来获得该工具。可以通过 clone（克隆）的方式获取 Capstone 的源码，命令如下：

```
git clone -b next https://github.com/capstone-engine/capstone
```

2. NDK 下载

下载 r22 版本 LTS 的 NDK，下载完成之后需要将其解压，可以使用 unzip 命令：

```
https://dl.google.com/android/repository/android-ndk-r22b-linux-x86_64.zip
```

3. 编译

编译说明位于 COMPILE.TXT 文件中，选择编译 Android 版本，编译说明如图 23-1 所示。

```
(5) Cross-compile for Android

To cross-compile for Android (smartphone/tablet), Android NDK is required.
NOTE: Only ARM and ARM64 are currently supported.

$ NDK=/android/android-ndk-r10e ./make.sh cross-android arm
or
$ NDK=/android/android-ndk-r10e ./make.sh cross-android arm64

Resulted files libcapstone.so, libcapstone.a & tests/test* can then
be used on Android devices.
```

图 23-1　编译说明

具体的编译命令如下：

```
NDK=/root/Desktop/NDK/android-ndk-r22 ./make.sh cross-android arm64
```

4. 头文件导入

编译完成之后有两个文件，其中一个是 libcapstone.so.5，将其放入 IDA 中查看，它本身就是一个 ELF 文件，提供了很多函数供我们使用，例如，如图 23-2 所示的 cs_disasm 函数，它是一个用于反编译的函数。

```
.text:00000000005DE310
.text:00000000005DE310   cs_disasm    EXPORT cs_disasm
.text:00000000005DE310                              ; DATA XREF: LOAD:00000000000053D↑o
.text:00000000005DE310                STP    X29, X30, [SP,#-0x20]!
.text:00000000005DE314                STR    X28, [SP,#0x10]
```

图 23-2　cs_disasm 函数

上述这个 SO 文件本身有很多 API 函数，但是在本章中使用到的关键函数就几个，如 cs_disasm 函数。使用这个 SO 文件需要用到头文件，它位于 capstone/include/capstone，需要将其放入新建的 capstone_test 项目中，如图 23-3 所示。

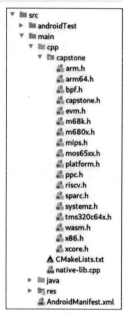

5. SO 文件更正

在源码编译中，其实会发生一个错误，但是依然会正常生成一个 libcapstone.so.5 文件，把这个文件放入 capstone_test 中，如图 23-4 所示。

图 23-3　放置头文件

图 23-4　libcapstone.so.5 的位置

然后在 CMakeLists.txt 中加入一段神秘的代码，如图 23-5 和图 23-6 所示。

```
set(CRYPTO_LIB_NAME capstone)
add_library(${CRYPTO_LIB_NAME} SHARED IMPORTED)
set_target_properties(${CRYPTO_LIB_NAME} PROPERTIES IMPORTED_LOCATION ${PROJECT_SOURCE_DIR}/../../../libs/lib${CRYPTO_LIB_NAME}.so.5)
```

图 23-5　神秘代码 1

```
target_link_libraries( # Specifies the target library.
                native-lib
                ${CRYPTO_LIB_NAME}
                # Links the target library to the log library
                # included in the NDK.
                ${log-lib} )
```

图 23-6　神秘代码 2

运行项目，程序直接崩溃，报错信息表明找不到 libcapstone.so.5 文件，如图 23-7 所示。报错的原因是 System.loadLibrary 在 loadlibrary 的时候，是根据 soname 来判断是否加载，我们需要进入 SO 文件内部更改其 soname。

```
dlopen failed: library "libcapstone.so.5" not found
```

图 23-7　报错代码

首先在 linker.cpp（在 Android 源码中查找）中查看 soname 的宏定义，如图 23-8 所示。14 的十六进制数对应为 E。

图 23-8　soname 的宏定义

将 SO 文件在 010 Editor 中打开，找到程序头表的动态段，并跳转到动态段的位置，根据 soname 的宏可以找到 soname 的相对偏移，如图 23-9 所示。

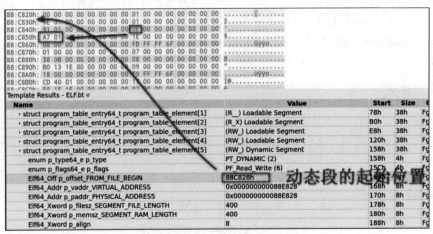

图 23-9　soname 相对偏移

然后根据上述方法找到字符串表的位置，soname 就是字符串，它在字符串表内，字符串表的宏定义为 5，如图 23-10 所示。

字符串表的偏移如图 23-11 所示。

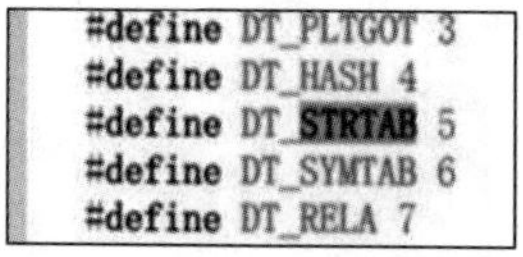

第 23 章　基于 Capstone 处理特殊指令　453

图 23-10　strtab 的宏定义

图 23-11　字符串表的偏移

使用十六进制加法找到 soname 的位置，计算结果如图 23-12 所示。

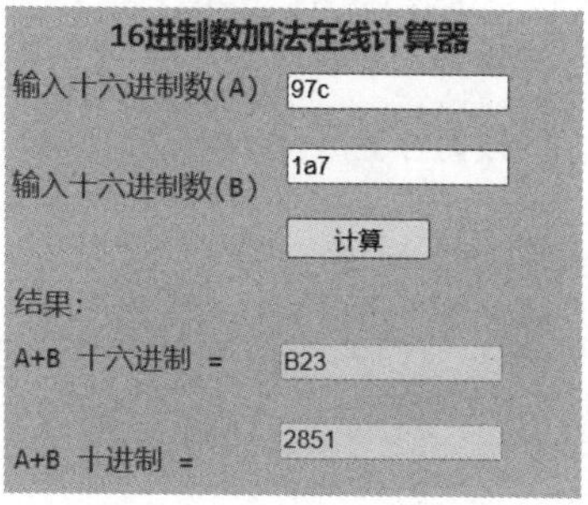

图 23-12　十六进制加法

跳转到 B23 之后，可以看到 soname，如图 23-13 所示，此时我们把 2E 35 置 0 即可。再次运行项目，发现可以正常运行了。

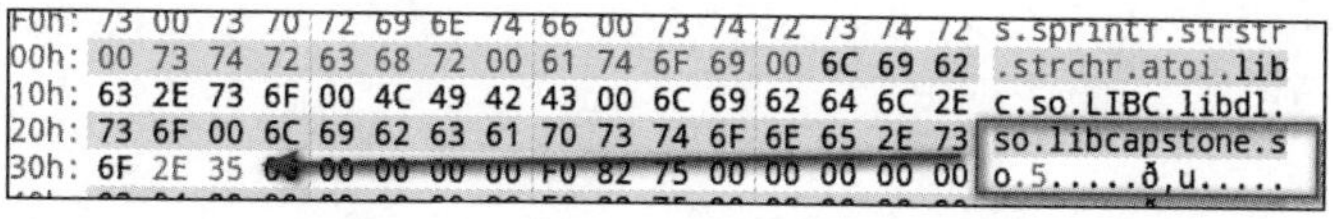

图 23-13　soname

23.2　Capstone 官方测试案例演示

网址 61（参见配书资源文件）为测试代码地址。

原始测试代码不适合直接在 Android 项目中编译运行，需要进行适当的改写，将测试案例移植到 Android 项目中。test 函数如代码清单 23-1 所示。移植过程中注释掉了项目自定义的 print_string_hex 函数和 print_insn_detail 函数，以及一些常规的错误检测。测试函数的主要功能是把函数内最开始的宏定义的十六进制转为汇编指令并执行。

代码清单 23-1　test 函数

```Java {.line-numbers}
// 全局变量
static csh handle;
// 平台结构体
struct platform {
    cs_arch arch;
    cs_mode mode;
```

```c
        unsigned char *code;
        size_t size;
        const char *comment;
    };

    static void test()
    {

#define ARM64_CODE "\x09\x00\x38\xd5\xbf\x40\x00\xd5\x0c\x05\x13\xd5\x20\x50\
x02\x0e\x20\xe4\x3d\x0f\x00\x18\xa0\x5f\xa2\x00\xae\x9e\x9f\x37\x03\xd5\xbf\x33\x03\xd5\
\xdf\x3f\x03\xd5\x21\x7c\x02\x9b\x21\x7c\x00\x53\x00\x40\x21\x4b\xe1\x0b\x40\xb9\x20\x0\
4\x81\xda\x20\x08\x02\x8b\x10\x5b\xe8\x3c"

        struct platform platforms[] = {
            {
                CS_ARCH_ARM64,
                CS_MODE_ARM,
                (unsigned char *)ARM64_CODE,
                sizeof(ARM64_CODE) - 1,
                "ARM-64"
            },
        };

        uint64_t address = 0x2c;
        cs_insn *insn;
        int i;
        size_t count;

    //    platforms

        for (i = 0; i < sizeof(platforms)/sizeof(platforms[0]); i++) {
            cs_err err = cs_open(platforms[i].arch, platforms[i].mode, &handle);
            // err 注释掉，忽略错误处理
    //        if (err) {
    //            printf("Failed on cs_open() with error returned: %u\n", err);
    //            abort();
    //        }

            cs_option(handle, CS_OPT_DETAIL, CS_OPT_ON);

            count = cs_disasm(handle, platforms[i].code, platforms[i].size, address, 0, &insn);
            if (count) {
                size_t j;

                __android_log_print(6,"roysue","****************\n");
                __android_log_print(6,"roysue","Platform: %s\n", platforms[i].comment);
                // 自定义函数先不进行处理
    //            print_string_hex("Code: ", platforms[i].code, platforms[i].size);
                __android_log_print(6,"roysue","Disasm:\n");
```

第 23 章　基于 Capstone 处理特殊指令

```
            for (j = 0; j < count; j++) {
                __android_log_print(6,"roysue","0x%" PRIx64 ":\t%s\t%s\n",
insn[j].address, insn[j].mnemonic, insn[j].op_str);
                // 自定义函数先不进行处理
//               print_insn_detail(&insn[j]);
            }
            __android_log_print(6,"roysue","0x%" PRIx64 ":\n", insn[j-1].address +
insn[j-1].size);

            // free memory allocated by cs_disasm()
            cs_free(insn, count);
        }
        // 错误不进行处理
//      else {
//          printf("****************\n");
//          printf("Platform: %s\n", platforms[i].comment);
//          print_string_hex("Code: ", platforms[i].code, platforms[i].size);
//          printf("ERROR: Failed to disasm given code!\n");
//          abort();
//      }

//      printf("\n");

        cs_close(&handle);
    }
}
```

通过测试案例，我们得知 Capstone 项目可以把自定义的字符串（十六进制）转换为汇编指令。日志输出结果如图 23-14 所示。

```
2022-04-27 11:29:29.802 5199-5199/? E/roysue: ****************
2022-04-27 11:29:29.802 5199-5199/? E/roysue: Platform: ARM-64
2022-04-27 11:29:29.802 5199-5199/? E/roysue: Disasm:
2022-04-27 11:29:29.802 5199-5199/? E/roysue: 0x2c: mrs    x9, midr_el1
2022-04-27 11:29:29.802 5199-5199/? E/roysue: 0x30: msr    spsel, #0
2022-04-27 11:29:29.802 5199-5199/? E/roysue: 0x34: msr    dbgdtrtx_el0, x12
2022-04-27 11:29:29.802 5199-5199/? E/roysue: 0x38: tbx    v0.8b, {v1.16b, v2.16b, v3.16b}, v2.8b
2022-04-27 11:29:29.802 5199-5199/? E/roysue: 0x3c: scvtf  v0.2s, v1.2s, #3
2022-04-27 11:29:29.802 5199-5199/? E/roysue: 0x40: fmla   s0, s0, v0.s[3]
2022-04-27 11:29:29.802 5199-5199/? E/roysue: 0x44: fmov   x2, v5.d[1]
2022-04-27 11:29:29.802 5199-5199/? E/roysue: 0x48: dsb nsh
2022-04-27 11:29:29.802 5199-5199/? E/roysue: 0x4c: dmb osh
2022-04-27 11:29:29.802 5199-5199/? E/roysue: 0x50: isb
2022-04-27 11:29:29.802 5199-5199/? E/roysue: 0x54: mul    x1, x1, x2
2022-04-27 11:29:29.802 5199-5199/? E/roysue: 0x58: lsr    w1, w1, #0
2022-04-27 11:29:29.802 5199-5199/? E/roysue: 0x5c: sub    w0, w0, w1, uxtw
2022-04-27 11:29:29.802 5199-5199/? E/roysue: 0x60: ldr    w1, [sp, #8]
2022-04-27 11:29:29.802 5199-5199/? E/roysue: 0x64: cneg   x0, x1, ne
2022-04-27 11:29:29.802 5199-5199/? E/roysue: 0x68: add    x0, x1, x2, lsl #2
2022-04-27 11:29:29.802 5199-5199/? E/roysue: 0x6c: ldr    q16, [x24, w8, uxtw #4]
2022-04-27 11:29:29.802 5199-5199/? E/roysue: 0x70:
```

图 23-14　测试函数日志输出

23.3 自定义汇编翻译函数

在代码清单 23-1 中，并没有传参进来，直接运行特定条件的代码，这样不是很智能。我们继续改写该代码，让其具有更高的可用性，以满足日常工作中的需求。我们的目标是通过此函数可以翻译一条指令，做到逐条翻译的效果。改写代码如代码清单 23-2 所示。这里不适合使用官方案例中的#define ARM64_CODE，而是申请一段内存空间来存放指令，去掉冗余操作。

代码清单 23-2　自定义汇编翻译函数

```Java {.line-numbers}
static void test2(unsigned int yy, uint64_t addr) {
    // 申请一段 8 字节内存，相当于 #define ARM64_CODE
    unsigned char *stc = static_cast<unsigned char *>(malloc(8));

    char code[40];

    memcpy(stc, &yy, 8);                              // 将 yy 复制到内存中
    csh handle;                                        // handle
    cs_insn *insn;                                     // 指令
    size_t count;                                      // size

    cs_open(CS_ARCH_ARM64, CS_MODE_ARM, &handle);
    cs_option(handle, CS_OPT_DETAIL, CS_OPT_ON);

    count = cs_disasm(handle, stc, sizeof(stc) - 1, addr, 0, &insn); //使用 stc, 不使用 platform
    if (count) {   // 如果可翻译
        __android_log_print(6, "r0ysue", "%s %s", insn[0].mnemonic, insn[0].op_str);
        // 操作, 操作码

        cs_free(insn, count);    // 释放
    }
    cs_close(&handle);           // 关闭
    free(stc);                   // 内存释放
}
```

指令翻译函数编写完成后，我们远程连接 IDA，随意找一条指令测试，如图 23-15 所示。

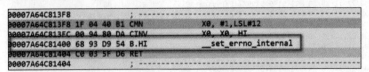

图 23-15　IDA 指令

在 JNI 静态注册的函数中调用，代码如代码清单 23-3 所示。

第 23 章 基于 Capstone 处理特殊指令

代码清单 23-3 函数调用

```C++ {.line-numbers}
extern "C" JNIEXPORT jstring JNICALL
Java_com_r0ysue_capstone_1test_MainActivity_stringFromJNI(
    JNIEnv *env,
    jobject /* this */) {
  std::string hello = "Hello from C++";

//    test();

    // 指令 -> 68 93 D9 54
    // addr -> 7A64C81400
    test2(0x54d99368, 0x7A64C81400);
}
```

运行项目，得到如图 23-16 所示的结果，可以看到成功翻译了一条指令。

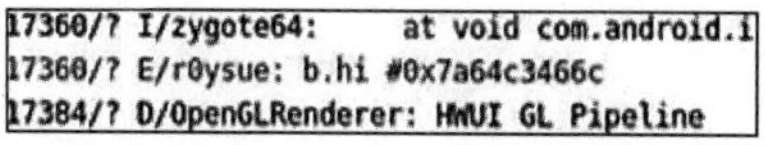

图 23-16 指令翻译结果

在 IDA 中跳转到对应的测试指令处，如图 23-17 所示，发现和 test2 翻译出来的地址是一样的，那么它的正确性就得到了验证。这样官方的案例就被我们改成了单条指令翻译的函数，它比较符合我们日常工作中的需求。

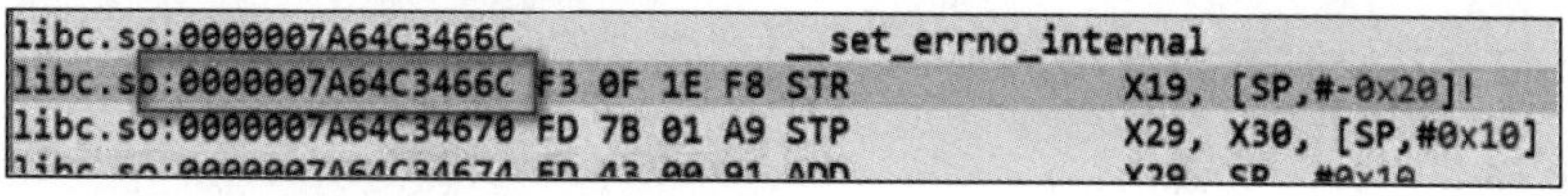

图 23-17 验证结果

继续改写 test2，让它变得更加智能一点，可以做到识别指令中是否存在跳转指令。这次测试 libart.so 中的指令，测试代码如代码清单 23-4 所示。

代码清单 23-4 测试 libart.so 中的指令

```C++ {.line-numbers}
    char line[1024];
    int *start;
    int *end;
    int n = 1;
    FILE *fp = fopen("/proc/self/maps", "r");
    while (fgets(line, sizeof(line), fp)) {
        if (strstr(line, "libart.so")) {
            __android_log_print(6,"r0ysue","%s", line);
            if (n == 1) {
                start = reinterpret_cast<int *>(strtoul(strtok(line, "-"), NULL, 16));
```

```cpp
                end = reinterpret_cast<int *>(strtoul(strtok(NULL, " "), NULL, 16));
            } else {
                strtok(line, "-");
                end = reinterpret_cast<int *>(strtoul(strtok(NULL, " "), NULL, 16));
            }
            n++;
        }
    }

    long *addr = reinterpret_cast<long *>((char *) start + 0x10D760);
    test2(*(int*)((char*)addr), reinterpret_cast<uint64_t>((char*)addr));
```

翻译结果如图 23-18 所示。

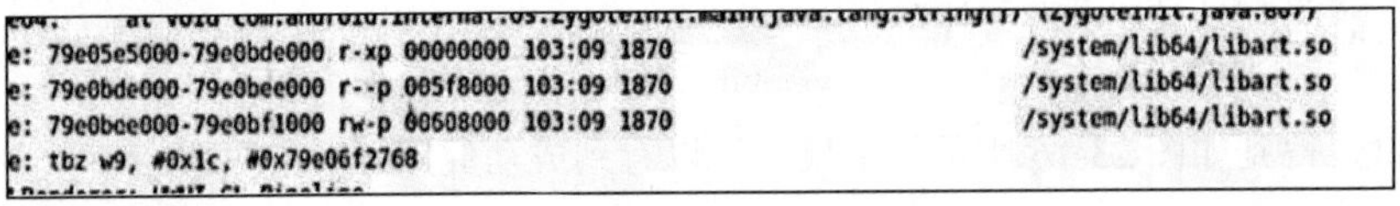

```
e: 79e05e5000-79e0bde000 r-xp 00000000 103:09 1870          /system/lib64/libart.so
e: 79e0bde000-79e0bee000 r--p 005f8000 103:09 1870          /system/lib64/libart.so
e: 79e0bee000-79e0bf1000 rw-p 00608000 103:09 1870          /system/lib64/libart.so
e: tbz w9, #0x1c, #0x79e06f2768
```

图 23-18　翻译结果

如图 23-19 所示，TBZ 的下一条指令是 BL 指令，我们更新一下 test2，让它有过滤的功能。

```
rt.so:00000079E06F275C LDR        W0, [X8]
rt.so:00000079E06F2760 TBZ        W9, #0x1C, loc_79E06F2768
rt.so:00000079E06F2764 BL         _ZN3art11ReadBarrier4MarkEPNS_6mirror6ObjectE ;
```

图 23-19　IDA 指令

更新的 test2 如代码清单 23-5 所示。

代码清单 23-5　test2 更新

```cpp {.line-numbers}
static void test2(unsigned int yy, uint64_t addr) {
    // 申请一段 8 字节内存，相当于 #define ARM64_CODE
    unsigned char *stc = static_cast<unsigned char *>(malloc(8));

    char code[40];

    memcpy(stc, &yy, 8);                              // 将 yy 复制到内存中
    csh handle;                                       // handle
    cs_insn *insn;                                    // 指令
    size_t count;                                     // size

    cs_open(CS_ARCH_ARM64, CS_MODE_ARM, &handle);
    cs_option(handle, CS_OPT_DETAIL, CS_OPT_ON);

    count = cs_disasm(handle, stc, sizeof(stc) - 1, addr, 0, &insn);  // 使用 stc，不
使用 platform
```

```
    if (count) { // 如果可翻译
        if(strstr(insn[0].mnemonic,"bl")&&!strstr(insn[0].mnemonic,"blr")){
            __android_log_print(6, "r0ysue", "%s %s", insn[0].mnemonic, insn[0].op_str);
                    // 操作，操作码
        }
        cs_free(insn, count);    // 释放
    }
    cs_close(&handle);           // 关闭
    free(stc);                   // 内存释放
}
```

再次运行，test2 程序就把 BL 指令识别出来了，如图 23-20 所示。

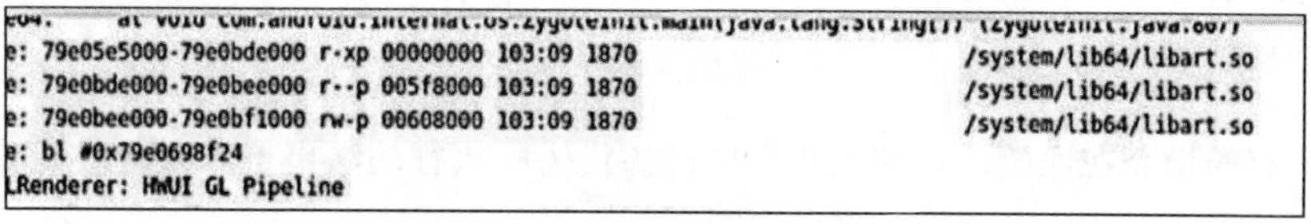
e: 79e05e5000-79e0bde000 r-xp 00000000 103:09 1870 /system/lib64/libart.so
e: 79e0bde000-79e0bee000 r--p 005f8000 103:09 1870 /system/lib64/libart.so
e: 79e0bee000-79e0bf1000 rw-p 00608000 103:09 1870 /system/lib64/libart.so
e: bl #0x79e0698f24
LRenderer: HWUI GL Pipeline

图 23-20

23.4　基于 Capstone 修正指令

23.4.1　指令修复的目的

为什么要对指令进行修复呢？这是因为部分指令移植到我们自己的代码中，执行的结果是不同的。例如在指令相同的情况下，偏移不同，结果就会不同，如图 23-21 和图 23-22 所示。

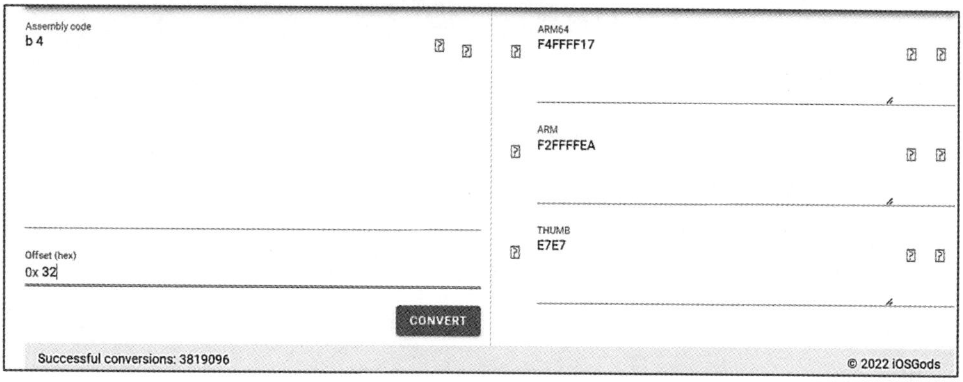
Assembly code
b 4
Offset (hex)
0x 32
CONVERT
ARM64
F4FFFF17
ARM
F2FFFFEA
THUMB
E7E7
Successful conversions: 3819096
© 2022 iOSGods

图 23-21　偏移为 0x32

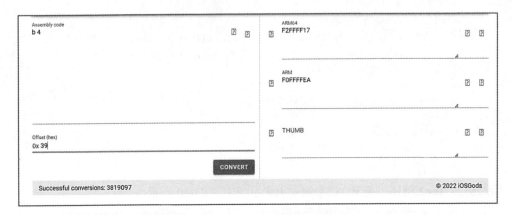

图 23-22　偏移为 0x39

为了让所有的指令在原始位置和在我们自己的代码中执行的结果是一致的，就需要进行指令修复。我们可以通过 Capstone 来解决这个问题。

23.4.2　修复指令的原理

指令修复的原理主要是进行等价替换，如图 23-23 所示。

图 23-23　b 指令等价替换

23.4.3　指令修复的种类

指令的种类有很多，不可能所有的指令都要修复，我们只演示常见指令的修复。此外，还要修复带偏移的指令，如 B 系列指令。

相对偏移指令（跳转指令）：

- B
- BCC
- BL
- CBZ
- TBNZ
- TBZ
- BR
- BLR

- RET

我们要修复的就是这些指令，指令修复的原理也就是等价替换原则，如把所有的 B 指令替换为 BR 指令。BR、BLR、RET 指令不用修复，因为它在原始代码位置和在我们的代码中的执行结果是一致的。

TBNZ 和 CBZ 指令的修复原理稍有不同，如图 23-24 所示。首先，我们需要给原指令赋一个初始值，然后跳转到中转代码，中转代码再跳到绝对地址的跳转点。如果条件不满足，就会按照原始的执行逻辑继续执行，跳过绝对地址的跳转。然而，这种转换会引发一个问题，如果跳转指令跳到了 Hook 的位置，我们就需要对指令进行特殊处理，将绝对地址指向原代码指令，进行减法替换。

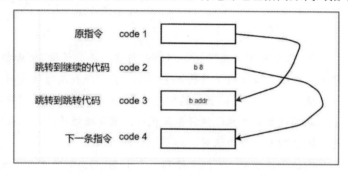

图 23-24　TBNZ 和 CBZ 指令修复原理图

23.5　本章小结

本章介绍了优秀的反汇编工具 Capstone，并详细讲解了它的一系列基础操作，包括如何把官网的源码编译为 SO 文件并导入工程中，以及解决 SO 文件不可用的方法。我们还演示了如何使用官方的测试案例，并对其进行了定制化开发，实现指令修复的功能。

第 24 章

inline Hook 框架集成

第 22 章介绍了使用 inline Hook 对普通函数进行 Hook 操作的过程，主要过程包括构建跳板函数，覆盖 4 条指令，获取跳板函数的地址，并跳转到跳板函数的地址，同时保存函数的返回地址。在跳板函数中，我们实现了重要的功能，如保存寄存器的内容、执行原始代码、恢复堆栈平衡，并最终复原 4 条指令，然后跳转回原指令的位置。

在这个过程中，可能会遇到一个小 BUG，那就是在 func 函数中执行复原代码时，CPU 的读写速度明显快于内存的读写速度，导致不能正确执行复原代码。为了解决这个问题，我们需要使用 sleep 函数来让整个过程延迟一点时间再进行。

本章将介绍的 inline Hook 集成框架是利用"以空间换时间"的办法来解决时间不匹配的问题，inline Hook 集成框架比第 22 章中的 inline Hook 先进许多。第 22 章只是对普通函数进行了简单的 Hook 操作，而本章介绍的集成框架更加综合和高级。

24.1 inline Hook 框架测试

新建一个项目，本项目需要将功能函数封装成为一个.cpp 文件，最终形成一个高可用的 Hook 框架。为了使这个框架有效，后续我们会将其集成到 r0Invoke 中，当然，集成也是一项比较耗费精力的工作，在集成的过程中，需要不断地处理未知的问题，对汇编的编译进行实时调节，否则可能导致运行逻辑跑偏。

该项目的最终效果是封装了 myinlinehook.h 和 myinlinehook.cpp 的文件，如图 24-1 所示。

第 24 章 inline Hook 框架集成

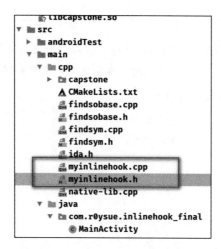

图 24-1 封装效果

编写两个测试函数 func、func1 来做测试,代码如代码清单 24-1 所示。这两个测试函数要替换 libart.so 中的两个连续的函数。

代码清单 24-1 测试函数

```
void func(void* a,void* b){
    __android_log_print(6,"r0ysue","%p",b);

}
void func1(void* a,void* b){
    __android_log_print(6,"r0ysue","i am from header%p",b);
}
```

编写一个 register_hook 函数,传入原始函数的地址和替换函数的地址。这就是整个集成框架的起点。

(1)寻找地址,这里就不需要寻找地址了,因为地址已经作为参数传入进来了,获取地址参数后先赋予地址访问权限,如代码清单 24-2 所示。

代码清单 24-2 赋予地址访问权限

```
void register_hook(void* address,void* func_replace){

    void* a=malloc(0x108);
    mprotect((void *) PAGE_START((long) a),PAGE_SIZE,PROT_WRITE | PROT_READ | PROT_EXEC);
    memcpy(a, (void*)myreplace, 0x108);
    __android_log_print(6,"r0ysue","%p",a);

    mprotect((void *) PAGE_START((long) address),PAGE_SIZE,
            PROT_WRITE | PROT_READ | PROT_EXEC);
```

}

（2）获取函数地址后，需要对参数进行保存。参数的定义可以是全局变量，也可以是局部变量，不会在执行的时候回填，因为在初始的时候就会回填，如代码清单24-3所示。

代码清单24-3　对参数进行保存

```
// 保存参数
// 不会在执行的时候回填，在初始的时候就会回填
// 全局变量、局部变量都可以
long old_code1 = *(int *) ((char *) address);
long old_code2 = *(int *) ((char *) address + 4);
long old_code3 = *(int *) ((char *) address + 8);
long old_code4 = *(int *) ((char *) address + 12);
```

（3）参数覆盖，如代码清单24-4所示。

代码清单24-4　参数覆盖

```
// 覆盖
int code1 = 0x58000050;    //ldr x16,8
int code2 = 0xd61f0200;
long code3 = reinterpret_cast<long>(myreplace);
long retaddress = reinterpret_cast<long>((char *) address + 0x10);
```

（4）因为这里我们要Hook多个函数，所以一个跳板是不够来支撑函数的跳转的。前面提到执行复原代码时，CPU的读写速度明显快于内存的读写速度，最终导致不能执行复原代码。这里采用一种方案，为myreplace函数配置一个模板，为每一个单独的inline Hook做一个跳板函数，这样就不会存在读写的问题了。经过模板的定义后，myreplace就成为一个裸函数了，这样就不用进行寄存器的恢复操作了。我们要进行指令修正，肯定要进行指令膨胀，更改为24条，如代码清单24-5所示。

代码清单24-5　指令膨胀

```
extern "C" void myreplace() {
    // 保存寄存器
    asm("sub SP, SP, #0x200");
    asm("mrs x16, NZCV");
    asm("stp X29, X30, [SP,#0x10]");
    asm("stp X0, X1, [SP,#0x20]");
    asm("stp X2, X3, [SP,#0x30]");
    asm("stp X4, X5, [SP,#0x40]");
    asm("stp X6, X7, [SP,#0x50]");
    asm("stp X8, X9, [SP,#0x60]");
    asm("stp X10, X11, [SP,#0x70]");
    asm("stp X12, X13, [SP,#0x80]");
    asm("stp X14, X15, [SP,#0x90]");
```

```
asm("stp X16, X17, [SP,#0x100]");
asm("stp X28, X19, [SP,#0x110]");
asm("stp X20, X21, [SP,#0x120]");
asm("stp X22, X23, [SP,#0x130]");
asm("stp X24, X25, [SP,#0x140]");
asm("stp X26, X27, [SP,#0x150]");

// 执行函数的偏移
asm("mov x0,x0");
asm("mov x0,x0");
asm("mov x0,x0");
asm("mov x0,x0");
asm("mov x0,x0");

asm("ldp X16, X17, [SP,#0x100]");
asm("ldp X28, X19, [SP,#0x110]");
asm("ldp X20, X21, [SP,#0x120]");
asm("ldp X22, X23, [SP,#0x130]");
asm("ldp X24, X25, [SP,#0x140]");
asm("ldp X26, X27, [SP,#0x150]");
asm("ldp X8, X9, [SP,#0x60]");
asm("ldp X10, X11, [SP,#0x70]");
asm("ldp X12, X13, [SP,#0x80]");
asm("ldp X14, X15, [SP,#0x90]");
asm("msr NZCV, x16");

asm("ldp X0, X1, [SP,#0x20]");
asm("ldp X2, X3, [SP,#0x30]");
asm("ldp X4, X5, [SP,#0x40]");
asm("ldp X6, X7, [SP,#0x50]");
asm("ldp X29, X30, [SP,#0x10]");
asm("add SP, SP, #0x200");

// 指令膨胀 4x6
// 指令恢复的偏移
asm("mov x0,x0");
asm("mov x0,x0");
asm("mov x0,x0");
asm("mov x0,x0");
asm("mov x0,x0");
asm("mov x0,x0");

asm("mov x0,x0");
asm("mov x0,x0");
asm("mov x0,x0");
asm("mov x0,x0");
```

```
    asm("mov x0,x0");
    asm("mov x0,x0");

    asm("mov x0,x0");
    asm("mov x0,x0");
    asm("mov x0,x0");
    asm("mov x0,x0");
    asm("mov x0,x0");
    asm("mov x0,x0");

    asm("mov x0,x0");
    asm("mov x0,x0");
    asm("mov x0,x0");
    asm("mov x0,x0");
    asm("mov x0,x0");
    asm("mov x0,x0");

    // 函数地址复原的偏移
    asm("ldr x16,8");
    asm("br x16");
    asm("mov x0,x0");
    asm("mov x0,x0");
}
```

（5）记住 3 个偏移，分别是执行函数的偏移、指令恢复的偏移和函数复原地址的偏移，如代码清单 24-6 所示。

代码清单 24-6　3 个偏移

```
// 记录执行函数的地址，把返回值函数填回
*(int *) ((char *) (a) + 0x44) = 0x58000070;
*(int *) ((char *) (a) + 0x44+4) = 0xd63f0200;
*(int *) ((char *) (a) + 0x44+8) = 0x14000003;
// 这里要使用 BL 指令，而不是 B 指令
*(long *) ((char *) (a) + 0x44+12) = reinterpret_cast<long>(func_replace);

// 记录执行指令的地址
*(int *) ((char *) (a) + 0x9c) = old_code1;
*(int *) ((char *) (a) + 0x9c+4) = old_code2;
*(int *) ((char *) (a) + 0x9c+8) = old_code3;
*(int *) ((char *) (a) + 0x9c+12) = old_code4;

// return 地址
*(long *) ((char *) (a) + 0x104) = retaddress;


*(int *) ((char *) address) = code1;
```

```
    *(int *) ((char *) address+4) = code2;
    *(int *) ((char *) address+8) = reinterpret_cast<long>(a);
```

（6）函数调用，开始 Hook，如代码清单 24-7 所示。

代码清单 24-7　开始 Hook

```
    initlinker();
    long* base= static_cast<long *>(findsobase("libart.so"));
    // 找地址
    int
off=findsym("/system/lib64/libart.so","_ZN3art11ClassLinker10LoadMethodERKNS_7DexFileER
KNS_21ClassDataItemIteratorENS_6HandleINS_6mirror5ClassEEEPNS_9ArtMethodE");
    long* addr= reinterpret_cast<long *>((char *) base + off );
    __android_log_print(6,"r0ysue","%p",addr);

    long* st= reinterpret_cast<long *>((char *) base + 0x10D760 + 0x25000);
    __android_log_print(6,"r0ysue","%p",st);

    // Hook 两个函数
    register_hook(st, (void*)(func));
    register_hook(addr,(void*)func1);
    // sleep 充足的时间连接 IDA
    //    sleep(20);
```

24.2　结合 Capstone 框架

在 24.1 节中，我们对普通函数的 inlinehook 做了测试，可以完美得到 Hook 的效果。但是，它还是不太智能，我们可以把第 23 章的 Capstone 框架也集成进来，可以根据不同的指令做出不同的动作，在指令恢复的部分加入遍历操作，然后将指令传入 Capstone 中，如代码清单 24-8 所示。

代码清单 24-8　结合 Capstone 框架

```
    for(int n=0x9c;n<=0x9c+72;n=n+24){

        test2(*(int*)((char*)addr+(n-0x9c)/6),
reinterpret_cast<uint64_t>((char*)addr+(n-0x9c)/6));

        if(isb==1){
            if((unsigned long)address-(unsigned long)addr<=0x10){
                address=((unsigned long)huifu+((unsigned long)address-(unsigned
long)addr)*6)+(unsigned long )0x9c;
            }

            *(int *) ((char *) (huifu) + n) = 0x58000050;
            *(int *) ((char *) (huifu) + n+4) = 0xd61f0200;
```

```c
                    *(long *) ((char *) (huifu) + n+8) = address;
//        00021FD6
            } else if(isbl==1){

                if((unsigned long)address-(unsigned long)addr<=0x10){
                    address=((unsigned long)huifu+((unsigned long)address-(unsigned long)addr)*6)+(unsigned long )0x9c;
                }

                *(int *) ((char *) (huifu) + n) = 0x58000070;
                *(int *) ((char *) (huifu) + n+4) = 0xd61f0200;
                *(int *) ((char *) (huifu) + n+8) = 0x14000003;
                *(long *) ((char *) (huifu) + n+12) = address;

        }else if(isbl==1){

                if((unsigned long)address-(unsigned long)addr<=0x10){
                    address=((unsigned long)huifu+((unsigned long)address-(unsigned long)addr)*6)+(unsigned long )0x9c;


                }

                *(int *) ((char *) (huifu) + n) = 0x58000070;
                *(int *) ((char *) (huifu) + n+4) = 0xd63f0200;
//        00023FD6
                *(int *) ((char *) (huifu) + n+8) = 0x14000003;
                *(long *) ((char *) (huifu) + n+12) = address;

        }else if(isb2==1){

                if((unsigned long)address-(unsigned long)addr<=0x10){
                    address=((unsigned long)huifu+((unsigned long)address-(unsigned long)addr)*6)+(unsigned long )0x9c;
                }

                *(int *) ((char *) (huifu) + n) =*(int*)((char*)addr+(n-0x9c)/6)&0xff00000f;
                *(int *) ((char *) (huifu) + n)=*(int *) ((char *) (huifu) + n)^0x40;
                *(int *) ((char *) (huifu) + n+4) = 0x14000005;
//        00023FD6
                *(int *) ((char *) (huifu) + n+8) = 0x58000050;
                *(int *) ((char *) (huifu) + n+12) = 0xd61f0200;
                *(long *) ((char *) (huifu) + n+16) = address;

        }else if(istb==1){
```

```
            if((unsigned long)address-(unsigned long)addr<=0x10){
                address=((unsigned long)huifu+((unsigned long)address-(unsigned
long)addr)*6)+(unsigned long )0x9c;
                    __android_log_print(6,"r0ysue","%p %p %p",address,addr,huifu);
            }

                        *(int *) ((char *) (huifu) + n)
=*(int*)((char*)addr+(n-0x9c)/6)&0xffff000f;
                        *(int *) ((char *) (huifu) + n)=*(int *) ((char *) (huifu) + n)^0x40;
            *(int *) ((char *) (huifu) + n+4) = 0x14000005;
//          00023FD6
            *(int *) ((char *) (huifu) + n+8) = 0x58000050;
            *(int *) ((char *) (huifu) + n+12) = 0xd61f0200;
            *(long *) ((char *) (huifu) + n+16) = address;

        }else {
            *(int *) ((char *) (huifu) + n) =*(int*)((char*)addr+(n-0x9c)/6);
        }
    }
```

24.3 本章小结

本章对之前学习的 inline Hook 框架进行了总结，并进行了优化，使其成为一个高可用的框架。在之前的 inline Hook 使用过程中，存在内存读写速度和 CPU 执行速度不匹配的问题。本章采用了"以空间换时间"的方法来解决这个问题。此外，还将在第 23 章中学习的 Capstone 框架集成到我们自己编写的 inline Hook 框架中，提高了整个框架的性能，实现了真正的高可用框架。

第 25 章

通杀的检测型框架 r0Invoke

什么是通杀？它是一种思路，可以通杀所有的一代、二代壳，也可以通杀所有的检测。但它只是一个框架，并不代表它可以一键解决所有的问题。本章通过 r0Invoke 教读者自己写一个这样的框架，好处是当我们在使用的过程中遇到问题时，能有思路去解决。

比如一个字符串定义错误，我们可以快速地定位到它，简单微调一下就能解决问题，这才是一种通杀。这就表明，我们对框架越了解，用起来就会越得心应手。

本章将通过以下有关 r0Invoke 的实际应用来认识这个框架：

（1）r0Invoke 牛刀小试脱壳 fulao2。

（2）在 FART 基础上进行脱壳操作。

（3）r0Invoke 进阶：跟踪所有运行在 ART 下的 Java 函数。

在 Java 层 Hook 多个函数，在 SO 文件中写会比较烦琐，所以为读者提供一个外挂，这个外挂会跟踪所有解释模式下的 Java 函数，跟踪它的执行流程。

（4）r0Invoke 主动调用 Native 函数并且修改参数。

某 App 的 SO 文件的主动调用，这个函数比较复杂，是一个 armVmp，它有一个参数 jobjectArray，我们可以尝试主动调用一下。

（5）r0Invoke Trace 高度混淆 OLLVM。

开发了一个追踪（Trace）的功能，与 Frida 的 Stalker 原理相似，但实现方法不一样，比 Stalker 更快、更准确，而且没有办法投毒，整个框架建立在 AOSP 系统上，而且不需要进行 root 操作。

25.1　r0Invoke 牛刀小试脱壳 fulao2

25.1.1　APK 静态分析

读过作者已经出版的《Frida 逆向与抓包实战》《Frida 逆向与协议分析》等图书的读者，都知道我们一直在和 fulao2 这个 App 对抗，有读者反应这个壳无论使用 Frida 还是 Xposed 都脱不掉，其实这是因为 fulao2 对壳做了一些小小的改动。我们先静态分析 APK，通过 jadx-gui 查看 APK，可以看到它是加壳的，如图 25-1 所示。

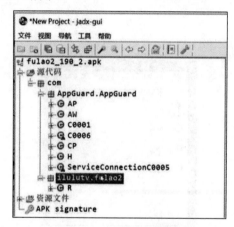

图 25-1　jadx-gui 反编译

然后查看它的 AndroidManifest.xml，如图 25-2 所示。

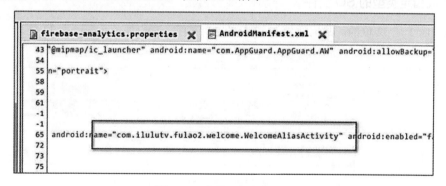

图 25-2　查看 MainActivity

可以看到这个 Activity 并没有在反编译中显示，同时也说明它加了壳。

25.1.2　使用 r0Invoke 脱壳

r0Invoke 是一个 Android 工程，使用方法是将编译后的 SO 文件推送（push）到手机中，进行动态加载。

具体的脱壳逻辑如下：

（1）把 SO 文件放到 SD 卡中，应用程序会自动加载 SD 卡的文件。
（2）任何 App 启动都会尝试去 SD 卡中加载，如果没有 SD 卡的权限，则会忽略这一步。
（3）在 ROM 中调用一个 system.loadlibrary，同 FART 一样的流程。
（4）Hook loadmethod 脱壳。

25.1.3　脱壳操作

脱壳步骤如下：

1. 对 SD 卡授权（见图 25-3）

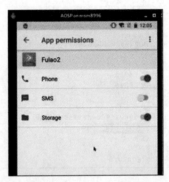

图 25-3　对 SD 卡授权

2. 生成指定类型的 SO 文件

因为 fulao2 是 32 位的，故需要导入 32 位的 SO 文件，我们把项目编译为 32 位，编译配置如图 25-4 所示。

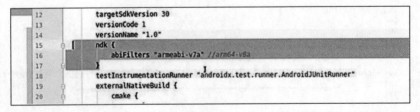

图 25-4　编译配置

3. Hook loadmethod 脱壳

需要特别说明的是，所有的 sdcard 加载的文件都会通过 loadmethod。

1）demo 演示

我们先通过一个 demo 来测试 r0Invoke 是否可以运行起来。测试程序是 getprocessname，通过这个程序来获取进程，代码如图 25-5 所示。

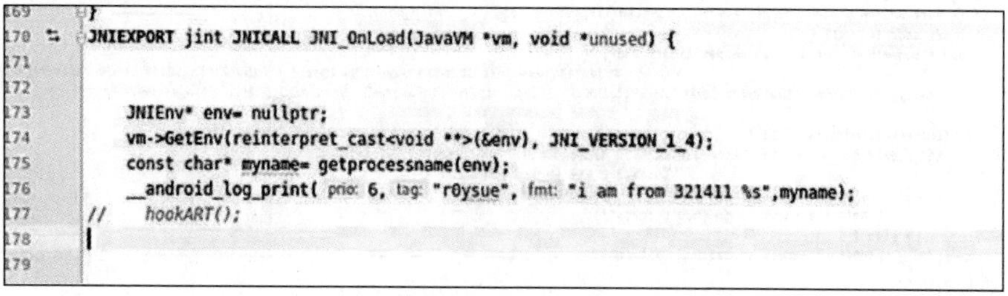

```
}
JNIEXPORT jint JNICALL JNI_OnLoad(JavaVM *vm, void *unused) {


    JNIEnv* env= nullptr;
    vm->GetEnv(reinterpret_cast<void **>(&env), JNI_VERSION_1_4);
    const char* myname= getprocessname(env);
    __android_log_print( prio: 6, tag: "r0ysue", fmt: "i am from 321411 %s",myname);
//    hookART();

```

图 25-5　测试 demo

编译 App，获取 SO 文件，并放入 sdcard 中，操作如图 25-6 所示。

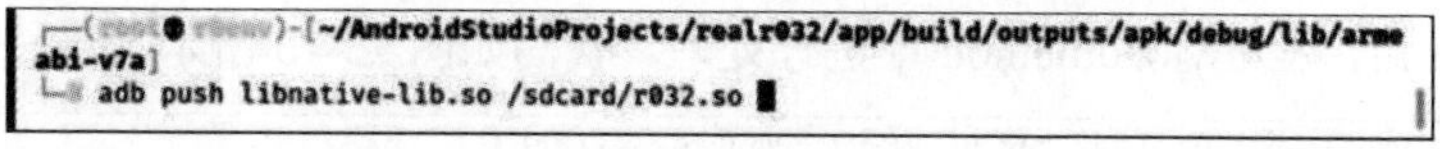

```
┌──(root● r0env)-[~/AndroidStudioProjects/realr032/app/build/outputs/apk/debug/lib/armeabi-v7a]
└─# adb push libnative-lib.so /sdcard/r032.so
```

图 25-6　push SO 文件

直接打开 fulao2，便可以得到注入结果，结果如图 25-7 所示。

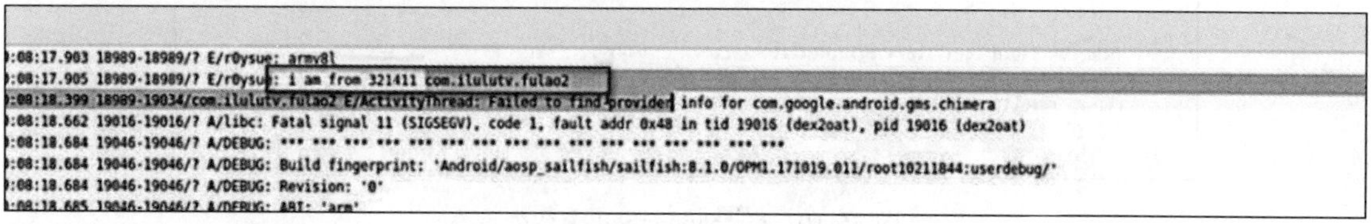

```
:08:17.903 18989-18989/? E/r0ysue: armv8l
:08:17.905 18989-18989/? E/r0ysue: i am from 321411 com.ilulutv.fulao2
:08:18.399 18989-19034/com.ilulutv.fulao2 E/ActivityThread: Failed to find provider info for com.google.android.gms.chimera
:08:18.662 19016-19016/? A/libc: Fatal signal 11 (SIGSEGV), code 1, fault addr 0x48 in tid 19016 (dex2oat), pid 19016 (dex2oat)
:08:18.684 19046-19046/? A/DEBUG: *** *** *** *** *** *** *** *** *** *** *** *** *** *** *** ***
:08:18.684 19046-19046/? A/DEBUG: Build fingerprint: 'Android/aosp_sailfish/sailfish:8.1.0/OPM1.171019.011/root10211844:userdebug/'
:08:18.684 19046-19046/? A/DEBUG: Revision: '0'
:08:18.685 19046-19046/? A/DEBUG: ABI: 'arm'
```

图 25-7　注入结果

2）脱壳核心原理

核心原理可从 LoadMethod 入手。我们可以到 Android 源码中追踪查看一下，如图 25-8 所示。

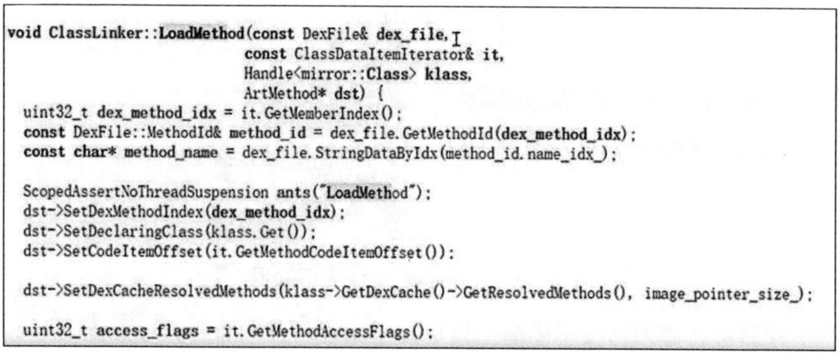

```
void ClassLinker::LoadMethod(const DexFile& dex_file,
                             const ClassDataItemIterator& it,
                             Handle<mirror::Class> klass,
                             ArtMethod* dst) {
  uint32_t dex_method_idx = it.GetMemberIndex();
  const DexFile::MethodId& method_id = dex_file.GetMethodId(dex_method_idx);
  const char* method_name = dex_file.StringDataByIdx(method_id.name_idx_);

  ScopedAssertNoThreadSuspension ants("LoadMethod");
  dst->SetDexMethodIndex(dex_method_idx);
  dst->SetDeclaringClass(klass.Get());
  dst->SetCodeItemOffset(it.GetMethodCodeItemOffset());

  dst->SetDexCacheResolvedMethods(klass->GetDexCache()->GetResolvedMethods(), image_pointer_size_);

  uint32_t access_flags = it.GetMethodAccessFlags();
```

图 25-8　源码追踪

我们根据参数 dex_file 来脱壳。

然后编写代码 Hook dex_file，代码如图 25-9 所示。

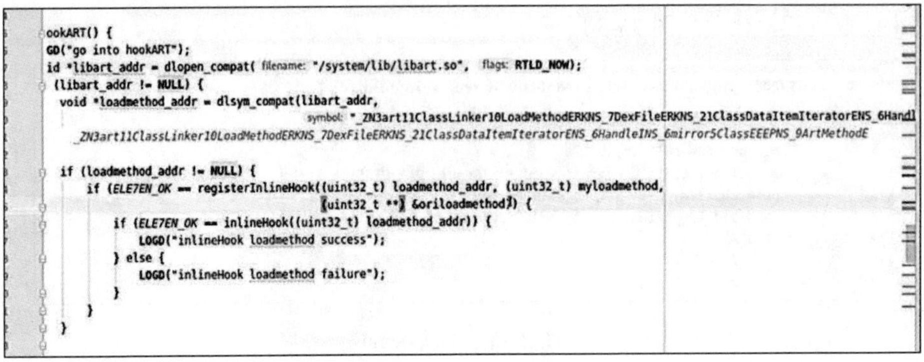

```
ookART() {
    GD("go into hookART");
    id *libart_addr = dlopen_compat( filename: "/system/lib/libart.so",  flags: RTLD_NOW);
    if (libart_addr != NULL) {
        void *loadmethod_addr = dlsym_compat(libart_addr,
                  symbol: "_ZN3art11ClassLinker10LoadMethodERKNS_7DexFileERKNS_21ClassDataItemIteratorENS_6Handl
                  _ZN3art11ClassLinker10LoadMethodERKNS_7DexFileERKNS_21ClassDataItemIteratorENS_6HandleINS_6mirror5ClassEEEPNS_9ArtMethodE
        if (loadmethod_addr != NULL) {
            if (ELE7EN_OK == registerInlineHook((uint32_t) loadmethod_addr, (uint32_t) myloadmethod,
                                                 (uint32_t **) &oriloadmethod)) {
                if (ELE7EN_OK == inlineHook((uint32_t) loadmethod_addr)) {
                    LOGD("inlineHook loadmethod success");
                } else {
                    LOGD("inlineHook loadmethod failure");
                }
            }
        }
    }
}
```

图 25-9 Hook dex_file

获取 dex_file 后，在 myloadmethod 中进行处理，代码如图 25-10 所示。

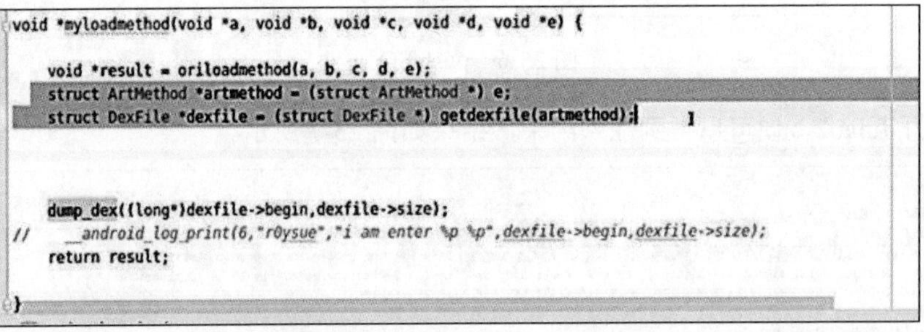

```
void *myloadmethod(void *a, void *b, void *c, void *d, void *e) {

    void *result = oriloadmethod(a, b, c, d, e);
    struct ArtMethod *artmethod = (struct ArtMethod *) e;
    struct DexFile *dexfile = (struct DexFile *) getdexfile(artmethod);

    dump_dex((long*)dexfile->begin,dexfile->size);
//      __android_log_print(6,"r0ysue","i am enter %p %p",dexfile->begin,dexfile->size);
    return result;
}
```

图 25-10 myloadmethod 函数处理

构造 ArtMethod 和 dex_file，获取拿到 dexfile 的 begin 和 size，传入 dump_dex 中，进行 DEX 的 dump 操作，代码如图 25-11 所示。

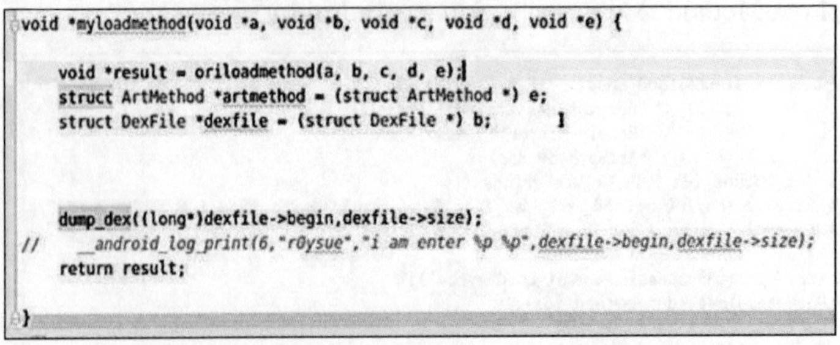

```
void *myloadmethod(void *a, void *b, void *c, void *d, void *e) {

    void *result = oriloadmethod(a, b, c, d, e);
    struct ArtMethod *artmethod = (struct ArtMethod *) e;
    struct DexFile *dexfile = (struct DexFile *) b;

    dump_dex((long*)dexfile->begin,dexfile->size);
//      __android_log_print(6,"r0ysue","i am enter %p %p",dexfile->begin,dexfile->size);
    return result;
}
```

图 25-11 myloadmethod 内部处理

这种脱壳方法目前是脱不掉这个壳的，这里其实还有一点小问题。先继续往下分析代码，在如图 25-12 所示的 dump_dex 函数中，把 DEX 指定到 sdcard 的目录中，给它一个读写权限，当然所有的 method 都会经过 classLinker，所以要判断一下，不然会重复写入。最后是一种常规文件的保存方式。

第 25 章　通杀的检测型框架 r0Invoke　475

```
void dump_dex(long *begin, int size) {
    char path[100];
    sprintf(path, fmt: "/sdcard/%d.dex", size);
    if (access(path, F_OK) != -1)
        return;

    mydex[n].begin=begin;
    mydex[n].size=size;
    int fp = open(path, flags: O_CREAT | O_APPEND | O_RDWR, modes: 0666);
//    syscall(__NR_write,fp,(void *) begin,size);
//    __android_log_print(6, "r0ysue", "i dump success %d.dex %x", size,begin);
```

图 25-12　dump_dex 代码 1

最后需要进行写入操作，然后打印 size 和魔术字段，代码如图 25-13 所示。

```
//    __android_log_print(6, "r0ysue", "i dump success %d.dex %x", size,begin);
      write(fp, (void *) begin, size);
      fsync(fp);
      close(fp);

//    n++;

      __android_log_print( prio: 6, tag: "r0ysue", fmt: "i dump success %d.dex %x", size,*begin);
}
```

图 25-13　dump_dex 代码 2

3）脱壳效果

我们来看一下脱壳效果，如图 25-14 所示，肉眼可见一些魔术字段，但是它真的把壳脱下来了吗？

魔术字段是正确的，没有问题，我们把脱出来的 DEX pull 到虚拟机中看一下效果。

```
mbstone written to: /data/tombstones/tombs
e: i dump success 3016944.dex a786564
e: i dump success 5039360.dex a786564
e: i dump success 9048832.dex a786564
e: i dump success 385040.dex a786564
e: i dump success 10181040.dex a786564
e: i dump success 140532.dex a786564
e: i dump success 405652.dex a786564
```

图 25-14　打印结果

使用 010 Editor 查看是否正常。如图 25-15 所示，可以看到是一些垃圾代码。

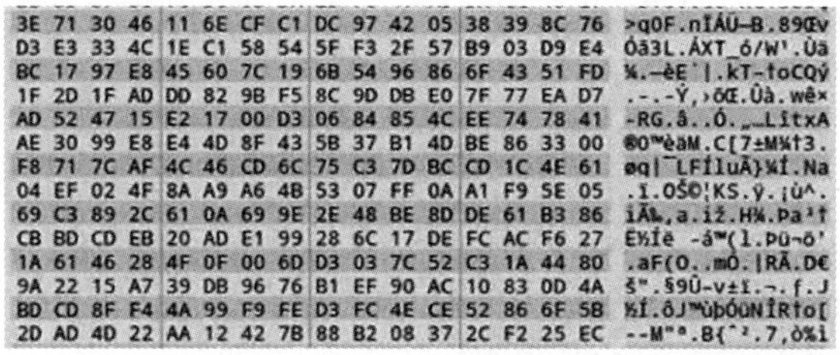

图 25-15　010 Editor 乱码

然后观察一下魔术头是否正确，如图 25-16 所示，可以看到并不是正确的魔术头。

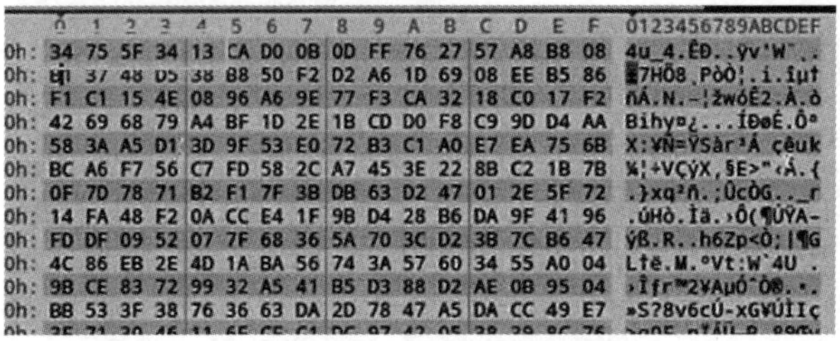

图 25-16　非正常魔术头

看似全脱出来了，其实都是假的，fulao2 其实挺狡猾的，给我们呈现假的东西。这是 Android log 和 write 的结果不匹配导致的，可以判断它对 write 做了 Hook 的处理，直接把这个 write 中的 begin 指向它准备好的 begin，这就是为什么打印的结果正确，但真实的脱壳却不是这样的，如图 25-17 所示。这不是什么高端的加密技术，就是简单的自定义 Hook，所以我们做一下微调。

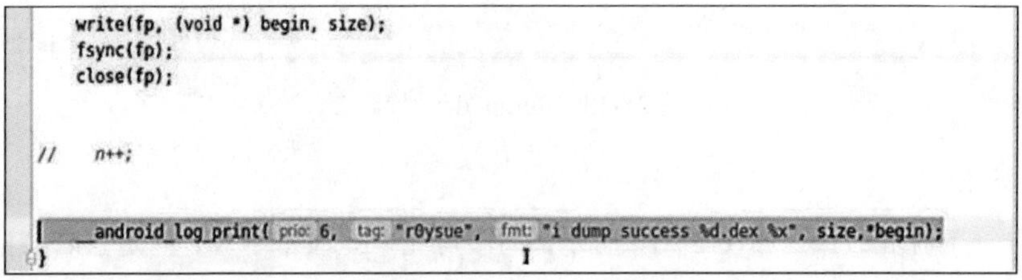

图 25-17　begin 指向问题

4）微调

我们把 write 函数转为 syscall 与 Xposed 进行比较，也不是很麻烦，将 APK 解压，获取 SO 文件，push 到 sdcard 中，转换如图 25-18 所示。

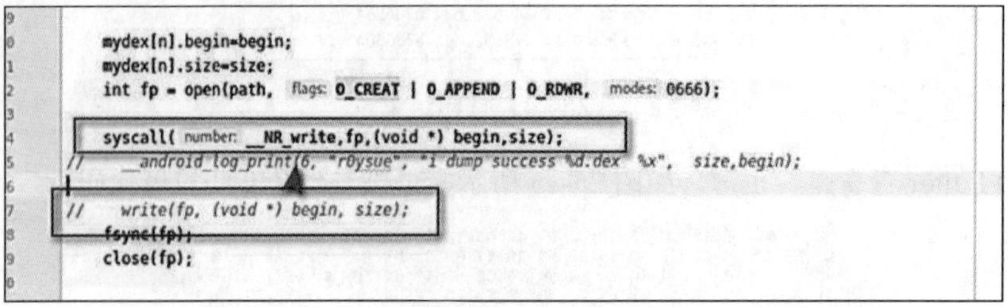

图 25-18　微调

再将编译解压后的 SO 文件 push 到手机中，重新运行 fulao2 App，打印结果如图 25-19 所示，打印没什么问题，看一下脱壳的真实效果。

在 010 Editor 中查看文件头，如图 25-20 所示。

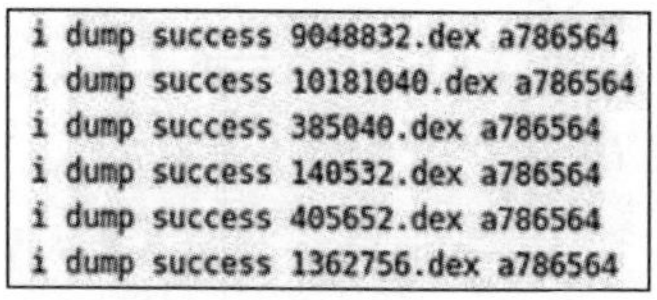

图 25-19 打印结果 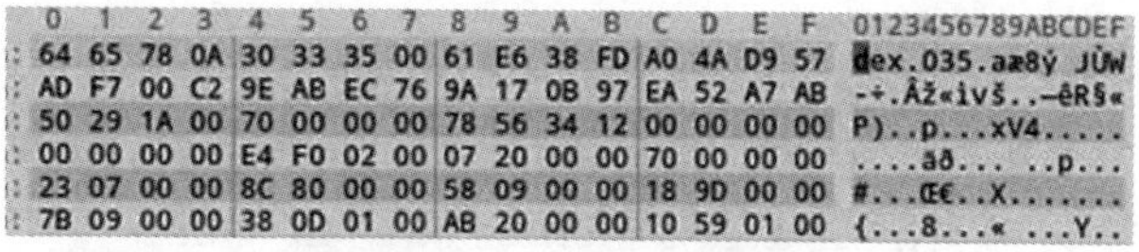   图 25-20 查看文件头

魔术字段没有问题，我们尝试使用 jadx-gui 看是否可以打开，效果如图 25-21 所示。

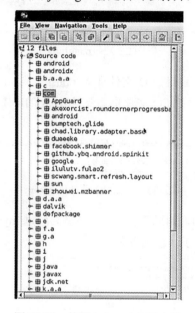

图 25-21 使用 jadx-gui 打开文件

fulao2 的壳不是抽取型的壳，其实这个壳还是很低端的，只是做了一点微调，市面上的脱壳机就搞不定它。只要我们理解了 ART，就能把这个壳脱下来，只要简单修改 write 函数即可。

25.2　r0Invoke 进阶：跟踪所有运行在 ART 下的 Java 函数

我们还可以使用 r0Invoke 来 Hook 多个 Java 函数，但 Hook 多个 Java 函数的效果并不理想，所以这里只 Hook 一个函数，通过编写一个辅助函数作为外挂，以此来查看函数的执行流程。

在这个案例中，我们使用 64 位应用的 App。内部实现在 libc 的 str 中，mainfun 是一个 inline Hook，这是编写的初版错码，写得不是很完整，如图 25-22 所示。当第二个参数等于 myblackinvoke 的时候，就打印第一个参数，第一个参数的内容是 Java 函数的名称和一个签名。

```
//
//    mainfun("_ZN3art9ArtMethod14RegisterNativeEPKvb", "libart.so",
//            reinterpret_cast<void *>(regist));

    //   myblackinvoke
  mainfun("strstr", "libc.so", (void* )(new_strstr));

    JNIEnv* env= nullptr;
    vm->GetEnv(reinterpret_cast<void **>(&env), JNI_VERSION_1_4);
const char* myname= getprocessname(env);
    __android_log_print( prio: 6, tag: "r0ysue", fmt: " i am from %s",myname);
```

图 25-22　mainfun

然后编译项目，解压 APK 后，将 SO 文件推送（push）到 sdcard 中，操作如图 25-23 所示，因为应用程序为 64 位，所以这里的 SO 文件也需要是 64 位的，本示例中的应用程序是随机找的，只要是 64 位的即可。

```
┌──(root㉿rOenv)-[~/AndroidStudioProjects/usehook/app/build/outputs/apk/debug/lib/arm64
-v8a]
└─$ adb push libnative-lib.so /sdcard/r0.so
```

图 25-23　push SO 文件到 sdcard 中

在解释器中做了插桩的处理，如果要解释语句，就要经过 myblackinvoke，如图 25-24 所示，这样我们就可以跟踪所有的流程了（包括签名）。

```
void new_strstr(const char* a,const char* b){
if(strcmp(b,"myblackinvoke")==0)
    __android_log_print( prio: 6, tag: "r0ysue", fmt: " method is  %s",a);
}
```

图 25-24　myblackinvoke

打印结果如图 25-25 所示。

```
method is  java.lang.reflect.Field com.alibaba.fastjson.util.b.a(java.lang.Class, java.lang.String, java.lang.reflect.Field[], ja
method is  java.lang.String java.lang.StringBuilder.toString()
method is  java.lang.reflect.Field com.alibaba.fastjson.util.b.a(java.lang.Class, java.lang.String, java.lang.reflect.Field[], ja
method is  java.lang.annotation.Annotation java.lang.reflect.Field.getAnnotation(java.lang.Class)
method is  void java.lang.Object.<init>()
method is  void libcore.reflect.AnnotationMember.<init>(java.lang.String, java.lang.Object, java.lang.Class, java.lang.reflect.Me
method is  java.lang.annotation.Annotation libcore.reflect.AnnotationFactory.createAnnotation(java.lang.Class, libcore.reflect.An
method is  void java.lang.reflect.Proxy.<init>(java.lang.reflect.InvocationHandler)
method is  java.lang.Object java.util.HashMap.get(java.lang.Object)
method is  int com.alibaba.fastjson.annotation.JSONField.ordinal()
method is  java.lang.Class java.lang.Object.getClass()
method is  java.lang.Object java.lang.reflect.Proxy.invoke(java.lang.reflect.Proxy, java.lang.reflect.Method, java.lang.Object[])
method is  boolean java.lang.Class.isAssignableFrom(java.lang.Class)
method is  com.alibaba.fastjson.serializer.SerializerFeature[] com.alibaba.fastjson.annotation.JSONField.serialzeFeatures()
method is  java.lang.Object java.util.HashMap.put(java.lang.Object, java.lang.Object)
method is  java.lang.Object java.lang.reflect.Proxy.invoke(java.lang.reflect.Proxy, java.lang.reflect.Method, java.lang.Object[])
method is  int com.alibaba.fastjson.serializer.SerializerFeature.of(com.alibaba.fastjson.serializer.SerializerFeature[])
```

图 25-25　打印结果

其实也可以做链路追踪，需要把所有的导出函数都 Hook 上，但是没有必要，链路追踪对内存的要求很高，也会浪费内存的资源。

25.3　r0Invoke 主动调用 Native 函数并且修改参数

这里的测试案例是一个非常难调用的函数，它有两个参数：一个是主动调用的调用号，另一个是 object 数组。这是一个 Native 函数，如图 25-26 所示。

```
mainfun("_ZN3art9ArtMethod14RegisterNativeEPKvb", "libart.so",
        reinterpret_cast<void *>(regist));
```

图 25-26　测试函数调用

对测试函数进行主动调用，然后注册到 myreplace 中，如图 25-27 所示。

```
void regist(void* a,void* b,int c){
    const char* ds=getartmethod((unsigned int *)a);
    if(strstr(ds, "doCommandNative")){
        st-b;
        long* d= reinterpret_cast<long *>(myreplace);
        __android_log_print( prio: 6, tag: "r0ysue", fmt: "%p",d);
        __asm__("str %[input_n], [X18,#0x28]\r\n"
        :[result_m] "=r" (d)
        :[input_n] "r" (d)
        );
        __android_log_print( prio: 6, tag: "r0ysue", fmt: "register %s",ds);
    }
}
```

图 25-27　注册

首先获取 ArtMethod 的 Methodname，如何获取呢？我们到源码中查看一下，如图 25-28 所示。第一个参数 ArtMethod 是函数名称，第二个参数是 native_method，使用要注册的函数把这个参数替换掉，就能实现动态注册函数替换。这里直接把第二个参数替换掉。

```
89      // Pop transition.
90      self->PopManagedStackFragment(fragment);
91    }
92
93    const void* ArtMethod::RegisterNative(const void* native_method, bool is_fast) {
94      CHECK(IsNative()) << PrettyMethod();
95      CHECK(!IsFastNative()) << PrettyMethod();
96      CHECK(native_method != nullptr) << PrettyMethod();
97      if (is_fast) {
98        AddAccessFlags(kAccFastNative);
99      }
00      void* new_native_method = nullptr;
01      Runtime::Current()->GetRuntimeCallbacks()->RegisterNativeMethod(this,
02                                                                      native_method,
03                                                                      /*out*/&new_native_method);
04      SetEntryPointFromJni(new_native_method);
05      return new_native_method;
06    }
```

图 25-28　RegisterName

在 myreplace 函数中，第一个参数是 JNI_onload，第二个参数是 jobject，第三个参数是调用编号，第四个参数是调用数组。直接用保存的地址来调用即可，如图 25-29 所示。

```
void* myreplace(void* a,void* b,int c,void* d){
    JNIEnv* env= static_cast<JNIEnv *>(a);
    jobject aa= static_cast<jobject>((docomm(st))(a, b, reinterpret_cast<void *>(c), d));


//    jobject st=static_cast<jobject>((docomm(st))(a, b, reinterpret_cast<void *>(yy), jk));

//__android_log_print(6,"r0ysue","dddddddddddd %p",st);

    if(aa!= nullptr&&c==0x28a1) {
```

图 25-29　myreplace 函数

使用 JNI 改变里面的一个参数，再做一次调用，如图 25-30 所示。

```
if(aa!= nullptr&&c==0x28a1) {
//    jclass jobjclass=env->FindClass("java/lang/String");
//    jobjectArray sg=env->NewObjectArray(5, jobjclass, nullptr);
//    env->SetObjectArrayElement(sg,0,env->NewStringUTF("{INPUT=YXUeAvXSKr0DAEMvjgqJg6G723867946}"));
    env->SetObjectArrayElement(static_cast<jobjectArray>(d), index: 2, value: env->NewStringUTF( bytes: "10"));
    jobject opp= static_cast<jobject>((docomm(st))(a, b, reinterpret_cast<void *>(c), d));
    char* change= const_cast<char *>(env->GetStringUTFChars(
            static_cast<jstring>(opp), isCopy: 0));
    change[20]='\0';
    __android_log_print( prio: 6, tag: "r0ysue", fmt: " 111111111111  %s",(const char*)change);
//    env->SetObjectArrayElement(sg,2,env->NewStringUTF("3"));
//    env->SetObjectArrayElement(sg,3,env->NewStringUTF("xxxxxxxxxxxxxxxxxxxxxxxxx"));
//    env->SetObjectArrayElement(sg,4,env->NewStringUTF("true"));
//
//    void* ccc= (docomm(st))(a, b, reinterpret_cast<void *>(c), sg);
```

图 25-30　再次调用

change 改变前的值和改变后的值分别如图 25-31 和图 25-32 所示。

```
199    //    jobjectArray sg=env->NewObjectArray(5, jobjclass, nullptr);
200    //    env->SetObjectArrayElement(sg,0,env->NewStringUTF("{INPUT=YXUeAvXSKr0DAEMvjgqJg6G723867946}"));
201        env->SetObjectArrayElement(static_cast<jobjectArray>(d), index: 2, value: env->NewStringUTF( bytes: "10"));
202        jobject opp= static_cast<jobject>((docomm(st))(a, b, reinterpret_cast<void *>(c), d));
203        char* change= const_cast<char *>(env->GetStringUTFChars(
204                static_cast<jstring>(opp), isCopy: 0));
205        change[20]='\0';
206        __android_log_print( prio: 6, tag: "r0ysue", fmt: " 111111111111  %s",(const char*)change);
207    //    env->SetObjectArrayElement(sg,2,env->NewStringUTF("3"));
```

图 25-31　改变后的 change

```
printobjearry(env,(jobjectArray)d);
change= const_cast<char *>(env->GetStringUTFChars(
        static_cast<jstring>(aa), isCopy: 0));
change[20]='\0';
__android_log_print( prio: 6, tag: "r0ysue", fmt: " ssssssss    int:%x    %s", c,(const char*)change);
```

图 25-32　原函数的 change

这个功能有着 Unidbg 的优点，不需要自己构造参数，我们可以通过 Hook 构造参数，慢慢修改。这里不具体介绍是哪个函数了，它是一个 ARM VMP，111111111111 是主动调用的结果，效果如图 25-33 所示。

```
oid V/HighwayApp: The address of obj delete invalid Event is 0x33a2
oid E/r0ysue:  111111111111  e7cef86fd0574030ea12     I
oid E/r0ysue:  ssssssss 0: {INPUT=YXUeAvXSKr0DAEMvjgqJg6G723867946}
oid E/r0ysue:  ssssssss 1: 23867946
oid E/r0ysue:  ssssssss 2: 10
oid E/r0ysue:  ssssssss 3:
oid E/r0ysue:  ssssssss 4: true
oid E/r0ysue:  ssssssss  int:28a1    2393e85cdd7ca90249a7
oid E/r0ysue:  aarch64
oid I/chatty:  uid=10079 com.lazada.android:channel identical 1 line
```

图 25-33　效果

最终结果如图 25-34 所示，此功能是基于 inline Hook 实现的，参数肯定是正确的。

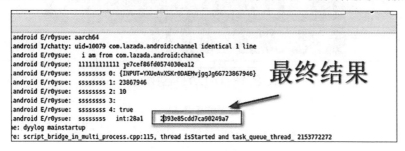

图 25-34　最终结果

> **注意**　如果在线程中调用，那么要对 env 进行处理。

r0Invoke 主动调用 Native 函数的优点如下：

（1）比 Frida 快。
（2）比 Frida 更难检测。
（3）不需要 root。

> **注意**　r0Invoke 主动调用 Native 函数比 Unidbg 简单，只要 Hook 出来，更改里面的参数即可。

25.4　r0Invoke Trace 高度混淆 OLLVM

本节介绍使用 r0Invoke Trace 高度混淆 OLLVM，此功能主要通过 Shell Code 编写完成，它的主要优点是速度快，比如 IDA 每分钟处理 1200 条指令，Unidbg 每分钟处理 1×10^5 条指令，不带寄存器，而 r0Invoke 20 秒可以处理 1.6×10^5 条指令，还带寄存器。

此外，该方法还有以下两个特点：

（1）基于 Capstone。

（2）无法投毒（无 root，无特征，无法被检测）。

下面我们来演示具体的操作过程。

演示的案例 App 是 3W 班的一道测试题目，这是一个 OLLVM 的混淆样本，题目混淆效果如图 25-35 所示。这里给大家看一下效果，使用 r0Invoke 20 秒就能解决。

图 25-35　OLLVM 混淆效果

上述案例没有对 sdcard 进行权限设置。这里采用另一种方式，即把 SO 文件导入应用目录中，赋予该目录权限，然后把 capstone.so 导入应用目录，具体操作如图 25-36 所示。接下来直接启动 App，即可开始测试。

```
com.ttxapps.wifiadb/
bullhead:/sdcard # cp libnative-lib.so /data/data/com.ex
com.example.classloadertest/          com.example.myapplication/
com.example.hellojni_sign2/
p libnative-lib.so /data/data/com.example.hellojni_sign2/r0.so
```

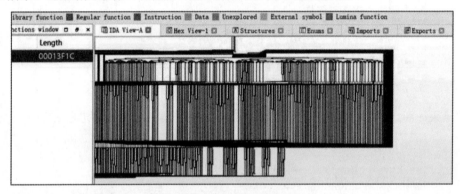

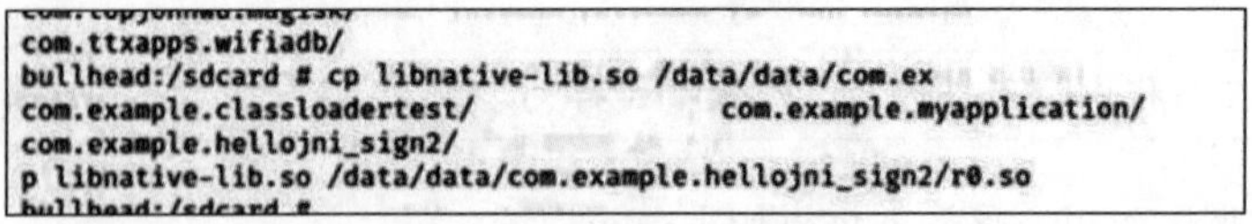

图 25-36　SO 文件导入操作

接下来进行代码讲解。

在本案例中，单独启动一个线程，代码如图 25-37 所示。

第 25 章　通杀的检测型框架 r0Invoke　483

```
//      mainfun("_ZN3art9ArtMethod14RegisterNativeEPKvb", "libart.so",
//              reinterpret_cast<void *>(regist));
    JNIEnv *env = nullptr;
    vm->GetEnv(reinterpret_cast<void **>(&env), JNI_VERSION_1_4);
    const char *myname = getprocessname(env);
    __android_log_print( prio: 6, tag: "r0ysue", fmt: " xxxxxxxxx i am from %s", myname);
//sleep(20);
    pthread_t thread;
    pthread_create(&thread, attr: nullptr, reinterpret_cast<void *(*)(void *)>(myin), nullptr);
}
```

图 25-37　启动一个线程

构造参数，如图 25-38 所示。

```
void myin() {
    sleep( seconds: 10);
    const char *a1 = "0123456789abcdef";
    int a2 = 16;
    const char *a3 = "www.pediy.com&kanxue";
    int a4 = 20;
    void *a5 = malloc( byte_count: 100);
    mprotect( addr: (void*)PAGE_START( x: (long)a1),PAGE_SIZE, prot: PROT_WRITE|PROT_READ|PROT_EXEC);
    mprotect( addr: (void*)PAGE_START( x: (long)a3),PAGE_SIZE, prot: PROT_WRITE|PROT_READ|PROT_EXEC);
//  char line[1024];
```

图 25-38　构造参数

获取首地址，进行主动调用，如图 25-39 所示。

```
    initlinker();
    long *start= static_cast<long *>(findsobase("libhello-jni.so"));
    __android_log_print( prio: 6, tag: "r0ysue", fmt: "xxxxxx %p", start);
    long *sub_13ce4 = reinterpret_cast<long *>((char *) start + 0x1494C4);

    mprotect( addr: (void *) PAGE_START( x: (long) sub_13ce4), PAGE_SIZE, prot: PROT_WRITE | PROT_READ | PROT_EXEC);

    __android_log_print( prio: 6, tag: "r0ysue", fmt: "xxxxxx %p", sub_13ce4);
    __android_log_print( prio: 6, tag: "r0ysue", fmt: "sssssss %p", jicunq);
    deal_with(sub_13ce4, end: reinterpret_cast<long *>((char *) sub_13ce4 + 0x13f1c));
    kk = (invoke) (c);
    __android_log_print( prio: 6, tag: "r0ysue", fmt: "xxxxxx %p", kk);
//  *(int*)((char *)sub_13ce4+n)=0x58000072;
//  *(int*)((char*)sub_13ce4+n+4)=0xd61f0240;
//  *(int*)((char*)sub_13ce4+n+8)=0x14000003;
//  *(long**)((char *)sub_13ce4+n+12)= reinterpret_cast<long *>(kk);
```

图 25-39　进行主动调用

最后展示最终效果，从开始到结束只用了 20 秒，如图 25-40 所示。

```
12:33:12.395 22963-22980/com.example.hellojni_sign2 E/r0ysue: xxxxxx success 0x0          开始
12:33:20.110 3192-3330/com.android.nfc E/NxpTml: _i2c_write() errno : 5
12:33:20.110 3192-3330/com.android.nfc E/NxpHal: PN54X - Error in I2C Write.....
12:33:20.111 3192-3332/com.android.nfc E/NxpHal: write error status = 0x1ff
12:33:20.111 3192-3332/com.android.nfc E/NxpHal: write_unlocked failed - PN54X Maybe in Standby Mode - Retry
12:33:21.373 3192-3330/com.android.nfc E/NxpTml: _i2c_write() errno : 5
12:33:21.374 3192-3332/com.android.nfc E/NxpHal: PN54X - Error in I2C Write.....           结束
12:33:21.374 3192-3332/com.android.nfc E/NxpHal: write error status = 0x1ff
12:33:21.375 3192-3311/com.android.nfc E/NxpHal: write_unlocked failed - PN54X Maybe in Standby Mode - Retry
12:33:32.219 22963-22980/com.example.hellojni_sign2 E/r0ysue: xxxxxx success 0x8ddf0095b8996c7e
```

图 25-40　最终效果 1

我们还可以把指令导出，看一下它有多少行，如图 25-41 所示。导出来为 $1.6×10^5$ 行代码，20 秒完成。笔者第一次用 IDA Trace，花费了将近 20 个小时，足以可见 r0Invoke 的强大威力。不过还是有一些小 BUG，有些指令回调的时候没有修正，不过它在测试环境下还是很准确的。它的原理是指令的翻译和指令的修正。速度快的原因是这个框架没有使用 C 语言，而是使用 Shell Code 来执行的。

```
162273    ldp x20, x19, [sp, #0x40]    x19 = 0x75b23ec4f0   x20 = 0x7
162274    ldp x22, x21, [sp, #0x30]    x21 = 0x75b23ec4f0   x22 = 0x5
162275    ldp x24, x23, [sp, #0x20]    x23 = 0x59b3   x24 = 0x75bec6c
162276    ldp x26, x25, [sp, #0x10]    x25 = 0x75b22f0000   x26 = 0x7
162277    ldp x28, x27, [sp], #0x60    x27 = 0x1   x28 = 0x75c99ee678
162278
```

图 25-41　最终效果 2

25.5　本章小结

本章介绍了一种新型的检测框架 r0Invoke。通过实际应用的演示，展示了 r0Invoke 在多个方面的功能，包括 fulao2 的脱壳效果、跟踪所有运行在 ART 下的 Java 函数、主动调用 Native 函数并修改参数，以及强大的追踪（Trace）功能。此外，r0Invoke 还可以使用 Shell Code 来快速追踪高度混淆的 OLLVM 代码。

第 26 章 SO 文件加载流程分析与注入实战

之前的几个章节讲述了 inline Hook 的原理和需要注意的一些问题。本章开启了新的篇章,研究了在 Android 中 SO 文件的加载流程。通过分析 SO 文件的加载流程,我们发现了 3 个主要的加载函数。从这 3 个加载函数中,我们可以衍生出不同的 SO 文件注入方式。

26.1 SO 文件的加载方式

新建一个以 C++为模板的工程,在 Java 代码中就会出现 SO 文件的加载代码,如代码清单 26-1 所示。

代码清单 26-1　SO 文件的加载代码

```
static {
        System.loadLibrary("native-lib");
}
```

我们先将 SO 文件的加载代码注释掉,编译 App,看看有什么结果。运行 App 后,在如图 26-1 所示的日志中可以看到报错内容为没有实现 stringFromJNI 方法。然后打开注释,再次编译运行,发现可以正常运行,说明这个位置就是 SO 文件加载的入口点,但是具体是怎么加载的,需要进入 Android 源码中查看。

```
2022-05-17 18:55:53.350 9297-9297/com.roysue.demoso E/AndroidRuntime: FATAL EXCEPTION: main
    Process: com.roysue.demoso, PID: 9297
    java.lang.UnsatisfiedLinkError: No implementation found for java.lang.String com.roysue.demoso.MainActivity.stringFromJNI() (tried
    Java_com_roysue_demoso_MainActivity_stringFromJNI and Java_com_roysue_demoso_MainActivity_stringFromJNI__)
        at com.roysue.demoso.MainActivity.stringFromJNI(Native Method)
        at com.roysue.demoso.MainActivity.onCreate(MainActivity.java:22)
        at android.app.Activity.performCreate(Activity.java:6975)
        at android.app.Instrumentation.callActivityOnCreate(Instrumentation.java:1213)
        at android.app.ActivityThread.performLaunchActivity(ActivityThread.java:2770)
        at android.app.ActivityThread.handleLaunchActivity(ActivityThread.java:2892)
        at android.app.ActivityThread.-wrap11(Unknown Source:0)
        at android.app.ActivityThread$H.handleMessage(ActivityThread.java:1593)
        at android.os.Handler.dispatchMessage(Handler.java:105)
        at android.os.Looper.loop(Looper.java:164)
        at android.app.ActivityThread.main(ActivityThread.java:6541) <1 internal call>
        at com.android.internal.os.Zygote$MethodAndArgsCaller.run(Zygote.java:240)
        at com.android.internal.os.ZygoteInit.main(ZygoteInit.java:767)
```

图 26-1　加载 SO 文件报错

同时，我们也知道了这行代码的作用是把 SO 文件加载到内存中，当 Java 代码中有 Native 函数的时候，就会到 SO 文件中寻找对应的符号。寻找的方式有以下两种：

- Java_com_roysue_demoso_MainActivity_stringFromJNI
- Java_com_roysue_demoso_MainActivity_stringFromJNI__

那么它是从哪里寻找的呢？在代码清单 26-1 中，加载的目录是一个相对路径，它其实是从 ClassLoader 中加载的，我们输出 ClassLoader 看一下它的结构，默认的 ClassLoader 是一个 PathClassLoader，输出结果如代码清单 26-2 所示。

代码清单 26-2　输出 ClassLoader

````
```Bash
dalvik.system.PathClassLoader[
  DexPathList[
    [zip file "/data/app/com.roysue.demoso-UFKjPdYe3BSYcIwgNKGalg==/base.apk"],
    nativeLibraryDirectories=[
      /data/app/com.roysue.demoso-UFKjPdYe3BSYcIwgNKGalg==/lib/arm64,
      /data/app/com.roysue.demoso-UFKjPdYe3BSYcIwgNKGalg==/base.apk!/lib/arm64-v8a,
      /system/lib64,
      /vendor/lib64
    ]
  ]
]
```
````

路径 /data/app/com.roysue.demoso-UFKjPdYe3BSYcIwgNKGalg==/lib/arm64 就是存放 SO 文件的位置，我们使用 adb 进入 Shell 查看一下，结果如图 26-2 所示。

图 26-2　SO 文件在 Android 系统中的位置

　　加载 SO 文件的方式有两种：一种是如代码清单 26-1 所示的相对路径加载，另一种是绝对路径加载，使用 System.load 函数加载，将绝对路径传入进去即可。但是加载的时候会被 SELinux 拦截，如果没有 root 权限，那么就不能关闭 SELinux，不能正常加载我们指定的 SO 文件。

26.2　SO 文件加载流程

　　26.1 节讲述了 SO 文件是从哪里开始加载的，但是加载的具体流程我们却不得而知。对此，我们需要深入 Android 源码详细探究。

　　先从 System.load 函数开始查找，如图 26-3 所示。

图 26-3　SO 文件加载入口

　　因为它加载的是绝对路径，所有比较好找，双击此函数，进入系统库中，如图 26-4 所示。

图 26-4　系统库定义处

　　继续往下追踪就追踪不到了，这时需要进入 Android 源码网站中寻找。

　　在图 26-4 中，可以看到 load 函数位于 System.java 文件中，我们就在此目录下查询，如图 26-5 所示。

```
AndroidXRef Oreo 8.1.0_r33

Home    Sort by: last modified time | relevance | path
Full Search [                    ]              In Project(s)  [select all] [invert selection]
Definition  [                    ]              art
Symbol     [                    ]               bionic
File Path  [System.java         ]               bootable
History    [                    ]               build
                                                compatibility
         [Search] [Clear] [Help]                cts

Searched path:"system . java" (Results 1 - 1 of 1) sorted by relevance
/libcore/ojluni/src/main/java/java/lang/
H A D    System.java
```

图 26-5 System.java

进入源码文件后，直接搜索 load，定位我们的目标函数，如图 26-6 所示。

```
1619      @CallerSensitive
1620      public static void load(String filename) {
1621          Runtime.getRuntime().load0(VMStack.getStackClass1(), filename);
1622      }
```

图 26-6 load 函数

传入的是文件的绝对路径，VMStack.getStackClass1()则是 PathClassloader。
继续追踪，会进入 doLoad 函数，如图 26-7 所示。

```
1070    private String doLoad(String name, ClassLoader loader) {
1071        // Android apps are forked from the zygote, so they can't have a custom LD_LIBRARY_PATH,
1072        // which means that by default an app's shared library directory isn't on LD_LIBRARY_PATH.
1073
1074        // The PathClassLoader set up by frameworks/base knows the appropriate path, so we can load
1075        // libraries with no dependencies just fine, but an app that has multiple libraries that
1076        // depend on each other needed to load them in most-dependent-first order.
1077
1078        // We added API to Android's dynamic linker so we can update the library path used for
1079        // the currently-running process. We pull the desired path out of the ClassLoader here
1080        // and pass it to nativeLoad so that it can call the private dynamic linker API.
1081
1082        // We didn't just change frameworks/base to update the LD_LIBRARY_PATH once at the
1083        // beginning because multiple apks can run in the same process and third party code can
1084        // use its own BaseDexClassLoader.
1085
1086        // We didn't just add a dlopen_with_custom_LD_LIBRARY_PATH call because we wanted any
1087        // dlopen(3) calls made from a .so's JNI_OnLoad to work too.
1088
1089        // So, find out what the native library search path is for the ClassLoader in question...
1090        String librarySearchPath = null;
1091        if (loader != null && loader instanceof BaseDexClassLoader) {
1092            BaseDexClassLoader dexClassLoader = (BaseDexClassLoader) loader;
1093            librarySearchPath = dexClassLoader.getLdLibraryPath();
1094        }
1095        // nativeLoad should be synchronized so there's only one LD_LIBRARY_PATH in use regardless
1096        // of how many ClassLoaders are in the system, but dalvik doesn't support synchronized
1097        // internal natives.
1098        synchronized (this) {
1099            return nativeLoad(name, loader, librarySearchPath);
1100        }
1101    }
1102
```

图 26-7 doLoad 函数

第 26 章　SO 文件加载流程分析与注入实战

在 doLoad 函数中，会再次对 loader 进行处理，然后进入 nativeLoad 函数。在 Java 函数中，找到 nativeLoad 函数，发现并没有对其进行处理，这时我们就需要进入 C 层去找它的实现。

C 层的实现如图 26-8 所示，它其实是一个 JNI 的动态注册，我们继续追踪，看它真实的实现位置。实现位置就是同文件中的 Runtime_nativeLoad，这是一种特殊的定义方式，再进入 JVM_NativeLoad 中继续追踪。

```
xref: /libcore/ojluni/src/main/native/Runtime.c
Home | History | Annotate | Line# | Navigate | Download            Search  □ only in Runtime.c
69
70  JNIEXPORT void JNICALL
71  Runtime_nativeExit(JNIEnv *env, jclass this, jint status)
72  {
73      JVM_Exit(status);
74  }
75
76  JNIEXPORT jstring JNICALL
77  Runtime_nativeLoad(JNIEnv* env, jclass ignored, jstring javaFilename,
78                     jobject javaLoader, jstring javaLibrarySearchPath)
79  {
80      return JVM_NativeLoad(env, javaFilename, javaLoader, javaLibrarySearchPath);
81  }
82
83  static JNINativeMethod gMethods[] = {
84    FAST_NATIVE_METHOD(Runtime, freeMemory, "()J"),
85    FAST_NATIVE_METHOD(Runtime, totalMemory, "()J"),
86    FAST_NATIVE_METHOD(Runtime, maxMemory, "()J"),
87    NATIVE_METHOD(Runtime, gc, "()V"),
88    NATIVE_METHOD(Runtime, nativeExit, "(I)V"),
89    NATIVE_METHOD(Runtime, nativeLoad,
90                  "(Ljava/lang/String;Ljava/lang/ClassLoader;Ljava/lang/String;)"
91                  "Ljava/lang/String;"),
92  };
93
94  void register_java_lang_Runtime(JNIEnv* env) {
95      jniRegisterNativeMethods(env, "java/lang/Runtime", gMethods, NELEM(gMethods));
96  }
97
```

图 26-8　nativeLoad 函数

继续追踪就会进入 LoadNativeLibrary 函数，它是一个 C 函数，如图 26-9 所示，但是它的函数实现比较长，追踪起来很麻烦，怎么办呢？其实我们只要跟着它的 library_path 参数追踪即可。

```
796  bool JavaVMExt::LoadNativeLibrary(JNIEnv* env,
797                                     const std::string& path,
798                                     jobject class_loader,
799                                     jstring library_path,
800                                     std::string* error_msg) {
801      error_msg->clear();
802
803      // See if we've already loaded this library.  If we have, and the class loader
804      // matches, return successfully without doing anything.
805      // TODO: for better results we should canonicalize the pathname (or even compare
806      // inodes). This implementation is fine if everybody is using System.loadLibrary.
807      SharedLibrary* library;
808      Thread* self = Thread::Current();
809      {
810        // TODO: move the locking (and more of this logic) into Libraries.
811        MutexLock mu(self, *Locks::jni_libraries_lock_);
812        library = libraries_->Get(path);
813      }
```

图 26-9　LoadNativeLibrary 函数

追踪 library_path 后，看哪些函数调用了这个参数，往下追踪，看到了核心的 OpenNativeLibrary 函数，它对 library_path 进行了处理，如图 26-10 所示。

```
void* handle = android::OpenNativeLibrary(env,
                                          runtime_->GetTargetSdkVersion(),
                                          path_str,
                                          class_loader,
                                          library_path,
                                          &needs_native_bridge,
                                          error_msg);
```

图 26-10　OpenNativeLibrary 函数

跟进去就到了 SO 文件的核心实现代码。这里 path 有两个处理分支，一个是 dlopen，另一个是 android_dlopen_ext。如果 class_loader 为空，就使用 dlopen（见图 26-11），如果 class_loader 不为空，就使用 android_dlopen_ext（见图 26-12）。这就是我们要 Hook 的一个点，Hook 这个点才能看到它的一个加载时机。

```
void* OpenNativeLibrary(JNIEnv* env,
                        int32_t target_sdk_version,
                        const char* path,
                        jobject class_loader,
                        jstring library_path,
                        bool* needs_native_bridge,
                        std::string* error_msg) {
#if defined(__ANDROID__)
  UNUSED(target_sdk_version);
  if (class_loader == nullptr) {
    *needs_native_bridge = false;
    return dlopen(path, RTLD_NOW);
  }

  std::lock_guard<std::mutex> guard(g_namespaces_mutex);
  NativeLoaderNamespace ns;

  if (!g_namespaces->FindNamespaceByClassLoader(env, class_loader, &ns)) {
    // This is the case where the classloader was not created by ApplicationLoaders
    // In this case we create an isolated not-shared namespace for it.
    if (!g_namespaces->Create(env,
                              target_sdk_version,
                              class_loader,
                              false /* is_shared */,
                              false /* is_for_vendor */,
                              library_path,
                              nullptr,
                              &ns,
                              error_msg)) {
      return nullptr;
    }
  }
```

图 26-11　SO 文件处理方式 dlopen

```
556    if (ns.is_android_namespace()) {
557      android_dlextinfo extinfo;
558      extinfo.flags = ANDROID_DLEXT_USE_NAMESPACE;
559      extinfo.library_namespace = ns.get_android_ns();
560
561      void* handle = android_dlopen_ext(path, RTLD_NOW, &extinfo);
562      if (handle == nullptr) {
563        *error_msg = dlerror();
564      }
565      *needs_native_bridge = false;
566      return handle;
567    } else {
568      void* handle = NativeBridgeLoadLibraryExt(path, RTLD_NOW, ns.get_native_bridge_ns());
569      if (handle == nullptr) {
570        *error_msg = NativeBridgeGetError();
571      }
572      *needs_native_bridge = true;
573      return handle;
574    }
```

图 26-12　SO 文件处理方式 android_dlopen_ext

每次加载 SO 文件，肯定会用到 dlopen 和 android_dlopen_ext，它们位于 libdl.so 中，使用 IDA 打开此文件，可以看到如图 26-13 所示的导出符号。我们使用 Frida Hook 这两个函数就能监控 System.load 和 System.loadlibrary 函数对 SO 文件的加载情况。这两个函数的加载流程其实是一样的。

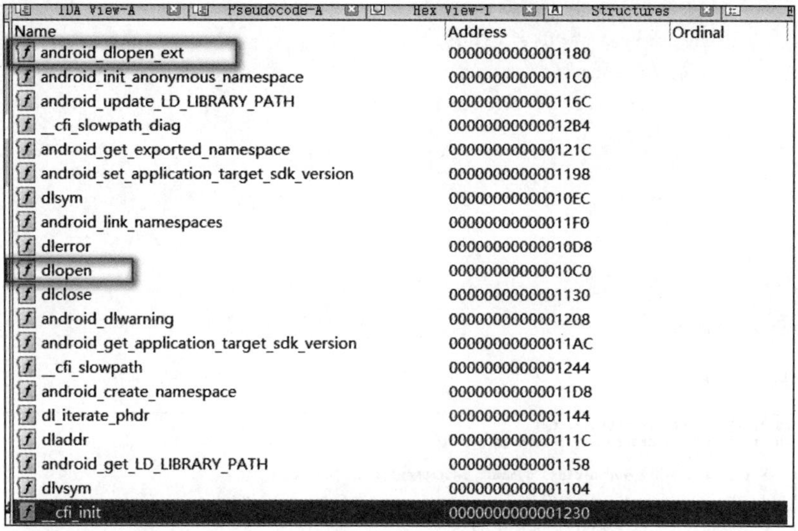

图 26-13　libdl.so

26.3　Frida Hook dlopen 和 android_dlopen_ext

使用 Frida 的 findExportByName 函数对 dlopen 和 android_dlopen_ext 进行 Hook 操作，并对 JNI 函数 stringFromJNI 进行 Hook 操作，如代码清单 26-3 所示。

代码清单 26-3　Hook 操作

```
function main(){
    var dlopen = Module.findExportByName(null,"dlopen")
    var android_dlopen_ext = Module.findExportByName(null,"android_dlopen_ext")

    Interceptor.attach(dlopen,{
        onEnter:function(args){
            console.log("dlopen => ",args[0].readCString())
        },onLeave:function(){

        }
    })

    Interceptor.attach(android_dlopen_ext,{
        onEnter:function(args){
```

```
                console.log("android_dlopen_ext => ",args[0].readCString())
            },onLeave:function(){
                var stringFromJNI =
Module.findExportByName("libnative-lib.so","Java_com_roysue_demoso_MainActivity_stringF
romJNI")
                Interceptor.attach(stringFromJNI,{
                    onLeave:function(args){
                        console.log("Enter stringFromJNI")
                    }
                })
            }
        })
    }

    setImmediate(main)
```

Hook 结果如图 26-14 所示。

```
┌──(root💀r0env)-[~/Desktop/workBench/tmp]
└─# frida -U -f com.roysue.demoso  -l invoke.js
     ____
    / _  |   Frida 12.8.0 - A world-class dynamic instrumentation toolkit
   | (_| |
    > _  |   Commands:
   /_/ |_|       help      -> Displays the help system
   . . . .       object?   -> Display information about 'object'
   . . . .       exit/quit -> Exit
   . . . .
   . . . .   More info at https://www.frida.re/docs/home/
Spawned `com.roysue.demoso`. Use %resume to let the main thread start executing!
[LGE Nexus 5X::com.roysue.demoso]-> %resume
[LGE Nexus 5X::com.roysue.demoso]-> android_dlopen_ext =>  /data/app/com.roysue.demoso-xOaCg-U2jbXJ7bR0Eii8bA==/lib/a
rm64/libnative-lib.so
Enter stringFromJNI
dlopen => /vendor/lib64/hw/gralloc.msm8992.so
dlopen => /vendor/lib64/egl/libEGL_adreno.so
dlopen => /vendor/lib64/hw/android.hardware.graphics.mapper@2.0-impl.so
dlopen => /vendor/lib64/hw/gralloc.msm8992.so
[LGE Nexus 5X::com.roysue.demoso]->
```

图 26-14　Hook 结果

以 spawn 模式 Hook 会导致线程中断，使用%resume 恢复线程即可。可以看到 SO 文件的加载路径，这也是我们日常工作中追踪 SO 文件的方式之一。

当然，也可以使用更复杂的 App 来查看脚本的效果，这里提供了一个加壳的 App，我们尝试用这个脚本进行 Hook，Hook 结果如图 26-15 所示。从 Hook 的结果中可以看到各种 SO 文件的加载情况，并看到 libjiagu_64.so 文件。同时，也可以发现 App 本身的 SO 文件都是通过 android_dlopen_ext 加载的，而系统中的 SO 文件是通过 dlopen 加载的。

第 26 章　SO 文件加载流程分析与注入实战

```
┌──(root💀r0env)-[~/Desktop/workBench/tmpDir/lesson21]
└─# frida -U -f com.meiliao.mlfsb -l invoke.js --no-pause
     ____
    / _  |   Frida 12.8.0 - A world-class dynamic instrumentation toolkit
   | (_| |
    > _  |   Commands:
   /_/ |_|       help      -> Displays the help system
   . . . .       object?   -> Display information about 'object'
   . . . .       exit/quit -> Exit
   . . . .
   . . . .   More info at https://www.frida.re/docs/home/
Spawned `com.meiliao.mlfsb`. Resuming main thread!
[LGE Nexus 5X::com.meiliao.mlfsb]-> android_dlopen_ext => /system/framework/oat/arm64/android.test.runner.odex
android_dlopen_ext => /data/app/com.meiliao.mlfsb-Cdn2cB-9vrpslzh1qRjF8A==/oat/arm64/base.odex
android_dlopen_ext => /data/data/com.meiliao.mlfsb/.jiagu/libjiagu_64.so
dlopen => liblog.so
dlopen => libz.so
dlopen => libc.so
dlopen => libm.so
dlopen => libstdc++.so
dlopen => libdl.so
dlopen => libjiagu_64.so
dlopen => libart.so
dlopen => libjiagu_64.so
dlopen => libjiagu_64.so
android_dlopen_ext => /system/framework/oat/arm64/android.test.runner.odex
dlopen => libjiagu_64.so
dlopen => libjiagu_64.so
android_dlopen_ext => /data/app/com.meiliao.mlfsb-Cdn2cB-9vrpslzh1qRjF8A==/lib/arm64/lib2D72A071.so
android_dlopen_ext => /data/app/com.meiliao.mlfsb-Cdn2cB-9vrpslzh1qRjF8A==/lib/arm64/libwind.so
android_dlopen_ext => /data/app/com.meiliao.mlfsb-Cdn2cB-9vrpslzh1qRjF8A==/lib/arm64/libBugly-ext.so
android_dlopen_ext => /data/app/com.meiliao.mlfsb-Cdn2cB-9vrpslzh1qRjF8A==/lib/arm64/libalicomphonenumberauthsdk-log-online-release_alijtca_plus.so
android_dlopen_ext => /data/app/com.meiliao.mlfsb-Cdn2cB-9vrpslzh1qRjF8A==/lib/arm64/libalicomphonenumberauthsdk_core.so
android_dlopen_ext => /data/app/com.meiliao.mlfsb-Cdn2cB-9vrpslzh1qRjF8A==/lib/arm64/libagora-rtc-sdk.so
dlopen => /data/app/com.meiliao.mlfsb-Cdn2cB-9vrpslzh1qRjF8A==/lib/arm64/libagora-crypto.so
android_dlopen_ext => /data/app/com.meiliao.mlfsb-Cdn2cB-9vrpslzh1qRjF8A==/lib/arm64/libfuai.so
android_dlopen_ext => /data/app/com.meiliao.mlfsb-Cdn2cB-9vrpslzh1qRjF8A==/lib/arm64/libCNamaSDK.so
dlopen => libEGL.so
```

图 26-15　复杂 App Hook

在 26.2 节，我们追踪到了 android_dlopen_ext，继续往下追踪看看还会进行哪些操作。继续追踪，到了如图 26-16 所示的位置。这时会发现再往下追踪线索就断了，这时需要进入 IDA 中查看它的实现。

```
151   void* android_dlopen_ext(const char* filename, int flag, const android_dlextinfo* extinfo) {
152     const void* caller_addr = __builtin_return_address(0);
153     return __loader_android_dlopen_ext(filename, flag, extinfo, caller_addr);
154   }
```

图 26-16　android_dlopen_ext

在 IDA 中，追踪到 __loader_android_dlopen_ext 就会停止，我们看一下它的汇编代码，如图 26-17 所示。它在 PLT 中，说明它的实现不在此 SO 文件中。

```
.plt:00000EB0  __loader_android_dlopen_ext                      ; CODE XREF: android_dlopen_ext+4↓p
.plt:00000EB0                  ADRL            R12, 0x3EB8
.plt:00000EB8                  LDR             PC, [R12,#(__loader_android_dlopen_ext_ptr - 0x3EB8)]!
```

图 26-17　__loader_android_dlopen_ext

然后在源码网站全局搜索 dlopen_ext，如图 26-18 所示。看到它在 dlfcn 中有声明，但是进入后并没有搜索到 __loader_android_dlopen_ext 的实现，因此我们转移战场，动态地链接到 IDA 上查看 SO 文件的加载。

使用 IDA 动态调试应用程序后，就能得到 android_dlopen_ext 的真实函数名称，调试结果如图 26-19 所示。定位到 __android_dlopen_ext 后，在 dlfcn 中继续追踪。

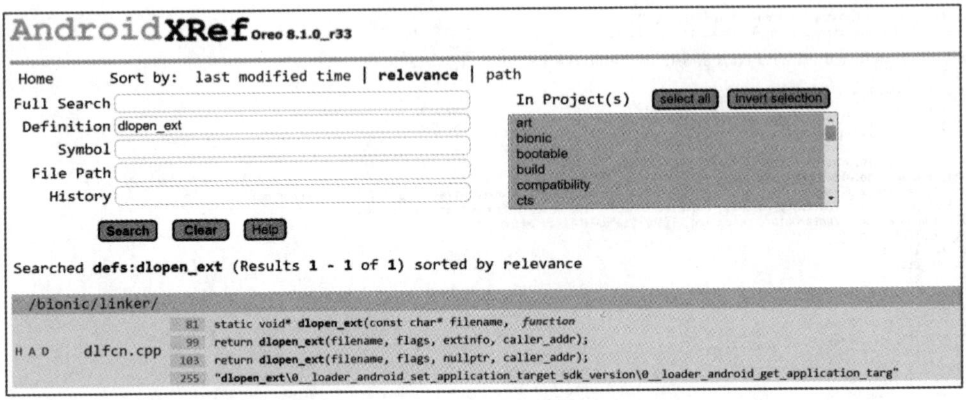

图 26-18 全局搜索

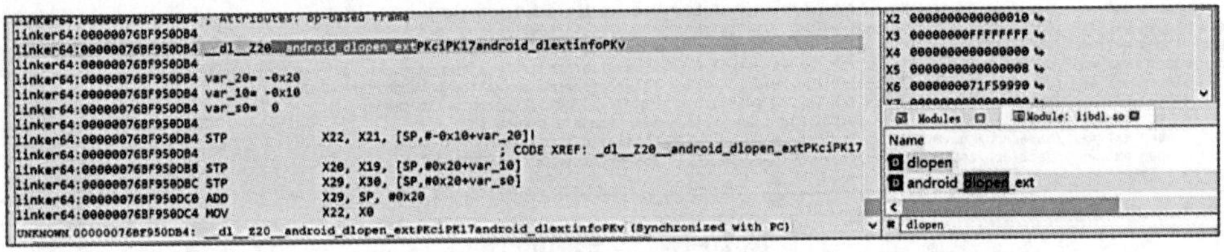

图 26-19 IDA 动态调试

定位到 __android_dlopen_ext 函数，如图 26-20 所示。在图中可以发现 dlopen 和 android_dlopen_ext 最终都使用了 dlopen_ext 函数。dlopen_ext 又调用了 do_dlopen，这时进入了 linker 的范畴，这就是 SO 文件装载链接的开始。

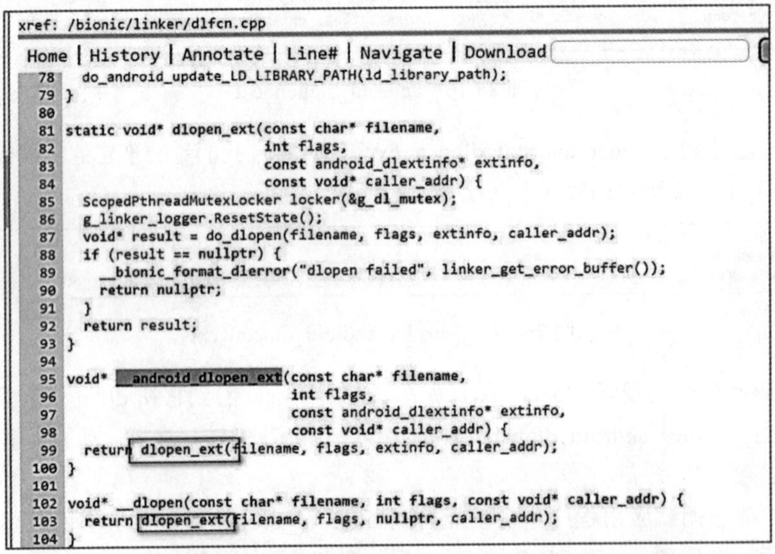

图 26-20 __android_dlopen_ext 函数

在 do_dlopen 中，我们跟着参数 name 继续追踪，name 被赋值给 translated_name，然后传入 find_library，如图 26-21 所示。

```
2012    ProtectedDataGuard guard;
2013    soinfo* si = find_library(ns, translated_name, flags, extinfo, caller);
2014    loading_trace.End();
```

图 26-21 find_library

获取 soinfo 指针后，在本函数中，还会通过 soinfo 指针调用 call_constructors 方法，如图 26-22 所示。

```
if (si != nullptr) {
  void* handle = si->to_handle();
  LD_LOG(kLogDlopen,
         "... dlopen calling constructors: realpath=\"%s\", soname=\"%s\", handle=%p",
         si->get_realpath(), si->get_soname(), handle);
  si->call_constructors();
  failure_guard.Disable();
  LD_LOG(kLogDlopen,
         "... dlopen successful: realpath=\"%s\", soname=\"%s\", handle=%p",
         si->get_realpath(), si->get_soname(), handle);
  return handle;
}
```

图 26-22 call_constructors 方法

在 call_constructors 方法中，核心处理位于 call_function 方法，如图 26-23 所示。传入了 3 个参数，分别是 DT_INIT、init_func_ 初始化方法以及相对路径。

```
418    // DT_INIT should be called before DT_INIT_ARRAY if both are present.
419    call_function("DT_INIT", init_func_, get_realpath());
420    call_array("DT_INIT_ARRAY", init_array_, init_array_count_, false, get_realpath());
```

图 26-23 call_function 方法

init_func_ 的定义位于 linker.cpp 文件中，如图 26-24 所示。它位于 DT_INIT 分支下，它把 load_bias 加一个值赋给了 init_func_。

```
3098   #endif
3099         case DT_INIT:
3100           init_func_ = reinterpret_cast<linker_ctor_function_t>(load_bias + d->d_un.d_ptr);
3101           DEBUG("%s constructors (DT_INIT) found at %p", get_realpath(), init_func_);
3102           break;
```

图 26-24 init_func_ 定义位置

接下来就会调用 call_array，把数组中的所有函数都调用一遍。追踪到这里，我们就能总结出在加载 SO 文件之前，会先调用 init 和 init_array 两个函数。

在执行完如图 26-10 所示的 OpenNativeLibrary 函数后，我们继续看 LoadNativeLibrary 中的实现。在继续追踪的过程中，会发现 library 调用了 FindSymbol 函数，获取 JNI_OnLoad 函数，如图 26-25 所示，并执行 JNI_OnLoad 函数。

```
915  bool was_successful = false;
916  void* sym = library->FindSymbol("JNI_OnLoad", nullptr);
917  if (sym == nullptr) {
918    VLOG(jni) << "[No JNI_OnLoad found in \"" << path << "\"]";
919    was_successful = true;
```

图 26-25 FindSymbol 函数

至此，整个执行流程基本就追踪完成了，执行顺序从上到下分别是：

（1）init。

（2）init_array。

（3）JNI_Onload。

这 3 个段我们可以自己编写代码来查看它们的执行顺序，当然它们的定义是固定的。这里编写代码来演示一下，如代码清单 26-4 所示，代码编写于 native-lib.cpp 文件中。

代码清单 26-4 查看执行顺序

```cpp
extern "C" void _init(void){
    __android_log_print(6,"r0ysue","I am from init");
}

__attribute__((constructor)) void a(void){
    __android_log_print(6,"r0ysue","I am from initArray1");
}

__attribute__((constructor)) void b(void){
    __android_log_print(6,"r0ysue","I am from initArray2");
}

JNIEXPORT jint JNICALL JNI_OnLoad(JavaVM *vm, void *unused) {
    __android_log_print(6,"r0ysue","i am from JNI_OnLoad");
    return JNI_VERSION_1_6;
}
```

输出结果如图 26-26 所示，可以看到 SO 文件加载的执行顺序。

```
2022-05-21 13:10:50.394 6269-6269/com.roysue.demoso E/r0ysue: I am from init
2022-05-21 13:10:50.394 6269-6269/com.roysue.demoso E/r0ysue: I am from initArray1
2022-05-21 13:10:50.394 6269-6269/com.roysue.demoso E/r0ysue: I am from initArray2
2022-05-21 13:10:50.394 6269-6269/com.roysue.demoso E/r0ysue: i am from JNI_OnLoad
```

图 26-26 输出结果

26.4 编译 AOSP 注入 SO

在 26.3 节中，我们讲述了 System.loadlibrary 把 SO 文件注入 App 中的过程，我们可以借鉴它的方式加载自己的 SO 文件。在编写代码之前，首先需要思考一个问题，在 init、init_array 中还是 JNI_OnLoad 中写注入代码呢？这里使用 JNI_OnLoad 来做注入的入口，它可以方便我们进行操作，因为它的第一个参数是 JavaVM，有了这个指针，我们就能操作一些基础的 API。

在 26.3 节中，我们也说明了 System.load 和 System.LoadLibrary 两个函数的区别，它们的功能都是加载 SO 文件，但是 System.load 函数可以加载绝对路径，我们就以它为例来进行测试。注入的方式有很多，以下是常见的 3 种方式：

（1）直接加载 sdcard 中的 SO 文件。
（2）加载私有目录的 SO 文件。
（3）编译 AOSP 来注入 SO 文件。

接下来主要介绍第 3 种方式。

26.4.1　直接加载 sdcard 中的 SO 文件

本小节先介绍第一种加载 SO 文件的方式，即直接加载 sdcard 中的 SO 文件。

要加载 sdcard 中的 SO 文件，需要在 AndroidManifest.xml 中赋予 sdcard 权限，具体如下：

```
<uses-permission android:name="android.permission.WRITE_EXTERNAL_STORAGE" />
<uses-permission android:name="android.permission.READ_EXTERNAL_STORAGE" />
```

下面介绍加载 SO 文件的具体操作。

新建一个空白模板的项目，我们通过空白的模板来实现 SO 文件的加载，如图 26-27 所示。

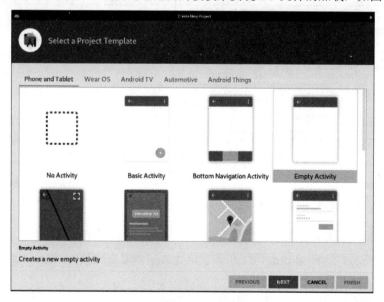

图 26-27　新建项目

把本章第一个项目中的 SO 文件推送（push）到 sdcard 目录下，如图 26-28 所示。

```
─(root㉿r0env)-[~/Desktop/workBench/tmpDir/lesson21/demoso/app/build/outputs/apk/debug/tm/lib/arm64-v8a]
└─# adb push libnative-lib.so /sdcard/r0.so
libnative-lib.so: 1 file pushed, 0 skipped. 231.8 MB/s (219264 bytes in 0.001s)
```

图 26-28 push SO 文件

编写加载代码，如图 26-29 所示。

```java
package com.roysue.dmeo;

import ...

public class MainActivity extends AppCompatActivity {

    @Override
    protected void onCreate(Bundle savedInstanceState) {
        super.onCreate(savedInstanceState);
        setContentView(R.layout.activity_main);

        System.load( filename: "/sdcard/r0.so");
    }
}
```

图 26-29 加载 SO 文件

注意，安装 App 后需要手动在设置中给 sdcard 赋予权限。

编译运行 App，发现会直接报错，如图 26-30 所示，报错的内容其实是我们在第 17 章中讲过的，dlopen 高版本无法加载白名单外 SO 文件的问题。但是第一种方案不可行，不采用这种方案。

```
2022-05-21 22:31:25.429 19853-19853/? E/AndroidRuntime: FATAL EXCEPTION: main
    Process: com.roysue.dmeo, PID: 19853
    java.lang.UnsatisfiedLinkError: dlopen failed: library "/sdcard/r0.so" needed or dlopened by "/system/lib64/libnativeloader.so" is not accessible for the namespace "classloader-namespace"
        at java.lang.Runtime.load0(Runtime.java:928)
        at java.lang.System.load(System.java:1621)
        at com.roysue.dmeo.MainActivity.onCreate(MainActivity.java:23)
        at android.app.Activity.performCreate(Activity.java:6975)
        at android.app.Instrumentation.callActivityOnCreate(Instrumentation.java:1213)
        at android.app.ActivityThread.performLaunchActivity(ActivityThread.java:2770)
        at android.app.ActivityThread.handleLaunchActivity(ActivityThread.java:2892)
        at android.app.ActivityThread.-wrap11(Unknown Source:0)
        at android.app.ActivityThread$H.handleMessage(ActivityThread.java:1593)
        at android.os.Handler.dispatchMessage(Handler.java:105)
        at android.os.Looper.loop(Looper.java:164)
        at android.app.ActivityThread.main(ActivityThread.java:6541) <1 internal call>
        at com.android.internal.os.Zygote$MethodAndArgsCaller.run(Zygote.java:240)
        at com.android.internal.os.ZygoteInit.main(ZygoteInit.java:767)
```

图 26-30 运行报错

26.4.2 加载私有目录的 SO 文件

本小节介绍第二种方式。

第二种方式是动态地将 sdcard 中的 r0.so 复制到 App 的私有目录，Android 系统会给它默认的权限，我们测试这种方案是否可行。

首先编写一个 Copy 文件的方法，如代码清单 26-5 所示。

代码清单 26-5 Copy 文件的方法

```java
// copy
public static void mycopy(String srcFileName,String destFileName){
    InputStream in = null;
    OutputStream out = null;
    try {
        in = new FileInputStream(srcFileName);
        out = new FileOutputStream(destFileName);
        byte[] bytes = new byte[1024];

        int i = 0;
        while((i=in.read(bytes)) != -1){
            out.write(bytes,0,i);
        }
    }catch (Exception e){
        e.printStackTrace();
    }finally {
        try{
            if(in != null){
                in.close();
            }
            if(out != null){
                out.flush();
                out.close();
            }
        }catch (Exception e) {
            e.printStackTrace();
        }
    }
}
```

主函数中的代码如代码清单 26-6 所示，将 sdcard 中的 r0.so 复制到应用程序的私有目录。

代码清单 26-6 主函数中的代码

```java
@Override
protected void onCreate(Bundle savedInstanceState) {
    super.onCreate(savedInstanceState);
    setContentView(R.layout.activity_main);

//      System.load("/sdcard/r0.so");

    Context context = this.getApplicationContext();
    String path = this.getPackageName();
    path = "/data/data/" + path + "/r0.so";
    File file1 = new File(path);
```

```
    mycopy("/sdcard/r0.so",path);


    Log.e("r0ysue",path);
}
```

编译运行 App 后，结果如图 26-31 所示，可以正常地打印目录。

```
2022-05-21 22:54:11.756 20473-20473/? E/r0ysue: /data/data/com.roysue.dmeo/r0.so
```

图 26-31　打印运行目录

我们再去 Shell 下查看它，如图 26-32 所示。在 Linux 系统中，直接赋予了该文件读写权限，因此我们可以正常打开它了。

```
bullhead:/data/data/com.roysue.dmeo # ls -l
total 224
drwxrws--x 2 u0_a105 u0_a105_cache   4096 2022-05-21 22:01 cache
drwxrws--x 2 u0_a105 u0_a105_cache   4096 2022-05-21 22:01 code_cache
-rw------- 1 u0_a105 u0_a105       219264 2022-05-21 22:54 r0.so
```

图 26-32　r0.so

在代码清单 26-6 中增加如下代码：

```
System.load(path);
```

运行结果如图 26-33 所示，这时就能正常打开 SO 文件了，SO 文件中的执行流程也被正常打印出来，因此这种方案是可行的。

```
2022-05-21 22:56:47.953 20601-20601/? E/r0ysue: I am from init
2022-05-21 22:56:47.953 20601-20601/? E/r0ysue: I am from initArray1
2022-05-21 22:56:47.953 20601-20601/? E/r0ysue: I am from initArray2
2022-05-21 22:56:47.955 20601-20601/? E/r0ysue: i am from JNI_OnLoad
```

图 26-33　运行结果

26.4.3　编译 AOSP 注入 SO 文件

使用编译 AOSP 的方式注入 SO 文件需要注意时机的问题，如果时机不对，注入就会失败。

这里选择 handlebindapplication 函数来注入 SO 文件，对应的实现如图 26-34 所示。在 makeApplication 处注入有点晚，因为已经构造完 Application，程序就要开始 attach 的操作，因此要选择一个更早的时机，我们选择在创建 ContextImpl 对象的时候进行 SO 文件的注入。

第 26 章　SO 文件加载流程分析与注入实战　501

```
private void handleBindApplication(AppBindData data) {
  //step 1: 创建LoadedApk对象
  data.info = getPackageInfoNoCheck(data.appInfo, data.compatInfo);
  ...
  //step 2: 创建ContextImpl对象;
  final ContextImpl appContext = ContextImpl.createAppContext(this, data.info);

  //step 3: 创建Instrumentation
  mInstrumentation = new Instrumentation();

  //step 4: 创建Application对象;在makeApplication函数中调用了newApplication，在该函数中又调用了app.attach(context),
  //在attach函数中调用了Application.attachBaseContext函数
  Application app = data.info.makeApplication(data.restrictedBackupMode, null);
  mInitialApplication = app;

  //step 5: 安装providers
  List<ProviderInfo> providers = data.providers;
  installContentProviders(app, providers);

  //step 6: 执行Application.Create回调
  mInstrumentation.callApplicationOnCreate(app);
}
```

图 26-34　handleBindApplication

源码位于 frameworks/base/core/java/android/app/ActivityThread.java 中，直接搜索 final ContextImpl appContext = ContextImpl.createAppContext(this, data.info); 就能找到目标代码，如图 26-35 所示。

```
                ii = new ApplicationPackageManager(null, getPackageManager());
                        .getInstrumentationInfo(data.instrumentationName, 0);
            } catch (PackageManager.NameNotFoundException e) {
                throw new RuntimeException(
                    "Unable to find instrumentation info for: " + data.instrumentationName);
            }

            mInstrumentationPackageName = ii.packageName;
            mInstrumentationAppDir = ii.sourceDir;
            mInstrumentationSplitAppDirs = ii.splitSourceDirs;
            mInstrumentationLibDir = getInstrumentationLibrary(data.appInfo, ii);
            mInstrumentedAppDir = data.info.getAppDir();
            mInstrumentedSplitAppDirs = data.info.getSplitAppDirs();
            mInstrumentedLibDir = data.info.getLibDir();
        } else {
            ii = null;
        }

        final ContextImpl appContext = ContextImpl.createAppContext(this, data.info);
        updateLocaleListFromAppContext(appContext,
                mResourcesManager.getConfiguration().getLocales());

        if (!Process.isIsolated()) {
            setupGraphicsSupport(appContext);
        }
```

图 26-35　更改源码位置

更改后的代码如图 26-36 所示。

```
        final ContextImpl appContext = ContextImpl.createAppContext(this, data.info);
// add

        ContextImpl  context = appContext;
        ActivityManager mAm = (ActivityManager) context.getSystemService("activity");
        String activity_packageName = mAm.getRunningTasks(1).get(0).topActivity.getPackageName();
        String tarPath = "/data/data/" + activity_packageName + "/r0.so";
        File file1 = new File(tarPath);
        mycopy("/sdcard/r0.so",tarPath);
        if(file1.exists()){
            System.load(tarPath);
        }
// add
```

图 26-36　更改后的代码

更改源码后，直接对源码进行编译，因为之前对系统源码编译过一次，再次编译只编译更改处的代码，这种编译方式称为增量编译，编译后的产物是 system.img。我们直接使用 fastboot flash system system.img 命令将 system.img 刷入手机，刷好后重启手机。

再次运行 App，运行结果与图 26-33 所示的结果一致。注意：运行前需要将第二种方式中注入 SO 文件的代码注释掉，做一个纯净的 App。

26.5　注入优化

在前面的讲解中，我们注入的 SO 文件使用的是 64 位的，并没有使用 32 位的 SO 文件来测试，接下来在 demoso 项目中，为 SO 文件中增加一点小功能，然后使用 32 位的 SO 文件进行注入测试。

这个小功能就是获取当前进程名称，通过前期调研，这里为读者演示 ActivityThread 中的 currentProcessName 函数来获取进程名称，因为它是一个静态方法，所以方便调用。除此之外，还需要获取一个 JNIEnv 指针方便我们调用。代码如代码清单 26-7 所示。

代码清单 26-7　获取进程名称

```
const char* getProcessName(JNIEnv* env){
    jclass ActivityThread = env->FindClass("android/app/ActivityThread");
    jmethodID currentProcess =
env->GetStaticMethodID(ActivityThread,"currentProcessName","()Ljava/lang/String;");
    jstring myProcessName =
static_cast<jstring>(env->CallStaticObjectMethod(ActivityThread,currentProcess));

    return env->GetStringUTFChars(myProcessName,0);
}

JNIEXPORT jint JNICALL JNI_OnLoad(JavaVM *vm, void *unused) {
```

第 26 章　SO 文件加载流程分析与注入实战　503

```
    JNIEnv* env;
    vm->GetEnv(reinterpret_cast<void **>(&env), JNI_VERSION_1_4);
    const char* myProcess = getProcessName(env);

    __android_log_print(6,"r0ysue","i am from JNI_OnLoad %s",myProcess);
    return JNI_VERSION_1_6;
}
```

编写完代码后，直接编译为 APK，然后将 APK 解压并把 SO 文件推送到 sdcard 中，我们先将 64 位的 SO 文件推送到 sdcard 中，看一下效果，如图 26-37 所示，可以正常获取当前进程的进程名。

```
1970-02-18 12:04:17.910 5928-5928/? I/zygote64: Late-enabling -Xcheck:jni
1970-02-18 12:04:18.127 5928-5928/? E/r0ysue: i am from JNI_OnLoad com.roysue.dmeo
```

图 26-37　64 位注入结果

因为刚才是直接编译生成 APK，所以可以直接通过 adb 命令在手机上以 32 位的形式安装。安装命令如下：

```
adb install --abi armeabi-v7a -r -t app-debug.apk
```

安装后点击 App 以运行之，不过运行后 App 直接崩溃并报错，如图 26-38 所示，这是因为 32 位的 App 和 64 位的 SO 文件不匹配所导致的。另外，在使用 ActivityThread 类时，还需要考虑其他 App 掺杂其中，因此需要将它们过滤掉。

```
1970-02-18 12:12:10.990 6099-6099/com.roysue.dmeo E/AndroidRuntime: FATAL EXCEPTION: main
    Process: com.roysue.dmeo, PID: 6099
    java.lang.UnsatisfiedLinkError:
        at java.lang.Runtime.load0(Runtime.java:928)
        at java.lang.System.load(System.java:1621)
        at android.app.ActivityThread.handleBindApplication(ActivityThread.java:5687)
        at android.app.ActivityThread.-wrap1(Unknown Source:0)
        at android.app.ActivityThread$H.handleMessage(ActivityThread.java:1663)
        at android.os.Handler.dispatchMessage(Handler.java:106)
        at android.os.Looper.loop(Looper.java:164)
        at android.app.ActivityThread.main(ActivityThread.java:6552)
        at java.lang.reflect.Method.invoke(Native Method)
        at com.android.internal.os.RuntimeInit$MethodAndArgsCaller.run(RuntimeInit.java:438)
        at com.android.internal.os.ZygoteInit.main(ZygoteInit.java:807)
```

图 26-38　运行结果：App 崩溃并报错

当前我们会遇到以下两个问题：

（1）在 32 位的情况下无法正常运行。
（2）过滤系统 App。

解决问题的代码如代码清单 26-8 所示，在源码的 ActivityThread 中编写代码，再次编译。

代码清单 26-8　解决问题的代码

```
ContextImpl context = appContext;
```

```java
        ActivityManager mAm = (ActivityManager) context.getSystemService("activity");
        String activity_packageName =
mAm.getRunningTasks(1).get(0).topActivity.getPackageName();
        if(activity_packageName.indexOf("com.android")<0){
            String targetPath = "/data/data/" + activity_packageName + "/r0.so";
            String targetPath2 = "/data/data/" + activity_packageName + "/r032.so";
            String targetPath3 = "/data/data/" + activity_packageName + "/r0capstone.so";
            File file1 = new File(targetPath);
            File file2 = new File(targetPath2);
            File file3 = new File(targetPath3);
            mycopy("/sdcard/r0.so",targetPath);
            mycopy("/sdcard/r032.so",targetPath2);
            mycopy("/sdcard/r0capstone.so",targetPath3);

            int perm = FileUtils.S_IRWXU | FileUtils.S_IRWXG | FileUtils.S_IXOTH;
            FileUtils.setPermissions(targetPath,perm,-1,-1);
            FileUtils.setPermissions(targetPath2,perm,-1,-1);
            FileUtils.setPermissions(targetPath3,perm,-1,-1);

            if(file1.exists()){
                Log.e("r0ysue",System.getProperty("os.arch"));
                if(System.getProperty("os.arch").indexOf("64")>=0){
                    System.load(targetPath3);
                    Log.e("r0ysue","sucess3");
                    System.load(targetPath);
                    Log.e("r0ysue","sucess1");
                    file1.delete();
                }else{
                    System.load(targetPath2);
                    file2.delete();
                }
            }
        }
    }
```

将 demoso 中的 32 位 SO 文件推送（push）到 sdcard 中，运行 demo 文件，此时就能在 32 位的 App 上注入 32 位的 SO 文件。

26.6 本章小结

在本章中，我们以 SO 文件的加载方式为入口点，追踪了 SO 文件的加载流程。在 Android 源码追踪的过程中，有两个关键函数起着重要的作用，它们分别是 dlopen 和 android_dlopen_ext。我们使用实际案例测试了 Frida 对这两个函数的 Hook 效果，并以此为基础学习了 SO 文件在加载之前做的 3 件事情。从中，我们引出了 SO 文件注入的方式。其中最主要的注入方式是通过编译 AOSP 来注入 SO 文件，此方法可以无须 root 权限即可完成注入操作。通过学习本章的内容，读者了解了 App 运行时必经的 SO 文件中的哪些函数，在注入 SO 文件时就会更加得心应手。